T0337927

NANOMEDICINE

NANOMEDICINE

DESIGN AND APPLICATIONS OF MAGNETIC NANOMATERIALS, NANOSENSORS AND NANOSYSTEMS

Vijay K. Varadan

Linfeng Chen

Jining Xie

WILEY

A John Wiley and Sons, Ltd., Publication

This edition first published 2008
© 2008 John Wiley & Sons, Ltd

Registered office
John Wiley & Sons Ltd, The Atrium, Southern Gate, Chichester, West Sussex, PO19 8SQ, United Kingdom

For details of our global editorial offices, for customer services and for information about how to apply for permission to reuse the copyright material in this book please see our website at www.wiley.com.

Library of Congress Cataloging-in-Publication Data

Nanomedicine: design and applications of magnetic nanomaterials, nanosensors, and nanosystems / Vijay K. Varadan ... [et al.].
        p.; cm.
    Includes bibliographical references and index.
    ISBN 978-0-470-03351-7 (cloth)
    1. Nanomedicine. 2. Ferromagnetic materials. 3. Nanostructured materials.
I. Varadan, V. K., 1943-
    [DNLM: 1. Nanostructures. 2. Biosensing Techniques. 3. Magnetics. 4.
Nanomedicine. QT 36.5 N18533 2008]
    R857.N34N355 2008
    610.28 – dc22
                            2008032218

A catalogue record for this book is available from the British Library.

ISBN 978-0-470-03351-7   (H/B)

Typeset in 10/12 Times by Laserwords Private Limited, Chennai, India
Printed and bound in Great Britain by CPI Antony Rowe, Chippenham, Wiltshire

Contents

Preface

Progress in nanoscience and nanotechnology has led to the development of a new field, nanomedicine, which is generally defined as the biomedical applications of nanoscience and nanotechnology. Nanomedicine stands at the boundaries between physical, chemical, biological and medical sciences, and the advances in nanomedicine have made it possible to analyze and treat biological systems at the cell and sub-cell levels, providing revolutionary approaches for the diagnosis, prevention and treatment of some fatal diseases. For example, the *US National Cancer Institute* expects that nanoscience and nanotechnology will be harnessed for the purposes of eliminating death and suffering from cancer. Many nanomedicine approaches are already quite close to fruition, and the *US Food and Drug Administration* has started to consider the complex issues related to the approval of nanomaterials, nanodevices and nanosystems, for human betterment.

Nanomagnetism is at the forefront of nanoscience and nanotechnology, and in the field of nanomedicine, magnetic nanomaterials are among the most promising nanomaterials for clinical diagnostic and therapeutic applications. The magnetic nanomaterials used for biomedical purposes generally include zero-dimensional nanospheres, one-dimensional nanowires and nanotubes, and two-dimensional thin films. Magnetic nanoparticles, mainly including nanospheres, nanowires and nanotubes, are widely used for labeling and manipulating biomolecules, targeting drugs and genes, magnetic resonance imaging, as well as hyperthermia treatment. Magnetic thin films are often used in the development of nanosensors and nanosystems for analyzing biomolecules and diagnosing diseases.

Due to the great market potential of nanomedicine, many universities, research institutions, hospitals, commercial companies and government organizations have spent a great deal of resources in the research of magnetic nanomaterials for biomedical applications, and amazing progress has been made in this field. Some magnetic nanoparticles and nanosensors are already commercially available. Some medical applications of magnetic nanoparticles and nanosensors are under clinical trials, and encouraging results have been reported.

Despite the rapid progress in nanomedicine, we are aware of the lack of good reference books in the field of magnetic nanomaterials and their biomedical applications. Though there are excellent reviews, book chapters, and books dealing with one or several topics in this field, a book containing a comprehensive coverage of up-to-date progress in this field is not available. The research in this field requires the collaboration between researchers from different disciplines, such as physics, chemistry, materials sciences, electrical engineering, biology and medicine. However, such cross-disciplinary cooperation is not easy. For example, the languages and the tools used by materials scientists are unfamiliar to many medical experts, and vice versa. Furthermore, the ways in which physicists and biologists, or chemists and cancer researchers, think about a biomedical problem may be totally different. Therefore a book, based on which the researchers with different backgrounds can communicate, is urgently needed.

Besides, due to the lack of a reference book that can provide a broad coverage and deep insight into this field, most research activities are based primarily on the information scattered throughout numerous reports and journals. Furthermore, because of the paucity of suitable textbooks, the training in this field is usually not systematic, and this is unfavorable for the further progress in this field.

This book aims to present a comprehensive treatment of this subject. It systematically discusses the synthesis techniques, the physical and chemical properties, and the working principles for biomedical applications of various types of magnetic nanomaterials. We aim to satisfy the need of a textbook for beginners and research students, and a reference book for professionals in this field. With such a book, beginners and research students can quickly obtain an overall picture of this field. Meanwhile, this book bridges the gaps between researchers from different disciplines, so that they can speak the same language, and get their ideas across to each other. The clinical doctors who are interested in this area will also find this book valuable.

The book mainly consists of three parts. The introductory part (Chapters 1 and 2) gives general information about magnetic nanomaterials and their biomedical applications, and provides the physical background for understanding and exploring the biomedical applications of magnetic nanomaterials. The second part (Chapters 3 to 7) deals with various types of magnetic nanoparticles and their biomedical applications. Chapters 3 and 4 discuss the synthesis, properties and biomedical applications of magnetic nanospheres. In Chapter 5, a special type of magnetic nanoparticle and magnetosomes that naturally exist in magnetotactic bacteria are discussed. Chapters 6 and 7 discuss the synthesis, properties and biomedical applications of nanowires and nanotubes, respectively.

The third part (Chapters 8 to 10) discusses the development of biosensors, biochips, and their biomedical applications, with emphases laid on the sensing effects of magnetic thin films. Chapter 8 discusses the development of magnetic biosensors widely used in biomedical tests, mainly including magnetoresistance-based sensors, Hall-effect sensors, sensors detecting magnetic relaxations, and sensors detecting susceptibilities of ferrofluids. Chapter 9 mainly discusses the development of magnetic biochips based on the magnetic biosensors discussed in Chapter 8. Chapter 10 discusses the typical biomedical applications of magnetic biosensor and biochip technologies. In these applications, magnetic biosensors and biochips are mainly used to detect the biomolecules labeled by magnetic nanoparticles. An outlook for the biomedical applications of magnetic biosensor and biochip technologies is made at the end of this chapter.

In this book, the interdisciplinary nature of nanomedicine is emphasized. We take bits and pieces from the contributing disciplines and integrate them in ways that produce a new conceptual framework. To make the book readable, the contents of the book are systematically and logically developed from the elementary level. Each chapter presents one of the major topics in the development of functional magnetic nanomaterials and their biomedical applications, and contains a brief introduction to the basic physical and chemical principles of the topic under discussion. Therefore, each chapter is a self-contained unit, from which readers can readily obtain comprehensive information on this topic. To provide an extensive treatment of each topic, we have condensed mountains of literature into a readable account within a reasonable size. Important references have been included for the benefit of the readers who wish to pursue further their interested topics in a greater depth.

In preparing the book, we have tried to emphasize the fundamental concepts. Though a considerable amount of the contents in this book is related to experimental details and results, we have tried to present the underlying sciences so that the readers can understand the process of applying fundamental concepts to design experiments and obtain useful results.

It should be indicated that, due to the rapid development of this field, and its interdisciplinary nature, a truly comprehensive coverage is difficult, and some important work in this field may have been missed. It is also difficult to always give proper credit to those who are the originators of new concepts and the inventors of new techniques. The summary and commentary we have written may not have grasped the essentials of the work under discussion. We hope this book does not have too many such errors, and we would appreciate it if readers could bring the errors they discover to our attention so that these can be corrected in future editions.

We would like to acknowledge the contributions made by the students and staff at the *Center of Excellence for Nano/Neuro Sensors and Systems (CENNESS)*, University of Arkansas, Fayetteville, including the contributions from Jose Abraham in Chapters 9 and 10.

This work is partially supported by the *National Science Foundation* NSF-EPSCoR Award No. EPS-0701890.

Vijay K. Varadan
Linfeng Chen
Jining Xie

About the Authors

Vijay K. Varadan is currently the Twenty-First Century Endowed Chair in Nano-and Bio-Technology and Medicine, and Distinguished Professor of Electrical Engineering and Distinguished Professor of Biomedical Engineering (College of Engineering) and Neurosurgery (College of Medicine) at University of Arkansas. He is also a Professor of Neurosurgery at the Pennsylvania State University College of Medicine. He joined the University of Arkansas in January 2005 after serving on the faculty of Cornell University, Ohio State University and Pennsylvania State University for the past 32 years. He is also the Director of the Center of Excellence for Nano-, Micro- and Neuro-Electronics, Sensors and Systems and the Director of the High Density Electronics Center. He has concentrated on the design and development of various electronic, acoustic and structural composites, smart materials, structures, and devices including sensors, transducers, Microelectromechanical Systems (MEMS), synthesis and large-scale fabrication of carbon nanotubes, NanoElectroMechanical Systems (NEMS), microwave, acoustic and ultrasonic wave absorbers and filters. He has developed neurostimulator, wireless microsensors and systems for sensing and control of Parkinson's disease, epilepsy, glucose in the blood and Alzheimer's disease. He is also developing both silicon and organic based wireless sensor systems with RFID for human gait analysis and sleep disorders and various neurological disorders. He is a founder and the Editor-in-Chief of the Journal of Smart Materials and Structures. He is the Editor-in-Chief of the Journal of Nanomedical Science in Engineering and Medicine. He is an Associate Editor of the Journal of Microlithography, Microfabrication and Microsystems. He serves on the editorial board of International Journal of Computational Methods. He has published more than 500 journal papers and 13 books. He has 13 patents pertinent to conducting polymers, smart structures, smart antennas, phase shifters, carbon nanotubes, an implantable device for Parkinson's patients, MEMS accelerometers and gyroscopes. He is a fellow of SPIE, ASME, Institute of Physics and Acoustical Society of America. He has many visiting professorship appointments in leading schools overseas.

Linfeng Chen received his B.Sc. degree in modern applied physics (major) and his B.Eng. degree in machine design and manufacture (minor) from the Tsinghua University, Beijing, China, in 1991, and he received his Ph.D. degree in physics from the National University of Singapore in 2001. From 1991 to 1994, he was an Assistant Lecturer in the Department of Modern Applied Physics, Tsinghua University. From 1994 to 1997, he was a Research Scholar with the Department of Physics, National University of Singapore. In 1997, he joined the Singapore DSO National Laboratories, as a project engineer, and two years later, he became a Member of Technical Staff. From 2001 to 2005, he was a research scientist at the Temasek Laboratories, National University of Singapore. Since July 2005, he has been a member of the research faculty at the High Density Electronics Center, University of Arkansas, Fayetteville. He is senior member of IEEE. His research interests mainly include microwave electronics, electromagnetic functional materials and magnetic nanomedicine.

Jining Xie received his Bachelor of Engineering degree in Chemical Engineering from Tsinghua University, Beijing, China in 1997 and received his Ph.D. degree in Engineering Science from the Pennsylvania State University, University Park in 2003. He continued as a postdoctoral research fellow at the Center for the Engineering of Electronic and Acoustic Materials and Devices (CEEAMD), Pennsylvania State University. In 2005, he joined the department of Electrical Engineering, University of Arkansas, Fayetteville, as a Research Assistant Professor. His research interests include chemical synthesis, surface modifications and characterizations of micro-/nano-materials and their applications in the field of biosensors, electrodes, neuron science, etc. He is a member of MRS and IEEE. He has published over 40 technical papers in journals and conferences. He also serves as a paper reviewer for eight journals.

1

Introduction

1.1 What is Nanoscience and Nanotechnology?

In the lexicology of science and technology, the prefix 'nano' refers to one-billionth of a unit. For example, one nanometer (nm) is one billionth (10^{-9}) of a meter. The nanometer scale is the natural spatial context for molecules and their interactions. Nanoscience and nanotechnology deal with the objects at the nanometer scale (National Science and Technology Council 1999a). The properties and functions of objects at the nanometer scale are significantly different from those at a larger scale. Generally speaking, nanoscience investigates the properties of materials at atomic, molecular and macromolecular scales, while nanotechnologies deal with the design, production and application of devices and systems by controlling their shapes and sizes at the nanometer scale (Royal Society and Royal Academy of Engineering 2004).

Biology is one of the most active fundamental sciences, and it is also a science that is the most visible to the public. The need for improvement in medicine for the treatment of disease or, in a general sense, for the amelioration, correction and prevention of dysfunction in health, will never disappear (Whitesides and Wong 2006). The combination of biology and medicine, generally referred to as 'biomedicine', represents a most exciting blend of science and technology. The nanoscale provides a junction for biomedicine and materials science and technology. This book discusses the developments in nanoscience and nanotechnology and their applications in biomedicine.

1.1.1 Nanoscale: Where Physical and Biological Sciences Meet

As indicated in Figure 1.1, nanoscale is generally defined as the range from the size of atoms up to 100 nanometers, and a nanomaterial is usually defined as a material whose smallest dimension is less than 100 nanometers (Yih and Wei 2005). In a general sense, nanomaterials include all the structures, devices and systems at nanoscale. In some cases the size limit of a nanomaterial can be extended up to 1000 nm, and the essential point is that a nanomaterial exhibits unique properties that are quite different from those at a larger scale.

In Figure 1.1, a lot of biological entities are within the range of nanoscale, such as proteins, antibodies, viruses and bacteria, and they are usually called biological nanomaterials. The special functions and properties of biological nanomaterials provide much inspiration for the design of non-biological nanomaterials; meanwhile, due to their suitable sizes, non-biological nanomaterials can be used to access or manipulate biological nanomaterials (Yih and Wei 2005). Nanomaterials with sizes smaller than 50 nm can get

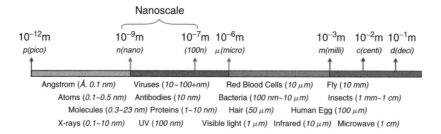

Figure 1.1 Nanoscale and typical materials whose dimension ranges are comparable to nanoscale. (Yih and Wei 2005)

inside most cells without difficulty. When nanomaterials with sizes smaller than 20 nm travel around the circulatory system of the body, they can move out of blood vessels. Therefore, after special treatments, nanomaterials are widely used as targeted drug delivery vehicles, which carry chemotherapeutic agents or therapeutic genes into the desired malignant cells while saving healthy cells. It should be noted that, in most of the technical literatures, nanomaterials are usually referred as non-biological nanomaterials, though biological entities and biological techniques have been widely used in the design and synthesis of non-biological nanomaterials.

The biological and physical sciences share a common interest in nanoscale, and the integration of biology and materials at the nanoscale has the potential to revolutionize many fields of science and technology. A vigorous trade across the borders of these areas exists in the development of new materials and tools, and the investigation of new phenomena. The advances in physical sciences offer materials useful in cell and molecular biology, and provide tools for characterizing cells and sub-cellular components; meanwhile the progress in biology provides a window for researchers to understand the most sophisticated functional nanostructures that have ever existed (Whitesides 2003).

1.1.2 Nanoscience

Nanoscience investigates those objects whose smallest dimensions range from several nanometers up to 100 nanometers (Royal Society and Royal Academy of Engineering 2004; Poole and Ownes 2003). As nanoscale may be the final engineering scale people have yet to master, nanoscience is regarded as a launch pad to a new technological era by many scientists and engineers (National Science and Technology Council 1999a).

Due mainly to the following two reasons, nanomaterials exhibit properties that are quite different from those of materials at large scales (Royal Society and Royal Academy of Engineering 2004; National Science and Technology Council 1999b). First, the surface areas of nanomaterials are much larger than those of the materials with the same mass but in a larger form. A larger surface area usually results in more reactive chemical properties, and also affects the mechanical or electrical properties of the materials. Second, nanomaterials are the natural home of quantum effects. At the nanoscale, quantum effects dominate the behaviors of a material, affecting its optical, electrical and magnetic properties.

1.1.2.1 Quantum Effect

To study the properties of the objects in the normal-sized realm, such as cars and houses, it is not usually necessary to use quantum mechanics, which is used by scientists to describe the properties of materials at the atom and electron levels. However, researchers in nanoscience are developing nanoscale building blocks, such as metallic and ceramic nanoparticles, and all-carbon nanotubes. These building blocks are hundreds of millions of times smaller than the bricks used for building houses and the tubes used for plumbing (National Science and Technology Council 1999a). Such nanoscale building blocks exhibit quantum effects.

Because the size of nanomaterials is close to the de Broglie wavelength of electrons and holes at room temperature, the states of free charge carriers in nanocrystals are quantized (Parak *et al.* 2003). For spherical nanocrystals, in which free electrons and holes are confined in all three directions, the movement of charge carriers is completely determined by quantum mechanics, and therefore the nanocrystals are often called quantum dots. Because of the similarity between the discrete energy levels of quantum dots and the discrete energy levels of atoms, quantum dots are often regarded as artificial atoms. Since the energy levels are determined by the size of the nanocrystal, they can be controlled by synthesizing nanocrystals of different diameters: the smaller a nanocrystal, the larger the spacing between its energy levels will be.

For semiconductor nanocrystals, their band gaps are size-dependent. If nanocrystals are excited optically, charge carriers are excited to upper energy levels. Fluorescent light will be emitted when the excited charge carriers fall back to the ground state. By controlled adjustment of the size during the synthesis of semiconductor nanocrystals basically all fluorescent colors in the visible region can be obtained, and there is no red-tail in the emission spectrum (Parak *et al.* 2003).

1.1.2.2 Surface Galore

For a given amount of material, the surface area compared to the volume increases when the particle size decreases, and thus the proportion of the constituent atoms at or near the surface increases. This feature is important because a lot of reactions occur at the surfaces of materials (National Science and Technology Council 1999a). For example, photosynthesis happens on the inside surfaces of cells, and catalysis happens on the surfaces of particles.

1.1.3 Nanotechnology

Nanotechnology usually refers to the capability of designing and controlling the structure of an object in the size range of nanometers. However, different researchers may have different opinions about what nanotechnology is, and it seems that the definitions of nanotechnology are as diverse as the applications of nanotechnology (Cao 2004; Malsch 2002; Taniguchi 1974). Some people consider the study of microstructures of materials using electron microscopy and the growth and characterization of thin films as nanotechnology. Other people consider a bottom-up approach in materials synthesis and fabrication, such as self-assembly or biomineralization to form hierarchical structures like abalone shell, to be nanotechnology. A drug-delivery system is a nanotechnology, and organizing molecules into functional complexes, for example a complex for delivering proteins to a certain position in the body, is also a nanotechnology. These definitions are true for certain specific research fields, but none of them covers the full spectrum of nanotechnology. The many diverse definitions of nanotechnology reflect the fact that nanotechnology covers a broad spectrum of research fields and requires true interdisciplinary and multidisciplinary efforts (Cao 2004). From the various definitions of nanotechnology listed above, we find that the only feature common to the diverse activities characterized as 'nanotechnology' is the tiny dimensions on which they operate (Royal Society and Royal Academy of Engineering 2004).

Generally speaking, nanotechnology can be understood as a technology of design, fabrication and applications of nanomaterials and nanostructures. Nanotechnology also includes a fundamental understanding of the physical properties and phenomena of nanomaterials and nanostructures. Study of fundamental relationships between physical properties and phenomena and material dimensions in the nanometer scale, is also referred to as nanoscience. To provide a more focused definition, nanotechnology deals with materials and systems whose structures and components possess novel and significantly improved physical, chemical and biological characteristics due to their nanoscale sizes (Cao 2004).

Though the word *nanotechnology* is new, the research on nanometer scale is not new. For example, the study of biological systems and the engineering of many materials such as colloidal dispersions, metallic quantum dots and catalysts have been in the nanometer regime for centuries. What is really new about nanotechnology is the combination of our capability of observing and manipulating materials at the nanoscale and our understanding of atomic scale interactions (Cao 2004). The invention and development of transmission electron microscopy (TEM), scanning tunneling microscopy (STM) and other scanning probe microscopy (SPM), such as atomic force microscopy (AFM), have opened up new possibilities for the characterization, measurement and manipulation of nanostructures and nanomaterials. Using these instruments, it is possible to study and manipulate the nanostructures and nanomaterials down to the atomic level.

After introducing the definition of nanotechnology by the US National Nanotechnology Initiative below, we discuss various nanotechnologies that often appears in the literature.

1.1.3.1 Definition by US National Nanotechnology Initiative

Though there are many different definitions for nanotechnology, most researchers follow the definition given by the US National Nanotechnology Initiative (NNI) (Alper 2005a). According to NNI, nanotechnology mainly involves three aspects: (i) research and technology development at the atomic, molecular or macromolecular levels; (ii) development and applications of structures, devices and systems with novel properties and functions due to their small and/or intermediate size; (iii) control and manipulation of materials at the atomic scale.

Encompassing science, engineering, and technology at the nanometer scale, nanotechnology involves imaging, characterizing, modeling and manipulating materials at the dimensions of roughly 1 to 100 nanometers. At this scale, the physical, chemical and biological properties of materials are fundamentally different from those of individual atoms, molecules and bulk materials (National Science and Technology Council 2004). By exploiting these novel properties, the main purpose of research and development in nanotechnology is to understand and create materials, devices and systems with improved characteristics and performances.

1.1.3.2 Natural Nanotechnology

Nanotechnology occurs in nature without intervention from human beings. The magic of natural nanotechnology encourages researchers in nanotechnology, including physicists, chemists, materials scientists, biologists, mechanical and electrical engineers and many other specialists, to learn from nature (National Science and Technology Council 1999a). In the following paragraphs, we discuss two examples of natural nanotechnology: chloroplast and F_1-ATPase complex.

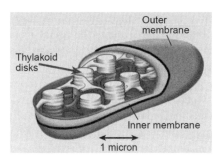

Figure 1.2 Basic structure of chloroplast. The nanoscale thylakoid disks arranged inside the stacked structures are the molecular machinery which convert light and carbon dioxide into biochemical energy. (National Science and Technology Council 1999a)

Photosynthesis is a biological way of capturing solar energy. Photosynthesis occurs in chloroplasts with nanometer and micrometer dimensions. Chloroplasts are a product of natural evolution. As shown in Figure 1.2, a chloroplast is a result of brilliantly engineered molecular ensembles, including light-harvesting molecules such as chlorophyll, arranged within the cell. These ensembles harvest the light energy by converting it into the biochemical energy stored in chemical bonds, which drives the biochemical machinery in plant cells (National Science and Technology Council 1999a).

F_1-ATPase complexes, approximately 10 nanometers across, are one of the greatest achievements of natural nanotechnology. They enable cells to produce the required biochemical fuel. As shown in Figure 1.3, F_1-ATPase complexes are molecular motors inside cells, and they continuously operate every time you move a muscle, or live another second. Each F_1-ATPase complex is a complex of proteins bound to the membranes of mitochondria, the bacteria-sized battery of a cell. F_1-ATPase complexes play a crucial role in synthesizing Adenosine triphosphate (ATP), which is the molecular fuel for cellular activity. Similar to fan motors, F_1-ATPase complexes can also make a rotary motion (National Science and Technology Council 1999a). The mechanism of rotary motion

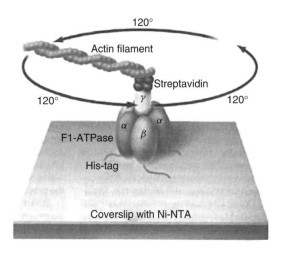

Figure 1.3 Basic structure of F_1-ATPase complex. (National Science and Technology Council 1999a)

seems to involve a sequential set of changes in conformation of the proteins driven by ions moving across the cell membrane. Though the rotary mechanism is yet to be further investigated, it certainly does not involve electrical current, magnetic fields or expansion of hot gasses in a cylinder (Whitesides 2003).

Natural nanotechnology provides much inspiration for researchers in nanotechnology. To learn from nature, it is necessary to have detailed knowledge about the basic biochemical and biophysical mechanisms at the level of the individual molecule, based on which the naturally occurring molecular machines operate. Knowledge about the working mechanisms of molecular machines is helpful for exploring new technological applications of the molecular machines and developing bio-mimic nanomaterials (Royal Society and Royal Academy of Engineering 2004).

1.1.3.3   Electronic Nanotechnology

Current nanotechnology is mainly driven by the continual shrinking of devices in the semiconductor industry and supported by the availability of characterization and manipulation techniques at the nanometer level (Cao 2004; Whitesides 2005). The continued decrease in device dimensions has followed the well-known Moore's law: the dimension of a device halves approximately every eighteen months. Electronic engineers have already shown how to extend existing methods for making microelectronic devices to new systems with wires and components whose dimensions are less than 100 nm. Scientists are currently working on molecular and nanoscaled electronics, which are constructed using single molecules or molecular monolayers. The continued size shrinkage of electronic circuits will sooner or later meet with the materials' fundamental limits imposed by thermodynamics and quantum mechanics.

1.1.3.4   Chemical Nanotechnology

For chemists, nanotechnology is not an entirely new field, as many existing chemical technologies employ nanoscale processes. Chemical catalysis is a typical example of chemical nanotechnology that has existed for more than a century. Catalysts accelerate numerous chemical transformations, such as the conversion of crude oil into gasoline and the conversion of small organic chemicals into drugs. Similarly, enzymes in cells are actually a kind of biological catalyst, and they organize and modulate the life chemistry of the body. It is expected that the nanoscale understanding of catalysis will lead to better and cleaner industrial processes.

1.1.3.5   Nanobiotechnology

Nanobiotechnology is a field that concerns the utilization of biological systems optimized through evolution, such as cells, cellular components, nucleic acids and proteins, to fabricate functional nanostructures and mesoscopic architectures comprised of organic and inorganic materials (Niemeyer and Mirkin 2004). Biological molecules can be harnessed for the creation of nanostructures, and can be used to assemble nanoscale building blocks based on the principle of molecular recognition (Parak *et al.* 2003). Many biological molecules, such as DNA, can bind to other molecules in a lock-and-key manner with very high selectivity and specificity. For example, the sequences of single-stranded oligonucleotides can be chosen so that they can bind to other partly complementary single-stranded oligonucleotides, causing the formation of complex patterns such as two-dimensional crystals or cubes.

The instruments originally developed for synthesizing and manipulating nanoscale materials have been refined and applied in the fundamental researches of biological activities (Niemeyer and Mirkin 2004). Nanotechnology has contributed important tools to investigate and manipulate biological nano-objects. One example is an atomic force microscope (AFM) based sensor, which detects biological molecules by binding them to a tiny cantilever, thereby tuning its resonance frequency. Another example is the miniaturization of a wide variety of laboratory apparatus to the size of a silicon chip. In this way the speed and throughput of classical biochemical methods, such as gel electrophoresis and polymerase chain reaction, can be increased (Parak *et al.* 2003).

1.1.3.6 Biomedical Nanotechnology

Nanomaterials and nanotechnology are widely applied in biomedicine, especially in the areas of biomedical diagnosis, drugs and prostheses and implants (Malsch 2002). The applications of biomedical nanotechnology generally fall into two categories: outside the body and inside the body. For applications outside the body, biosensors and biochips have been used to analyze blood and other biological samples. For applications inside the body, researchers are working on targeted drugs delivery, implantation of insulin pumps and gene therapy. In addition, great achievements have been made on the prostheses and implants that include nanostructured materials.

1.1.3.7 Cancer Nanotechnology

Cancer is one of the leading causes of death in developed countries. Conventional treatments, including surgery, radiation, chemotherapy and biological therapies (immunotherapy) are limited by the accessibility of the tumor, the risk of operating on a vital organ, the spread of cancer cells throughout the body and the lack of selectivity toward tumor cells (Arruebo *et al.* 2007). Nanotechnology can provide a better chance of survival. Cancer nanotechnology is actually a kind of biomedical nanotechnology. As more and more attention is paid to the diagnosis and therapy of cancers using nanotechnology, so cancer nanotechnology becomes a special branch in nanotechnology.

Cancer nanotechnology includes varieties of materials and techniques that are used for solving various problems. The research activities in cancer nanotechnology generally fall into seven categories. The first is the development of early imaging agents and diagnostic techniques for detecting cancers at their earliest, pre-symptomatic stage. Second is the development of techniques that can provide on-site assessments of the effects of the therapies. Third is the development of targeting devices that can bypass biological barriers and accurately deliver therapeutic agents to the tumor sites. Fourth, the development of agents that can be used to monitor predictive molecular changes and to prevent pre-cancerous cells from becoming malignant ones. Fifth is the development of surveillance systems for detecting the mutations that could trigger the cancer process and also for detecting genetic markers indicating a predisposition to cancers. Sixth is the development of methods for controlling cancer symptoms that badly affect quality of life. Seventh is the development of techniques helping researchers to rapidly identify new targets for clinical treatment and forecast possible side effects and drug resistance.

There are two major trends in cancer nanotechnology research (Alper 2005a). One trend is the development of multi-functional nanomaterials than can be used to simultaneously image a tumor and deliver drugs to the tumor. This may be the most radical improvement that nanotechnology can make for cancer treatment. The other trend in cancer nanotechnology is to dose a tumor with many drugs simultaneously, not just with one drug. In this way, the drug resistance problem, which is one of the most vexing problems in cancer treatment, could be solved. Usually, the drug resistance of a cancer cell is due to its ability to pump out the anti-cancer drugs once they are delivered into the cell. However, by delivering an agent that can inhibit the pumping at the same time as the anti-cancer drugs are delivered to cancer cells, the problem of drug resistance may disappear.

1.1.4   Typical Approaches for Synthesis of Nanomaterials

The approaches for synthesis of nanomaterials are commonly categorized into top-down approach, bottom-up approach and hybrid approach.

1.1.4.1   Top-down Approach

Generally speaking, the top-down approach is an extension of lithography. This approach starts with a block of material, and reduces the starting material down to the desired shape in nanoscale by controlled etching, elimination and layering of the material (Cao 2004). Owing to the advancement of the semiconductor industry, the top-down approach for the fabrication of nanomaterials is a well developed method.

One problem with the top-down approach is the imperfection of the surface structure (Cao 2004). The conventional top-down techniques, such as lithography, may cause severe crystallographic damage to the processed patterns, and some uncontrollable defects may also be introduced even during the etching steps. For example, a nanowire fabricated by lithography usually contains impurities and structural defects on the surface. As the surface over volume ratio in nanomaterials is very large, such imperfections may significantly affect the physical properties and surface chemistry of the nanomaterials.

Regardless of the surface imperfections and other defects, the top-down approach is still important for synthesizing nanomaterials. However, it should be noted that, in the quest for miniaturization, top-down lithographic approaches for creating nanomaterials are approaching the fundamental limitations. Even cutting-edge electron beam lithography cannot create structures smaller than 10 nm (Darling and Bader 2005). Besides, lithographic techniques are usually expensive, and their productivities are usually low.

1.1.4.2   Bottom-up Approach

In a bottom-up approach, materials are fabricated by efficiently and effectively controlling the arrangement of atoms, molecules, macromolecules or supramolecules (Luo 2005). The bottom-up approach is driven mainly by the reduction of Gibbs free energy, so the nano-materials thus produced are in a state closer to a thermodynamic equilibrium state. The synthesis of large polymer molecules is a typical example of the bottom-up approach, where individual building blocks, monomers, are assembled into a large molecule or polymerized into bulk material. Crystal growth is another example of the bottom-up approach, where growth species – either atoms, or ions or molecules – assemble in an orderly fashion into the desired crystal structure on the growth surface (Cao 2004).

The concept and practice of a bottom-up approach have existed for quite a while, and this approach plays a crucial role in the fabrication and processing of nanomaterials. The nanostructures fabricated in the bottom-up approach usually have fewer defects, a more homogeneous chemical composition and better short and long range ordering.

1.1.4.3 Hybrid Approach

Though both the top-down and bottom-up approaches play important roles in the synthesis of nanomaterials, some technical problems exist with these two approaches. For the top-down approach, the main challenge is how to accurately and efficiently create structures which are becoming smaller and smaller; while for the bottom-up approach, the main challenge is how to fabricate structures which are of sufficient size and amount to be used as materials in practical applications. The top-down and bottom-up approaches have evolved independently. It is found that, in many cases, combining top-down and bottom-up methods into a unified approach that transcends the limitations of both is the optimal solution (Royal Society and Royal Academy of Engineering 2004; Darling and Bader 2005). A thin film device, such as a magnetic sensor, is usually developed in a hybrid approach, since the thin film is grown in a bottom-up approach, whereas it is etched into the sensing circuit in a top-down approach.

1.1.5 Interdisciplinarity of Nanoscience and Nanotechnology

Nanoscience and nanotechnology are highly interdisciplinary, encompassing aspects of physics, chemistry, biology, materials science and engineering, and medicine (Hurst *et al.* 2006). Due to their interdisciplinarity, nanoscience and nanotechnology have brought about cooperation between scientists and engineers with different backgrounds to share their expertise, instruments and techniques (Royal Society and Royal Academy of Engineering 2004). The evolutionary developments within different areas in the investigation of materials that are becoming smaller and smaller have contributed to the rapid progress in nanoscience and nanotechnology, and meanwhile nanoscience and nanotechnology benefit not only the electronics industry, but also the chemical and space industries, as well as medicine and health care.

In the following subsections we concentrate on the roles of three disciplines in the research of nanoscience and nanotechnology: chemistry, physics and biology and medicine.

1.1.5.1 Chemistry

Chemistry plays a leading role in nanotechnology, and in a sense, chemistry is the ultimate nanotechnology. The opportunities for chemistry to make important contributions to nanoscience abound, and three promising areas include synthesis of nanomaterials, molecular mechanisms in nanobiology, and risk assessment and evaluation of safety (Whitesides 2005). Chemistry is unique in the sophistication of its ability to synthesize new forms of matter. In making new forms of matter by joining atoms and groups of atoms together with bonds, chemistry contributes to the invention and development of materials whose properties depend on nanoscale structure. Meanwhile, chemistry makes unique contributions to the study of the molecular mechanisms of functional nanostructures in biology, such as the light-harvesting apparatus of plants, ATPases, the ribosome and the structures that package DNA, ultimately the cell. Furthermore, analyzing the risks of nanomaterials to health and the environment requires cooperation across various disciplines, including chemistry, physiology, molecular medicine and epidemiology.

1.1.5.2 Physics

Compared with bulk materials, materials at the nanoscale exhibit quite different properties, and physics studies the underlying mechanisms of the changes of properties due to the size changes. Physicists are investigating the special mechanical, thermal, electrical and optical properties of various types of nanomaterials, such as quantum dots and hybrid thin films, and most of the researches involves quantum mechanics.

Among various categories of nanomaterials, magnetic nanomaterials exhibit unique size dependence of magnetic properties in the nanoscale, and knowledge of these properties is essential for the design and modifications of magnetic nanomaterials and for the development their specific applications. The research on nanomagnetism, magnetism in nanomaterials, has been among the most challenging topic in nanoscience and nanotechnology (Himpsel *et al.* 1998).

1.1.5.3   Biology and Medicine

It is a main trend that biomedical and physical sciences share a common interest in nanomaterials, and the conventional borders of these areas are disappearing. The union of biologists and physicians with engineers and materials scientists is encouraging (Ritchie 2005). As few biologists and physicians know much about engineering and materials science, and even fewer engineers and materials scientists know much about medicine, the potential for such a union seems boundless.

Nanofabrication can provide analytical tools for investigating biomolecules as well as for exploring the interior structure and function of cells (National Science and Technology Council 1999b). In return, biology is clearly having an equally significant impact on nanoscience and nanotechnology. Methods in biology can be used to make nanomaterials that are difficult or impossible to be fabricated by synthetic means (Taton 2003). Due to the evolution of billions of years, organisms of all types are equipped with numerous nanomachines, such as DNA that can be used for information-storage and chloroplasts that capture the solar energy (Service 2002; National Research Council 1994).

Researchers in the field of nanoscience and nanotechnology are seeking practical help from biology. One of the most attractive features of biological systems is that an organism has the capability to produce extremely complex molecules, for example DNA and proteins, with atomic precision. The powerful biomachinery can further arrange different organisms into a complicated system. However, synthesized nanomaterials, for example carbon nanotubes and metal nanoparticles, do not have similarly efficient guiding mechanisms. Besides, it is very difficult to handle and manipulate nanomaterials using the traditional methods. Inspired by the discoveries in biology, researchers in the field of nanoscience and nanotechnology are trying to use the molecular toolbox in biology, for the synthesis of functional nanomaterials (Service 2002).

Researchers attempt to combine the capability of assembling complex structures in biology and the ability of developing functional devices in nanoscience and nanotechnology. Such a combination is helpful for the development of a variety of novel structures and devices (Service 2002). Williams *et al.* (2002) demonstrated the bioelectronic assembly, as shown in Figure 1.4. In this example, the assembly of the carbon nanotubes into molecular-scale electronic devices is based on the selective binding capabilities of peptide nucleic acid (PNA). Similar to DNA, PNA consists of a series of nucleotide bases (A's, T's, G's and C's) that selectively bind to one another. However, in a PNA, the backbone of sugar and phosphate groups in DNA is substituted with more stable links based on peptides. Due to this substitution PNAs can endure higher temperatures and stronger solvents, often used in chemical and biological processing, than DNAs.

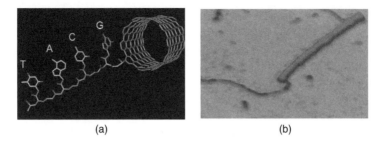

(a)   (b)

Figure 1.4 Bioelectronic assembly of a carbon nanotube. (a) Schematic drawing, and (b) micrograph. (Service 2002)

1.2 Magnets and Nanometers: Mutual Attraction

As physicists and materials scientists are becoming increasingly interested in the properties of magnetic materials on the nanometer scale, biology is benefiting from nanomagnets. Magnetic nanomaterials are quite different from other nanomaterials, because the fundamental properties of magnets are defined at nanometer length scales (Koltsov and Perry 2004). Nanomagnets can measure anything from just under a micron to a few nanometers in size, and have applications that range from medical imaging and drug delivery to sensors and computing. As will be discussed in later chapters, magnetic nanomaterials are widely used in biology and medicine.

Nanomagnetism is at the frontiers of nanoscience and nanotechnology, and magnetic nanomaterials are among the most promising nanomaterials for clinical diagnostic and therapeutic applications. Nanomagnetism basically involves studying how such ferromagnetic materials behave when they are geometrically restricted in at least one dimension. One of the central topics in nanomagnetism is the superparamagnetic properties of magnetic nanomaterials; in most cases, what makes ferromagnets useful in biomedical applications is their superparamagnetic properties. More discussions about nanomagnetism can be found in Chapter 2.

1.3 Typical Magnetic Nanomaterials

The magnetic nanomaterials used in biology and medicine generally fall into three categories: zero dimensional nanomaterials such as nanospheres; one-dimensional nanomaterials such as nanowires and nanotubes; and two-dimensional nanomaterials such as thin films. Usually, all the nanospheres, nanorods, nanowires and nanotubes are called nanoparticles, among which, nanorods, nanowires and nanotubes are high aspect-ratio nanoparticles.

In most of the biomedical applications, magnetic nanoparticles are suspended in appropriate carrier liquids, forming magnetic fluids, also called ferrofluids. The properties of ferrofluids are discussed in Chapter 2. In most of the biomedical applications, magnetic thin films are usually fabricated into magnetic biosensors or biochips, by etching them into certain patterns to perform specific functions. As discussed in detail in Chapter 8, most of the magnetic biosensors fabricated from magnetic thin films detect the stray magnetic fields from magnetic nanoparticles that are attached to biomolecules.

1.3.1   Nanospheres

Among the three types of magnetic nanoparticles, magnetic nanospheres are most widely used in biomedicine. To realize their biomedical applications, the magnetic nanospheres should be stably suspended in the carrier liquid, and they should also carry out certain biomedical functions. The magnetic material most often used is iron oxides, usually in the form of magnetite (Fe_3O_4) or maghemite (γ-Fe_2O_3), and the carrier liquids are usually water, kerosene or various oils.

Figure 1.5 shows the basic structures of magnetic nanospheres used for biomedical applications. Due to their small size, the magnetic nanoparticles in carrier liquids neither form sediment in the gravitational field or in moderate magnetic field gradients, nor do they agglomerate due to magnetic dipole interaction. However, a stable suspension can only be achieved if the particles are protected against agglomeration due to the van der Waals interaction. Usually this protection can be achieved by two approaches (Could 2004; Odenbach 2004). One is the electric charge stabilization. In this approach, a thin layer of gold is coated on the surface of the nanospheres. Meanwhile, the thin gold layer can also serve as an ideal base on which chemical or biological agents can be functionalized, as shown in Figure 1.5(a). These molecules generate a repulsive force, preventing the particles from coming into contact and thus suppressing the destabilizing effect of the van der Waals interaction. In practical applications, these two approaches are often used in combination for the majority of ferrofluids, since this allows the synthesis of suspensions which are stable over years (Could 2004). Usually, iron oxide nanoparticles used in biomedical applications are coated with gold or silica, and subsequently functionalized, for example, with antibodies, oligonucleotides or peptide ligands.

One major trend in the research of nanospheres is the development of multi-functional nanospheres, and one typical approach for realizing the multi-functionality is to functionalize a magnetic nanosphere with different functional groups. Figure 1.6 illustrates a multifunctional magnetic nanosphere. The targeting ligands and delivery peptides conjugated on the nanoparticle surface are used for targeting the nanosphere to the desired location and for delivering the drug respectively, and the sensor and reporter functional group is for checking the effects of drug delivery.

Another approach to performing multiple functions simultaneously is to use multicomponent nanomaterials, such as core–shell, alloyed and striped nanoparticles (Hurst *et al.* 2006). Researchers hope to design multicomponent nanostructures and exploit their

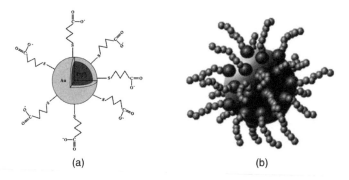

(a)   (b)

Figure 1.5   Two approaches for protection of nanospheres. (a) Electric charge stabilization (Could 2004); (b) organic molecule stabilization. (Odenbach 2004)

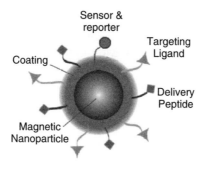

Figure 1.6 Illustration of a multifunctional magnetic nanosphere, modified from Could (2004)

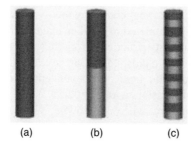

Figure 1.7 Configurations of three types of nanowires. (a) Single-segment nanowire; (b) two-segment nanowire; (c) multilayer nanowire. (Sun *et al.* 2005). Reprinted from IBM Sun, L., Hao, Y., Chien, C.L. and Searson, P.C. (2005) Tuning the properties of magnetic nanowires, *Journal of Research and Development*, **49**(1), 79–102, with permission from IBM

inherent multiple functionalities for use in many novel applications. For example, in a core–shell system, the favorable properties of the core are maintained, while the shell functions to provide additional stabilization, passivation or chemical functionality.

1.3.2 Nanorods and Nanowires

Nanorods and nanowires are straight solid one-dimensional high aspect ratio nanomaterials. Usually, a nanowire has a higher aspect ratio than a nanorod; however there is no strict standard by which we can differentiate a nanorod and a nanowire. A nanorod named by some researchers may be called a nanowire by other researchers, and vice versa. Therefore, in this book, we do not distinguish between nanorods and nanowires, calling them both nanowires.

As shown in Figure 1.7(a), in most cases nanowires are cylindrical in shape with a radius in the range from 5 to 500 nm, and length up to about 100 μm. The elongated structure of nanowires may result in inherent chemical, electrical, magnetic and optical anisotropy that can be exploited for interactions with cells and biomolecules in fundamentally new ways (Bauer *et al.* 2004). Although a majority of the magnetic carriers currently used for biomedical applications are magnetic nanospheres, nanowires are an alternative type of nanoparticles with considerable potential. Due mainly to the following attractive properties, many efforts are being made to explore the applications of nanowires in biomedicine (Hultgren *et al.* 2005).

1.3.2.1 Biocompatibility

Biocompatibility is one of the most important considerations in the development of biomedical applications of nanomaterials. Most of the magnetic nanowires are compatible with living cells. They can be functionalized with biologically active molecules, and they do not disrupt normal cell functions, such as cell proliferation and adhesion, and gene expression (Hultgren *et al.* 2005).

1.3.2.2 Magnetization

Magnetic nanoparticles can be used to apply forces to biological systems. Therefore, they can be used in the magnetic separation of biological entities, and the targeted delivery of drugs and genes, and they can also be used to study mechanotransduction at the cellular level (Hultgren *et al.* 2005). Usually, magnetic nanowires are made of solid metals, and thus they may possess large magnetic moments per volume of material. Therefore magnetic nanowires can be manipulated at weaker magnetic fields.

It should be noted that, due to its structural properties, the magnetization of a nanowire is quite different from that of a nanosphere (Sun *et al.* 2005). The magnetization of a sphere under an external magnetic field is independent of the direction of the applied magnetic field. However, it is easier to magnetize a nanowire along its axis than perpendicular to its axis. Furthermore, when a nanowire is placed between a north pole and a south pole, outside the nanowire, the magnetic field lines emanating from the north pole end at the south pole. While inside the nanowire, the magnetic field lines are in the direction from the north pole to the south pole, and thus in the opposite direction of the magnetization of the material. Therefore, the magnetic field inside the nanowire tends to demagnetize the nanowire, and this field is usually called the demagnetizing field.

1.3.2.3 Controllable Dimensions

As the geometry of nanoparticles may have biomedical effects, it is desirable to tune the size and shape of nanoparticles to study the mechanisms of their interactions with cells. Magnetic nanowires are often synthesized by the template method. In this method, the diameter and length of magnetic nanowires are independently controllable. The diameter of nanowires can be controlled in the range from nanometer to micrometer, by using templates with different pore diameters. Meanwhile, the length of magnetic nanowires can be controlled by using templates with different thicknesses.

1.3.2.4 Multi-segment Structure

As shown in Figures 1.7(b) and (c), different types of materials, for example, magnetic and non-magnetic, can be selectively electrodeposited along the nanowire axis, resulting in multi-segment nanowires. By precisely modulating the composition along the axis of a nanowire, the architecture and magnetic properties of the nanowire can be precisely controlled (Sun *et al.* 2005). Both single-segment and multiple-segment magnetic nanowires have been used in biomedicine. To optimize their biomedical applications, the magnetic properties of magnetic nanowire, for example, the Curie temperature, the easy magnetization axis, the saturation magnetic field, the saturate magnetization, remanent magnetization and coercivity, can be adjusted to meet special biomedical application requirements.

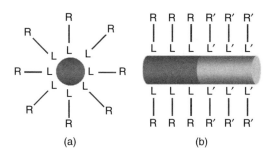

Figure 1.8 Functionalization of a nanosphere (a) and a two-segment nanowire (b). L and L′ represent two different types of ligands, and R and R′ represent two different types of functional groups. (Sun *et al.* 2005). Reprinted from IBM Sun, L., Hao, Y., Chien, C.L. and Searson, P.C. (2005) Tuning the properties of magnetic nanowires, *Journal of Research and Development*, **49**(1), 79–102, with permission from IBM

1.3.2.5 Multi-functionality

In biomedical applications, such as drug and gene delivery, magnetic nanowires are usually functionalized with biologically active molecules. As shown in Figure 1.8, compared to nanospheres, multi-functionality can be more easily realized on multi-segment nanowires (Sun *et al.* 2005, Reich *et al.* 2003).

Based on surface coordination chemistry, using ligands that bind selectively to different segments of a multi-component wire, spatially modulated multiple functionalization can be realized in the wire. Figure 1.8(b) shows a two-segment nanowire functionalized with ligands whose headgroups (L and L′) selectively bind to desired segments and whose tail groups (R and R′) will target two different biomolecules. The control of selective functionalization is crucial to the realization of the multi-functionality of nanowires. For example, for a Ni-Au nanowire, based on the differences in surface chemistry of the gold and nickel segments, different molecules can be bound to different segments of the nanowire, so different segments can be arranged to carry out different tasks.

1.3.3 Nanotubes

The magnetic nanotubes discussed in the literature can be classified into three types. The first type is non-magnetic nanotubes, such as carbon nanotubes, whose inner void is filled with magnetic nanomaterials. This type of magnetic nanotube can also be taken as a magnetic nanowire covered with a layer of non-magnetic material. The second type is non-magnetic nanotubes whose walls are deposited with magnetic nanomaterials. The third type is nanotubes whose whole structure is made of magnetic materials. In this book, we concentrate on the third type of magnetic nanotubes.

As shown in Figure 1.9(a), a magnetic nanotube is the hollow counterpart of a magnetic nanowire. Similar to magnetic nanowires, magnetic nanotubes have high aspect ratio, and usually have much stronger magnetization than magnetic nanospheres.

Due to their unique structural properties, nanotubes are ideal for the realization of multifunctionality. As shown in Figure 1.9(b), as a nanotube has distinctive inner and outer surfaces, the inner surface and the outer surface of a nanotube can be functionalized to perform different biomedical functions. Depending on the inner diameter of a nanowire, the inner empty space of the nanowire can be used to capture, concentrate and release biological entities ranging in size from small molecules to large proteins. The outer surface

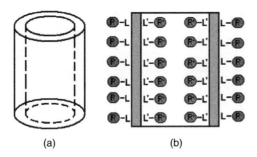

(a)   (b)

Figure 1.9 The basic structure of a magnetic nanotube (a), and its multi-functionalization (b)

of a nanotube is often functionalized with environment-friendly molecules or probing molecules to a specific target (Son *et al.* 2005).

Template method is often used in the synthesis of magnetic nanotubes. Using this method, the three major structural parameters can be independently controlled. The outer diameter of a nanotube is mainly determined by the pore diameter of the template, and the inner diameter of a nanotube can be controlled by the concentration of the solution and the filtration speed. The length of a nanotube is mainly determined by the thickness of the template.

1.3.4 Thin Films

Magnetic thin films are actually sheets of magnetic material with thicknesses usually less than one hundred nanometers. Magnetic thin films can have single-layer, or multilayer structures, and they can be single-crystal, polycrystalline or amorphous. One important property of a magnetic thin film is that its electrical resistance usually changes when an external magnetic field is applied, and the change of electrical resistance due to the application of external magnetic field is called magnetoresistance. For certain multilayer structures composed of alternating ferromagnetic and non-magnetic layers, the resistance of a multilayer thin film drops dramatically as a magnetic field is applied, and this phenomenon is called giant magnetoresistance (GMR) (Binasch *et al.* 1989; Baibich *et al.* 1988).

Figure 1.10 shows the change in room-temperature resistance ΔR *vs* in-plane magnetic field for two polycrystalline multilayers of Fe/Cr and Co/Cu, respectively (Parkin 1995). The resistance is measured with the current in the plane of the layers, and the magnetic field is applied orthogonal to the current also in the plane of the layers. The variation in resistance is related to a change in the relative orientation of neighboring ferromagnetic Fe or Co layers with applied magnetic field. The resistance is higher when adjacent magnetic layers are aligned anti-parallel to one another, as compared with parallel alignment, shown schematically in Figure 1.10.

Since the discovery of the GMR effect in magnetic multilayer systems, many biosensors employing this effect have been developed. As shown in Figure 1.11, GMR sensors could be employed for the detection of biomolecules that have been tagged with magnetic nanoparticles. GMR sensors are favored over competing optical detection schemes due to their higher sensitivity, lower background, compact size and easy integrability with existing semiconductor electronics (Li *et al.* 2004). GMR sensors with minimum dimensions below 100 nm have been fabricated using electron beam lithography (Wood 2005). Such technology has implications in many areas of biological and medical research, including disease detection, treatment and prevention.

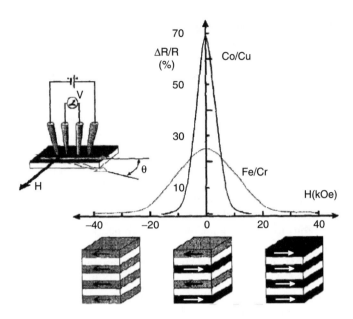

Figure 1.10 Room-temperature resistance *vs* in-plane magnetic field curves for polycrystalline Fe/Cr and Co/Cu multilayers. The measurement geometry is shown schematically in the top left corner. The magnetic state of the antiferromagnetically coupled multilayers is shown schematically in the lower portion of the figure for large negative, zero, and large positive magnetic fields (Parkin 1995)

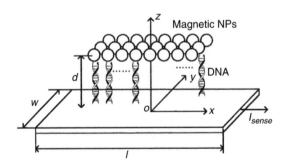

Figure 1.11 Schematic view of a magnetic nanoparticle array immobilized on a magnetoresistance sensor through DNA hybridization (modified from Li *et al.* 2004)

1.4 Nanomedicine and Magnetic Nanomedicine

The integration of biological and physical sciences at nanoscale has impacts in many areas of science and technology. One area that is particularly promising is the biomedical applications of magnetic nanomaterials.

1.4.1 Inspiration from Nature

It is well known that the most advanced nanoscale machines that have ever existed are cells regulating and controlling biological systems, and understanding cells is one of the great unanswered questions in science (Royal Society and Royal Academy of Engineering 2004; Whitesides 2005). A cell is the smallest and most fundamental unit, from which the rest is built. It is a system of molecules and remarkable nanoscale 'machines' – functional molecular aggregates of great complexity. Understanding these molecular nanostructures in their full, mechanistic, molecular complexity is vital to understanding the cell. Proteins are one typical example of molecular nanostructures. They have numerous highly specific functions, and take part in almost all the biological activities, such as sensory, metabolic, information and molecular transport processes. However, compared to an individual cell, a single molecule bio-nanodevice, such as a protein, only occupies about one-millionth to one-billionth of the volume. Therefore, in the biological world, there are innumerable biological nanoscale structures, devices and machines, which the researchers in the field of nanoscience and nanotechnology may be interested to investigate and emulate. The methods emerging from this research will help us to move closer to understanding human life and health, and thus towards nanomedicine, which has attracted a lot of attention and has become a fast growing field.

It should be noted that current materials used in biomedicine have only a small fraction of the sophistication of the naturally occurring materials that they replace. In general, synthetic materials are poor or partial replacements for naturally occurring ones. However, synthetic materials are more satisfactory when they serve a function that does not exist in nature. For example, delivery of a drug by erosion of a polymeric matrix *in vivo*; and alteration of the relaxation time of protons in water to improve contrast in magnetic resonance imaging (Whitesides and Wong 2006).

1.4.2   What is Nanomedicine?

Nanomedicine stands at the boundaries between the physical, the chemical and the biological sciences. It originated from the imaginative idea that robots and other related machines at the nanometer scale could be designed, fabricated and introduced into the human body for repairing malignant cells at the molecular level. According to its original vision, nanomedicine is a process including the diagnosis, treatment and prevention of diseases and traumatic injuries, and the preservation and improvement of human health, using molecular tools and molecular knowledge of the human body (Freitas 2005).

The progress in both nanoscience and nanotechnology makes nanomedicine practical. From a technical viewpoint, nanomedicine consists of the applications of particles and systems at the nanometer scale for the detection and treatment of diseases at the molecular level, and it plays an essential role in eliminating suffering and death from many fatal diseases, such as cancer (Yih and Wei 2005; National Institute of Health and National Cancer Institute 2004). Based on nanofabrication and molecular self-assembly, various biologically functional materials and devices, such as tissue and cellular engineering scaffolds, molecular motors and biomolecules, can be fabricated for sensor, drug delivery and mechanical applications (Royal Society and Royal Academy of Engineering 2004).

Nanomedicine has obvious advantages (Whitesides and Wong 2006). First, nanoparticles are potentially invaluable tools for investigating cells because of their small size. Second, as their size can be controlled, from that of large molecules to that of small cells, the ability of nanoparticles to escape the vasculature *in vivo* can also be controlled. Third, because of their small size, nanoparticles can circulate systemically in the bloodstream and thus serve in roles such as magnetic resonance enhancement, iron delivery for the production of red blood cells and drug delivery to improve the availability of serum-insoluble drugs.

1.4.3 Status of Nanomedicine

Nanomedicine has developed in numerous directions, and it has been fully acknowledged that the capability of structuring materials at the molecular scale greatly benefits the research and practice of medicine. The investigation of fundamental problems regarding the biocompatibility of nanomaterials has been initiated both theoretically and experimentally. The complicated issues related to the future approval of nanomedical materials by the US Food and Drug Administration are extensively discussed. It seems that preparations are being made for our society to deploy nanomedicine for human betterment (Freitas 2005).

However, nanomedicine is a long-term expectation. Before nanomedicine can be used in clinics, fundamental mechanisms of nanomedicine should be fully investigated, and clinical trials and validation procedures should be strictly conducted. Though it is possible that some biological entities, such as proteins, DNA and other bio-polymers, could be directly used for biosensor applications, nevertheless some serious issues, such as biocompatibility and robustness, may hinder the progress of these efforts. Though in many areas, such as disease diagnosis, targeted drug delivery and molecular imaging, clinical trials of some nanomedicine products are being made, the clinical applications of these techniques, which require rigorous testing and validation procedures, may not be realized in the near future (Royal Society and Royal Academy of Engineering 2004).

At all events, it should be noted that although the applications of nanomaterials in biology and medicine are in an embryo stage, it is the great promise of nanomedicine that has inspired researchers to extensively investigate the interfaces between nanotechnology, biology and medicine (Satyanarayana 2005).

1.4.4 Magnetic Nanomedicine

Magnetic nanomedicine is growing rapidly, and there is already a broad range of applications including cell separation, biosensing, studies of cellular function, as well as a variety of potential medical and therapeutic uses. The magnetic nanomaterials used in magnetic nanomedicine can be generally classified into magnetic thin films and magnetic nanoparticles which include nanospheres, nanowires and nanotubes. Magnetic thin films are often used in the development of high sensitivity and high accuracy magnetic sensors and biochips, which are important for the detection of biological entities bound with magnetic nanoparticles. The magnetic nanoparticles used in magnetic nanomedicine usually consist of a single magnetic species and a suitable coating to allow functionalization with bioactive ligands. Magnetic nanoparticles have attractive advantages in biomedical applications as discussed below (Pankhurst *et al.* 2003).

1.4.4.1 Easy Detection

As almost all biological entities are non-magnetic, magnetic nanoparticles in biological systems can be easily detected and traced. One typical example is the enhancement of the signal from magnetic resonance imaging (MRI) using magnetic nanoparticles. In this technique, a subject is placed in a large, external magnetic field and then exposed to a pulse of radio waves. Changes to the spin of the protons in water molecules are measured after the pulse is turned off. Tiny differences in the way that protons in different tissues behave can then be used to build up a 3D image of the subject (Koltsov and Perry 2004).

1.4.4.2 Magnetic Manipulation

Magnetic nanoparticles will rotate under an external uniform magnetic field, and will make translational movements under an external magnetic field gradient. Therefore, magnetic nanoparticles, or magnetically tagged molecules, can be manipulated by applying an external magnetic field. This is important for transporting magnetically tagged drug molecules to diseased sites. The magnetic manipulation of magnetic nanowires and nanotubes is important for applying forces to biological entities, and for nanowires or nanotubes to get into biological entities.

1.4.4.3 Energy Transfer

Magnetic nanoparticles can resonantly respond to a time-varying magnetic field, transferring energy from the exciting magnetic field to the nanoparticles and the tagged biological entities. This property has been used in hyperthermia treatment of cancer tumors (Pankhurst *et al.* 2003).

1.5 Typical Biomedical Applications of Functional Magnetic Nanomaterials

The advances in nanoscience and nanotechnology result in a variety of biomedical applications for magnetic nanomaterials. Most of the biomedical applications of magnetic nanomaterials are based on the specific characteristics of magnetic nanomaterials and the benevolent relationships between magnetic fields and biological systems. The strength of magnetic field required for manipulating magnetic nanoparticles does not have harmful effects on biological tissues, and the biotic environment does not interfere with the magnetism of magnetic particles (Could 2004). Though magnetic nanoparticles with different compositions, sizes and shapes have been developed for biomedical applications, the most frequently used magnetic materials are two types of iron oxide particles, maghemite (γ-Fe_2O_3) and magnetite (Fe_3O_4). In some applications of magnetic nanoparticles, magnetic beads of micrometer size, consisting of magnetic nanoparticles, are used.

As shown in Figure 1.12, the biomedical applications of magnetic nanoparticles can be generally classified into diagnosis and therapy (Pankhurst 2003; Arruebo *et al.* 2007). Magnetic thin films are often used in the development of biosensors and biochips, which play crucial roles in magnetic diagnosis. We briefly discuss below the diagnostic and therapeutic applications of magnetic nanoparticles, and the magnetic biosensors and biochips based on magnetic thin films. In the final part, the trends of magnetic nanomaterials in biomedicine are outlined.

1.5.1 Diagnostic Applications of Magnetic Nanoparticles

Magnetic nanomaterials have been widely used in the diagnosis of diseases. As diseases can be detected at the cell or molecular level, many diseases can be diagnosed at very early stages, much earlier than the disease symptoms appear. This is especially important for some fatal diseases, such as cancer. We discuss below several typical examples of diagnostic applications of magnetic nanoparticles.

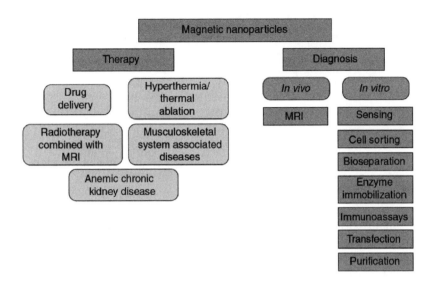

Figure 1.12 Biomedical applications of magnetic nanoparticles. (Arruebo *et al.* 2007)

1.5.1.1 Magnetic Separation

Separation of specific biological entities from their environments is very important in biochemical analysis and disease diagnosis. Magnetic nanoparticles usually exhibit superparamagnetic behaviors at room temperature. They have strong magnetization under an external magnetic field, but do not have remanent magnetism once the magnetic field is moved away. This magnetic on/off switching behavior is extremely helpful for magnetic separation (Could 2004; Pankhurst 2003).

Magnetic separation is a well-developed method and can be used as an alternative to the conventional centrifugal separation method. A typical magnetic separation procedure mainly consists of four steps, and iron oxide particles are often used in magnetic separation (Could 2004). In the first step, magnetic microbeads are made by encasing iron oxide nanoparticles in a biocompatible coating, and the surface of the microbeads is functionalized with a special biological or chemical agent that can selectively bind to the target cells or molecules to be separated. In the second step, the functionalized microbeads are added to the solution containing the target cells or molecules, and the target cells or molecules are subsequently bound to the magnetic microbeads. In the third step, a permanent magnet is placed at the side of the solution, inducing a magnetic moment on the magnetic microbeads and establishing a magnetic field gradient which drives the magnetized microbeads to move along the field lines. Finally, the magnetized microbeads cluster together near the magnet, and thus the target cells or molecules bound to the magnetic microbeads are separated.

1.5.1.2 Medical Imaging

One of the most attractive advantages of the magnetic resonance imaging (MRI) technique is that this technique is non-invasive. The MRI technique is extensively used in diagnosing diseases, making pre-surgical assessment and monitoring the therapy effects. A lot of efforts are being made to enhance its resolution and contrast. One way of boosting the MRI signal is to use contrast agents made of magnetic nanoparticles. Superparamagnetic iron oxide particles coated with a suitable chemically neutral material to prevent them from reacting with body fluids are often used as contrast agents. These magnetic particles are usually injected into the bloodstream and travel to different organs depending on their size. Therefore, by selecting particles of particular sizes, researchers can study specific parts of the body (Koltsov and Perry 2004).

1.5.1.3 Targeted Detection

The targeted detection technique can be used for detecting extremely early signs of disease (Could 2004; Pray 2005). Usually, a tumor with a diameter of 10 mm has more than one hundred million tumor cells. Using the targeted detection technique, a cancer cluster with about 10–100 cancer cells could be detected, so cancers could be diagnosed at a very early stage of malignancy. It is expected that the targeted detection technique can be further improved to detect individual cancer cells before the formation of a cluster. This technique can also be used in AIDS research. To detect HIV viruses, ferromagnetic nanoparticles are usually coated with gold, and tagged with an HIV antibody via an Au-S covalent bond. Such nanoparticle probes could sensitively detect very small amounts of viral particles that could not be detected by the conventional AIDS diagnosing techniques.

It should be noted that before the targeted detection technique can be used in clinics, some practical problems should be addressed (Could 2004). For example, it should be investigated whether a small number of malignant cells can produce signals that are strong enough to be reliably detected by magnetic probes, whether the magnetic probes can reach the intended targets, and whether the signal is truly caused by the particulate clustering due to the probe–target binding.

1.5.2 Therapeutic Applications of Magnetic Nanoparticles

Using magnetic nanomaterials, therapeutic treatments could be performed at cell or molecular level, and therefore the therapy efficiency will be greatly improved, and the side effects will be greatly decreased. In the following subsections, we discuss five typical therapeutic applications of magnetic nanoparticles.

1.5.2.1 Hyperthermia Treatment

As some cancer cells are more susceptible to high temperatures than normal cells, such cancerous cells can be treated thermally. Therefore, by increasing the temperature of the tissue to above 42 °C, the cells could be selectively destroyed. To achieve this, a dose of magnetic nanoparticles could be injected into a region of malignant tissue, after which an alternating magnetic field could be applied to the magnetic nanoparticles. If the field is sufficiently strong and of optimum frequency, the magnetic nanoparticles will absorb energy and heat the surrounding tissue, affecting only the infected cells (Pankhurst et al. 2003).

However, this method still remains problematic for clinical use for several reasons (Koltsov and Perry 2004). In hyperthermia treatments, high magnetic fields are required for this technique to be effective. It is also difficult to localize enough magnetic particles in the cancerous region because the body's main defense system, the reticuloendothelial system, engulfs and removes any inert materials, so special coatings for the nanoparticles should be used to overcome this problem. Furthermore, to ensure the nanoparticles have a suitable Curie temperature, above which they no longer absorb energy, the size, shape and physical properties of the magnetic nanomaterials should be optimized.

1.5.2.2 Targeted Drug Delivery

As shown in Figure 1.13, the effectiveness of a drug is related to the drug dose over a certain dose range, and usually this relationship changes in the higher dose range. Often, toxicity sets in before the saturation limit is approached, and the therapeutic window with acceptable side effects is often narrow. The task of drug targeting is to push the systemic toxicity/side effect threshold to extremes such that the therapeutic window widens enough to cover the dose-response space up to the limit of local (target site) saturation (Plank *et al.* 2003b; Neuberger *et al.* 2005).

Magnetic nanoparticles can be used in targeted drug delivery. Functionalized magnetic nanoparticles used in drug delivery may contain drugs that are expected to be released into malignant cells and on-board sensors and actuators that control and regulate the drug release. As shown in Figure 1.14, in a drug delivery process, the drug molecules are bound to magnetic carriers. The process of drug localization is based on the competition between forces exerted on the carriers by blood compartment, and magnetic forces generated from the magnet. When the magnetic forces exceed the linear blood flow rates in arteries ($10 \, \text{cm} \, \text{s}^{-1}$) or capillaries ($0.05 \, \text{cm} \, \text{s}^{-1}$), the magnetic carriers are retained at the target site and maybe internalized by the endothelial cells of the target tissue. The use of nanoparticles favors the transport through the capillary systems of organs and tissues avoiding vessel embolism (Tartaj *et al.* 2003). Once the drugs/carriers are concentrated at

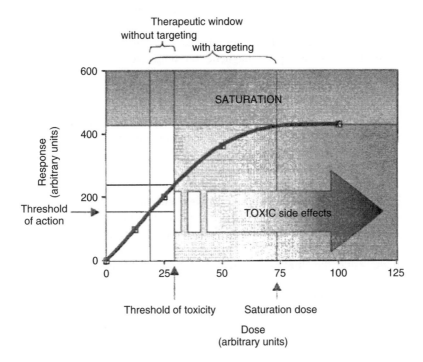

Figure 1.13 Toxic side effects often restrict the possibility of exploiting the full dose-response range of a drug up to (local) saturation levels. One objective of targeting is achieving target site saturation levels while pushing the non-target side toxicity threshold to higher doses. In this manner, the therapeutic window widens enough to achieve maximum local effect. Shown is a hypothetical dose-response relationship with arbitrary toxicity and saturation levels just to illustrate the potential of drug targeting. (Plank *et al.* 2003b)

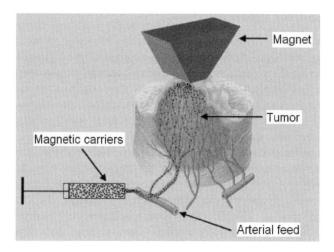

Figure 1.14 Schematic representation of the magnetically driven transport of drugs to a specific region. A catheter is inserted into an arterial feed to the tumor and a magnetic stand is positioned over the targeted site. (Alexiou *et al.* 2005)

the diseased site, the drugs are released from the carriers through modulation of magnetic field, enzymatic activity or changes in physiological conditions such as pH, osmolality or temperature. The released drugs then enter into the malignant cells. The treatment efficacy can be significantly improved by the magnetic 'tag-drag-release' process, and the required doses are simultaneously reduced.

The use of magnetic nanoparticles for drug delivery has significant advantages, such as the ability to target specific locations in the body, the reduction of the quantity of drug needed to attain a particular concentration in the vicinity of the target and the reduction of the concentration of the drug at non-target sites minimizing severe side effects (Arruebo *et al.* 2007). The advantages of targeted drug delivery make this method attractive for cancer treatment. Conventional chemotherapy has severe side effects as the agents used to kill cancerous tumors also kill healthy cells. In a targeted delivery approach, the chemotherapy agents are attached to magnetic nanoparticles, and the chemotherapy agents are then pulled towards malignant cells by a magnetic field focused on the target tumor. After the chemotherapy agents are aggregated around the tumor, the drug can be released by hitting the aggregate with an RF pulse, and therefore the drug concentration in the tumor is high, while the drug concentration at the rest of the body is relatively low (Could 2004).

To realize effective drug delivery, special attention should be paid to the choice of magnetic particles (Could 2004). Using a larger magnetic particle, a stronger magnetic force against blood flow can be exerted on the particle. Moreover, the particles should be small enough to avoid the risk of clogging small capillaries, whose diameter is about several micrometers. Magnetic nanoparticles for targeted drug delivery should also be completely biocompatible. It has been verified that iron oxide particles are non-toxic, and can be eventually used in the formation of blood hemoglobin. However, for gold-coated ferromagnetic particles, the situation is somewhat complicated. The small amount of gold may pass through the body eventually, and the iron will be metabolized. Cobalt is more stable than iron, and so it is easier for fabrication. However, cobalt is not suitable for *in vivo* applications due to its toxicity.

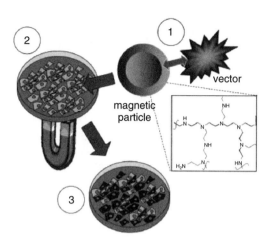

Figure 1.15 Procedure of magnetofection. Gene vectors are associated with magnetic particles coated with polyelectrolytes. Cells are cultured with the vector-magnetic particles under an external magnetic field. As the magnetic field attracts the vector-magnetic particles toward the cells, almost all the cells get in contact with vectors, and a high percentage of cells are rapidly transfected. (Plank 2003a)

1.5.2.3 Magnetofection

With advances in molecular biology and the sequencing of the human genome, gene therapy is playing a pivotal role in the treatment of genetic diseases. Gene therapy involves the introduction of healthy copies of mutated or absent genes into target cells so as to promote the expression of normal protein and to restore correct cellular function (Mehier-Humbert and Guy 2005). DNA drugs provide new hope in treating, preventing and controlling disease, almost certainly improving overall human health. Although non-viral DNA delivery systems have great therapeutic and prophylactic potential, their clinical utility has been limited by three major barriers (Luo 2005): inefficient uptake by the cell, insufficient release of DNA within the cell and ineffective nuclear targeting and transport. As the size of most cells is in the micrometer range, and the space inside a cell is extremely crowded, ideal DNA delivery systems must be in the nanometer range where they interact with or are delivered into a cell. Nanoscale science and nanotechnology provide more flexibility and precision to control the structure and composition of delivery systems, thus promising approaches for more effective and specific non-viral DNA delivery systems (Scherer *et al.* 2002; Plank *et al.* 2003a).

The magnetofection method is inspired by the method of targeted drug delivery using magnetic nanoparticles. Figure 1.15 illustrates the procedure of magnetofection. In this method, gene vectors are associated with superparamagnetic nanoparticles. Under the influence of magnetic gradient fields, the magnetic particles accumulate on the target cells. Due to the magnetic force exerted upon gene vectors, the entire vector dose is rapidly concentrated on the cells, and almost all the cells are in contact with a significant dose of gene vectors. The magnetic field may further push the magnetic particles associated with gene vectors across the plasma membrane and into the cell, and DNA is then released into the cytoplasm.

It should be noted that magnetofection is a method that enhances standard transfection procedures using viral or non-viral vectors. The magnetic field itself is ineffective unless the DNA vectors are complicated with magnetic particles. Magnetofection is applicable

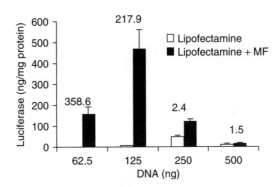

Figure 1.16 Luciferase expression in cultured HUVEC, using Lipofectamine with or without magnetofection (MF). Numbers above the bars represent the n-fold increase achieved by magnetofection compared to the conventional non-viral transfection technique. (Mehier-Humbert and Guy 2005)

to both viral and non-viral gene vectors, and it can be used in targeted gene delivery *in vitro* as well as *in vivo* (Plank 2003a). Magnetofection is a simple method. This method only requires a magnetic plate, and no expensive instruments are needed. However the magnetic plate should be specially designed so that the heterogeneous magnetic field generated by the plate can satisfy three requirements. First, the magnetic field should effectively magnetize the magnetic nanoparticles in solution. Second, the magnetic field should have a very strong gradient to attract the nanoparticles. Third, the magnetic field should cover the whole surface of the plate (Mehier-Humbert and Guy 2005).

As shown in Figure 1.16, magnetofection allows an increase in transfection efficiency by up to several hundred-fold (Mehier-Humbert and Guy 2005). In the experiment, the cell cultures subjected to magnetofection are placed on Nd-Fe-B magnetic plates for 15 minutes. Using a gene reporter encoding for Green Fluorescent Protein (GFP), transfection rates of about 40 % are achieved in human umbilical vein endothelial cells (HUVEC), which are well known as being difficult to transfect.

The magnetofection technique opens a novel perspective on gene delivery (Plank 2003a; Mehier-Humbert and Guy 2005; Scherer *et al.* 2002). Firstly, this technique can be used to rapidly and efficiently introduce nucleic acids into primary cells, and this will greatly improve the speed and efficiency of many gene-related tests, such as the examination of gene function, the identification of nucleic acids with therapeutic potential and the assessment of risks associated with the transfer of genetic material into cells. The high efficiency of magnetofection is mainly due to the quick sedimentation of the full vector dose on the target cells. In a magnetofection procedure, all the cells can be bound with vector particles in several minutes, while in a conventional transfection procedure, it takes several hours to achieve the same frequency of vector–target cell contact. Secondly, the magnetofection technique saves material. By using this technique, very low vector doses and extremely short incubation times are needed to achieve high transgene expression levels. In this way, the possible toxicity to the cells due to the transfection process can be avoided. Thirdly, it allows magnetic field-guided targeting. The force exerted on a magnetic particle in liquid suspension is related to the magnetic flux density, the magnetic field gradient and the particle volume. This provides flexibilities for optimizing these parameters to achieve the best effects. Fourthly, this method greatly profits from the fact that the individual modules of a system can be optimized independently and variants can be assembled in a combinatorial manner, thus facilitating optimization towards specific applications. The size and surface chemistry of magnetic particles can be tailored to meet specific demands on physical and biological characteristics; also the linkage between vector and magnetic particle can be designed accordingly.

There are several perspectives to the future use of magnetofection (Scherer *et al.* 2002). For *in vitro* application, magnetofection is particularly useful in the transfection/transduction of difficult-to-transfect/transduce cells, and it is an ideal research tool when the available vector dose, the process time and the sustainable costs are limited. Magnetofection is also a very good choice for *in vivo* gene- and nucleic acid-based therapies which require local treatments. With a simple magnetofection set-up, gene delivery to surgically accessible sites such as gut, stomach and vasculature can be greatly improved and further local applications can be achieved.

1.5.2.4 Mechanical Forces on Cells

The mechanical forces on cells play a vital role in controlling the forms and functions of cells, and the processes of many diseases (Could 2004). Generally speaking, the forces applied on biological entities generally fall into two types: pulling (tensional) forces or twisting (torsional) forces. The tensional force often results from the magnetic separation process, where the functionalized magnetic particles pull their attached targets along the external magnetic field gradient. The torsional force often results from coated ferromagnetic beads, which are first magnetized by a strong and short magnetic pulse, and subsequently subjected to a weak but constant magnetic field whose direction is perpendicular to the direction of the pulsed field. The ferromagnetic beads try to realign along the direction of the constant magnetic field, and the movement of the beads results in a twisting force on the surfaces of cells associated to the beads. The information about the stiffness and viscoelasticity of the surrounding cells could be derived from the extent of the rotation. This procedure can be improved to study the effects of the mechanical force inside individual cells. Many efforts have been made to investigate how external mechanical forces are transferred across specific cell receptors and how this translates into the changes in intracellular biochemistry and gene expression.

1.5.2.5 Detoxification

Magnetic nanoparticles can be used in the treatment after a poison gas attack. The detoxification of contaminated personnel and environment is based on the magnetic 'tag and drag' mechanism, which is often used in the targeted drug delivery (Could 2004). As shown in Figure 1.17, in the detoxification process, the magnetic nanoparticles, which are specially functionalized for catching the toxin to be detoxified, are injected into the body of the patient, and then an external magnetic field is applied to concentrate the toxin-tagged particles and draw them out of the patient's body. To achieve efficient detoxification, the magnetic moment of the magnetic particles should be high enough, so that the magnetic particles attached with toxin molecules can be quickly concentrated and drawn out of the body by an externally applied magnetic field gradient.

1.5.3 Magnetic Biosensors and Biochips Based on Magnetic Thin Films

Nanotechnologies have been used in the development of high sensitivity, high accuracy and high spatial resolution sensors. A sensor usually consists of a power supply, a sensing action element which converts the detected property into an electrical signal and a reporting unit which transmits the sensing signal to a remote detector. Using nanotechnologies, sensors can be made as small as possible so that they are minimally invasive. Secondly,

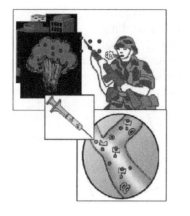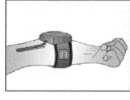

Figure 1.17   Mechanism of detoxification using magnetic nanoparticles. (Could 2004)

using nanotechnologies, the sensing element can be made to be very specific and accurate. Along with decrease of the sensor dimension, the sensing area also decreases, making increasing demands on sensitivity. This may require detection at the single molecule level, which is close to the limit of the length scale in nanotechnology. Therefore, nanoscience and nanotechnologies are expected to help the design and production of smaller, cheaper sensors with increasing selectivity (Royal Society and Royal Academy of Engineering 2004).

Nanosensors have been widely used in various areas, for example monitoring the quality of drinking water, detecting and tracking environmental pollution and checking food edibility. Nanosensors can also be used to achieve individualized healthcare with greater safety and security (Royal Society and Royal Academy of Engineering 2004).

Magnetic thin films have been used in the development of biosensors and biochips for biomedical applications. Most nanosensors developed from magnetic thin films are based on the magnetoresistance effect. The development of magnetoresistance nanosensors, along with other types of biosensors, is discussed in detail in Chapter 8. Here we briefly discuss biosensors and sensor arrays.

1.5.3.1   Biosensors

Biosensors usually detect the stray magnetic fields of magnetic nanoparticles which are attached to biological entities (Could 2004; Han 2006; Rife *et al.* 2003). In a magnetoresistive biosensor detection scheme, to perform genetic screening, iron oxide nanoparticles are functionalized, for example, with streptavidin, so that they can bind with targets containing biotinyl groups. Meanwhile, a surface is coated with biomolecular probes that can bind with complementary target species. As shown in Figure 1.18, the target hybridizes with the immobilized molecular probe, and links to the functionalized iron oxide magnetic label. The stray magnetic field of the captured magnetic nanoparticle is detected by the magnetic sensor beneath the functionalized surface.

1.5.3.2   Sensor Arrays

The efficiency of bio-analysis can be greatly improved by arraying biosensors. Generally speaking, there are two approaches to arraying biosensors. One approach is to array many sensors with the same functionality. Using such arrays, many bio-analyses of the same

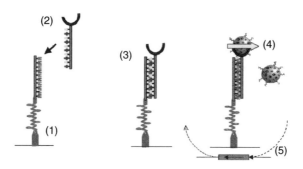

Figure 1.18 Detection scheme of a magnetoresistive (MR) sensor. (1) Immobilized probe DNA; (2) biotinilated target DNA; (3) biotinilated target DNA hybridizes with probe DNA; (4) magnetic label bound to biotinilated target; (5) MR sensor detects the fringe field from the magnetic label. (Could 2004)

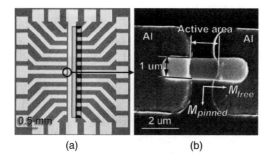

Figure 1.19 (a) Optical microscope image of a 1 μm wide spin valve sensor array and (b) scanning electron microscope image of a 1 μm wide sensor strip (center). The dark bar in the center of image (a) is the photoresist passivation layer. (Li *et al.* 2003)

type can be conducted concurrently. Figure 1.19 shows an array of 11 spin valve sensors (Li *et al.* 2003).

The other approach is to array different types of sensor together, resulting in multi-sensing chips, as shown in Figure 1.20. Using such an array, different bio-analysis on a sample could be simultaneously conducted using the different sensors on the array (Could 2004; Rife *et al.* 2003).

1.5.4 Trends of the Biomedical Applications of Magnetic Nanomaterials

Though great progress has been made in the applications of magnetic nanoparticles in biomedicine (Pankhurst 2003), many challenges remain. To realize the full potential of magnetic nanomaterials, researchers within different disciplines, such as physics, chemistry, materials science, electric and electronic engineering, biology and medicine, should work together. We discuss below two main trends in the biomedical applications of magnetic nanomaterials: multi-functional nanomaterials and lab-on-a-chip devices.

1.5.4.1 Multi-functional Nanomaterials

One of the main trends in the research of magnetic nanoparticles for biomedical applications is the development of magnetic nanoparticles that can perform more than one

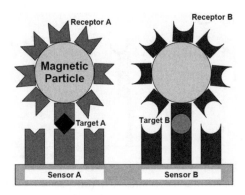

Figure 1.20 Detection of two different types of targets using an array with two types of sensors. (Rife *et al*. 2003). Reprinted from: Rife, J. C., Miller, M. M., Sheehan, P. E. Tamanaha, C. R. Tondra, M. Whitman, L. J. (2003). Design and performance of GMR sensors for the detection of magnetic microbeads in biosensors, *Sensors and Actuators A*, **107**, 209–218, with permission from Elsevier

function. A great deal of effort is being made to build a multi-functional nanoplatform that can be used to create desired multi-functional nanoparticles for diagnostic and/or therapeutic applications, as shown in Figure 1.21.

Figure 1.21(a) schematically shows a multi-functional nanoparticle for cancer therapy. The cancer cell targeting component ensures the nanoparticle reaches the cancer cells. The drug delivery indicator controls and reports the drug delivery. The cell death sensor reports whether the cancer cell has been killed, and the contrast agent checks the therapeutic effects. It should be noted that there is a long way to go to develop such a multi-functional nanoparticle, and at present it is still at an early stage (National Institute of Health and National Cancer Institute 2004). As shown in Figure 1.21(b), a multi-functional nanoplatform needs to be developed for the development of multi-functional nanoparticles. Such a multi-functional nanoplatform should include targeting unit, report unit, therapeutic unit, imaging unit and so on. Much effort needs to be made in developing such units, and constructing the nanoplatform using these units.

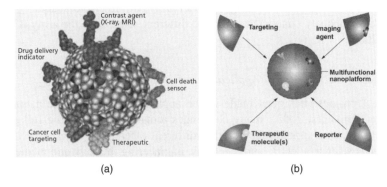

(a)   (b)

Figure 1.21 Development multi-functionality of particles for nanomedicine. (a) An example of multi-functional nanoparticle; (b) a multi-functional nanoplatform for the development of multi-functional nanoparticles. (National Institute of Health and National Cancer Institute 2004)

1.5.4.2 Microfluidics and Labs-on-chips

Many efforts have been made in the development of magnetoelectronic tools that can precisely detect and manipulate individual cells and biomolecules. The development of such magnetoelectronic tools is based on microfluidics, which deal with the knowledge and techniques for manipulating, investigating and utilizing tiny fluid volumes in a controlled way (Satyanarayana 2005). The microfluidic devices are usually built on microchips with channels conducting liquid under pressure or with an applied electrical current (Alper 2005b). This technology is expected to revolutionize many fields, especially chemistry, biology and medicine. In technical literatures, many words have been coined to describe such microfluidic devices: lab-on-a-chip device, micro total analysis system (μTAS), miniaturized analysis system, microfluidic system and nanofluidic system. In the following discussion, we use lab-on-a-chip and labs-on-chips.

A lab-on-a-chip device is a combination and integration of fluidic, sensor and detection elements to perform a complete sequence of chemical reactions or analyses, including sample preparation, mixing, reaction, separation and detection. In the example shown in Figure 1.22, microfluidic components with different functions are connected with each other to form an integrated system; therefore multiple functionalities can be realized on a single chip (Liu *et al.* 2004). In biomedical applications, the liquid in the channels may contain small particles, such as proteins, DNA or even single cells, so that changes during disease development can be monitored (Alper 2005b).

Lab-on-a-chip technology can be regarded as an interface between the nanoscale and the macro-world. It can be used for handling nanoparticles, cells or nanobarcodes, and for manipulating cellular machinery. However, it should be noted that the structures in a lab-on-a-chip device are not necessarily nanostructures, and they may be in the micrometer to even millimeter range. But, on the other hand, by integrating nanostructures, nanocoatings, nanoactuators and nano-detection and measurement tools such as nano-electrodes and nano-optics, more and more powerful microfluidic platforms are being developed.

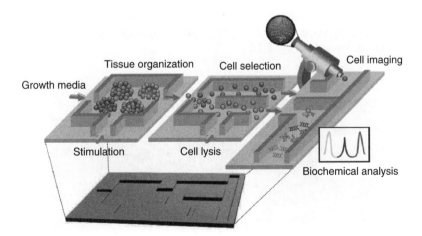

Figure 1.22 Tissue organization, culture and analysis in a lab-on-a chip device. (El-Ali *et al.* 2006)

Lab-on-a-chip technology exhibits both technical and economical advantages (Niemeyer and Mirkin 2004; Alper 2005b). First, as the surface to volume ratio of the fluids in lab-on-a-chip devices is extremely large, the surface effects dominate volume effects. This leads to well-defined flow characteristics: the flow is strictly laminar and turbulence only appears in very limited regions around sharp edges, the equilibrium conditions can be reached much faster and the capillary forces may advantageously be used for fluid transport. Meanwhile, the large surface implies a high reaction efficiency, as the surface areas which may be coated with catalysts or enzymes are large compared to the reaction volume. Second, the small sample volumes involved are of enormous advantage especially for highly parallel applications, like array devices used in genomics, proteomics and drug discovery. This fact is especially significant when analyzes are expensive or scarce. Furthermore, due to the small sample volumes needed, minimally invasive methods are sufficient for taking samples, for example blood or interstitial fluids. Third, lab-on-a-chip devices can be used to measure individual biomolecules, such as a single enzyme or a single piece of DNA. Because many biology effects are related to the behaviors of individual molecules, using this capability, we can investigate the topics that have been very difficult study, especially those related to rare biochemical and genetic events. Fourth, thousands of analyses can be performed concurrently by multiplexing individual microfluidic devices, or combining them in parallel. The integration and the mass-fabrication capabilities of micro-fabrication technology make the application of labs-on-chips economically attractive.

Lab-on-a-chip technologies show great application promise in cancer diagnosis (Satyanarayana 2005). As cancer is a very complicated fatal disease, it is necessary to identify quickly the mutations that may predispose one to cancer, and to study how the communications between cancer cells are made to cause the disease. The less invasive procedures and testing methods based on microfluidic technology can show early evidence of disease, and can be used to understand the circumstances that foster disease. This technique can provide genetic and proteomic information at the single-cell level which is valuable for the diagnosis and treatment of the disease. Furthermore, by using microfluidic devices, both the time and the amount of biological sample needed to conduct a large number of tests can be greatly reduced. As shown in Figure 1.23, in a lab-on-a-chip device which is slightly larger than a dime (with a diameter of 18 mm), thousands of experiments can be run concurrently, and this technique can be used to define the genetic bases of diseases, such as cancer (Balagadde *et al.* 2005).

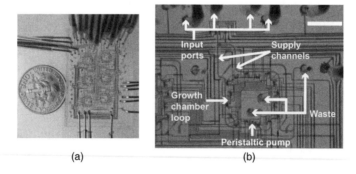

(a)   (b)

Figure 1.23   (a) Optical micrograph of six microchemostats that operate in parallel on a single chip; (b) optical micrograph of a single microchemostat and its main components. Scale bar, 2 mm. (Balagadde *et al.* 2005)

Though many problems for the design and fabrication of microfluidic devices have been solved, such as on-chip preparation of samples, functional complexity, integration level and fabrication cost (Liu *et al.* 2004), challenges still exist in the development of microfluidic devices. First, microfluidics is a new field of science and technology. When the dimensions of a system become smaller and smaller, the system behaviors may change greatly. More fundamental studies in fluid flows at the micro- and nanoscale are needed (Alper 2005b). Second, to satisfy the requirements of complex analysis, it is necessary to increase the integration density and the functionalities of lab-on-a-chip devices. Finally, before lab-on-a-chip devices can be widely used, they should be developed at lower cost, and in addition they should be portable and easy to operate.

References

Alexiou, C., Jurgons, R., Schmid, R., Erhardt, W., Parak, F., Bergemann, C. and Iro, H. (2005). Magnetic drug targeting – a new approach in locoregional tumortherapy with chemotherapeutic agents: experimental animal studies, *HNO*, **53**, 618–22.

Alper, J. (2005a). Why nano?, *Monthly Feather*, NCI Alliance for Nanotechnology in Cancer, September 2005.

Alper, J. (2005b). Chips ahoy – the lab-on-a-chip revolution is near, *Monthly Feather*, NCI Alliance for Nanotechnology in Cancer, April 2005.

Arruebo, M., Fernández-Pacheco, R., Ricardo Ibarra, M. and Santamaría, J. (2007). Magnetic nanoparticles for drug delivery, *Nanotoday*, **2** (3), 22–32.

Baibich, M.N., Broto, J.M., Fert, A., Nyuyen van Dau, F., Petroff, F., Eitenne, P., Creuzet, G., Friederich, A. and Chazelas, J. (1988). Giant magnetoresistance of (001) Fe/(001) Cr magnetic superlattices, *Physical Review Letters*, **61**, 2472–5.

Balagadde, F.K., You, L., Hansen, C.L., Arnold, F.H. and Quake, S.R. (2005). Long-term monitoring of bacteria undergoing programmed population control in a microchemostat, *Science*, **309**, 137–40.

Bauer, L.A., Birenbaum, N.S. and Meyer, G.J. (2004). Biological applications of high aspect ratio nanoparticles, *Journal of Materials Chemistry*, **14**, 517–26.

Binasch, G., Grunberg, P., Saurenbach, F. and Zinn, W. (1989). Enhanced magnetoresistance in layered magnetic structures with antiferromagnetic interlayer exchange, *Physical Review B*, **39**, 4828–30.

Cao, G. (2004). *Nanostructures and Nanomaterials: Synthesis, Properties, and Applications*, Imperial College Press, London.

Could, P. (2004). Nanoparticles probe biosystems, *Materials Today*, **7** (2), 36–43.

Darling, S.B. and Bader, S.D. (2005). A materials chemistry perspective on nanomagnetism, *Journal of Materials Chemistry*, **15**, 4189–95.

El-Ali, J., Sorger, P.K. and Jensen, K.F. (2006). Cells on chips, *Nature*, **442**, 403–11.

Freitas, R.A. Jr. (2005). What is nanomedicine?, *Nanomedicine: Nanotechnology, Biology, and Medicine*, **1**, 2–9.

Han, S.J., Xu, L., Yu, H., Wilson, R.J., White, R.L., Pourmand, N. and Wang, S.X. (2006). CMOS integrated DNA microarray based on GMR sensors, *Electron Devices Meeting, IEDM '06 International*, 1–4.

Himpsel, F.J., Ortega, J.E., Mankey, G.J. and Willis, R.F. (1998). Magnetic nanostructure, *Advances in Physics*, **47** (4), 511–97.

Hultgren, A., Tanase, M., Felton, E.J., Bhadriraju, K., Salem, A.K. Chen, C.S. and Reich, D.H. (2005). Optimization of yield in magnetic cell separations using nickel nanowires of different lengths, *Biotechnology Progress*, **21**, 509–15.

Hurst, S.J., Payne, E.K., Qin, L. and Mirkin, C.A. (2006). Multisegmented one-dimensional nanorods prepared by hard-template synthetic methods, *Angewandte Chemie Int. Ed.*, **45**, 2672–92.

Koltsov, D. and Perry, M. (2004). Magnets and nanometres: mutual attraction, *Physics World*, **17**(7), 31–5.

Li, G.X., Wang, S.X. and Sun, S.H. (2004). Model and experiment of detecting multiple magnetic nanoparticles as biomolecular labels by spin valve sensors, *IEEE Transactions on Magnetics*, **40**(4), 3000–2.

Li, G.X., Joshi, V., White, R.L., Wang, S.X., Kemp, J.T., Webb, C., Davis, R.W. and Sun, S.H. (2003). Detection of single micron-sized magnetic bead and magnetic nanoparticles using spin valve sensors for biological applications, *Journal of Applied Physics*, **93**(10), 7557–9.

Liu, R.H., Yang, J., Lenigk, R., Bonanno, J. and Grodzinski, P. (2004). Self-contained, fully integrated biochip for sample preparation, polymerase chain reaction amplification, and DNA microarray detection, *Analytical Chemistry*, **76**, 1824–31.

Luo, D. (2005). Nanotechnology and DNA Delivery, *MRS Bulletin*, **30**(9), 654–8.

Malsch, I. (2002). Biomedical applications of nanotechnology, *The Industrial Physicist*, **8**(3), 15–17.

Mehier-Humbert, S. and Guy, R.H. (2005). Physical methods for gene transfer: improving the kinetics of gene delivery into cells, *Advanced Drug Delivery Reviews*, **57**, 733–53.

National Institute of Health and National Cancer Institute. (2004). *Cancer Nanotechnology*, US Department of Health and Human Services.

National Science and Technology Council (2004). *National Nanotechnology Initiative Strategic Plan*, Executive Office of the President of the United States, Washington, DC.

National Science and Technology Council (1999a). *Nanotechnology: Shaping the World Atom by Atom*, Executive Office of the President of the United States, Washington, DC.

National Science and Technology Council (1999b). *Nanotechnology Research Directions: IWGN Workshop Report– Vision for Nanotechnology R&D in the Next Decade*, Executive Office of the President of the United States, Washington, DC.

National Research Council (1994). *Hierarchical Structures in Biology as a Guide for New Materials Technology*, National Academy Press, Washington DC.

Neuberger, T., Schopf, B., Hofmann, H., Hofmann, M. and von Rechenberg, B. (2005). Superparamagnetic nanoparticles for biomedical applications: Possibilities and limitations of a new drug delivery system, *Journal of Magnetism and Magnetic Materials*, **293**, 483–96.

Niemeyer, C.M. and Mirkin, C.A. (2004). *Nanobiotechnology: Concepts, Applications and Perspectives*, WILEY-VCH, Chichester.

Odenbach, S. (2004). Recent progress in magnetic fluid research, *Journal of Physics: Condensed Matter*, **16**, R1135–R1150.

Pankhurst, Q.A., Connolly, J., Jones, S.K. and Dobson, J. (2003). Applications of magnetic nanoparticles in biomedicine, *Journal of Physics D: Applied Physics*, **36**, R167–R181.

Parak, W.J. Gerion, D., Pellegrino, T., Zanchet, D., Micheel, C., Williams, S.C., Boudreau, R., Le Gros, M.A., Larabell, C.A. and Alivisatos, A.P. (2003). Biological applications of colloidal nanocrystals, *Nanotechnology*, **14**, R15–R27.

Parkin, S.S.P. (1995). Giant magnetoresistance in magnetic nanostructures, *Annual Review in Materials Science*, **25**, 357–88.

Plank, C., Schillinger, U., Scherer, F., Bergemann, C., Remy, J.S., Krotz, F., Anton, M., Lausier, J. and Rosenecker, J. (2003a). The magnetofection method: using magnetic force to enhance gene delivery, *Biological Chemistry*, **384**, 737–47.

Plank, C., Anton, M., Rudolph, C., Rosenecker, J. and Krötz, F. (2003b). Enhancing and targeting nucleic acid delivery by magnetic force, *Expert Opinion on Biological Therapy*, **3**(5), 745–58.

Poole, C.P. Jr. and Ownes, F.J. (2003). *Introduction to Nanotechnology*, John Wiley & Sons, Inc., Hoboken.

Pray, L.A. (2005). *Molecular Diagnostics: New Growth, New Markets*, Cambridge Healthtech Advisors, Waltham.

Ritchie, R. (2005). Whither 'nano' or 'bio'?, *Materials Today*, **8**(12), 72.

Reich, D.H., Tanase, M., Hultgren, A., Bauer, L.A., Chen, C.S. and Meyer, G.J. (2003). Biological applications of multifunctional magnetic nanowires, *Journal of Applied Physics*, **93**(10), 7275–80.

Rife, J.C., Miller, M.M., Shaheen, P.E., Tamanaha, C.R., Tondra, M. and Whitman, L.J. (2003) Design and performance of GMR sensors for the detection of magnetic microbeads in biosensors,, *Sensors and Actuators A*, **107**, 209–218.

Royal Society and Royal Academy of Engineering (2004). *Nanoscience and nanotechnologies: opportunities and uncertainties*, Royal Society and Royal Academy of Engineering, London.

Satyanarayana, M. (2005). Microfluidics: the flow of innovation continues, *Monthly Feather*, NCI Alliance for Nanotechnology in Cancer, August 2005.

Scherer, F., Anton, M., Schillinger, U., Henke, J., Bergemann, C., Kruger, A. and Gansbacher, B. (2002). Magnetofection: enhancing and targeting gene delivery by magnetic force *in vitro* and *in vivo*, *Gene Therapy*, **9**, 102–9.

Service, R.F. (2002). Biology offers nanotechs a helping hand, *Science*, **298**, 2322–3.

Son, S.J., Reichel, J., He, B., Schuchman, M. and Lee, S.B. (2005). Magnetic nanotubes for magnetic-field-assisted bioseparation, biointeraction, and drug delivery, *Journal of the American Chemical Society*, **127**, 7316–17.

Sun, L., Hao, Y., Chien, C.L. and Searson, P.C. (2005). Tuning the properties of magnetic nanowires, *Journal of Research and Development,* **49**(1), 79–102.

Taniguchi, N (1974). On the basic concept of nanotechnology, Proceedings of the International Congress on Production Engineering, Japanese Society of Precision Engineering, Tokyo, Japan.

Tartaj, P., Morales, M.D., Veintemillas-Verdaguer, S., Gonzalez-Carreno, T. and Serna, C.J. (2003) The preparation of magnetic nanoparticles for applications in biomedicine, *Journal of Physics D- Applied Physics*, **36**, R182–R197.

Taton, T.A. (2003). Bio-nanotechnology two-way traffic, *Nature Materials*, **2**, 73–4.

Whitesides, G.M. and Wong, A.P. (2006). The Intersection of Biology and Materials Science, *MRS Bulletin*, **31**(1), 19–27.

Whitesides, G.M. (2005). Nanoscience, nanotechnology, and chemistry, *Small*, **1**(2), 172–9.

Whitesides, G.M. (2003). The 'right' size in nanobiotechnology, *Nature Biotechnology*, **21**(10), 1161–5.

Williams, K.A., Veenhuizen, P.T.M., de la Torre, B.G., Eritja, R. and Dekker, C. (2002). Carbon nanotubes with DNA recognition, *Nature*, **420**(6917), 761.

Wood, D.K., Ni, K.K., Schmidt, D.R. and Cleland, A.N. (2005). Submicron giant magnetoresistive sensors for biological applications, *Sensors and Actuators A* **120**, 1–6.

Yih, T.C. and Wei, C. (2005). Nanomedicine in cancer treatment, *Nanomedicine: Nanotechnology, Biology, and Medicine*, **1**, 191–2.

2

Physical Background for the Biomedical Applications of Functional Magnetic Nanomaterials

2.1 Requirements for Biomedical Applications

Magnetic nanomaterials have been widely used for diagnostic and therapeutic applications. In biomedical applications, magnetic nanoparticles are usually in the form of magnetic beads made by embedding magnetic nanoparticles in a suitable matrix. Usually magnetic beads are tailor-made to meet the requirements for specific applications (Gijs 2004). In the design and fabrication of magnetic beads, the biocompatibility, biodegradability and stability should be taken into full consideration, and meanwhile the shape and size of the magnetic beads should be controlled. The physical properties of the beads, such as the content of iron oxide, which determine the magnetic behavior of the beads, need to be carefully designed. Meanwhile, suitable bead surface modifications on magnetic beads are needed to allow covalent bonding or simple unspecific adsorption of biomolecules.

2.1.1 Magnetic Particles and Ferrofluids

The magnetic nanomaterials in most of the biomedical applications are in the form of ferrofluids, which are stable dispersions of magnetic beads in an organic or aqueous carrier medium. Ferrofluids are also called magnetic bead solutions or magnetic fluids. In order to obtain a stable dispersion of magnetic particles in an aqueous medium, the characteristics of the particle surface have to be tailored to the medium. Particle-solvent interactions and interparticle repulsions must be strong enough to overcome Van der Waals attraction between the particles and magnetic attraction in the case of particles with a permanent magnetic moment (Connolly and St Pierre 2001).

2.1.1.1 Magnetic Particles

Magnetic particles are magnetically responsive solid phases, and they could be nanoparticles or aggregates of nanoparticles of micro- to nanometer size. Under an external magnetic field, magnetic particles will rotate. To move the particles in a preferred direction of space, an inhomogeneous magnetic field should be used. The magnetic force acting on such particles in a liquid suspension is proportional to the magnetization of the particle, the magnetic flux density and the magnetic field gradient (Plank 2003a). The most frequently used magnetic particles are ferrites, having the general composition MFe_2O_4 (with M being a bivalent metal cation such as Ni, Co, Mg or Zn and including magnetite Fe_3O_4) and maghemite Fe_2O_3 (Halbreich *et al.* 1998).

Magnetic nanoparticles offer attractive advantages for their applications in biomedicine (Pankhurst *et al.* 2003). First, their sizes could be controlled in the range of a few nanometers to tens of nanometers, so they could be smaller than or comparable to a cell (10–100 μm), a virus (20–450 nm), a protein (5–50 nm) or a gene (2 nm wide and 10–100 nm long). Therefore they can get close to, or enter a biological entity of interest. If coated with suitable biological molecules, they can interact with or bind to a biological entity, providing a controllable means of tagging for the biological entity. Second, magnetic nanoparticles can be manipulated by an external magnetic field gradient. In combination with the intrinsic penetrability of magnetic fields into human tissue, this 'action at a distance' opens up many applications involving the transport and immobilization of magnetic nanoparticles, or of magnetically tagged biological entities. Therefore, they can be used to deliver a package, such as an anticancer drug, to a targeted region of the body, such as a tumor. Third, magnetic nanoparticles can resonantly respond to a time-varying magnetic field, transferring energy from the exciting field to the nanoparticles and therefore the nanoparticles can heat up. Therefore, magnetic nanoparticles can be used as hyperthermia agents, delivering toxic amounts of thermal energy to targeted bodies such as tumors; or as chemotherapy and radiotherapy enhancement agents.

In the development of magnetic nanoparticles, three aspects should be taken into consideration (Parak *et al.* 2003). First, magnetic nanoparticles should be crystalline, and each particle consist of only one domain. Second, the size distribution of the nanoparticles should be as narrow as possible. Third, all the magnetic nanoparticles in a particular sample should have a unique and uniform shape. Besides spherical nanoparticles which are most widely used, nanoparticles with more complex geometries, such as nanowires and nanotubes, are also often used.

2.1.1.2 Ferrofluids

In most of the biomedical applications of magnetic nanoparticles, they are dispersed in a suitable liquid, and stabilized in a way that prevents agglomeration, forming a colloidal solutions. Ferrofluids were initially produced by grinding large particles in suitable organic solvents und sieving, while now most of the ferrofluids are prepared chemically.

To obtain a stable ferrofluid in physiological media at neutral pH and appropriate ionic strength, the particle surface should be functionalized. Usually these particles are coated with dextran, albumin or synthetic polymers such as methacrylates and organosilanes, and the effector, usually an antibody, is attached to the particle through a covalent bond to the coating Polymer. The stability of the coating determines the stability of the effector-particle complex (Halbreich *et al.* 1998).

2.1.2 Biocompatibility and Chemical Stability

The above discussion indicates that the magnetic nanomaterials in biomedical applications are usually in the form of ferrofluids. Usually ferrofluids are water colloids of magnetic nanomaterials. The ferrofluids and the magnetic nanomaterials in the ferrofluids should be biocompatible and chemically stable. To achieve this, the size of the magnetic nanomaterials should be suitably chosen, and the surfaces of the magnetic nanoparticles should be functionalized to ensure that magnetic nanoparticles can be stably suspended in the fluids and can be chemically stable so that they are biocompatible.

2.1.3 Magnetic Properties

One of the key points in biomedical applications is the remote control or manipulation of biological entities or drugs. Usually this is realized using external magnetic fields. Therefore the magnetic properties of the magnetic nanoparticles attached to biological entities or drugs are very important. Usually magnetic nanomaterials have superparamegnetic properties. Detailed discussion on the magnetic properties of magnetic nanomaterials can be found in the second section of this chapter.

2.1.4 Physical Properties

To realize the manipulation of ferrofluids and perform desired biomedical functions, the physical properties of ferrofluids and magnetic nanoparticles should satisfy special requirements. The typical physical properties of ferrofluids which play important roles in the biomedical applications mainly include viscosity and magnetic relaxation. In some biomedical applications, especially in cases where magnetic nanoparticles are used as tools to apply mechanical forces on biological entities, we should also consider the mechanical properties of the magnetic nanomaterials. As shown in Figure 2.1, magnetic nanowires may penetrate the cell membranes, and get inside the cells, for drug or gene delivery, or cell separations. In these cases, the magnetic nanowires should have sufficient mechanical strength.

2.2 Fundamentals of Nanomagnetism

Nanomagnetism have been under intensive investigation for decades (Darling and Bader 2005; Himpsel *et al.* 1998). In this section, we aim to outline the fundamental magnetic properties of magnetic nanomaterials and discuss the theoretical frameworks for analyzing these properties. We lay our emphasis on the fundamentals of nanomagnetism related to their biomedical applications and we will concentrate on the magnetism of magnetic

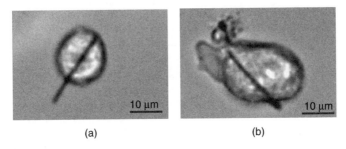

(a)   (b)

Figure 2.1 Optical images of trypsinized 3T3 cells penetrated by nanowires. (a) a cell (average diameter 15 μm) with 22-μm nanowire. (b) a cell (average diameter 23 μm) with 22 μm nanowire. (Hultgren *et al.* 2004)

nanospheres. Special attention will be paid to nanospheres with strict spherical structures, and the effects of shape variations, such as triangular, square and pentagonal, on the magnetic properties can be found in Cowburn (2000). The magnetic properties of magnetic nanowires, magnetic nanotubes and magnetic thin films are discussed in Chapters 6, 7 and 8 respectively.

2.2.1 Basic Concepts of Nanomagnetism

Before introducing the basic concepts of nanomagnetization, we give some explanations about magnetic units, as the mix of units in technical literatures may confuse or mislead readers with different backgrounds. In discussing the engineering applications of magnetic materials, SI units are usually used. However, in more fundamental research, cgs units are often used. In the cgs unit system, magnetization, M, is measured in emu/cm^3 (electromagnetic units), magnetic field strength, H, is measured in Oersted (Oe), and the permeability of vacuum is taken as one. The relationship between magnetic induction (**B**), magnetic field strength (**H**) and magnetization (**M**) is given by $\mathbf{B} = \mathbf{H} + 4\pi\mathbf{M}$, with **B** in gauss. Scholten (1995) provides a convenient overview of the various magnetic unit conventions.

2.2.1.1 Classification of Nanostructure Morphologies

The nanostructure morphology of magnetic nanomaterials can be classified according to the relationship between nanostructure and magnetic properties. Here we discuss a classification method, which emphasizes the physical mechanisms responsible for the behavior of magnetic nanomaterials (Leslie-Pelecky and Rieke 1996). As shown in Figure 2.2, there are four general classifications of magnetic nanostructured materials ranging from non-interacting particles (type A) in which the magnetization is determined strictly by size effects, to fine-grained nanostructures, in which interactions dominate the magnetic properties. At one extreme, denoted type A, are the systems consisting of isolated

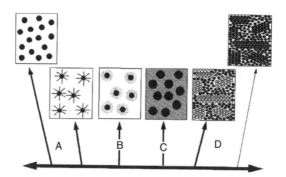

Figure 2.2 Classification of nanostructure morphologies. In a type A material, the inter-particle spacing is large enough to approximate the particles as non-interacting. Ferrofluids, in which magnetic particles are surrounded by a surfactant preventing interactions, are a subgroup of Type A materials. Type B materials are ultra-fine particles with a core–shell morphology. Type C nanocomposites are composed of small magnetic particles embedded in a chemically dissimilar matrix. The matrix may or may not be magnetoactive. Type D materials consist of small crystallites dispersed in a non-crystalline matrix. In such a nanostructure, the nanoparticles may be in a distinct phase from that of the matrix, while in an ideal case, both the nanoparticles and the matrix are made of the same material. (Leslie-Pelecky and Rieke 1996)

particles with nanoscale diameters. The magnetic properties of these non-interacting systems can be derived strictly from the reduced size of the components, and there is no contribution from interparticle interactions. At the other extreme, denoted type D, are bulk materials with nanoscale structure. In a bulk material, a significant fraction of the volume (up to 50%) is composed of grain boundaries and interfaces, and the interactions between the nanoparticles dominate the magnetic properties of a bulk material. The length scale of the interactions is critically dependent on the characteristics of the interphase, and may span many grains. As the magnetic properties of type D bulk materials are dominated by the interactions and grain boundaries, the magnetic behavior of these materials cannot be predicted simply by applying theories for polycrystalline materials with reduced length scales. In most cases, the magnetic behavior of a magnetic nanomaterial is a result of contributions from both interaction and size effects. Type B and type C are two typical intermediate forms. Type B particles are ultra-fine particles with a core–shell morphology. The presence of a shell can help prevent particle-particle interactions, but the interactions between the core and the shell may affect the overall magnetic performance of the material. In some cases, the shells themselves can be magnetic. Type C nanocomposite materials consist of two chemically dissimilar materials: magnetic particles distributed throughout a matrix. The magnetic interactions between the magnetic particles are mainly determined by the volume fraction of the magnetic particles and the character of the matrix (Leslie-Pelecky and Rieke 1996).

2.2.1.2 Fundamental Magnetic Lengths

One of the fundamental motivations for the investigation of magnetic nanomaterials is the impressive change of magnetic properties that occurs when the critical length governing some phenomenon is comparable to the size of the nanoparticle. Changes in the magnetization of a material occur via activation over an energy barrier, and each physical mechanism responsible for an energy barrier has an associated length scale. The fundamental magnetic lengths for magnetic materials mainly include the crystalline anisotropy length, l_K, the applied field length, l_H, and the magnetostatic length, l_S, as defined below (Leslie-Pelecky and Rieke 1996):

$$l_K = \sqrt{J/K} \tag{2.1}$$

$$l_H = \sqrt{2J/HM_S} \tag{2.2}$$

$$l_S = \sqrt{J/2\pi M_S^2} \tag{2.3}$$

where K is the anisotropy constant of a bulk material due to the dominant anisotropy, and J is the exchange within a grain. If there is more than one type of barrier, the magnetic properties of the magnetic material are dominated by the shortest characteristic length. The fundamental magnetic lengths for most magnetic materials are on the order of 1–100 nm. For example, at 1000 Oe and room temperature, nickel has lengths $l_S \approx 8$ nm, $l_K \approx 45$ nm and $l_H \approx 19$ nm.

2.2.1.3 Hysteresis Loops

Most of the magnetic properties of a material can be derived from its hysteresis loop (Leslie-Pelecky and Rieke 1996). Figure 2.3 schematically illustrates a magnetization *vs* field (M–H) hysteresis loop. When the external magnetic field is sufficiently large, all the

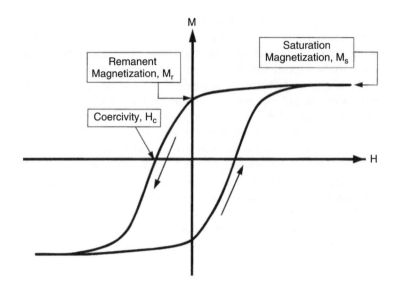

Figure 2.3 A typical magnetization *vs* field (M–H) hysteresis loop of a ferromagnetic material. Several important parameters are shown: saturation magnetization M_s, remanent magnetization M_r, and coercivity H_c. (Leslie-Pelecky and Rieke 1996)

spins within a magnetic material align with the applied magnetic field. In this state, the magnetization of the material achieves its maximum value, and this value is called the saturation magnetization, M_s. When the external magnetic field becomes weaker, the spins in the material cease to be aligned with the external magnetic field, so the total magnetization of the material decreases. For a ferromagnetic material, when the external magnetic field decreases to zero, the material still has a residual magnetic moment, and the value of the magnetization at zero field is called the remanent magnetization, M_r. The remanence ratio is defined as the ratio of the remanent magnetization to the saturation magnetization, M_r/M_s, which varies from 0 to 1. To bring the material back to zero magnetization, a magnetic field in the negative direction should be applied, and the magnitude of the field is called the coercive field, H_c. The re-orientation and growth of spontaneously magnetized domains within a magnetic material depends on both microstructural features such as vacancies, impurities or grain boundaries, and intrinsic features such as magnetocrystalline anisotropy as well as the shape and size of the particle. In most cases, the hysteresis loop of a magnetic material should be experimentally measured using, for example, a vibrating sample magnetometer (VSM) or superconducting quantum interference device SQUID magnetometer, and it is not possible to predict a priori what the hysteresis loop will look like.

Materials with different magnetic properties have different shapes of hysteresis loops. Figure 2.4 shows a schematic diagram of a blood vessel into which some magnetic nanoparticles have been injected, and the magnetic properties of both the injected particles and the ambient biomolecules in the blood stream (Pankhurst *et al.* 2003). Generally speaking, the blood vessel and the biomaterials in the blood vessel are either diamagnetic or paramagnetic, while the injected magnetic particles are either ferromagnetic or superparamagnetic, depending on their sizes.

When a magnetic field of strength **H** is applied to a magnetic material, the individual atomic moments in the material contribute to the overall magnetic induction **B** of the

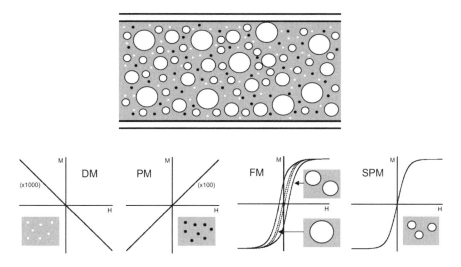

Figure 2.4 A hypothetical situation in which ferromagnetic particles with a range of sizes from nanometer up to micron scale are injected into a blood vessel. The magnetic responses associated with different classes of magnetic material are illustrated by their corresponding M–H curves. The biomaterials in the blood vessel are diamagnetic (DM) or paramagnetic (PM). Depending on their sizes, the injected particles can be ferromagnetic (FM) or superparamagnetic (SPM). Ferromagnetic materials can be multi-domain (- - - - in FM diagram) or single-domain (_____ in FM diagram). (Pankhurst *et al.* 2003)

material:

$$\mathbf{B} = \mu_0(\mathbf{H} + \mathbf{M}) \tag{2.4}$$

where μ_0 is the permeability of free space, and the magnetization $\mathbf{M} = \mathbf{m}/V$ is the magnetic moment per unit volume, where $\mathbf{m}$ is the magnetic moment on a volume V of the material.

Broadly, all materials can be regarded as magnetic materials because all the materials respond to magnetic fields to some extent. However, they are usually classified based on their volumetric magnetic susceptibility, χ, describing the relationship between the magnetic field $\mathbf{H}$ and the magnetization $\mathbf{M}$ induced in a material by the magnetic field:

$$\mathbf{M} = \chi\mathbf{H} \tag{2.5}$$

In the SI unit system, χ is dimensionless, while both M and H are expressed in Am^{-1}. Most materials display little magnetism, and these are classified either as paramagnets or diamagnets. The χ value for paramagnets is usually in the range of 10^{-6} to 10^{-1}, while the χ value for diamagnets is usually in the range -10^{-6} to -10^{-3}. Negative χ value of diamagnets indicates that in such materials, the magnetization $\mathbf{M}$ and the magnetic field $\mathbf{H}$ are in opposite directions. However, some materials exhibit ordered magnetic states, and they are usually classified as ferromagnets, ferrimagnets and antiferromagnets. The prefixes of these names refer to the nature of the coupling interactions between the electrons within the material (Pankhurst *et al.* 2003). Such couplings may lead to large spontaneous magnetizations, and this is the reason why ordered magnetic materials usually have much larger χ values than paramagnetic or diamagnetic materials.

It should be noted that the susceptibility in ordered materials also depends on applied magnetic field **H**. This magnetic field gives rise to the characteristic sigmoidal shape of the M–H curve, with **M** approaching a saturation value at high magnetic field. In ferromagnetic and ferrimagnetic materials, hysteresis loops can be observed, as shown in Figure 2.3. The shape of a hysteresis loop is partly determined by the particle size. A particle in the order of micron size or more usually has a multi-domain structure. As it is easy to make the domain walls move, the hysteresis loop of such particles is narrow. In a smaller particle, the single-domain structure leads to a broad hysteresis loop. When particle size become even smaller, in the order of tens of nanometers or less, superparamagnetism can be found. The magnetic moment of a superparamagnetic particle as a whole is free to fluctuate in response to thermal energy, while the individual atomic moments maintain their ordered state relative to each other. As shown in Figure 2.4, the M–H curve of a superparamagnetic particle is anhysteretic, but still sigmoidal (Pankhurst *et al.* 2003).

2.2.1.4   Magnetic Anisotropy

Most materials contain some type of anisotropy affecting their magnetization behaviors. The magnetic anisotropy of a material can be modeled as uniaxial in character and represented by (Leslie-Pelecky and Rieke 1996):

$$E = KV \sin^2 \theta \tag{2.6}$$

where K is the effective uniaxial anisotropy energy per unit volume, θ is the angle between the moment and the easy axis, and V is the particle volume. In the following, we discuss two general types of magnetic anisotropy: bulk anisotropy and surface anisotropy.

(1)   Bulk Anisotropy

The most common types of anisotropy include crystal anisotropy, shape anisotropy, stress anisotropy, externally induced anisotropy and exchange anisotropy (Leslie-Pelecky and Rieke 1996). The two typical anisotropies in nanostructured materials are crystalline anisotropy and shape anisotropy.

Arising from spin-orbit coupling, magnetocrystalline anisotropy is specific to a given material and independent of particle shape. Magnetocrystalline anisotropy energetically favors alignment of the magnetization along a specific crystallographic direction, and this direction is called the easy axis of the material. The magnetocrystalline anisotropy properties of typical ferromagnetic materials are listed in Table 2.1. As the coercivity is proportional to the anisotropy constant, high-anisotropy materials are attractive candidates for high-coercivity applications (Leslie-Pelecky and Rieke 1996).

Table 2.1   Magnetocrystalline anisotropy of typical ferromagnetic materials at room temperature

Material	Easy axis	Magnitude at room temperature (erg/cm^3)
Hcp Co	c axis	7×10^6
cubic Fe	<100>	8×10^5
cubic Ni	<111>	5×10^4

Due to averaging over all orientations, a polycrystalline specimen with no preferred grain orientation does not have net crystal anisotropy. However, a nonspherical polycrystalline sample possesses shape anisotropy (Leslie-Pelecky and Rieke 1996). For example, it is easier to magnetize a cylindrical sample along its long axis than along its short axes. Samples with symmetric shapes, such as a sphere, have no net shape anisotropy. Shape anisotropy causes large coercive forces. An increase of the aspect ratio from 1.1 to 1.5 in single-domain iron particles with easy axis aligned along the field quadruples the coercivity, and a further increase in the aspect ratio to 5 produces another doubling of the coercivity.

Besides crystalline anisotropy and shape anisotropy, there are other types of bulk anisotropies, such as stress anisotropy and exchange anisotropy (Leslie-Pelecky and Rieke 1996). Stress anisotropy is caused by external or internal stresses due to rapid cooling, application of external pressure, etc., and may also be induced by annealing in a magnetic field, plastic deformation or ion beam irradiation. Exchange anisotropy occurs when a ferromagnetic material is in close proximity to an antiferromagnetic material or ferromagnetic material. Magnetic coupling at the interface of the two materials can create a preferential direction in the ferromagnetic phase, which takes the form of a unidirectional anisotropy. This type of anisotropy is often observed in type B particles shown in Figure 2.2, when an antiferromagnetic or ferrimagnetic oxide forms around a ferromagnetic core.

(2) Surface Anisotropy

Usually the anisotropy of fine metallic and oxide particles increases as the size decreases, and the increase of anisotropy is mainly due to the contribution of surface anisotropy. Surface anisotropy has a crystal-field nature which comes from the symmetry breaking at the boundaries of the particle. The structural relaxation yielding the contraction of surface layers and the existence of some degree of atomic disorder and vacancies induce local crystal fields with predominant axial character normal to the surface, which may produce easy-axis or easy-plane anisotropies (Batlle and Labarta 2002). The axis of the local crystal field, $\hat{n}$, can be evaluated from the dipole moment of the nearest-neighbor atomic positions with respect to the position of a given surface atom as:

$$\hat{n}_i \propto \sum_{j}^{nn} (P_j - P_i) \tag{2.7}$$

where P_i is the position of the ith atom and the sum extends to the nearest neighbors of this atom. Because some of the neighbors at the surface are missing, $\hat{n}_i$ is not zero and directed approximately normal to the surface. The effect of these local fields can be modeled by adding to the Hamiltonian a term in the form of KS_ς^2, where S_ς is the component of the spin along a vector normal to the surface, and $K < 0$ corresponds to the easy-axis case and $K > 0$ to the easy plane one (Batlle and Labarta 2002). If $|K|$ is comparable to the ferromagnetic exchange energies, we can obtain spin configurations similar to those shown in Figure 2.5. Such configurations are due to the competition between surface anisotropy and ferromagnetic alignment. Generally speaking, the surface anisotropy makes the surface layer of a particle magnetically harder than the core of the particle.

The existence of boundaries also causes lattice deformations at the surface, and thus results in strains at the surface. Through magnetostriction effects, such strains induce an additional surface anisotropy (Batlle and Labarta 2002). Usually thin films exhibit strong strain anisotropy because stresses always exist at the interface between substrate and film

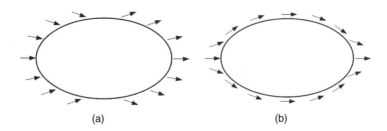

Figure 2.5   Surface spin arrangement of a ferromagnetic particle with a surface anisotropy of the form KS_s^2. Case (a) corresponds to $K < 0$ (radial) and case (b) corresponds to $K>0$ (tangential). (Batlle and Labarta 2002)

due to lattice mismatch. But the strain energy for most of the nanoparticles is weak, and so the contribution of the strain energy could be neglected, especially for free particles. It should be noted that as surface anisotropy is normal to surface, the surface anisotropy of a spherical particle would be averaged to zero based on symmetry arguments.

In analyzing the anisotropy of magnetic nanomaterials, it is necessary to include the contributions from both the core and the surface. The core contributes the bulk anisotropy, and the surface contributes the surface anisotropy. Usually an effective anisotropy energy per unit volume, K_{eff}, is used. For a spherical particle the K_{eff} is give by (Batlle and Labarta 2002):

$$K_{eff} = K_b + \frac{6}{d}K_S \qquad (2.8)$$

where K_b is the bulk anisotropy energy per unit volume, K_s is the surface density of anisotropy energy and d is the diameter of the particle.

2.2.1.5   Single-domain Particles

A domain is a group of spins whose magnetic moments are in the same direction, and in the magnetization procedure, they act cooperatively. In a bulk material, domains are separated by domain walls, which have a characteristic width and energy associated with their formation and existence. The movement of domain walls is a primary means of reversing magnetization and a major source of energy dissipation.

Figure 2.6 schematically shows the relationship between the coercivity in particle systems and particle sizes (Leslie-Pelecky and Rieke 1996). In a large particle, energetic considerations favor the formation of domain walls, forming a multi-domain structure. The magnetization of such a particle is realized through the nucleation and motion of these walls. As the particle size decreases toward a critical particle diameter, D_c, the formation of domain walls becomes energetically unfavorable. So there is no domain wall in such a particle, and this particle is called a single-domain particle. For a single-domain particle, the magnetization procedure is realized through the coherent rotation of spins. The particles with size close to D_c usually have large coercivities. As the particle size is much smaller than D_c, the spins in this particle are affected by thermal fluctuations, and such a single-domain particle is usually called a superparamagnetic particle, which will be discussed in Section 2.2.2.

Frenkel and Dorfman (1930) theoretically predicted the existence of single-domain particles. The D_c values for some typical magnetic materials of spherical shape are listed in Table 2.2. It should be noted that particles with significant shape anisotropy usually have

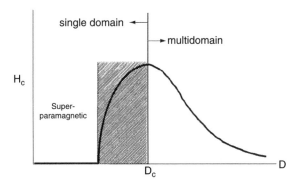

Figure 2.6 The relationship between the coercivity in ultrafine particle systems and particle sizes. (Leslie-Pelecky and Rieke 1996)

Table 2.2 Critical single-domain sizes, D_c, for spherical particles with no shape anisotropy. Adapted from Leslie-Pelecky, D.L. and Rieke, R.D. (1996) Magnetic Properties of Nanostructured Materials. Chemistry of Materials, **8**(8), 1770–83.

Material	Dc (nm)
Co	70
Fe	14
Ni	55
Fe_3O_4	128
γ-Fe_2O_3	166

a larger effective critical single-domain diameter than corresponding spherical particles (Leslie-Pelecky and Rieke 1996).

2.2.1.6   Time Dependence of Magnetization

The time dependence of magnetization of a material is important for its engineering applications and for investigating the fundamental mechanisms of magnetism. The variation of magnetization with time of a magnetic material can be described by:

$$\frac{dM(t)}{dt} = -\frac{M(t) - M(t = \infty)}{\tau} \qquad (2.9)$$

where $M(t = \infty)$ is the magnetization at the equilibrium state, and τ is a characteristic relaxation time given by:

$$\tau = \frac{1}{f_0} \exp\left(\frac{\Delta E}{kT}\right) \qquad (2.10)$$

For a uniaxial anisotropy, the energy barrier, ΔE, is equal to the product of the anisotropy constant and the volume. In most cases, f_0 is often taken as a constant of value $10^9 \, s^{-1}$. As the behavior of τ is dominated by the exponential argument, the accurate value of f_0 is usually not necessary. However, the particle size greatly affects the relaxation time. If we choose typical values $f_0 = 10^9 \, s^{-1}$, $K = 10^6 \, erg/cm^3$, and $T = 300 \, K$, the relaxation time of a particle with diameter of 11.4 nm is 0.1 s, while the relaxation time of a particle with diameter of 14.6 nm is $10^8 \, s$ (Leslie-Pelecky and Rieke 1996).

If all components of a system have the same relaxation time, Equation (2.9) offers the simplest solution. However this assumption is not applicable to real systems because of the distribution of energy barriers in real systems. The energy barrier distribution may be related to the variations of a lot of parameters, such as particle sizes, anisotropies or compositional inhomogeneity, and the distribution of energy barriers causes a distribution of relaxation times. If the distribution of energy barriers is constant, the magnetization decays logarithmically (Leslie-Pelecky and Rieke 1996):

$$M(t) = M(t = 0) - S \ln(t) \qquad (2.11)$$

where the magnetic viscosity, S, is related to the energy barrier distribution. If the distribution of energy barriers is constant, deviations from the ln(t) behavior can be observed. To keep Equation (2.11) applicable, the magnetic viscosity, S, should be accordingly modified (Leslie-Pelecky and Rieke 1996).

2.2.2 Superparamagnetism

Neel (1949) theoretically demonstrated that H_c approaches zero when particles become very small because the thermal fluctuations of very small particles prevent the existence of a stable magnetization. This is a typical phenomenon of superparamagnetism. There are two experimental criteria for superparamagnetism (Bean and Jacobs 1956). First, the magnetization curve exhibits no hysteresis, and second the magnetization curves at different temperatures must superpose in a plot of M vs H/T. Figure 2.7 shows the magnetization curves of iron amalgam on H/T bases. Measurements were made at 77 K and 200 K respectively, and the magnetization curves at 77 K and 200 K superpose each other. The imperfect H/T superposition may be due to a broad distribution of particle sizes, changes in the spontaneous magnetization of the particle as a function of temperature or anisotropy effects.

The basic mechanism of superparamagnetism is based on the relaxation time τ of the net magnetization of a magnetic particle (Brown 1963):

$$\tau = \tau_0 \exp \left(\frac{\Delta E}{k_B T} \right) \qquad (2.12)$$

where ΔE is the energy barrier to moment reversal, and $k_B T$ is the thermal energy. For non-interacting particles the pre-exponential factor τ_0 is in the order of $10^{-10} - 10^{-12} \, s$ and only weakly dependent on temperature. The energy barrier has several origins, including both intrinsic and extrinsic effects such as the magnetocrystalline and shape anisotropies, respectively (Pankhurst et $al.$ 2003). However, in the simplest cases, it is given by $\Delta E = KV$, where K is the anisotropy energy density and V is the particle volume. For small particles, ΔE is comparable to $k_B T$ at room temperature, so superparamagnetism is important for small particles.

It should be noted that, for a given material, the observation of superparamagnetism is dependent not only on temperature, but also on the measurement time τ_m of the experimental technique used (Pankhurst et $al.$ 2003). As shown in Figure 2.8, if $\tau \ll \tau_m$, the flipping is fast relative to the experimental time window and the particles appear to be

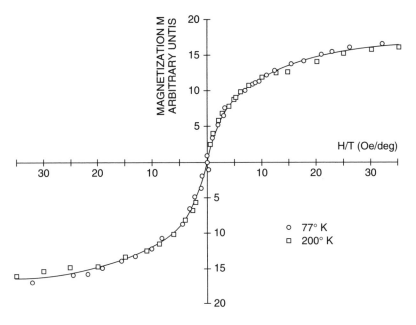

Figure 2.7 Magnetization curves on H/T bases, as a demonstration of the superparamegnetism of iron amalgam. (Bean and Jacobs 1956)

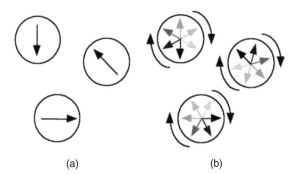

(a)   (b)

Figure 2.8 Observation of superparamagnetism. The circles depict three magnetic nanoparticles, and the arrows represent the net magnetization direction in these particles. In case (a), at temperatures well below the measurement-technique-dependent blocking temperature T_B of the particles, or for relaxation times τ (the time between moment reversals) much longer than the characteristic measurement time τ_m, the net moments are quasi-static. In case (b), at temperatures well above T_B, or for τ much shorter than τ_m, the moment reversals are so rapid that in zero external field the time-averaged net moment on the particles is zero. (Pankhurst *et al.* 2003)

paramagnetic; while if $\tau \gg \tau_m$, the flipping is slow, and such a state is called a blocked state. In a block state, the quasi-static properties of the material can be observed. The blocking temperature T_B can be obtained by assuming $\tau = \tau_m$. In typical experiments, the measurement time τ_m can range from the slow timescale of 10^2 s for DC magnetization, and medium timescale of 10^{-1}–10^{-5} s for AC susceptibility, through to the fast timescale of 10^{-7}–10^{-9} s for 57Fe Mossbauer spectroscopy.

2.2.3 Nanoparticle Assemblies

In nanoparticle assemblies, the interactions between the particles should be taken into consideration. Stoner-Wohlfarth theory and Holz-Scherer theory are well-acknowledged theories addressing magnetic nanoparticle assemblies.

2.2.3.1 Stoner-Wohlfarth Theory

Stoner-Wohlfarth theory describes the behavior of an assembly of single-domain, non-interacting particles with uniaxial anisotropy (Stoner and Wohlfarth 1948). This theory analyzes how different types of anisotropies affect the magnetic properties, including the coercivity and remanence, of fine particles. One important conclusion about the remanence behavior can be derived from the Stoner-Wohlfarth theory (Wohlfarth 1958):

$$\frac{M^{DCD}(H)}{M(\infty)} = \left(1 - 2\frac{M^{IRM}(H)}{M(\infty)}\right) \tag{2.13}$$

where M^{DCD} is the DC demagnetization remanence, M^{IRM} is the isothermal remanent magnetization, and $M(\infty) = M^{DCD}(H = \infty) = M^{IRM}(H = \infty)$.

As most of real materials could not strictly satisfy the non-interacting assumption required by Equation (2.13), the deviation from Equation (2.13) can be used to investigate the interactions between the particles in real materials. Kelly $et\ al.$ (1989) suggested the difference term, $\Delta M(H)$, defined by:

$$\Delta M(H) = \frac{M^{DCD}(H)}{M(\infty)} - \left(1 - 2\frac{M^{IRM}(H)}{M(\infty)}\right) \tag{2.14}$$

Positive values of $\Delta M(H)$ indicate the presence of stabilizing (ferromagnetic) inter-actions, while negative values indicate demagnetizing interactions (Leslie-Pelecky and Rieke 1996). Equation (2.14) is applicable to strongly interacting systems.

2.2.3.2 Holz-Scherer Theory

Holz and Scherer (1994) developed a micromagnetic theory that can explicitly analyze the behavior of the type D nanostructures in Figure 2.2. Using the micromagnetic lengths given by Equations (2.1) and (2.3), Holz-Scherer theory assumes that crystallographic correlations in the space of the nanocrystals are on the order of the nanocrystallite size, d_{NC}. Under the condition $l_K \gg d_{NC} \gg l_S$, crystal anisotropy plays no role in the magnetic properties, and the magnetic properties are dominated by the magnetostatic and exchange energies. Table 2.3 compares the magnetic lengths of the three elemental ferromagnets and a rare-earth alloy, and these parameters are calculated by using Equations (2.1) and (2.3). Table 2.3 indicates that the ratios of the anisotropy to magnetostatic lengths for Ni and Fe are larger than those for Co and SmCo$_5$, so this theory has more chance of applying to Ni and Fe than to Co or SmCo$_5$ (Leslie-Pelecky and Rieke 1996).

2.2.4 Colloidal Magnetic Nanoparticles

Various types of magnetic complex fluids have been artificially prepared for different purposes. Depending on the dimensions of the magnetic particles, these fluids can be magnetic fluids (ferrofluids), magneto-rheological suspensions or other complex fluids between ferrofluids and magneto-rheological suspensions. The ferromagnetic particles in

Table 2.3 Magnetic lengths for ferromagnetic materials. Adapted from Leslie-Pelecky, D.L. and Rieke, R.D. (1996) Magnetic Properties of Nanostructured Materials. Chemistry of Materials, **8**(8), 1770–83.

Material	l_K/l_S
Fe	7.1
Ni	6.0
Co	2.0
SmCo$_5$	1.4

ferrofluids, usually magnetic ferrite, have a diameter of about 10 nm, and are often covered with a surfactant layer, forming a core–shell structure. The colloid stability is ensured by the shell layer together with thermal agitation. The ferrofluids exhibit a very small magneto-rheological effect. The ferromagnetic particles in magneto-rheological fluids are usually made of iron, and their diameters are about a few micrometers (Popa *et al.* 2005).

Magnetic fluids have been widely used in biological applications, and their magnetic properties and micro-structural characteristics should be carefully investigated. In the following subsections, we discuss the magnetic properties of core–shell nanoparticles and the micro-structural characterization of magnetic suspensions.

2.2.4.1 Core–shell Nanoparticles

Generally speaking, a core–shell nanoparticle consists of a magnetic core encapsulated in a protective shell that is usually biocompatible. The applications of magnetic nanoparticles can be significantly extended by the introduction of core–shell structure (Darling and Bader 2005). There are two general approaches for the development of core–shell structures. With the first approach, the core is magnetic while the shell is nonmagnetic; and using the second approach, the core and the shell are made of magnetic materials of different hardness.

A magnetic core–shell nanoparticle using the first approach can be formed by surrounding the core by an oxide shell which is a natural result of exposure to environmental oxygen. Such core–shell nanoparticles can also be synthesized chemically through tuning the dimensions of both layers and the interface between them. In such a system, the core material is ferromagnetic while the shell is antiferromagnetic, such as colloidal Co/CoO and CoNi/(CoNi)O. As shown in Figure 2.9(a), a shifted hysteresis loop can be observed due to the exchange anisotropy caused by the interfacial couplings. This effect has been used to control the magnetization of devices, such as a spin valve sensor, via the giant magnetoresistance effect. Though the exchange bias was discovered a long time ago (Meiklejohn and Bean 1957), the microscopic mechanisms of exchange bias are yet to be fully understood.

Most of the functionalized magnetic nanoparticles used in biomedical applications are developed in this approach. Biomedical applications require rigorous surface functionalizations to make the particles invisible to the reticulo-endothelial system of the body, and to prevent aggregation that would inhibit the transport of particles through the body. By coating a magnetic core with a thin layer of gold, ligands with various functionalities can be introduced through Au-thiol chemistry, meanwhile maintaining the magnetic utility of the core. Much progress has been made in using iron oxide nanoparticles capped with dextran, polyethylene glycol, polyethylene oxide and other brush polymer coatings (Darling and Bader 2005).

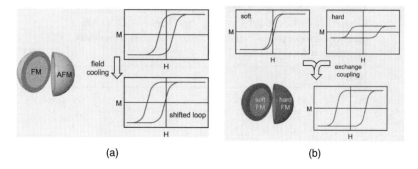

Figure 2.9 (a) M–H hysteresis loops of an exchange bias system. The center of the loop is shifted to the side after field cooling from above the Neel temperature due to coupling between antiferromagnetic and ferromagnetic layers; (b) M–H hysteresis loops of an exchange spring system. Coupling of the soft and hard ferromagnets leads to an enhanced energy product with both large coercivity and large magnetization. (Darling and Bader 2005)

Using the other approach, one ferromagnetic material is grown onto another ferromagnetic material. Figure 2.9(b) shows a core–shell structure consisting of a soft ferromagnetic core covered by a hard ferromagnetic shell. Due to the magnetic exchange coupling between the core and the shell, such a system combines a large coercive field and large magnetization. In such a system, the interphase coupling between the soft and hard ferromagnetic materials can be tuned (Zeng *et al.* 2004a; 2004b; 2002), and extremely strong permanent magnets can be achieved by optimizing the interphase coupling (Darling and Bader 2005).

Though there are many problems to be addressed for single-component and basic core–shell systems, and many applications could be developed based on single-component and basic core–shell systems, researchers are becoming interested in the synthesis and characterization of 'onion' particles, which are composed of many layers (Darling and Bader 2005). The properties of multi-layer core–shell particles could be further tailored to meet special application requirements.

2.2.4.2 Micro-structural Characterization of Magnetic Suspensions

The micro-structural properties of magnetic suspensions can described by peak (P) curves, the first derivative (relative to the magnetic field strength) of the magnetization curves relative to the saturation magnetization. By numerical derivation of the hysteresis curves, P curves can provide knowledge about the microstructure of the magnetic particles in a complex fluid (Popa *et al.* 2005).

Figure 2.10 shows the M–H hysteresis curves of three different magnetic suspensions. Some information about the microstructure of the suspensions can be obtained using the first derivative of the magnetization relative to the magnetic field strength H. The definition of the P_X value (in m/kA) of the hysteresis curve at the point X (having coordinates H_X and M_X) is given by (Popa *et al.* 2005):

$$P_X = \frac{d(M_X/M_S)}{dH_X} \tag{2.15}$$

where M_s is the saturate magnetization.

Figure 2.11 shows three P curves corresponding to three M–H hysteresis curves shown in Figure 2.10. The diameters of the magnetic particles in the three magnetic suspensions

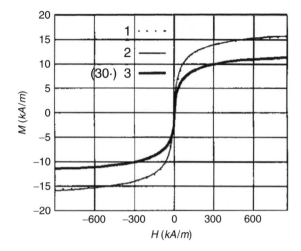

Figure 2.10 Hysteresis curves for three different magnetic suspensions. The saturation magnetizations for the three magnetic suspensions are $M_{s1} = 15.6$ kA/m; $M_{s2} = 15.9$ kA/m; and $M_{s3} = 0.38$ kA/m. For visibility M_3 was multiplied by a factor of 30. (Popa *et al.* 2005)

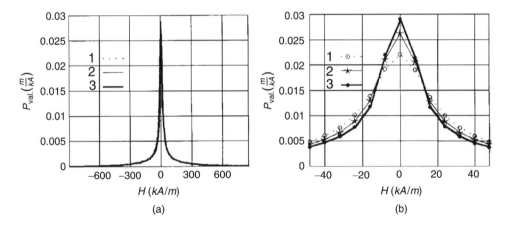

Figure 2.11 (a) P curves for the ferrofluids from Figure 2.10; (b) enlargement of P curves in the low field region. (Popa *et al.* 2005)

are about 10.5, 8.5 and less than 5 nm respectively. Figure 2.11 indicates that the height of the P curves for small magnetic particles is greater than that for large magnetic particles, and the spread of the P curves for large magnetic particles is wider than that for small magnetic particles. Among various parameters derived from the P curves, the amplitude of the P value at the point H = 0, denoted by P(0), is an important characteristic parameter describing the micro-structural properties of a magnetic suspension (Popa *et al.* 2005).

P values are related to the microstructure through a previous calibration. With a previous calibration of P curves using spherical particles, the information about the presence, dimension, modification or orientation of the magnetic particles can be obtained. From Equation (2.15), we can find that P values are the same whether the volume magnetization or the mass magnetization are used. Further, the use of P curves has additional advantages (Popa *et al.* 2005). P values do not depend on the accuracy of sample mass or volume measurement, the accuracy of the magnetometer y-axis calibration and the accuracy of the saturation magnetization (M_s) measurement. Using P curves, the living biological material could be investigated without extracting a sample. By placing a branch of a plant between the poles of the magnetometer and gently vibrating this branch, the P curves for the plant branch could be extracted.

2.2.5   Heating Mechanisms for Hyperthermia

From open M–H curves due to hysteresis as shown in Figure 2.3, we find that external energy is required to overcome the barrier to domain wall motion imposed by the intrinsic anisotropy and microstructural impurities and grain boundaries in the material (Pankhurst *et al.* 2003). This energy is related to the area enclosed by the hysteresis loop, and in the magnetization procedure, it is delivered by the externally applied magnetic field. If a time-varying magnetic field is continuously applied to a ferromagnetic or ferrimagnetic material, there is a constant flow of energy into that material, and this energy will be transferred into thermal energy. This is the physical basis of hyperthermia treatments. A similar argument about energy transfer can be made for superparamagnetic materials. For superparamagnetic materials, external energy is required to coherently align the particle moments to achieve the saturated state. In the following subsections, we discuss the heat generating mechanisms associated with the two types of magnetic materials used for hyperthermia treatments: ferromagnetic or ferrimagnetic single-domain or multi-domain particles, and superparamagnetic (SPM) particles.

2.2.5.1   Ferromagnetic or Ferrimagnetic (FM) Particles

When FM particles are exposed to a time-varying magnetic field, the amount of heat generated per unit volume is determined by the area of the hysteresis loop multiplied by the frequency (Pankhurst *et al.* 2003):

$$P_{FM} = \mu_0 f \oint H dM \qquad (2.16)$$

It should be noted that other possible mechanisms for magnetically induced heating such as eddy current heating and ferromagnetic resonances are ignored in Equation (2.16). The magnetic particles in hyperthermia treatment are much too small and the AC field frequencies much too low for the generation of any substantial eddy currents. For these magnetic particles, their ferromagnetic resonance effects, which are discussed in Section 2.3, can only be observed at frequencies much higher than the frequencies generally considered appropriate for hyperthermia treatments.

For FM particles much larger than the SPM size limit, the integral in Equation (2.16) does not have implicit frequency dependence, therefore P_{FM} can be determined from the hysteresis loops obtained by quasi-static measurements using a VSM or SQUID magnetometer (Pankhurst *et al.* 2003).

In principle, sufficient hysteresis heating of the FM particles could be obtained by using strongly anisotropic magnets such as Nd-Fe-B or Sm-Co. However, in practical hyperthermia treatments, fully saturated loops cannot be used because of the constraints on the amplitude of magnetic field that can be used. In these cases, minor (unsaturated) loops are used, producing heat at reduced levels (Pankhurst *et al.* 2003). According to Equation (2.16), it can be seen that the maximum realizable P_{FM} should involve a rectangular hysteresis loop. Such a hysteresis loop could only be achieved with an ensemble of uniaxial particles perfectly aligned with H, but it is very difficult to achieve such a configuration *in vivo*. For most cases, FM particles are randomly aligned, and in this case, the most heating that can be expected is around 25 % of the ideal maximum.

2.2.5.2 Superparamagnetic Particles

For hyperthermia treatments, superparamagnetic nanoparticles are suspended in water or a hydrocarbon fluid, resulting in ferrofluids. When the magnetic field applied to a ferrofluid is removed, the magnetization of the ferrofluid relaxes back to zero due to the ambient thermal energy of its environment. Two mechanisms are responsible for this relaxation: Brownian rotation which is the physical rotation of the particles themselves within the fluid, and Neel relaxation which is the rotation of the atomic magnetic moments within the particle. There is a relaxation time corresponding to each of these processes. The Brownian relaxation time τ_B is mainly determined by the hydrodynamic properties of the fluid; while the Neel relaxation time τ_N depends on the magnetic anisotropy energy of the superparamagnetic particles relative to the thermal energy (Pankhurst *et al.* 2003). Both of the relaxation times τ_B and τ_N depend on particle size. While, for a given size, the Brownian relaxation usually occurs at a lower frequency than the Neel relaxation. The magnetic relaxation will be discussed in details in Section 2.3.

The heating of superparamagnetic particles by AC magnetic fields can be analyzed using the Debye model originally developed to describe the dielectric dispersion in polar fluids. Due to the finite rate of change of M in a ferrofluid, magnetic moment M will lag behind H. For small field amplitudes, the response of a ferrofluid to an AC field can be described by its complex susceptibility $\chi = \chi' + i\chi''$. The real part χ' corresponds to energy storage, while the imaginary part χ'' corresponds to energy dissipation which generates heat (Pankhurst *et al.* 2003):

$$P_{SPM} = \mu_0 \pi f \chi'' H^2 \tag{2.17}$$

where f is the frequency of the applied AC magnetic field. Equation (2.17) means that if **M** lags **H**, there is a positive conversion of magnetic energy into internal energy.

2.2.5.3 Specific Absorption Rate

The heat generation from magnetic particles is usually described in terms of the specific absorption rate (SAR) in the unit of Wg^{-1}, and the values of P_{FM} and P_{SPM} can be obtained by multiplying the SAR by the density of the particles. This parameter can be used to compare the efficacies of magnetic particles with different sizes (Pankhurst *et al.* 2003). It should be noted that as the saturation of most real FM materials requires applied field strengths of ca $100\,kAm^{-1}$ or more, due to the operational constraint of ca $15\,kAm^{-1}$ in hyperthermia treatments, only minor hysteresis loops can be utilized, resulting in low SARs. However, superparamagnetic materials can generate useful heating at lower magnetic field strengths.

2.3 Magnetic Relaxation of Ferrofluids

The complex susceptibility, $\chi(\omega) = \chi'(\omega) - i\chi''(\omega)$, of magnetic fluids is crucial for understanding the dynamic behavior of these colloidal suspensions. From the frequency dependence of complex susceptibility, various relaxation mechanisms, both Brownian and Neel, as well as ferromagnetic resonance, can be identified and investigated, and meanwhile the macroscopic and microscopic properties of the fluids can also be determined, including the mean particle radius, the mean value of anisotropy field, the gyromagnetic constant, γ, and the damping constant, α (Fannin 2002; Chung *et al.* 2005).

This dynamical susceptibility of a magnetic fluid depends on the relaxation mechanisms and the corresponding relaxation times. There are two distinct mechanisms by which the magnetization of ferrofluids may relax after an applied field has been removed: the physical rotation of the particles within the fluid, and the rotation of the magnetic moments within the particles (Chung *et al.* 2005; Fannin 2002). The first mechanism is called Brownian relaxation with relaxation time τ_B and the second is called Neel relaxation with relaxation time τ_N (Brown 1963). The relaxation times τ_B and τ_N depend on the particle diameter differently. Moreover, under the application of an external DC magnetic field, a magnetic fluid will exhibit ferromagnetic resonance.

2.3.1 Debye Theory

The theory developed by Debye for analyzing the dielectric dispersion in dipolar fluids can be used to understand the frequency dependence of the complex susceptibility of magnetic fluids. Debye's theory is applicable for spherical magnetic particles when the dipole-dipole interaction energy, U, is smaller than the thermal energy kT. According to Debye's theory, the frequency dependence of complex susceptibility $\chi(\omega)$ can be described by (Fannin 2002):

$$\chi(\omega) = \chi_\infty + \frac{\chi_0 - \chi_\infty}{1 + i\omega\tau} \qquad (2.18)$$

where the relaxation time τ is given by:

$$\tau = \frac{1}{\omega_{max}} = \frac{1}{2\pi f_{max}} \qquad (2.19)$$

f_{max} is the frequency where $\chi''(\omega)$ reaches its maximum value, χ_0 is the static suscepti-bility value, and χ_∞ is susceptibility value at very high frequency.

Figure 2.12 shows the frequency dependence of complex susceptibility according to Debye's theory. We find that $\chi'(\omega)$ falls monotonically with the increase of frequency, and that the maximum possible value of the absorption component, $\chi''(\omega)$, is equal to half that of the $\chi'(\omega)$ component (Fannin 2002).

2.3.2 Magnetic Relaxations of Magnetic Fluids

For magnetic nanoparticles in fluids, if the diameter of the magnetic core D_c is suffi-ciently small, and the double shell thickness of the nonmagnetic shell D_s is sufficiently large, the magnetic interactions between the particles can be neglected, independent of their concentration in the fluid. The magnetic susceptibility depends on the orientation of the external magnetic field with respect to the particle anisotropy axis, and the over-all magnetic susceptibility consists of longitudinal ($\chi_\parallel$) and transverse susceptibility ($\chi_\perp$) components. For a fluid with the particle anisotropy axes randomly distributed, the overall susceptibility is given by (Chung *et al.* 2005; Fannin 2002):

$$\chi(\omega) = \frac{1}{3}\left[\chi_\parallel(\omega) + 2\chi_\perp(\omega)\right] \qquad (2.20)$$

As the characteristic relaxation time for transverse relaxation is about 0.1 ns, if the susceptibility is measured in a frequency range of about $10-10^4$ Hz, $\chi_\perp(\omega)$ can be taken as a purely real quantity. Therefore, for $\omega/2\pi < 10$ kHz, we have:

$$Im\chi(\omega) = \frac{1}{3}Im\chi_\parallel(\omega) = \frac{1}{3}\chi_\parallel(0)\frac{\omega\tau_\parallel}{1 + (\omega\tau_\parallel)^2} \qquad (2.21)$$

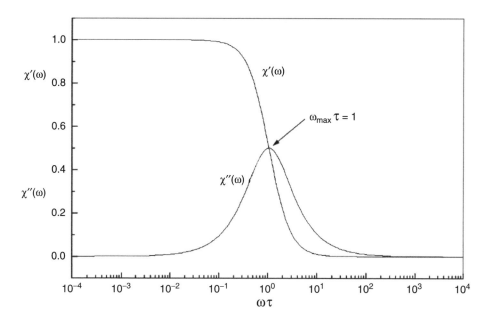

Figure 2.12   Debye plot of $\chi'(\omega)$ and $\chi''(\omega)$ against $\omega\tau$. (Fannin 2002). Reproduced by permission of Springer Science and Business Media

where the effective longitudinal relaxation time $\tau_{//}$ is given by (Chung *et al.* 2005; Fannin 2002):

$$\tau_{//} = \frac{\tau_N \tau_B}{\tau_N + \tau_B} \tag{2.22}$$

The Brownian relaxation time is given by:

$$\tau_B = \frac{3\eta V}{k_B T} \tag{2.23}$$

where η is the viscosity of liquid, and V is the particle effective hydrodynamic volume ($V > V_m$ due to a surfactant layer coating) (Chung *et al.* 2005).

Depending strongly on the relative energy barrier ($a = \Delta E/k_B T$), the longitudinal relaxation time can vary from several ns to several years. For uniaxial, uniform particles, the energy barrier is $\Delta E = K_1 V_m$, where K_1 is the uniaxial anisotropy constant and V_m is the magnetic particle volume. For a large energy barrier ($a > 2$) and zero magnetic field, the Neel relaxation time can be expressed as (Brown 1963):

$$\tau_N = \frac{1}{2}\pi^{1/2}\tau_D a^{-3/2} \exp(a) \tag{2.24}$$

where the free diffusion time τ_D is given by:

$$\tau_D = \frac{M_S V_m}{2\alpha\gamma k_B T} \tag{2.25}$$

where α is the damping constant and γ is the gyromagnetic ratio. For a small energy barrier ($a \ll 1$) and intermediate energy barrier ($a \sim 1$), some approximations for the function $\tau_N/\tau_D = F(a)$ can be used (Chung *et al.* 2005).

In a real magnetic fluid, magnetic nanoparticles have a size distribution. Therefore we need to introduce particle diameter distribution functions for the magnetic core $P(D_c)$ and the nonmagnetic shell $P(D_s)$, or the total (hydrodynamic) diameter $P(D)$ with $D = D_c + D_s$, and the effective susceptibility is an integral over the particle size distribution (Chung *et al.* 2005):

$$\text{Im}\chi(\omega) = \frac{1}{3}\hat{\chi}_0 \int_0^\infty dD_c \int_0^\infty dD P(D_c)P(D)D_c^6 \times \frac{\omega\tau(D_c, D)}{1 + [\omega\tau(D_c, D)]^2} \qquad (2.26)$$

where:

$$\hat{\chi}_0 = \frac{nM_S^2(\pi/6)^2}{3k_B T} \qquad (2.27)$$

here n is the particle concentration per unit volume, and the effective relaxation time is given by:

$$\tau(D_c, D) = \frac{\tau_N^{-1}(D_c)\tau_B^{-1}(D)}{\tau_N^{-1}(D_c) + \tau_B^{-1}(D)} \qquad (2.28)$$

The absorption susceptibility Im $\chi(\omega)$ in Equation (2.26) for distributed core diameters has two maximum values. Usually the maximum value at the low frequency is mainly due to the Brown relaxation, while the maximum value at the high frequency is mainly due to the Neel relaxation (Chung *et al.* 2005). The particle core diameter D_c and the total diameter D can be estimated from the following relationship (Kotitz *et al.* 1995):

$$\omega_{m1}\tau_B(D) = 1, \omega_{m2}\tau_N(D_c) = 1 \qquad (2.29)$$

where ω_{m1}, ω_{m2} are the frequencies of the low-frequency and high-frequency maximum values of the absorption susceptibility Im $\chi(\omega)$, respectively.

Figure 2.13(a) shows the calculated dynamical susceptibility for magnetite (Fe_3O_4) particles with the log-normal particle size distribution $P(D_c)$. Figure 2.13(b) shows the dependence of the low-frequency peak position ω_{m1} on the particle shell diameter D_s. The positions of both the low-frequency peak and the high-frequency peak depend on the width of the size distribution, and the peak positions can change over one order of magnitude by changing the size distribution. Meanwhile, it should be noted that, even for a very broad size distribution, the peak positions still strongly depend on the shell diameter as shown in Figure 2.13(b).

2.3.3 Ferromagnetic Resonances

In the GHz frequency range, the character of the susceptibility dispersion changes from relaxation to resonance. The component $\chi_\perp(\omega)$ in Equation (2.20) is associated with resonance. Due to the resonance, the value of $\chi'(\omega)$ has a change in sign at the angular frequency, ω_{res}, given by (Fannin 2002):

$$\omega_{res} = \gamma H_A \qquad (2.30)$$

where γ is the gyromagnetic constant, and H_A is the value of the anisotropy field.

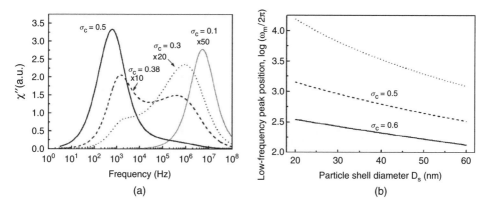

Figure 2.13 (a) Im χ (ω) as a function of frequency calculated for different size dispersions σ_c for the magnetic core of magnetite nanoparticles. Parameters for the calculation are $M_s = 480\,G$, $K_1 = 5 \times 10^4\,erg/cm^3$, $T = 300\,K$, $\alpha = 0.01$, $\eta = 0.01\,erg\,s/cm^3$, $D_c^m = 10\,nm$, and $D_s = 40\,nm$ ($\sigma_s = 0$). (b) The dependence of the low-frequency peak position of Im χ (ω) on the shell diameter D_s. Except for D_s, the parameters are the same as those in (a). The solid line corresponds to $\sigma_c = 0.6$, the dashed line corresponds to $\sigma_c = 0.5$. The dotted line is for the case without any size dispersion. (Chung *et al.* 2005)

The resonant component of the susceptibility, $\chi_\perp(\omega)$, may be described by (Coffey *et al.* 1993):

$$\chi_\perp(\omega) = \chi_\perp(0)\frac{1 + i\omega\tau_2 + \Delta}{(1 + i\omega\tau_2)(1 + i\omega_\perp) + \Delta} \tag{2.31}$$

where $\chi_\perp(0)$ is the static transverse susceptibility, and the concepts of $\tau_\perp$, Δ and τ_2 are defined in (Coffey *et al.* 1993). By combining Equations (2.20) and (2.31), the overall frequency-dependent susceptibility can be obtained (Fannin 2002):

$$\chi(\omega) = \chi'(\omega) - i\chi''(\omega) = \frac{1}{3}\left[\frac{\chi_\parallel(0)}{1 + i\omega\tau_\parallel} + 2\chi_\perp(0)\frac{1 + i\omega\tau_2 + \Delta}{(1 + i\omega\tau_2)(1 + \omega\tau_{\perp eff}) + \Delta}\right] \tag{2.32}$$

It should be noted that in Equation (2.32), $\tau_\perp$ is replaced by $\tau_{\perp eff} = \tau_\perp\tau_B/(\tau_\perp + \tau_B)$ to include the effects of Brownian relaxation.

Figure 2.14 shows the susceptibility spectra for a 760 G suspension of magnetite in isopar M subjected to a polarizing field in the range 0–100 kAm^{-1} (Fannin 2002), obtained by means of the transmission line technique (Chen *et al.* 2004). It can be seen that for the unpolarized case, resonance occurs at a frequency $f_{res} = 1.6\,GHz$, while the maximum $\chi''(\omega)$ of the loss-peak occurs at a frequency $f_{max} = 1.0\,GHz$. When the polarizing field, H, is increased to 100 kAm^{-1}, f_{res} and f_{max} increase 5.0 GHz and 4.7 GHz, respectively.

Figure 2.15 shows the relationship between the resonant frequency f_{res} and the polarizing field H. According to the relationship $\omega_{res} = 2\pi f_{res} = \gamma(H + \overline{H}_A)$, the mean value of the anisotropy field $\overline{H}_A$ can be determined from the intercept of the plot. Based on the value of $\overline{H}_A$, the mean value of anisotropy constant, $\overline{K}$, and bulk, M_s, can be obtained. Further, the magneto-mechanical ratio, γ, can be found from the slope of the plot (Fannin 2002).

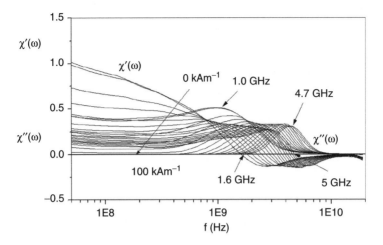

Figure 2.14 Spectra of $\chi'(\omega)$ and $\chi''(\omega)$ over the frequency range 50 MHz to 18 GHz for 17 values of polarizing field in the range 0–100 kA/m. (Fannin 2002). Reproduced by permission of Springer Science and Business Media

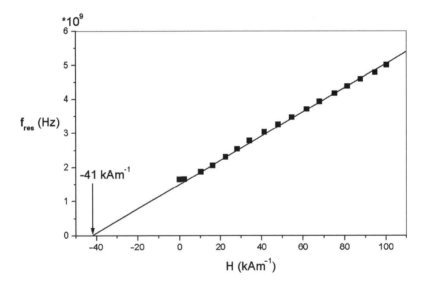

Figure 2.15 The relationship between f_{res} and H (Fannin 2002). Reproduced by permission of Springer Science and Business Media

The value of damping constant α can be estimated by fitting the susceptibility measurement data to theoretical susceptibility profiles based on Equation (2.32), as shown in Figure 2.16. Based on the damping constant α and the other macroscopic and microscopic properties of the particles, the 'magnetic viscosity', $a_m = M_s/6\alpha\gamma$, and the exponential prefatory, $\tau_0 = M_s/2\alpha\gamma K$, of Neel's expression for τ_N can be obtained (Fannin 2002).

2.3.4 Characterization of the Electromagnetic Responses of Ferrofluids

The characterization of the electromagnetic responses of ferrofluids is very important for the study of the relaxation mechanism of ferrofluids. Various methods have been

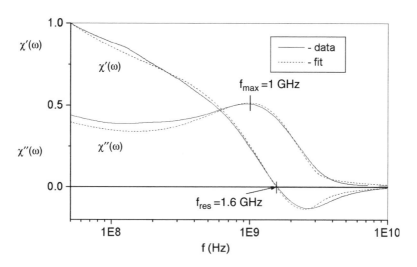

Figure 2.16 Fitting of the measurement results of $\chi'(\omega)$ and $\chi''(\omega)$ to Equation (2.32), for unpolarized plot in Figure 2.14. (Fannin 2002). Reproduced by permission of Springer Science and Business Media

developed for this purpose. Generally speaking, these methods should cover a wide frequency range, so that both the Brownian and Neel relaxations can be studied. However, practically, the methods used in the study of ferrofluids can be generally classified into low-frequency methods and high-frequency methods. Usually the working frequency range of a low-frequency method is below 50 MHz, and a low-frequency method is mainly for Brownian relaxation. While the working frequency range of a high-frequency method is usually from 50 MHz to 20 GHz, and a high-frequency method is mainly for Neel relaxation and ferromagnetic resonance.

In the following section we discuss a typical low-frequency method and a typical high-frequency method. It should be noted that most of the methods for the characterization of complex susceptibility χ of ferrofluids measure the relative complex permeability μ_r, and the complex susceptibility is obtained from the following relationship:

$$\mu_r = \mu_r' - i\mu_r'' = 1 + \chi = 1 + \chi' - i\chi'' \tag{2.33}$$

In this section, we will also discuss a method for the study of the Faraday effect of ferrofluids, and introduce magnetorelaxometry of ferrofluids.

2.3.4.1 Toroidal Method

The complex relative permeability of a fluid, $\mu_r(\omega)$, at a particular frequency, $\omega/2\pi$, can be derived from the increase of impedance, $Z(\omega)$, of a toroidal coil when it is filled with the fluid. This method requires a considerable quantity of fluid, and usually ferrofluids are difficult and costly to prepare in large amounts. Fannin $et\ al.$ (1986) developed a technique which can be used to measure complex susceptibility of samples with volumes not greater than $30\,\mu l$.

As shown in Figure 2.17, the test cell consists of a slot of length λ_1 cut in a toroid with rectangular cross-section A and mean magnetic length λ_2. The toroid is made of a material with very high permeability $\mu_r = \mu_r' - i\mu_r''$. Due to the existence of the slot, the

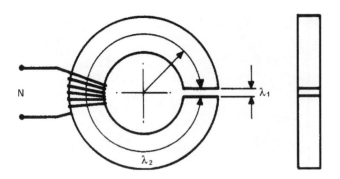

Figure 2.17 Toroidal test cell for susceptibility measurement. (Fannin *et al.* 1986)

effective permeability μ_{eff} of the toroid becomes

$$\mu_{\mathrm{eff}} = \mu_r / (1 + \gamma \mu_r / \mu_1) \qquad (2.34)$$

where $\gamma = \lambda_1 / \lambda_2$. Equation (2.34) is based on several assumptions: there is no leakage flux, the magnetic flux density is uniform over the cross-section, and there is no the fringing effect near the slot.

The impedance, $Z(\omega)$, across the N turns wound on the toroid, is given by:

$$Z(\omega) = R_w + i\omega \mu_{eff} L_0 \qquad (2.35)$$

where R_w is the resistance of the winding, and:

$$L_0 = \mu_0 N^2 A / \lambda_2 (1 + \gamma) \qquad (2.36)$$

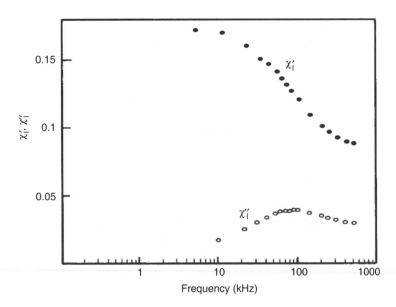

Figure 2.18 Frequency dependence of the susceptibility of cobalt ferrite dispersed in toluene. (Fannin *et al.* 1986)

When the slot is empty, its impedance is given by:

$$Z_e = R_w + i\omega\mu_r L_0/(1 + \gamma\mu_r) \tag{2.37}$$

When it is filled with a magnetic fluid, the impedance of the toroid becomes:

$$Z_f = R_w + i\omega\mu_r L_0/(1 + \gamma\mu_r/\mu_1) \tag{2.38}$$

The complex permeability of the fluid filled in the slot can be derived from Equations (2.37) and (2.38):

$$\mu_1 = -\frac{(Z_e - R_w)(Z_f - R_w)}{(Z_f - R_w)(i\omega L_0 + R_w - Z_e) - (Z_e - R_w)i\omega L_0} \tag{2.39}$$

Therefore the values of χ_1' and χ_1'' can be obtained from the measurements of Z_e, Z_f and R_w based on Equations (2.39) and (2.33). Figure 2.18 shows the susceptibility spectra of cobalt ferrite dispersed in toluene measured using this method. This method is suitable for low-frequency measurements.

2.3.4.2 Short-circuited Coaxial Line Method

Many methods have been developed for the characterization of ferrofluids at high frequencies (Chen *et al.* 2004; Usanov *et al.* 2001; Hrianca and Malaescu 1995). We concentrate below on the short-circuited coaxial line method based on the transmission line theory (Fannin *et al.* 1995).

As shown in Figure 2.19, the input impedance of a transmission line at a distance x from a terminating load, Z_R, is given by:

$$Z_{in} = \frac{V_{in}}{I_{in}} = \frac{V_R \cosh(\gamma x) + Z_0 I_R \sinh(\gamma x)}{I_R \cosh(\gamma x) + (V_R/Z_0) \sinh(\gamma x)} \tag{2.40}$$

where Z_0 is the characteristic impedance of the transmission line, and $\gamma = \alpha + i\beta$ is the propagation constant with α the attenuation coefficient, $\beta = 2\pi/\lambda$ the phase change coefficient, and λ the operating wavelength in the line.

If the transmission line has a very low loss ($\alpha \approx 0$), Equation (2.40) becomes:

$$Z_{in} = Z_0 \frac{Z_R + i Z_0 \tan(\beta x)}{Z_0 + i Z_R \tan(\beta x)} \tag{2.41}$$

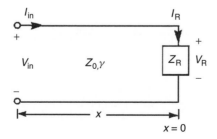

Figure 2.19 The equivalent circuit of a transmission line terminated in a load impedance Z_R

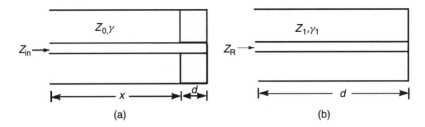

Figure 2.20 (a) The model of a coaxial transmission line terminated in a toroidal sample with a short circuit; (b) the model of a short-circuited toroidal sample

If the load is a short circuit, i.e. $Z_R = 0$, Equation (2.41) can be written as:

$$Z_{in} = i Z_0 \tan(\beta x) \tag{2.42}$$

Consider a short-circuited, air-filled coaxial transmission line, terminated with a toroidal sample of material with thickness d, as shown in Figure 2.20(a). The characteristic impedance of the air-filled line is Z_0, and its propagation constant is γ. As shown in Figure 2.20(b), the shorted sample-filled section in Figure 2.20(a) can be modeled as a shorted transmission line with input impedance Z_R, characteristic impedance Z_1 and propagation constant γ_1. In this case, the general equation, Equation (2.40), should be used because the line cannot be assumed to be lossless.

As the line is shorted, the input impedance, Z_R, is given by:

$$Z_R = Z_1 \tanh(\gamma_1 d) \tag{2.43}$$

The intrinsic impedance Z of a medium with absolute values of complex permeability, μ_1, can be written as:

$$Z = i\omega\mu_1/\gamma \tag{2.44}$$

And the characteristic impedance, Z_1, of a coaxial line containing such a medium as its dielectric is:

$$Z_1 = \frac{1}{2\pi} Z \ln\left(\frac{b}{a}\right) \tag{2.45}$$

where b is the radius of the outer conductor and a is the radius of the inner conductor of the coaxial line. By combining Equations (2.43), (2.44) and (2.45), we have:

$$Z_R = A \frac{i\omega\mu_1}{\gamma_1} \tanh(\gamma_1 d) \tag{2.46}$$

with

$$A = \frac{1}{2\pi} \ln\left(\frac{b}{a}\right) \tag{2.47}$$

If the sample depth is much less than the wavelength of electromagnetic radiation in the sample medium ($\gamma_1 d \ll 1$), we can assume that $\tanh(\gamma_1 d) \approx \gamma_1 d$. In this case,

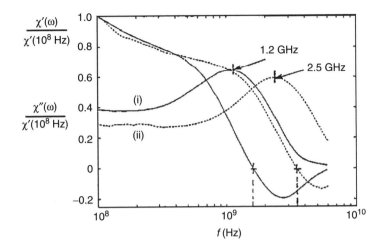

Figure 2.21 A normalized plot of $\chi'(\omega)$ and $\chi''(\omega)$ against f (Hz) for sample (i) magnetite in isopar M, and sample (ii) cobalt in toluene. (Fannin *et al.* 1995)

Equation (2.46) becomes:

$$Z_R = i A \omega \mu_1 d \tag{2.48}$$

By substituting Equation (2.48) for Z_R in Equation (2.41), we have:

$$Z_{in} = Z_0 \frac{i A \omega \mu_1 d + i Z_0 \tan(\beta x)}{Z_0 - A \omega \mu_1 d \tan(\beta x)} \tag{2.49}$$

Equation (2.49) can be re-written in terms of the permeability of the sample, μ_1:

$$\mu_1 = Z_0 \frac{Z_{in} - i Z_0 \tan(\beta x)}{i A \omega d Z_0 + Z_{in} A \omega d \tan(\beta x)}. \tag{2.50}$$

By eliminating the constant factor A, which is determined by the dimensions of the coaxial line as indicated in Equation (2.47), we can get:

$$\mu_r = \mu_r' - i \mu_r'' = \frac{\mu_1}{\mu_0} = \frac{\lambda}{2\pi d} \cdot \frac{(Z_{in}/Z_0) - i \tan(\beta x)}{(Z_{in}/Z_0) \tan(\beta x) + i} \tag{2.51}$$

Therefore, the relative complex permeability, and thus the complex susceptibility, can be obtained from the impedance measurements of the short-circuited line.

Figure 2.21 shows typical measurement results obtained using this method (Fannin *et al.* 1995). For both samples, the $\chi''(\omega)$ component does not change up to a frequency of about 400 MHz, while the corresponding $\chi'(\omega)$ component decreases at this frequency range. This is mainly due to the contribution of the relaxational Neel components to the susceptibility. When the frequency is higher than 400 MHz, a relaxation to resonance transition occurs with the $\chi'(\omega)$ components becoming negative at approximately 1.7 and 3.6 GHz respectively.

It should be noted that this method is based on the assumption that the sample depth is much less than the wavelength of the electromagnetic radiation in the sample medium. To obtain more accurate results, or measure samples at higher frequency ranges, one needs to use the transmission/reflection methods, or modified reflection methods (Chen *et al.* 2004). In a transmission/reflection method, the fluid under test is placed in a transmission line; from the transmission and reflection coefficients of the segment of the transmission line filled with the fluid, both complex permittivity and complex permeability of the sample can be obtained. One typical modified reflection method which is useful for characterization of fluids is a two-thickness reflection method. In this method, two complex reflection coefficients from two shorted samples with different lengths are measured, and the complex permittivity and complex permeability of the samples are calculated from the two measured complex reflection coefficients.

2.3.4.3   Faraday Effect in Magnetic Fluids

The Faraday effect could be used to study the structure and properties of magnetic colloids. Maiorov (2002) has developed a method for observing the Faraday effect in magnetic fluids in the GHz frequency range. The effect is observed when a microwave in mode H_{11} propagates in the circular waveguide with a magnetic fluid, and a DC magnetic field is applied along the waveguide.

Figure 2.22 schematically shows the experimental set-up (Maiorov 2002). A circular metallic waveguide with 23 mm diameter passes through the electric coil system which provides the external magnetic field, and the magnetic fluid sample in a dielectric container is placed inside the circular waveguide. A DC coil provides a DC magnetic field for magnetization of the sample, while an AC coil provides a weak alternating magnetic field with a frequency of about 100 Hz for modulation of the polarization, which is essential for angular sensitivity of the microwave detector head. A standard microwave generator operating at 10 GHz provides an electromagnetic wave for the circuit waveguide, and the H_{11} mode is established in the waveguide. The polarization of the electromagnetic wave at the outlet of the waveguide is measured using a rotating detector head.

Figure 2.23 shows the angular shift of the polarization plane of the incident wave as a function of the applied magnetic field. The two magnetic fluid samples in this experiment are 200 mm long. Magnetic fluids consist of magnetite core nanoparticles suspended in tetradecane, and oleic acid is used as a surfactant. The diameter of magnetic particles d in magnetic fluids is the same for both samples, and the difference of magnetic saturation

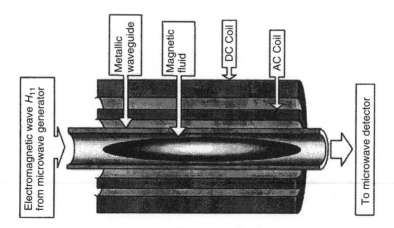

Figure 2.22   Measurement of Faraday effects of magnetic fluids. (Maiorov 2002)

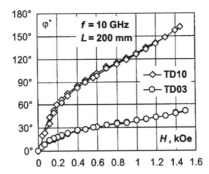

Figure 2.23   Faraday rotation curves for the magnetite colloids in tetradecane. (Maiorov 2002)

M_s for these samples is approximately 3. Figure 2.23 shows slight hysteresis, which may be related to the relaxation of the sample at the changing magnetic field. It seems that the time given for a relaxation, several seconds for each step, is too short for the magnetic particles to achieve their equilibrium states.

2.3.4.4   Magnetorelaxometry of Ferrofluids

Most of the methods for the detection of labeled substances, such as fluorescence or enzymatic reactions, usually require additional separation and cleaning steps to eliminate the unbound labels. Magnetorelaxometry (MRX) allows the discrimination of the signals from the bound and unbound magnetic nanoparticles used to label substances, without further separation and cleaning steps. MRX is the measurement of the relaxation of magnetic nanoparticles (MNPs) after the magnetizing field H_{mag} is switched off. The relaxation of these magnetic nanoparticles may be Brownian relaxation or Neel relaxation. For movable particles, both the Brownian relaxation and Neel relaxation exist, while for immovable particles, only Neel relaxation can be observed. As shown in Figure 2.24, because of the different relaxation times, it is possible to distinguish between the signals of movable and immovable particles (Warzemann *et al.* 1999).

Magnetorelaxometry can be used for the detection of biomolecules using different approaches. If the MNPs with a suitable core size are immobilized by binding to a specific biomolecule, they exhibit Neel relaxation, but the unbound MNPs exhibit Brownian relaxation. Therefore the bound and unbound MNPs can be distinguished by their different relaxation times, as indicated in Figure 2.24. If the MNP marker binds to a mobile biomolecule in suspension, its hydrodynamic volume will be increased, resulting in an increased Brownian relaxation time. Depending on whether it exceeds the corresponding Neel time constant or not, MNPs bound to biomolecules can either be distinguished from unbound ones by their increased Brownian relaxation time or by the change of Brownian to Neel relaxation (Ludwig 2004).

Magnetorelaxometry is mainly performed with superconducting quantum interference devices (SQUIDS), the most sensitive magnetic field sensors with magnetic field noise values of the order of one $fT/Hz^{1/2}$. To suppress the disturbances from remote sources an unshielded measurement environment, hardware gradiometer systems, is often used (Schambach *et al.* 1999). In a measurement set-up, shown in Figure 2.25, the sample is magnetized in the x-direction, whereas the planar $\partial B_Z/\partial x$ SQUID gradiometer is located above it in the x-y plane. If the contributions by parasitic areas are neglected, the magnetizing field does not penetrate the pick up areas of the SQUID antenna, and thus does not couple flux into the gradiometer antenna.

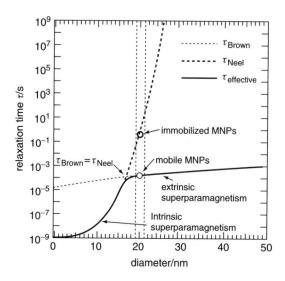

Figure 2.24 Relaxation times of Brownian and Neel relaxation *vs* MNP core diameter. In the calculations a shell thickness of 15 nm, K = 20 kJ/m^3 and T = 300 K are assumed. For core diameters below about 15 nm Neel relaxation dominates independent of whether Brownian rotation is suppressed or not. For a core diameter of 20 nm (indicated by circles) the MNP relaxes via Brownian rotation if it is mobile and via Neel relaxation if it is immobilized. (Ludwig 2004)

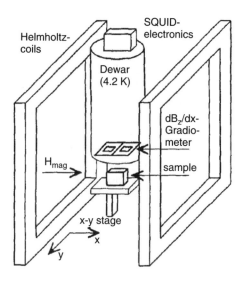

Figure 2.25 Principle sketch of magnetorelaxometry based on SQUID gradiometer. (Warzemann *et al.* 1999)

In a relaxation measurement, the sample is exposed to a magnetic field H$_{mag}$ for a defined time t$_{mag}$, and then the field is switched off. After a small delay time the SQUID gradiometer measures a time-dependent gradient signal $\partial B_z / \partial x$ caused by the sample magnetization. Figure 2.26 shows a typical relaxation signal $(\partial B_z / \partial x)(t)$ of a sample of dried ferrofluid, and only pure Neel relaxation can be observed. In the measurement, the sample was located 16 mm beneath the SQUID and was magnetized with a field of 1 mT for 1 s.

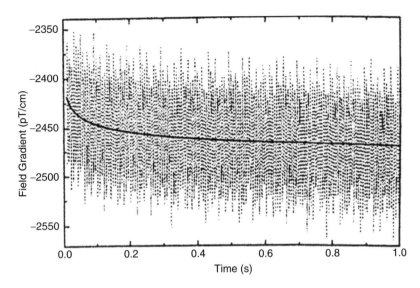

Figure 2.26 Measured (dotted curve) and fitted (full curve) MRX signal $(\partial B_z/\partial x)(t)$ of a sample with dried ferrofluid containing 3×10^{-8} mol (1.7 µg) iron. (Warzemann *et al.* 1999)

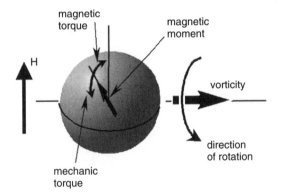

Figure 2.27 The origin of a field dependent increase of viscosity in a ferrofluid. (Odenbach 2004)

2.4 Magnetorheology of Ferrofluids

2.4.1 Effects of Magnetic Field on Ferrofluid Viscosity

It is well known that the viscous behavior of a ferrofluid will be changed under the influence of a magnetic field, and the effects of the magnetic field on the viscosity of a ferrofluid are dependent on the relative angle between the magnetic field direction and the vorticity of the flow (Odenbach 2004). As shown in Figure 2.27, in a ferrofluid flow, the rotation axis of the magnetic particles is aligned with the vorticity of the flow. If a magnetic field is applied to the ferrofluid flow, the magnetic moments of the particles will tend to align with the magnetic field direction. If the magnetic moments of the magnetic particles are fixed inside the particles, the rotation caused by the viscous friction will lead to a disalignment of the magnetic moments of the magnetic particles and the externally applied magnetic field. The magnetic torque resulted from this disalignment counteracts

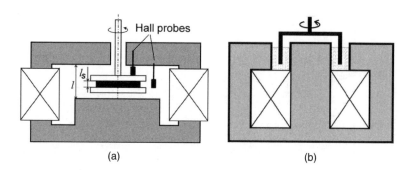

Figure 2.28 Magnetic circuits for rheometry study. (a) Parallel plate geometry; (b) cylindrical Couette geometry. (Bossis 2002)

the mechanic torque due to the shear flow. Therefore the free rotation of the particles will be hindered and, macroscopically, the viscosity of the fluid will be increased. If the magnetic field and the vorticity of the fluid are in the same direction, the rotation of the particle will not cause a disalignment of magnetic moment and field direction, and thus no magnetic torque will counteract the free rotation of the particles. In this case, the viscosity of the fluid will not be affected by the magnetic field.

2.4.1.1 Ferrofluid Viscosity in Absence of Magnetic Field

The viscosity of a fluid is related to the temperature in the fluid, interactions between particles, the concentration, shape and dimension of the particles, and viscosity of the carrier liquid. As the relationships between the viscosity and these aspects are quite complicated, there is no general formula for the viscosity of suspensions, in particular for magnetic fluids. However, many approximate models have been proposed, each of which is based on specific considerations and therefore with limited applicability (Vekas *et al.* 2000). In the following, we discuss several typical models often used in the analysis of the viscosity of ferrofluids in the absence of a magnetic field.

For isotropic diluted suspensions with non-interacting spherically shaped particles, the viscosity is given by Einstein's relationship (Vekas *et al.* 2000):

$$\eta = \eta_0(1 + 2.5\Phi_h) \tag{2.52}$$

where η_0 is the viscosity of the carrier liquid, 2.5 is the shape factor for spherical particles, and $\Phi_h = p\Phi$ is the hydrodynamic volume fraction. Φ is the volume fraction of dispersed particles and p a proportionality factor between Φ and Φ_h, which could be a fit parameter.

Vand's model considers the hydrodynamical interactions and the collisions between particles, but the Brownian motion is neglected (Vand 1948). This model can be used for a large range of volume fractions. According to Vand's model, the viscosity is given by:

$$\eta = \eta_0 \exp\left(\frac{k_1\Phi_h + r_2(k_2 - k_1)\Phi_h^2}{1 - Q\Phi_h}\right) \tag{2.53}$$

where $Q = 0.609$ is a constant related to the hydrodynamical interactions, $k_1 = 2.5$ is the shape factor for isolated spherical particles, $k_2 = 3.175$ is the shape factor for the ensemble of two colliding particles, and $r_2 = 4$ is a time constant for binary collisions.

Another well-known model is Krieger-Dougherty's model, which is valid for a large range of volume fractions (Barnes *et al.* 1989). According to this model, the viscosity is given by:

$$\eta = \eta_0 \left(1 - \frac{\Phi_h}{\Phi_x} \right)^{[\eta]\Phi_x} \tag{2.54}$$

where Φ_x is the maximum packing fraction, and usually it refers to the fraction at which η becomes infinity, and $[\eta]$ is the so-called intrinsic viscosity, and actually it is the shape factor of particles; for example, for spherical particles, it equals 2.5.

2.4.1.2 Ferrofluid Viscosity in Presence of Magnetic Field

When a magnetic field is applied to a magnetic fluid, its rheological properties will be changed. For a diluted ferrofluid where there is no interaction between particles, the Brownian rotation of the particles during the flow is hindered by the magnetic field, so its effective viscosity is increased. For a Couette flow, the viscosity of the fluid is given by (Vekas *et al.* 2000):

$$\eta(H) = \eta(0) \left[1 + \frac{3}{2} \Phi_h \frac{\xi - \tanh \xi}{\xi + \tanh \xi} \sin^2 \beta \right] \tag{2.55}$$

where $\eta(0)$ is the ferrofluid viscosity at zero field, and β is the angle between the vorticity and the magnetic field. It should be noted that Equation (2.55) is based on two assumptions: the particles are identical, and the magnetization has a Langevin behavior:

$$M_L = M_s L(\xi) \tag{2.56}$$

where M_s is the saturate magnetization and $L(\xi) = \coth \xi - 1/\xi$ is the Langevin function. The Langevin parameter ξ is given by $\xi = \mu_0 mH/(k_B T)$ where μ_0 is the magnetic permeability of vacuum, m is the magnetic moment of a particle, H is the applied field, k_B is Boltzmann's constant and T is the absolute temperature.

For magnetic fluids with high volume fractions, the situations are much more complicated than that for a diluted ferrofluid. The interactions between the magnetic particles and especially the cluster formation greatly increase the viscosity. The severe aggregation of magnetic particles may make the behaviors of the magnetic fluids non-Newtonian (Vekas *et al.* 2000).

2.4.2 Rheometers for the Study of Magnetorheology Fluids

With some modifications, conventional rheometers can be used to study the effect of a magnetic field on the rheology of magnetorheology fluids. Figure 2.28 shows two typical arrangements for the study of the rheology of magnetorheology fluids (Bossis 2002). In Figure 2.28(a), the field is parallel to the axis of rotation and the polar pieces have the same axis of symmetry. In Figure 2.28(b), the geometry is a cylindrical Couette cell and the field is perpendicular to the lines of flow everywhere. In this case, the radial field is not constant throughout the gap. The magnetic field on the inner wall of the yoke can be 25 % higher than the magnetic field on the outer wall. This will attract the particles on the inner wall, and so the apparent viscosity of the fluid will be decreased. Moreover, to avoid the magnetic saturation of the central iron rod, which may introduce serious end effects, the height of the cylinder must be less than its radius (Bossis 2002).

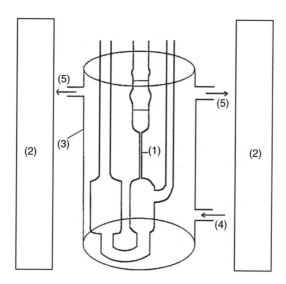

Figure 2.29 Experimental set-up for measurement of magneto-viscosity with capillary viscosimeter. (1) Capillary; (2) magnetic pole; (3) cylinder; (4) input port; (5) output port. (Vekas *et al.* 2000)

Vekas *et al.* (2000) proposed another method for the study of magnetoviscosity. As shown in Figure 2.29, an Ubbelohde capillary viscosimeter is placed between the plane-parallel pole pieces of an electromagnet, which produce a uniform magnetic field at the place where the capillary is placed. The viscosimeter is concentrically fixed inside the cylinder, so that the magnetic field is perpendicular to the flow direction. The thermostabilization of the ferrofluid is achieved using a thermostat with liquid circuit. The liquid enters at the bottom input port of the support cylinder and goes out at the top output ports. The kinematic viscosity v is measured. The shear viscosity is given by $\eta = v\rho$, where the mass density ρ was measured with the aid of a picnometer. The use of the Ubbelohde capillary viscosimeter has the advantages of easy adaptability of the device to magnetorheological measurements and simple flow geometry.

Figure 2.30 shows an example of magnetorheological measurement results using the set-up shown in Figure 2.29. At low fields the increase in viscosity was not obvious. At 0.04 T, a strong increase in viscosity was observed and each new measurement gave a higher viscosity. This shows that the stability of the magnetic liquid was lost and the aggregate formation continued at constant magnetic field. The viscosity further increased at higher fields, but at about 0.07 T, the viscosity decreased after successive measurements at constant field. Further on, the viscosity decreased even with the increase in the magnetic field. The decrease of viscosity may be related to structural transitions, in particular phase separation, occurring due to the drop-like condensation of primary aggregates under the action of the field, followed by the sedimentation of the largest aggregates. Therefore, the liquid which arises in the capillary is less and less concentrated (Vekas *et al.* 2000).

2.5 Manipulation of Magnetic Particles in Fluids

Manipulation of magnetic particles in fluids is crucial for the biomedical applications of magnetic particles, such as magnetic separation and drug delivery. In the following, we

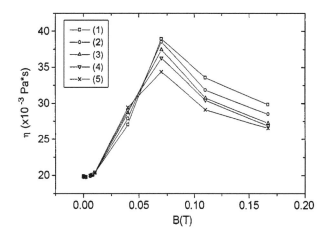

Figure 2.30 Successive measurements of the effective viscosity versus magnetic flux density. The sample is prepared by the chemical coprecipitation method, and its carrier liquid is transformer oil (TR30). Magnetite particles are stabilized with technical grade oleic acid (TOA). TOA contains about 70 % oleic acid, while the remaining 30 % consists of saturated or nonsaturated acids with shorter molecules. (Vekas *et al.* 2000)

discuss the general magnetic properties of magnetic particles often used in biomedical applications, then analyze the forces on magnetic particles by external magnetic fields, and after that, we discuss the basic requirements about the forces on magnetic particles for their biomedical applications.

2.5.1 Magnetic Nanoparticles and Microparticles

Both magnetic nanoparticles and microparticles are used in biomedical applications. As shown in Figure 2.31(a), a typical magnetic nanoparticle consists of a magnetic core with diameter ϕ, covered by a non-magnetic shell for selectively binding the target biological entities, for example a specific cell, protein or DNA sequence. The magnetization curve shown in Figure 2.31(b) indicates that an ensemble of such superparamagnetic particles is hysteresis-free. So the magnetic interaction between magnetic nanoparticles can be easily switched on and off, and this characteristic is important for their biomedical applications (Gijs 2004). For example, the biological entities tagged by superparamagnetic particles can be concentrated or removed from a matrix by applying an external magnetic field, but they quickly decompose after the field is removed, as shown in Figure 2.31(c). Magnetic nanoparticles have additional advantages. Magnetic nanoparticles cause the minimum disturbance to the tagged biological entities, and magnetic nanoparticles have a large surface-to-volume ratio for chemical binding (Gijs 2004).

Magnetic microparticles usually have diameters in the range of 0.5 to 5 μm. As shown in Figure 2.32(a), there are two types of magnetic microparticles. One type of magnetic microparticles has a single magnetic core, while the other type has a core composed of multiple nanoparticles embedded in a non-magnetic matrix, and there are more or less magnetic interactions between the nanoparticles in the core. As shown in Figure 2.32(b), magnetic microparticles usually have a multi-domain structure and their M–H curves exhibit hysteresis. After the magnetization field is removed, magnetic microparticles have a remnant magnetization M_{rem}, which results in the clustering of magnetic particles, as

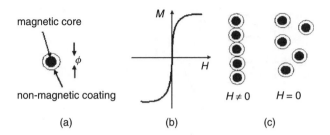

Figure 2.31   (a) A spherical nanoparticle consisting of a magnetic core with diameter ϕ and a non-magnetic shell. (b) M–H curve of an ensemble of magnetic nanoparticles. (c) A magnetic nanoparticle superstructure in the presence of a magnetic field H. After the field is removed, the superstructure decomposes into single particles. (Gijs 2004). Reproduced by permission of Springer Science and Business Media

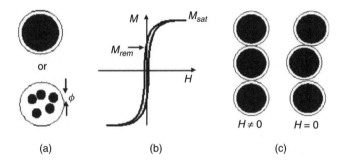

Figure 2.32   (a) A spherical magnetic microparticle with a single internal magnetic core or consisting of multiple nanometer-sized cores of diameter ϕ; (b) hysteresis loop of an ensemble of ferromagnetic particles; (c) microparticle superstructure in the presence of a magnetic field H. After the field is removed, the particles keep a remnant moment and the superstructure does not decompose. (Gijs 2004). Reproduced by permission of Springer-Verlag

shown in Figure 2.32(c). Therefore, when magnetic microparticles are exposed to an external magnetic field, they acquire a magnetic dipole moment and coalesce. Due to the interaction between the magnetic dipoles, the magnetic microparticles will form a supraparticle structure, consisting of chain-like columnar structures along the field direction. A lot of parameters affect the shape of the supraparticle structure, such as the particle concentration and the applied magnetic field (Gijs 2004).

2.5.2   Forces on Magnetic Particles by Magnetic Fields

The manipulation of magnetic particles by external magnetic fields is required by most of the biomedical applications. To manipulate magnetic particles, it is important to recognize that a magnetic field gradient is required to exert a translation force; while a uniform field gives rise to a torque, but no translational action (Pankhurst 2003).

The magnetic force acting on a point-like magnetic dipole **m** is given by:

$$\mathbf{F}_m = (\mathbf{m} \cdot \nabla)\mathbf{B} \tag{2.57}$$

From a geometrical view, this force is the differentiation of magnetic inductance **B** with respect to the direction of magnetic dipole **m**. For example, if $\mathbf{m} = (0, 0, m_z)$ then

$\mathbf{m} \cdot \nabla = m_z(\partial/\partial z)$. In this case, if there is a field gradient in $\mathbf{B}$ in the z-direction, a force will be applied on the dipole. For a magnetic particle suspended in water, the total moment on the particle is given by $\mathbf{m} = V_m\mathbf{M}$, where V_m is the volume of the particle and $\mathbf{M}$ is its volumetric magnetization. The volumetric magnetization is given by $\mathbf{M} = \Delta\chi\mathbf{H}$, where $\Delta\chi = \chi_m - \chi_w$ is the effective susceptibility of the magnetic particle relative to the water (Pankhurst 2003). In the case of a dilute suspension of particles in pure water, we can assume that $\mathbf{B} = \mu_0\mathbf{H}$. Therefore, Equation (2.57) can be rewritten as:

$$\mathbf{F}_m = \frac{V_m\Delta\chi}{\mu_0}(\mathbf{B} \cdot \nabla)\mathbf{B} \tag{2.58}$$

If there are no time-varying electric fields or currents in the medium, we have the Maxwell equation $\nabla \times \mathbf{B} = 0$, based on which we obtain the following mathematical identity:

$$\nabla(\mathbf{B} \cdot \mathbf{B}) = 2\mathbf{B} \times (\nabla \times \mathbf{B}) + 2(\mathbf{B} \cdot \nabla)\mathbf{B} = 2(\mathbf{B} \cdot \nabla)\mathbf{B} \tag{2.59}$$

By combining Equations (2.58) and (2.59), we obtain (Pankhurst 2003):

$$\mathbf{F}_m = V_m\Delta\chi\nabla\left(\frac{1}{2}\mathbf{B} \cdot \mathbf{H}\right) = V_m\Delta\chi\nabla\left(\frac{B^2}{2\mu_0}\right) \tag{2.60}$$

Equation (2.60) indicates that the magnetic force on a magnetic particle is related to the differential of the magnetostatic field energy density, $(1/2)\mathbf{B} \cdot \mathbf{H}$. If $\Delta\chi > 0$, the magnetic force will be in the direction of the steepest ascent of the energy density scalar field. It is the basis for many biomedical applications of magnetic particles, such as magnetic separation and drug delivery (Pankhurst 2003).

2.5.3 Mechanism of Magnetic Manipulation

In many biomedical applications, the magnetically labeled biological entities are separated from a liquid solution by passing the fluid through a region where there is a magnetic field gradient that can produce a magnetic force to immobilize the biological entities. For *in vivo* applications, the magnetically labeled biological entities are transported through the blood flow and locally retained to a destination under the application of external magnetic fields. To achieve effective manipulation, the magnetic force given by Equation (2.60) should overcome the hydrodynamic drag force acting on the biological entities in the flowing solution, as given by:

$$F_d = 6\pi\eta R_m \Delta v \tag{2.61}$$

where η is the viscosity of the medium surrounding the biological entity, for example blood, R_m is the radius of the biological entity and $\Delta v = v_m - v_w$ is the difference in velocities of the biological entity and the blood.

For most biomedical applications, the effects of the buoyancy force on the motion of biological entities can be neglected (Pankhurst 2003). Therefore, by equating the hydrodynamic drag and magnetic forces, we obtain the velocity of the biological entity relative to the carrier fluid:

$$\nabla v = \frac{\xi}{\mu_0}\nabla(B^2) = \frac{R_m^2\Delta\chi}{9\mu_0\eta}\nabla(B^2) \tag{2.62}$$

where ξ is the magnetophoretic mobility of the particle, which describes how manipulable a particle is. Generally speaking, due to their larger size, magnetic microspheres have much larger magnetophoretic mobility than nanoparticles. (Pankhurst 2003).

In most cases, permanent magnets, rather than coils, are used to perform magnetic manipulation (Gijs 2004). Typically, a permanent magnet has a magnetic induction $B_m = 0.5-1\,T$ and a field gradient $\nabla B \approx B_m/w$, where w is the geometrical dimension of the permanent magnet. The magnetic field generated by a current-fed coil is much smaller: a flat millimeter-size coil with ten windings and a current of $0.5-1\,A$ generates a magnetic induction of $1-10\,mT$, at least 100 times smaller than the permanent magnet, and the gradient is also a factor of 100 times lower. Therefore, the force given by Equation (2.60) is a factor of 10^4 larger when using a permanent magnet rather than a coil.

2.6 Interactions Between Biological Nanomaterials and Functionalized Magnetic Nanoparticles

Magnetic nanoparticles have been widely used in biomedicine. With the development of nanotechnology, it possible to produce, characterize and specifically tailor the functional properties of magnetic nanoparticles for biomedical applications (Berry 2005). The further improvement of biomedical applications of magnetic nanomaterials requires a multidisciplinary approach involving expertise from many different fields, including drug delivery, polymer chemistry, physics, biochemistry, engineering and so on (Labhasetwar 2005; Spatz 2004). It is becoming more and more important to study the interactions between biological nanomaterials and functionalized magnetic nanoparticles.

As discussed earlier, most magnetic particles for biomedical applications comprise of a magnetite Fe_3O_4, or maghemite γ-Fe_2O_3 core, covered with a biocompatible shell. Iron oxide nanoparticles are stable and not toxic, and the use of iron oxide nanoparticles has the additional advantage that the body can process excess iron. In the human body, iron is stored primarily in the core of the iron storage protein ferritin. The iron contained in endosomes and lysosomes is metabolized into elemental iron and oxygen by hydrolytic enzymes. Iron homeostasis is well controlled by adsorption, excretion and storage. Following the administration of iron nanoparticles, iron in the body can be processed. But it must be kept in mind that, though iron is important for almost all living tissues, it has a rather limited bioavailability, and in some situations it can also be toxic to cells (Berry 2005).

In biomedical applications, magnetic nanoparticles may be injected intravenously, and transported by the blood circulation, and they can also be injected directly into the general area where the treatment is desired. Whichever approach is used, aggregation of the magnetic particles should be avoided. Actually, magnetic nanoparticles diffuse through the intercellular space in the body. For example, with the aid of blood pressure gradients, ferritin particles with a diameter of $9\,nm$ can diffuse rapidly through intercellular spaces. Generally speaking, nanoparticles with about $5-20\,nm$ diameter should be ideal for most of the therapies (Berry 2005).

Despite their various purposes, the effectiveness of biomedical applications is controlled by cell–particle interaction, which results in either particle attachment to the cell membrane or particle uptake into the cell body, as shown in Figure 2.33. These two responses determine the success of a particular application, and many efforts have been made to exploit and enhance these cell–particle interactions (Berry 2005). In the following section, we discuss the role of materials chemistry in cell–particle interactions, followed by targeting to cell receptors and targeted cell uptake. In the final part, the interactions between magnetic nanoparticles and living cell membranes will be discussed.

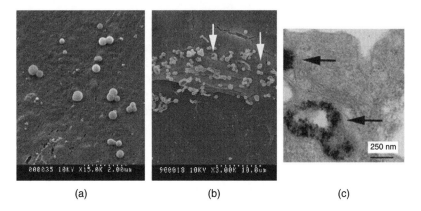

(a)   (b)   (c)

Figure 2.33 Cell–particle interaction depends on particle attachment to the cell membrane and possible subsequent uptake into the cell body. (a) Scanning electron micrograph (SEM) images illustrating insulin-derivatised nanoparticle attachment onto a fibroblast cell membrane; (b) SEM evidence of F-actin protrusions in specific areas on the cell membrane indicative of endocytosis (arrows), in response to albumin-derivatised nanoparticles; (c) transmission electron micrograph of identical particles located in a vesicle. (Berry 2005)

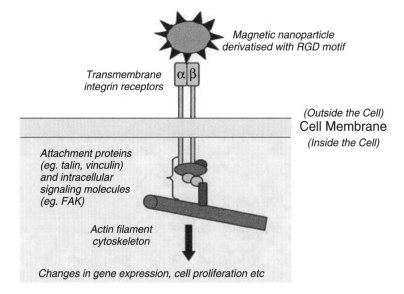

Figure 2.34 Possible interaction of RGD-derivatised magnetic nanoparticles with transmembrane integrin receptors, highlighting the range of subsequent cell responses. (Berry 2005)

2.6.1 Surface Coating

Due to the progress in biotechnology and materials science, magnetic colloids have been widely used in biomedical applications. The colloidal stability not only depends on particle size, but also on the charge and surface chemistry, which provide steric and coulombic repulsions (Berry 2005). If they can successfully avoid the immune system, magnetic particles can get close to, and interact with, the interested biological entities, due to their controllable sizes.

Magnetic nanoparticles are physiologically well tolerated (Berry 2005). After magnetic nanoparticles are injected into the bloodstream, they are coated by the components in the bloodstream, such as plasma proteins. Such an opsonisation process makes the particles recognizable by the reticulo-endothelial system (RES), the body's major defense system. The RES is a diffuse system of specialized phagocytic cells, associated with the connective tissue framework of the liver, spleen and lymph nodes. The macrophages of the liver, spleen and circulation, play an important role in removing the opsonised particles. Therefore, the surfaces of the magnetic nanoparticles for biomedical applications should be modified to ensure that they are biocompatible and stable to the RES.

Generally speaking, hydrophobic particles are quickly coated with plasma components after injection and thus rapidly removed from the circulation, while hydrophilic particles can resist this coating process and will stay in the circulation for a longer time. Therefore, to evade the RES, the particles should be sterically stabilized with a layer of hydrophilic polymer chains. The most common coatings are derivatives of dextran, polyethylene glycol, polyethylene oxide, poloxamers and polyoxamines. However, it should be noted that, though great progress has been made, it is still impossible to completely evade the RES by coating these nanoparticles (Berry 2005).

2.6.2 Targeting to Cell Receptors

By conjugating nanoparticles with functional groups that permit specific recognition of cell types, nanoparticles can be targeted to a specific tissue or cell type in the body. This has been used in active targeting, which is based on the use of ligands that can bind to a cell surface receptor. A wide variety of ligands have been used to target various types of cell surface biomarkers, such as antibodies, peptides, polysaccharides and drugs (Berry 2005).

This approach can be used for magnetic drug delivery systems. For example, magnetic particles have been used to target cytotoxic drugs to sarcoma tumors implanted in rat tails. In this experiment, the magnetic particles were conjugated with starch and anionic phosphate groups, so that cationic binding to the positively charged amino sugars of epirubicin could be achieved. Human trials, for example, intravenous infusion of the chemically bound drug, have also been undertaken. During infusion, a magnetic field was built up as close to the tumor as possible. However, several problems need to be solved before this technique could be successfully used in humans, such as poor targeting in deep tissue, and poor retention of the particles after the external magnetic field is removed (Berry 2005).

Cell adhesion sequences are a class of cell surface targeting molecules under intense research. Cells in the body reside in a three-dimensional environment termed the extracellular matrix (ECM). Cells attach to the ECM through transmembrane receptors that recognize and bind to particular amino acid sequences from the proteins in the ECM. Among various sequences that have been investigated, the RGD amino acid sequence (Arg-Gly-Asp) which is a recognized adhesion motif located in ECM proteins such as fibronectin, is most studied. As shown in Figure 2.34, improvement on the use of such sequences may enhance cell targeting and adhesion to specific receptors, and also cell behavior via activation of specific signaling cascades (Berry 2005).

Many efforts have been made to capitalize on particle attachment to membrane receptors (Berry 2005). A magnetic field could be used to mechanically stimulate cells by applying a twisting motion to the attached magnetic particles (Cartmell et al. 2003), and many cell functions, such as cell growth, proliferation, protein synthesis and gene expression, can be regulated by mechanical forces (Eastwood et al. 1998; Zhu et al. 2000). In

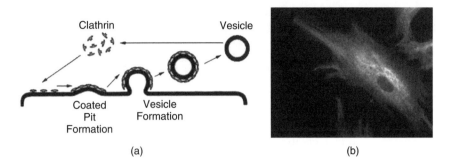

Figure 2.35   (a) The involvement of clathrin in pit formation; (b) a fluorescence image representing clathrin localization in a fibroblast after 30 minutes incubation with dextran-derivatised magnetic nanoparticles. (Berry 2005)

addition, magnetic twisting cytometry (MTC) has been developed to study the mechanical properties of a cell by applying a torque to the magnetic particle that is tightly bound to the cell surface (Mijailovich *et al.* 2002). This work is based on the fact that the external mechanical forces are transmitted across the cell membrane to the cytoskeleton by trans-membrane receptors, such as integrins (Ingber 1997; Chen and Ingber 1999). By using MTC techniques, researchers can make baseline measurements of cell deformability, for example in terms of stiffness, creep response and elastic modulus (Bausch *et al.* 1999).

2.6.3 Targeted Cell Uptake

One main requirement of some biomedical applications of magnetic nanoparticles, such as MRI, hyperthermia treatment and gene therapy, is that the cells efficiently capture the particles. In most cases, particles are captured through endocytosis, a procedure consisting of several main steps including membrane invaginations, clathrin-coated pit formation, coated pit sequestration, detachment of the newly formed vesicle via action of the small GTPase dynamic and finally movement of this new endocytic compartment away from the plasma membrane into the cytosol (Berry 2005). Figure 2.35 schematically shows the process of pit formation. Researchers are trying to understand the complex protein machinery involved in the vesicle formation during endocytosis, and how it interconnects functionally with the cortical cytoskeleton underlying the plasma membrane. The work in this field will be helpful for the optimization of cell loading with magnetic nanoparticles.

DNA transfection delivery represents an exciting development in medical treatment; however this method currently suffers from poor efficacy. The low efficiency of DNA delivery from outside the cell to inside the nucleus is a natural consequence of its multistep process, involving traveling across the cell membrane, through the cell body, and entering the nucleus. The true benefits of gene therapy cannot be realized until the current gene delivery systems are improved or new vectors are developed. The main obstacle in the field is the inefficient delivery of genes because of short *in vivo* half-life, lack of cell-specific targeting and particularly low transfection efficiencies (Berry 2005).

Many efforts have been made to increase the efficiency of DNA delivery (Berry 2005). Cell-penetrating peptides, defined as peptides with a maximum of 30 amino acids that are able to translocate across cell membranes in a non-endocytic fashion are used for this purpose (Lundberg and Langel 2003). Using these peptides, the cellular delivery of conjugated molecules such as nucleic acids, full-length proteins and nanoparticles, can be realized. The attachment of a cell-penetrating peptide, HIV-1 tat peptide, to dextran cross-linked iron oxide nanoparticles may increase the particle uptake into lymphocytes over 100-fold (Allport and Weissleder 2001; Wunderbaldinger *et al.* 2001).

2.6.4   Interactions Between Magnetic Nanoparticles and Cell Membranes

The above discussions indicate that the cell membranes play an important role in the interactions between magnetic nanoparticles and cells. Two approaches have been used to study the interaction between magnetic nanoparticles and cell membranes. One approach is to use phospholipid bilayers to mimic cellular membrane, and with the other approach, living cell membranes are used.

2.6.4.1   Phospholipid Bilayers

Physical model systems using phospholipid bilayers mimicking a cellular membrane have been used to study the interactions between magnetic nanoparticles and cell membranes (Fabre *et al.* 1990; Hare *et al.* 1995; Spoliansky *et al.* 2000). The magnetic nanoparticles adsorbed on the membranes can rotate in the presence of a magnetic field, resulting in a distortion of the membrane structure. A change in birefrigence can be determined using a polarizing microscope (Fabre *et al.* 1990) or a polarized He–Ne laser (Spoliansky *et al.* 2000), whereas X-ray scattering is used to measure the layer separation and the rotation angle of the layer molecules due to the magnetic field (Hare *et al.* 1995).

2.6.4.2   Living Cell Membranes

A study of living cells is needed to determine the effectiveness of the biomedical applications of magnetic nanoparticles. Koh *et al.* (2005) studied the cellular membranes of living cells, *Escherichia coli* (*E. coli*), instead of artificial phospholipid bilayers or membranes of lipids extracted from the cells. *E. coli* cells belong to gram negative bacteria, which have an outer membrane as well as a plasma membrane.

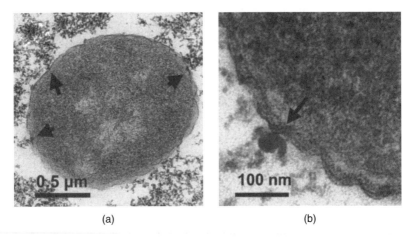

(a)   (b)

Figure 2.36 Transmission electron microscope images showing that *E. coli* have contacts with the SiO$_2$/γ-Fe$_2$O$_3$ particles. Arrows point out contacting parts between the cells and the magnetic particles. (a) An *E. coli* cell surrounded by the magnetic particles. (b) A magnetic particle in contact with the *E. coli* membrane. (Koh *et al.* 2005)

Figure 2.36 shows two TEM images of *E. coli* interacting with magnetic particles. In the experiments, *E. coli* cells are grown in the absence and presence of $SiO_2/\gamma\text{-}Fe_2O_3$ composite particles, respectively. Magnetic fields up to 423 mT were applied on the cells, to observe the effect of the magnetic particles and the magnetic field strengths. *In situ* X-ray scattering from continuous cultures gave significant differences, depending on whether or not the magnetic particles were included in the cell cultures. X-ray scattering results from continuous cultures of *E. coli* showed two peaks, a sharp peak at $q = 0.528\,\text{Å}^{-1}$ and a diffuse one at $q = 0.612\,\text{Å}^{-1}$. When magnetic particles were added and the applied magnetic field strength was increased from 227 to 298 mT, even to 362 mT, the sharp peak was shifted to the smaller side of q, while the diffuse peak did not change. A critical magnetic field strength was found at 362 mT, where the sharp peak disappeared (Koh *et al.* 2005).

References

Allport, J.R. and Weissleder, R. (2001) In vivo imaging of gene and cell therapies. *Experimental Hematology*, **29** (11), 1237–46.

Batlle, X. and Labarta, A. (2002) Finite-size effects in fine particles: magnetic and transport properties. *Journal of Physics D: Applied Physics*, **35**, R15–R42.

Bausch, A.R. Hellerer, U. Essler, M. Aepfelbacher, M. and Sackmann, E. (2001) Rapid stiffening of integrin receptor-actin linkages in endothelial cells stimulated with thrombin: A magnetic bead microrheology study. *Biophysical Journal*, **80** (6), 2649–57.

Bausch, A.R., Moller, W. and Sackman, E. (1999) Measurement of local viscoelasticity and forces in living cells by magnetic tweezers. *Biophysical Journal*, **76** (1), 573–9.

Bean, C.P. and Jacobs, I.S. (1956) Magnetic granulometry and super-paramagnetism. *Journal of Applied Physics*, **27**, 1448–52.

Berry, C.C. (2005) Possible exploitation of magnetic nanoparticle–cell interaction for biomedical applications. *Journal of Materials Chemistry*, **15**, 543–7.

Bossis, G., Volkova, O., Lacis, S. and Meunier, A. (2002) Magnetorheology: fluids, structures and rheology. *Lecture Notes in Physics*, **594**, 202–30.

Brown, W.F. Jr. (1963) Thermal fluctuations of a single-domain particle. *Physical Review*, **130**, 1677–86.

Cartmell, S.H., Magnay, J., Dobson, J. and El Haj, A.J. (2003) Effects of mechanical force application on 3D bone tissue engineered constructs using magnetic microparticles. *European Cells & Materials Journal*, **6** (supp 2), 7.

Chen, C.S. and Ingber, D.E. (1999) Tensegrity and mechanoregulation: from skeleton to cytoskeleton. *Osteoarthritis Cartilage*, **7**, 81–94.

Chen, J.X., Fabry, B., Schiffrin, E.L. and Wang, N. (2001). Twisting integrin receptors increases endothelin-1 gene expression in endothelial cells. *American Journal of Physiology – Cell Physiology*, **280**, C1478–84.

Chen, L.F., Ong, C.K., Neo, C.P., Varadan, V.V. and Varadan, V.K. (2004) *Microwave Electronics: Measurement and Materials Characterization*, John Wiley & Sons, Ltd, Chichester.

Chung, S.H., Hoffmann, A., Guslienko, K., Bader, S.D., Liu, C., Kay, B., Makowski, L. and Chen L. (2005) Biological sensing with magnetic nanoparticles using Brownian relaxation (invited). *Journal of Applied Physics*, **97**, 10R101.

Chung, S.H. Hoffmann, A. Bader, S.D. Liu, C. Kay, B. Makowski, L. and L. Chen (2004) Biological sensors based on Brownian relaxation of magnetic nanoparticles. *Applied Physics Letters*, **85** (14), 2971–3.

Coffey, W.T., Kalmykov, Y.P. and Massawe. E.S. (1993) The effective eigenvalue method and its application to stochastic problems in conjunction with the nonlinear Langevin equation, in *Advances in Chemical Physics: Modern Nonlinear Optics*, (eds Evans, M.W. and Kielich, S.), vol. 85 part 2, Wiley Interscience, New York.

Connolly, J. and St Pierre T.G. (2001) Proposed biosensors based on time-dependent properties of magnetic fluids. *Journal of Magnetism and Magnetic Materials*, **225**, 156–60.

Cowburn, R.P. (2000) Property variation with shape in magnetic nanoelements. *Journal of Physics D: Applied Physics*, **33**, R1–R16.

Darling, S.B. and Bader, S.D. (2005) A materials chemistry perspective on nanomagnetism. *Journal of Materials Chemistry*, **15**, 4189–95.

Eastwood, M., McGrouther, D.A. and Brown, R.A. (1998) Fibroblast responses to mechanical forces, *Proceedings of the Institution of the Mechanical Engineers Part H – Journal of Engineering in Medicine*, **212**, 85–92.

Fabre, P., Casagrande, C., Veyssie, M., Cabuil, V. and Massart, R. (1990) Ferrosmectics – a new magnetic and mesomorphic phase. *Physical Review Letters*, **64**, 539–42.

Fannin, P.C. (2002) Magnetic spectroscopy as an aide in understanding magnetic fluids. *Lecture Notes in Physics*, **594**, 19–32.

Fannin, P.C., Giannitsis, A.T. and Charles, S.W. (2001) Frequency-dependent loss tangent and power factor of magnetic fluids. *Journal of Magnetism and Magnetic Materials*, **226**, 1887–9.

Fannin, P.C., Relihan, T. and Charles, S.W. (1995) Investigation of ferromagnetic resonance in magnetic fluids by means of the short-circuited coaxial line technique. *Journal of Physics D: Applied Physics*, **28**, 2003–6.

Fannin, P.C., Scaife, B.K. and Charles, S.W. (1986) New technique for measuring the complex susceptibility of ferrofluid. *Journal of Physics E: Scientific Instruments*, **19**, 238–9.

Frenkel, J. and Dorfman, J. (1930) Spontaneous and induced magnetisation in ferromagnetic bodies. *Nature*, **126**, 274–5.

Gijs, M.A.M. (2004) Magnetic bead handling on-chip: new opportunities for analytical applications, *Microfluid and Nanofluid*, **1**, 22–40.

Halbreich, A., Roger, J., Pons, J.N., Geldwerth, D., Da Silva, M.F., Roudier, M. and Bacri, J.C. (1998) Biomedical applications of maghemite ferrofluid. *Biochimie*, **80**, 369–90.

Hare, B.J., Prestegard, J.H. and Engelman, D.M. (1995) Small angle x-ray scattering studies of magnetically oriented lipid bilayers, *Biophysical Journal*, **69**, 1891–6.

Himpsel, F.J., Ortega, J.E., Mankey, G.J. and Willis, R.F. (1998) Magnetic nanostructures, *Advances in Physics*, **47** (4), 511–97.

Holz, A. and Scherer, C. (1994) Topological theory of magnetism in nanostructured ferromagnets, *Physical Review B*, **50**, 6209–32.

Hrianca, I. and Malaescu, I. (1995) The rf magnetic permeability of statically magnetized ferrofluids, *Journal of Magnetism and Magnetic Materials*, **150**, 131–6.

Hultgren, A., Tanase, M., Chen, C.S. and Reich, D.H. (2004) High-yield cell separations using magnetic nanowires. *IEEE Transactions on Magnetics*, **40** (4), 2988–90.

Ingber, D.E. (1997) Tensegrity: the architectural basis of cellular mechanotransduction. *Annual Review of Physiology*, **59**, 575–99.

Kelly, P.E., O'Grady, K., Mayo, P.I. and Chantrell, R.W. (1989) Switching mechanism in cobalt-phosphorus thin-films. *IEEE Transactions on Magnetics*, **25**, 3880–3.

Koh, I., Cipriano, B.H., Ehrman, S.H., Williams, D.N., Holoman, T.R.P. and Martínez-Miranda, L.J. (2005) X-ray scattering study of the interactions between magnetic nanoparticles and living cell membranes. *Journal of Applied Physics*, **97**, 084310.

Kotitz, R., Fannin, P.C. and Trahms, L. (1995) Time-domain study of Brownian and Neel relaxation in ferrofluids. *Journal of Magnetism and Magnetic Materials*, **149**, 42–6.

Labhasetwar, V. (2005) Nanotechnology for drug and gene therapy: the importance of understanding molecular mechanisms of delivery. *Current Opinion in Biotechnology*, **16**, 674–80.

Leslie-Pelecky, D.L. and Rieke, R.D. (1996) Magnetic properties of nanostructured materials. *Chemistry of Materials*, **8** (8), 1770–83.

Ludwig, F., Mauselein, S., Heim, E. and Schilling, M. (2005) Magnetorelaxometry of magnetic nanoparticles in magnetically unshielded environment utilizing a differential fluxgate arrangement. *Review of Scientific Instruments*, **76**, 106102.

Ludwig, F., Heim, E. and Schilling, M. (2004) Magnetorelaxometry of magnetic nanoparticles – a new method for the quantitative and specific analysis of biomolecules. *Proceedings of the 4th IEEE Conference on Nanotechnology*, 245–8.

Lundberg P. and Langel, U. (2003) A brief introduction to cell-penetrating peptides. *Journal of Molecular Recognition*, **16**, 227–33.

Maiorov, M.M. (2002) Faraday effect in magnetic fluids at a frequency 10 GHz. *Journal of Magnetism and Magnetic Materials*, **252**, 111–13.

Mehier-Humbert, S. and Guy, R.H. (2005) Physical methods for gene transfer: improving the kinetics of gene delivery into cells. *Advanced Drug Delivery Reviews*, **57**, 733–53.

Meiklejohn, W.H. and Bean, C.P. (1957) New magnetic anisotropy. *Physical Review*, **105** (3), 904–13.

Mijailovich, S.M., Kojic, M., Zivokovic, M., Fabry, B. and Fredberg, J.J. (2002) A finite element model of cell deformation during magnetic bead twisting. *Journal of Applied Physiology*, **93**, 1429–36.

Odenbach, S. (2004) Recent progress in magnetic fluid research. *Journal of Physics: Condensed Matter*, **16**, R1135–R1150.

Pankhurst, Q.A., Connolly, J., Jones, S.K. and Dobson, J. (2003) Applications of magnetic nanoparticles in biomedicine. *Journal of Physics D: Applied Physics*, **36**, R167–R181.

Parak, W.J., Gerion, D., Pellegrino, T., Zanchet, D., Micheel, C., Williams, S.C., Boudreau, R., Le Gros, M.A., Larabell, C.A. and Alivisatos, A.P. (2003) Biological applications of colloidal nanocrystals. *Nanotechnology*, **14**, R15–R27.

Plank, C., Schillinger, U., Scherer, F., Bergemann, C., Rémy, J.S., Krötz, F., Anton, M., Lausier, J. and Rosenecker, J. (2003a) The magnetofection method: using magnetic force to enhance gene delivery. *Biological Chemistry*, **384**, 737–47.

Plank, C., Anton, M., Rudolph, C., Rosenecker, J. and Krötz, F. (2003b) Enhancing and targeting nucleic acid delivery by magnetic force. *Expert Opinion on Biological Therapy*, **3** (5), 745–58.

Popa, N.C., Siblinib, A. and Nader, C. (2005) P curves for micro-structural characterization of magnetic suspensions. *Journal of Magnetism and Magnetic Materials*, **293**, 259–64.

Schambach, J., Warzemann, L. and Weber, P. (1999) SQUID gradiometer measurement system for magnetorelaxometry in a disturbed environment. *IEEE Transactions on Applied Superconductivity*, **9** (2), 3527–30.

Scherer, F., Anton, M., Schillinger, U., Henke, J., Bergemann, C., Kruger, A., Gansbacher, B. and Plank, C. (2002) Magnetofection: enhancing and targeting gene delivery by magnetic force in vitro and in vivo. *Gene Therapy*, **9**, 102–9.

Scholten, P.C. (1995) Which SI. *Journal of Magnetism and Magnetic Materials*, **149** (1–2), 57–9.

Spatz, J.P. (2004) Cell–Nanostructure Interactions, in *Nanobiotechnology*, (eds Christof Niemeyer, Chad Mirkin), WILEY-VCH Verlag GmbH & Co. KGaA, Weinheim.

Spoliansky, D., Ponsinet, V., Ferré, J. and Jamet, J.P. (2000) Magneto-optical study of the orientation confinement of particles in ferrolyotropic systems. *European Physical Journal E*, **1**, 227–35.

Stoner, E.C. Wohlfarth, E.P. (1948) A mechanism of magnetic hysteresis in heterogeneous alloys. *Philosophical Transactions of the Royal Society of London Series A – Mathematical and Physical Sciences*, **240**, 599–642.

Vand, V. (1948) Viscosity of solutions and suspensions. 1. Theory, *Journal of Physical and Colloid Chemistry*, **52** (2), 277–99.

Vekas, L., Rasa, M. and Bica, D. (2000) Physical properties of magnetic fluids and nanoparticles from magnetic and magneto-rheological measurements. *Journal of Colloid and Interface Science*, **231**, 247–54.

Usanov, D.A., Skripal, A.V., Skripal, A.V. and Kurganov, A.V. (2001) Determination of the magnetic fluid parameters from the microwave radiation reflection coefficients. *Technical Physics*, **46** (12), 1514–17.

Warzemann, L., Schambach, J., Weber, P., Weitschies, W. and Kotitz, R. (1999) LTS SQUID gradiometer system for *in vivo* magnetorelaxometry. *Superconductor Science and Technology*, **12**, 953–5.

Weber, P., Romanus, E., Warzemann, L., Prass, S., Groβ, C., Hückel, M., Bräuer, R. and Weitschies, W. (2000) Spatially resolved relaxation measurements of magnetic nanoparticles as a novel tool for in vivo imaging. *Proceedings of the 3rd International Conference on the Scientific and Clinical Applications of Magnetic Carriers, Rostock*.

Wohlfarth, E.P. (1958) Relations between different modes of acquisition of the remanant magnetization of ferromagnetic particles. *Journal of Applied Physics*, **29**, 595–6.

Wunderbaldinger, P., Josephson, L. and Weissleder, R. (2001) Tat peptide directs enhanced clearance and hepatic permeability of magnetic nanoparticles. *Bioconjugate Chemistry*, **13**, 264–8.

Zeng, H., Li, J., Wang, Z.L., Liu J.P. and Sun, S. (2004a) Bimagnetic core/shell FePt/Fe_3O_4 nanoparticles. *Nano Letters*, **4**, 187–90.

Zeng, H., Sun, S., Li, J., Wang Z.L. and Liu, J.P. (2004b) Tailoring magnetic properties of core/shell nanoparticles. *Applied Physics Letters*, **85** (5), 792–4.

Zeng, H., Li, J., Liu, J.P., Wang Z.L. and Sun, S. (2002) Exchange-coupled nanocomposite magnets by nanoparticle self-assembly. *Nature*, **420**, 395–8.

Zhu, C., Bao, G. and Wang, N. (2000) Cell mechanics: Mechanical response, cell adhesion, and molecular deformation. *Annual Review of Biomedical Engineering*, **2**, 189–226.

3

Magnetic Nanoparticles

3.1 Introduction

Investigations on magnetic nanoscaled materials have continued for several decades. The multidisciplinary studies on magnetic nanoparticles and other nanostructure combine a broad range of synthesis and characterization techniques from physics, chemistry and materials science and engineering. The aims of these studies are to understand the synthesis mechanisms, provide information about both the structural and magnetic properties of the magnetic nanomaterials and explore their potential applications which will be beneficial to our daily lives eventually. Magnetic materials have been widely used as the recording media. As the grain size of advanced magnetic recording media is shrinking to nanodimensions, their magnetic properties depend strongly upon the size of their nanocrystals. Superparamagnetic properties result at the smallest sized nanocrystals. Hence, the study of nanoscale magnetic domains is not only scientifically fundamental but also technically necessary.

In this chapter, the basics of nanomagnetics will first be presented followed by a review on the synthesis and functionalization of magnetic nanoparticles. Generally speaking, nanoparticles can be classified into nanospheres, nanowires and nanotubes. The magnetic nanoparticles discussed in this chapter and Chapter 4 are mainly magnetic nanospheres. The properties and biomedical applications of magnetic nanowires and nanotubes are discussed in Chapter 6 and Chapter 7, respectively.

3.2 Basics of Nanomagnetics

3.2.1 Classification of Magnetic Nanoparticles

A classification of nanostructured magnetic morphologies was desirable because of the correlation between nanostructure and magnetic properties. Among many schemes proposed by various researchers, we have chosen here the following classification, which was designed to emphasize the magnetic behavior-related physical mechanisms (Leslie-Pelecky and Rieke 1996). The classification is illustrated in Figure 3.1. Type A is denoted for systems consisting isolated particles with nanoscale diameters. Since the interparticle interactions can be ignored for these systems, their unique magnetic properties are completely attributable to the isolated components with their reduced sizes. Another type, type D, is assigned to bulk materials with nanoscale structure. This type is featured

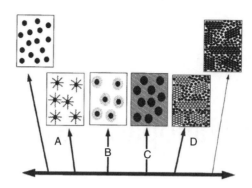

Figure 3.1 Schematic presentation of different types of magnetic nanostructured materials. (Leslie-Pelecky and Rieke 1996)

by a significant fraction (up to 50 %) of the sample volume composed of grain boundaries and interfaces. Compared with type A systems, the interparticle interactions cannot be ignored and the bulk magnetic properties for type D are indeed dominated by the interactions. It is believed that the length scale of the interactions can span up to many grains and is critically related to the interphase characteristics. Because of the existence of the interactions and grain boundaries, the magnetic behaviors of type D nanostructures cannot be predicted theoretically simply by considering only the polycrystalline materials with reduced length scales. Other than type A and type D, intermediate forms such as core–shell nanoparticles (type B) and nanoparticle-based nanocomposites (type C) are classified, as shown in Figure 3.1. In type B, the shells on magnetic nanoparticles, which may not be magnetic themselves, are usually used to reduce interparticle interactions. For type C systems, the magnetic properties of nanocomposites are determined by the faction of magnetic nanoparticles as well as the characteristics of the matrix material.

3.2.2 Single-domain Particles

Single-domain and multidomain are important for ultrafine magnetic particles. Domain walls have a characteristic width and energy associated with their formation and existence. They separate domains – groups of spins all pointing in the same direction and acting cooperatively. Reversing magnetization is primarily achieved by the motion of domain walls. Figure 3.2 illustrates the dependence of coercivity on particle size by an experimental investigation. Multidomain is the case for large particles in which domain walls form energy-favorably. As the particle size decreases below a critical diameter, D_c, single-domain particles form where the formation of domain walls becomes energetically unfavorable. Thus, magnetization reversal cannot be obtained readily leading to larger coercivities because of the lack of nucleation and motion of the domain walls. If the particle size continues to decrease, the spins are increasingly influenced by thermal fluctuations and this phenomenon is called superparamagnetism. The estimated single-domain diameter for some materials in the shape of spherical particles is listed in Table 3.1.

3.2.3 Superparamagnetism

It has been shown theoretically that, for very small magnetic particles, as the thermal fluctuation can prevent the existence of a stable magnetization, coercivity H_c approaches zero. This superparamagnetism has two experimental criteria which are no hysteresis

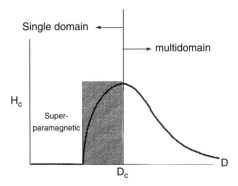

Figure 3.2 Qualitative illustration of the coercivity behavior in the function of particle sizes in ultrafine particle systems. Adapted from Leslie-Pelecky, D.L. and Rieke, R.D. (1996). Magnetic Properties of Nanostructured Materials, *Chemistry of Materials*, **8**(8), 1770–83

Table 3.1 Estimated values of single-domain sizes for spherical nanoparticles without shape anisotropy. Reproduced by permission of American Chemical Society. Adapted from Leslie-Pelecky, D.L. and Rieke, R.D. (1996). Magnetic Properties of Nanostructured Materials, *Chemistry of Materials*, **8**(8), 1770–83.

Material	D_{crit} (nm)	material	D_{crit} (nm)
Co	70	Fe_3O_4	128
Fe	14	γ-Fe_2O_3	166
Ni	55		

for the magnetization curve and overlapping of the magnetization curves at different temperatures. Possible reasons for imperfect superposition could be anisotropy effects, a wide distribution of particle sizes, and changes of spontaneous particle magnetization with temperatures. The width and mean particle size of superparamagnetic particles can be obtained by determining the magnetization as a function of field. It is necessary to point out that this method can only be used for weakly interacting systems where the interparticle interactions are not considered.

3.3 Synthesis Techniques

Since nanoparticles/nanocrystals often exhibit novel electrical, chemical, magnetic and optical properties, which are unique compared with those of their bulk counterparts, extensive efforts have been made to develop uniform nanometer sized particles for both technological and fundamental scientific importance. For instance, magnetic nanoparticles/nanocrystals with 2–20 nm dimensions can be used in multi-terabit in high density magnetic storage devices (Hyeon 2003). These ultrafine magnetic nanoparticles also have broad applications in magnetic ferrofluids, contrast enhancement in magnetic resonance imaging (MRI), highly active catalysts, magnetic refrigeration systems, magnetic carriers for drug targeting, hyperthermia treatment, etc. Here we will focus on colloidal chemical synthesis. Efforts have been made to synthesize different types of magnetic nanoparticles/nanocrystals using colloidal chemical synthetic approaches. Generally, in a colloidal system, separated nanoparticles with ultra-fine dimensions are stabilized by adding surfactant reagents resulting in a uniform suspension in an aqueous or organic solvent. Magnetic colloidal systems consist of magnetic nanoparticles/nanocrystals. As mentioned above, these colloidal systems are examples of type A, where contributions from interparticle interactions are negligible and their unique magnetic behaviors are attributed to their reduced size. The surfactant coating plays an important role in colloidal systems to prevent magnetic nanoparticles clustering owing to steric repulsion. At the same time, assisted by surfactant molecules, reactive species can be added or removed during the synthesis of magnetic nanoparticles. In addition, these surfactant-attached magnetic nanoparticles can be suspended in various aqueous or organic solvents uniformly and powder forms can be obtained again by removing the solvent.

3.3.1 Chemical Methods

Chemical methods, with their straightforward nature, have been widely used to prepare nanostructured materials. An obvious merit of chemical methods is their potential for large-scale production of high pure nanoparticles/nanocrystals. Different particle dimensions, ranging from micrometers to nanometers, can be achieved by careful control of the competition between nucleation and growth during the chemical synthesis.

Several synthetic procedures have been reported for synthesizing monodisperse nanoparticles. Generally, all the chemical methods of dispersing monodisperse particles share the same growth process which is a sudden burst of nucleation process followed by slow growth in a controlled manner. Compared with conventional wet-synthetic methods using ionic precursors in water or another polar solvent, the hot-injection solvothermal method represents a promising technique, in which neutral organometallic precursors are used in a coordinating alkyl solution with a high boiling point (Donegá *et al.* 2005). In a typical hot injection synthetic procedure, as shown in Figure 3.3, the rapid injection of the reagents (often organometallic compounds) into a hot solution with surfactants causes a simultaneous formation of many nuclei (Hyeon 2003). Alternatively, the synthesis can be achieved by adding reducing agents to solutions containing metal salts at a high temperature. In another synthesis procedure, chemical precursors can be stirred at room temperature followed by slow heating in a controlled manner to generate nuclei. Then, reactive species are added for particle growth. High temperature Oswalt ripening is generally used to increase particle size, where smaller nanocrystals dissolve and precipitate on nanoparticles with

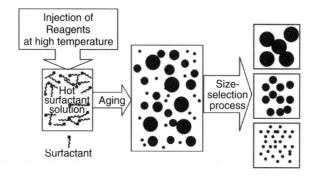

Figure 3.3 Schematic presentation of synthesis procedures of monodisperse magnetic nanoparticles by the injection of precursors into a solution containing surfactants at high temperature followed by a post treatment process such as aging and a size control process. (Hyeon 2003)

larger sizes. The nanoparticles' growth can be stopped by promptly dropping the temperature. These synthetic procedures usually produce nanoparticles with a narrow particle size distribution ($\sigma \sim 10\%$). Controlling the size dimension of synthetic nanoparticles is always desirable. Systematically optimizing the reaction conditions including reaction temperature, reaction time, different surfactants, and reagent concentrations has been demonstrated to control the particle sizes. In general, larger particles are obtained with a longer reaction time and higher reaction temperature. This can be explained as a longer reaction time results in a generation of more monomeric species, and a higher reaction temperature leads to an increased rate of reaction. Also, it is possible to further narrow the particle size distribution below $\sigma = 5\%$ by additional size-selection processes, such as adding a relatively poor solvent to deposit the particles in larger diameters. It is true that, when adding a poor solvent to a suspension of nanoparticles with various dimensions, bigger particles aggregate and precipitate first due to the strong van der Waals attraction among them. Centrifugation is always the method used for precipitate separation. The size selection process can be repeated for the precipitate nanoparticles which will lead to narrower size distribution.

Another synthesis technique is the so-called reverse micelle method for nanoparticle synthesis. It was discovered that reverse micelles can serve as templates for nanoparticle formation. Usually, the micelles were obtained by adding surfactants, which is critical for the shape of the formed nanoparticle. For instance, in the case of champagne cork-shaped surfactants (branched hydrocarbon chains and small polar heads) reverse micelles with spherical shapes are formed. As illustrated in Figure 3.4, the inner core is created by the head groups and the chains form the outer surface (Pileni 2003). This spherical reverse micelle is also termed a water-in-oil droplet. In contrast to normal micelles, the dimension of reverse micelles increases linearly with the amount of water added to the system, hence the diameter of the synthetic nanoparticles can be controlled. More interestingly, spherical reverse micelles can be used as variably sized nanoreactors. Experiments have revealed that two reverse micelles will exchange their water content and reform into two distinct micelles when they meet. Chemical reactions in nanoreactors can be achieved by dissolving two reagents, A and B, in two micellar solutions. When mixing two solutions, A and B are in contact and react because of the exchange process occurring in the reverse micelles system. Since this synthesis technique is rather general, a broad range of spherical nanomaterials, including semiconductors, metals, oxides and alloys can be prepared. In particular, this method is ideal for the synthesis of some alloy nanoparticles which cannot be produced by other synthesis methods. Another merit of this reverse micelles method is its high purity. It was noted that pure metal nanocrystals without any detectable oxide could be obtained by using reverse micelles as nanoreactors.

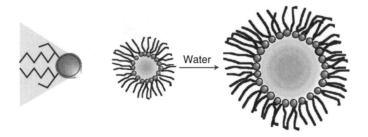

Figure 3.4 Schematic of generalized synthesis of nanoparticles in aqueous solutions by a reverse micelles method. (Pileni 2003)

3.3.2 Biological Methods

Since nanomaterials have comparable dimensions to biological aggregates, bio-related synthesis methods have been explored for novel nanoparticle synthesis. In biological methods, synthesis and assembly of crystalline inorganic materials can be regulated by biological organisms under environmentally benign conditions and desired chemical compositions and phases can be achieved. For example, the nucleation of semiconducting nanoparticles can be initiated in the presence of viruses expressing material-specific peptides. Other examples are the use of porous protein crystals, manipulation of bacteria to produce oxide nanoparticles and selection of metal-specific polypeptides from combinatorial libraries (Reiss *et al.* 2004).

In biological methods, biological entities usually serve as templates for nanoparticle formation. In all cases, the biological entities were used not only to encapsulate the nanoparticles, but to strictly regulate the dimension of the crystals. To prepare magnetic nanoparticles, ferritin can be used which consists of 24 nearly identical subunits. Self-assembly of ferritin will form a spherical cage with a 7.5–8.0 nm-diameter cavity, which can be used for the biological storage of iron in the form of ferrihydrite, an iron (III) oxy-hydroxide. The protein cage is able to withstand relatively high temperatures for biological systems (up to 65 °C) and various pH values (~4.0–9.0) for certain periods of time. Therefore this protein template is quite strong and will not cause any significant disruption of the quaternary structure.

A protein, named 'Apoferritin', has been used as a generic reaction container for the synthesis of a variety of materials, such as iron and manganese oxides (Mayes *et al.* 2003). Apoferritin proteins can be prepared by demineralizing native ferritin (lyophilized

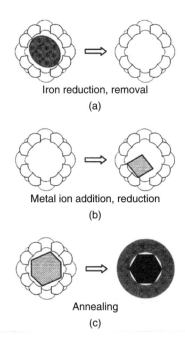

Iron reduction, removal

(a)

Metal ion addition, reduction

(b)

Annealing

(c)

Figure 3.5 Schematic diagram of the synthesis of nanomagnets by a biological method; (a) reduction of amorphous ferrihydrite followed by removal from ferritin proteins to form apoferritin; (b) reconstitution of Co and Pt ions, and chemical reduction to form a full metal alloy core; (c) annealing to form the L1 phase encased in a carbonized matrix. (Mayes *et al.* 2003)

horse spleen ferritin) through reductive dissolution. Apart from regular magnetic nanoparticles, this protein has also been used for the production of superparamagnetic magnetite and semiconducting cadmium sulphide nanoparticles. A good example of using apoferritin protein for nanoparticle synthesis is Co–Pt nanoparticle dispersion. The synthesis includes a coordination of metal ions within the cavity of the protein, followed by their chemical reduction, as illustrated in Figure 3.5 (Mayes *et al.* 2003). This two-step synthesis process needs to be repeated multiple times so that nanoparticles continually grow and fill the cavity of proteins resulting in a high yield. Then nanoparticles inside the protein cavity can be obtained from the dispersion by filtration and L1 phase nanoparticles formed after thermal annealing.

3.4 Synthesis of Magnetic Nanoparticles

Nanoscaled magnetic particles have been used in ferrofluids, refrigeration systems, etc., and their potential applications are storage devices, medical imaging, catalysis, etc (Morales *et al.* 1999). In view of the property characterization and application exploration, synthesis of various high pure magnetic nanoparticles is highly desirable. Here we will discuss representative synthesis of a number of magnetic nanoparticles.

3.4.1 Synthesis of Magnetic Monometallic Nanoparticles

The magnetic properties of the magnetic nanoparticles are determined not only by particle size but also the precise crystal structure and the presence of defects. Transition metals, such as Ni, Fe and Co, have strong magnetic properties. Though magnetic monometallic nanoparticles are preferred, their applications are mainly hindered by their difficult processing. For transition metal nanoparticles, stable colloids are extremely difficult to obtain because of their strong magnetic interactions which result in uncontrolled agglomeration of the particles. Thus, the size-selection process, which is necessary for many applications, cannot be performed readily. In addition, magnetic monometallic nanoparticles are air sensitive and their handling is complicated. In this regard, use of the metal oxide nanoparticles with weaker magnetic properties has been prompted in many applications. To overcome the problem, magnetic monometallic nanoparticles coated with a thin protective layer have been investigated for certain applications.

It has been documented that cobalt nanocrystals possess extraordinary structural, electronic, magnetic and catalytic characteristics which are size-dependent. In particular, the magnetization relaxation time of Co nanoparticles is exponentially dependent on their volume. Intensive studies have revealed their application in magnetic storage (Puntes *et al.* 2001). Because of their high surface tension and high electron affinity owing to the partially filled d-band, the strong van der Waals interactions among polarizable metallic nanoparticles, and interactions among magnetic dipoles, isolating Co nanoparticles is difficult. Fortunately, several research groups have made major progress and isolated Co nanoparticles have been achieved.

Sun's group reported their synthesis of monodisperse cobalt nanoparticles with the capability of precisely controlling the nanocrystal diameter ranging from 2 to 11 nm and size distribution ($\sigma < 7\%$ std. dev.) (1999). They also prepared ordered thin films (superlattices, colloidal crystals) from these monodisperse cobalt nanocrystals. To prevent unnecessary oxidation and irreversible aggregation, cobalt nanocrystals are sheathed with a robust organic ligand shell. The phase of the prepared Co nanocrystals was studied and was found to be different from two known stable forms, face-centered-cubic (fcc) and hexagonal-close-packed (hcp), of bulk elemental cobalt. In practice, a combination of hcp and fcc structures is observed for many bulk cobalt materials due to the low energy stacking faults. In solution-based chemical synthesis, since the reaction is not thermo-dynamically controlled, metastable crystal phases of nanocrystals can be obtained. This is exactly the reason why the synthesized Co nanocrystals have an unusual symmetry structure, which is different from hcp and fcc structures. This novel form of cobalt is similar to the β phase of Mn, and has been designated as the epsilon phase of cobalt ε-Co. Compared with hcp phase, ε-Co displays a lower magnetocrystalline anisotropy. Transformation of ε-Co to hcp phase can be realized by thermal annealing at 300 °C. Experimental measurements also proved that ε-Co nanocrystals are soft magnetic mate-rials with low coercivity. Their low coercivities, coupled with their reduced magnetic dipole interaction, offer several practical benefits for applications.

Sun's Co nanocrystal synthesis was conducted using standard airless synthesis processes and commercially available chemical reagents. First, tetrahydrofuren (THF) superhydride (LiBEt$_3$H) solution was mixed with dioctylether and evaporating THF under vacuum. The obtained dioctylether solution of superhydride was injected into a hot CoCl$_2$ dioctylether solution (200 °C) with oleic acid (octadec-9-ene-1-carboxylic acid, CH$_3$(CH$_2$)$_7$CH = CH(CH$_2$)$_7$COOH) and trialkylphosphine (PR$_3$, R = n-C$_4$H$_9$, or n-C$_8$H$_{17}$). Reaction started and reduction led to a simultaneous formation of numerous Co nuclei. As the reaction was last at 200 °C, Co nuclei grew into nanometer-sized, single crystalline Co parti-cles. The dimension of the nanocrystals can be controlled by a careful selection of the alkylphosphines used in the reaction. It was found that short chain alkylphosphines (e.g., tributylphosphine) resulted in large particles due to a faster growth, while addition of longer chain alkylphosphines (e.g., trioctylphosphine) produce smaller Co nanocrystals due to the slower particle growth. For example, bulky P(C$_8$H$_{17}$)$_3$ produces particles rang-ing 2–6 nm and less bulky P(C$_4$H$_9$)$_3$ led to larger particles (7–11 nm). At the end of the reaction, quenching was performed to stop the growth of the particles and organi-cally stabilized cobalt nanocrystals were obtained. To get monodisperse nanocrystals, the as-prepared cobalt nanocrystals were dispersed in a solvent (e.g. aliphatic, aromatic or chlorinated solvents) and can be precipitated by short chain alcohols. This is because the attractive forces between particles are strongly dependent on their sizes, and the addition of an alcohol induces aggregation of the large particles before the small ones precipi-tate. By carefully controlling the reaction and performing size-selective precipitation, Co nanocrystals of extremely narrow size distributions ($\sigma < 7\%$) in diameter were isolated (Sun and Murray 1999). Figure 3.6a and 3.6b show TEM images of the synthesized Co nanocrystals after separation procedures. It is obvious that cobalt nanocrystals with a mean diameter of 6 and 9 nm are nearly monodisperse and the organic ligand shell helps each crystal separate from its neighbor. Hysteresis measurements at 5 K was performed on a ε-Co superlattice sample on silicon nitride before and after annealing at 300 °C, as shown in Figure 3.6(c–e). The change of its coercivity indicates a phase transformation occurring during the annealing.

Alivisatos' research group has also reported their synthesis of cobalt nanocrystals by injection of an organometallic reagent into a surfactant solution at high temperature with an inert gas protection (Puntes et al. 2001). Figure 3.7a shows a TEM image of synthe-sized spherical iron nanoparticles with diameters of 2 nm. To prevent the agglomeration of particles as well as passivate them against oxidation, a dense-packed monolayer of coordinating ligand was coated on the nanoparticles. The surfactant served to control the shape and dimension of the nanoparticles, which grew from many small metal nuclei instantly formed during the hot injection. Meanwhile, the surface tension of the nanopar-ticles could be lowered through charge transfer, hence modified growth was achieved.

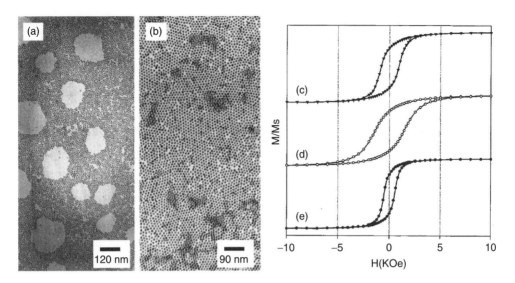

Figure 3.6 TEM image of as-prepared Co nanoparticles with (a) 6 nm and (b) 9 nm in diameter by superhydride reduction. A comparison of the magnetization curves of Co nanoparticles from (c) 9 nm ε-Co nanoparticles, (d) hcp cobalt nanoparticles, and (e) fcc+hcp Co nanoparticles measured at 5 K. (Sun and Murray 1999)

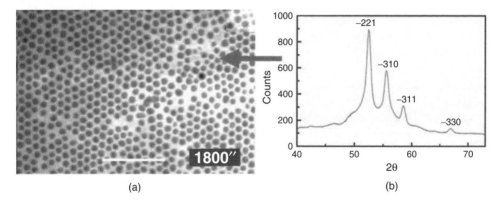

Figure 3.7 (a) TEM image of prepared Co nanoparticles (scale bar: 100 nm); (b) XRD pattern of the Co nanoparticles confirming a ε-Co crystal structure. (Puntes *et al.* 2001)

The experimental results indicated that the ratio of surfactant to precursor determined the size of the synthesized nanoparticles. A mixture of trioctylphosphine oxide (TOPO) and oleic acid was found to be the best surfactant which effectively regulates the growth rates of different planes to form cobalt rods. TOPO plays several roles in cobalt crystal growth. First, TOPO is able to enhance the atom exchange process between nanoparticles, which is necessary for both kinetic control and dimension distribution. Secondly, TOPO helps the decomposition of Co nanoparticles to form monomeric species. In addition, the concentration of TOPO is directly proportional to the length of the nanorods grown in the synthesis. The use of oleic acid was also necessary for the growth of cobalt nanocrystals with narrow distribution of particle sizes and shapes. Similarly to Sun's work mentioned above, the cobalt nanocrystals synthesized by this method showed complex cubic primitive

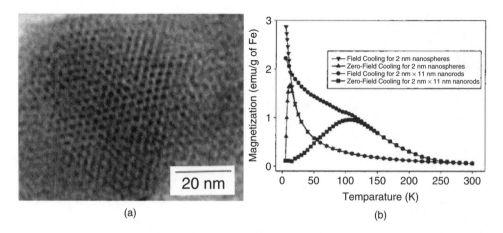

(a)   (b)

Figure 3.8 (a) TEM image of spherical iron nanoparticles with 2 nm diameter; (b) curves of normalized magnetization by mass as the temperature changes for synthesized nanoparticles and nanorods. (Park *et al.* 2000)

structure, ε-Co, which is the favorable crystalline structure in nanoparticles synthesized by a wet chemical method. The X-ray diffraction pattern of cobalt nanocrystals is shown in Figure 3.7b. An average diameter of 7.3 nm for nanoparticles can be estimated from the XRD peaks by Scherrer's formula. Both high-resolution TEM and dark-field TEM confirmed the defect-free single crystalline phase of cobalt nanoparticles.

The hot injection method has also been used for iron nanoparticle synthesis. Park *et al.* prepared iron nanospheres via a thermal decomposition of $Fe(CO)_5$, an organometallic precursor, in the presence of a particular surfactant (2000). In their synthesis, trioctylphosphine oxide (TOPO) was also used as the surfactant. 0.2 ml $Fe(CO)_5$ was mixed with 5.0 g TOPO at 340 °C under an argon protection followed by aging for 30 minutes at 320 °C. The synthesis temperature, 340 °C, is higher than the decomposition temperature of $Fe(CO)_5$, so that a complete decomposition can be achieved and metallic iron atoms can form as nucleis for nanoparticle growth. Black precipitation was observed when the reaction mixture was added into acetone for the quenching purpose. After filtering and washing, the resulting powders consisting of spherical nanocrystals were able to be dispersed uniformly in pyridine resulting in a clear homogeneous solution. Figure 3.8a is a transmission electron micrograph image showing nanoparticles with uniform diameter (around 2 nm) packing closely. Unlike the single-crystalline phase of cobalt nanoparticles, the as-prepared iron nanoparticles are amorphous according to the electron diffraction pattern obtained from TEM operation. Interestingly, it was noticed that the shape of nanoparticles can be tuned by adjusting the surfactant used in the synthesis. For example, by using didodecyldimethylammonium bromide (DDAB) as the surfactant, nanorods can be prepared, induced by an irreversible and strong binding between the surfactant molecules and the central part of the growing nanocrystals. During the fusion process for nanoparticles, since the central regions were occupied by DDAB, the nanospheres would only bind on the edge. Consequently, a catenated structure was generated and unidirectional nanorods formed. A superconducting quantum interference device (SQUID) was used to measure the magnetic properties of iron nanoparticles, both spherical particles (2 nm in diameter) and nanorods (2 nm × 11 nm). The results are shown in Figure 3.8b. Figure 3.8b also shows the temperature-dependent magnetization in an applied magnetic field of 100 Oe at temperatures ranging from 5 to 300 K using field-cooling (FC) procedures and zero-field cooling (ZFC).

3.4.2 Synthesis of Magnetic Alloy Nanoparticles

Other than magnetic monometallic nanoparticles, magnetic alloy nanoparticles have also attracted lots of interests. A good example is FePt. Owing to its rather high uniaxial magnetocrystalline anisotropy ($K_u \cong 7 \times 10^6$ J/m^3) coupled with its excellent chemical inertness, FePt is the material used as the permanent magnet in versatile applications. It is believed that FePt nanoparticles may be valuable in ultrahigh-density magnetic recording applications in the future because their large anisotropy constant and nanoscale dimension provide a high magnetic stability for individual particles.

A number of research groups have reported their syntheses of magnetic alloy nanoparticles by chemical methods. To synthesize FePt nanoparticles, Sun *et al.* used a diol or polyalcohol (e.g. ethylene glycol or glycerol) to reduce metal salts. This method is named the 'polyol process' (Sun *et al.* 2000). In a typical synthesis, two reactions including the reduction of Pt(acac)$_2$ (acac = acetylacetonate, CH$_3$COCHCOCH$_3$) and the decomposition of Fe(CO)$_5$ at high temperature were initiated simultaneously by oleic acid and oleyl amine, a combination of surfactants. At high reaction temperature, Pt metals from Pt(acac)$_2$ and Fe metals from Fe(CO)$_5$ reacted and monodisperse FePt nanoparticles formed. The composition of alloy nanoparticles can be tailored readily by adjusting the molar ratio of precursor reagents, the iron carbonyl and the platinum salt. Fe$_{48}$Pt$_{52}$ particles were obtained by setting the molar ratio of Fe(CO)$_5$ to Pt(acac)$_2$ as 3:2, while Fe$_{52}$Pt$_{48}$ and Fe$_{70}$Pt$_{30}$ nanoparticles were yielded with 2:1 and 4:1 molar ratios, respectively. The strategy of tuning the particle size is to add more reagents to enlarge the existing seeds, by which nanoparticles with an average diameter of 3 to 10 nm can be prepared. Due to the surfactants used in the preparation, colloids with isolated monodisperse FePt are quite stable and oxidation is avoided for FePt nanoparticles protected by the surfactant layer. To prepare a sample for TEM observation, a

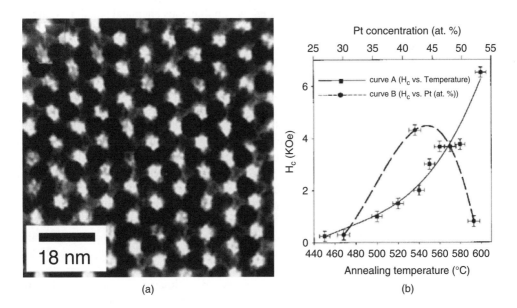

(a)   (b)

Figure 3.9 (a) TEM image of 3-dimensional assembly made from as-synthesized Fe$_{50}$Pt$_{50}$ nanoparticles with 6 nm diameter. The TEM sample was prepared by dropping a hexane/octane suspension onto a silica-coated copper TEM grid; (b) coercivities of Fe$_{50}$Pt$_{50}$ nanoparticles in function of annealing temperature and Pt concentration, measured by a SQUID on 4 nm Fe$_{52}$Pt$_{48}$ nanoparticles. (Sun *et al.* 2000)

drop of the FePt colloid was spread on a copper TEM grid followed by a slow evaporation of the carrier solvent. TEM observation revealed a three-dimensional (3D) superlattice which is composed of monodisperse FePt nanoparticles with $\sigma \leq 5\%$ in diameter. As shown in a TEM image, Figure 3.9a, monodisperse $Fe_{50}Pt_{50}$ particles (about 6 nm in diameter) are assembled in a hexagonal densely packed 3D array. The surfactants, which are oleyl amine and oleic acid, serve as the capping groups and maintain a nearest neighbor spacing about 4 nm between nanoparticles. This interparticle distance can be adjusted by using shorter-chain capping groups, instead of long-chain ones. Magnetic property measurement of 4 nm FePt particles was conducted on a superconducting quantum interference device. The results indicated that the as-prepared nanoparticle assemblies were superparamagnetic with a coercivity about 0 Oe at room temperature. Magnetic studies measured in a 10 Oe magnetic field at 5–400 K with the standard zero-field cooling and field cooling procedures suggested temperature-dependence of the magnetization. It was revealed that the superparamagnetic behavior of FePt nanoparticles has a blocking temperature around 20–30 K. This low blocking temperature may be caused by the low magnetocrystalline anisotropy of the fcc structure. High anisotropy fct phase FePt nanoparticles can be prepared by thermal annealing, after which they are transformed into room temperature nanoscale ferromagnets. In order to tailor the coercivity of these ferromagnets, several synthesis parameters need to be adjusted carefully, such as the annealing temperature, nanoparticle dimension and the elemental composition of FePt alloys. Figure 3.9b shows the dependence of coercivity on the annealing temperature and the elemental composition. No significant difference appears between out-of-plane and in-plane coercivities and hysteresis characteristics, which are attributed to the random orientation of individual FePt nanocrystals' easy axis. Additionally, the measurement by superconducting quantum interference device (SQUID) magnetometry indicated the largest coercivity for Fe_xPt_{1-x} when x varied from about 0.52 to 0.60.

Synthesis of $Co_{48}Pt_{52}$ and $Fe_{50}Pd_{50}$ nanoparticles has been demonstrated by Chen et al. (2002a). $Co_{48}Pt_{52}$ nanoparticles were prepared by simultaneous chemical reactions of platinum acetylacetonate and cobalt tricarbonyl nitrosyl. In a typical synthesis, 0.50 mmol platinum acetylacetonate, $Pt(acac)_2$, and 1.5 mmol 1,2-hexadecanediol were dissolved completely in 20 ml dioctyl ether under nitrogen protection at 100 °C. Then, 1.0 mmol $Co(CO)_3NO$, 0.50 mmol oleic acid and 0.50 mmol oleylamine were added to the solution in a three-necked bottle. The mixture was refluxed for 30 minutes at the boiling point of dioctyl ether, which is 286 °C. After cooling to room temperature, 20 ml ethanol was added to the resulting black dispersion and initiated a precipitation of nanoparticles. The last step was to isolate nanoparticles by centrifuging and washing with ethanol. Similarly, $Fe_{50}Pd_{50}$ nanoparticles were prepared by the simultaneous chemical reactions of palladium acetylacetonate and iron pentacarbonyl. To prepare FePd nanoparticles, 0.50 mmol palladium acetylacetonate and 1.5 mmol 1,2-hexadecanediol were dissolved in 20 ml of dioctylether under N_2 protection followed by an addition of 0.75 mmol iron pentacarbonyl, 0.50 mmol oleic acid and 0.50 mmol oleylamine. The mixture was heated to 286 °C and refluxed for 30 minutes. The same precipitation and separation processes were applied to obtain FePd nanoparticles. The TEM images of as-prepared $Co_{48}Pt_{52}$ and $Fe_{50}Pd_{50}$ nanoparticles are shown in Figure 3.10a and 3.10b, respectively. From the TEM images, the average diameters of $Co_{48}Pt_{52}$ and $Fe_{50}Pd_{50}$ nanoparticles can be estimated to be 7 nm and 11 nm, respectively. Both nanoparticles can be dispersed in hydrocarbon solvents. Magnetic characterization revealed that the as-prepared $Co_{48}Pt_{52}$ nanoparticles were superparamagnetic, while the as-prepared $Fe_{50}Pd_{50}$ nanoparticles exhibited ferromagnetic properties.

By using a similar synthesis strategy, $Fe_xCo_yPt_{100-x-y}$ alloy nanoparticles can be prepared (Chen and Nikles 2002b). Basically, the synthesis process includes simultaneous reductions of both cobalt acetylacetonate and platinum acetylacetonate, as well as the

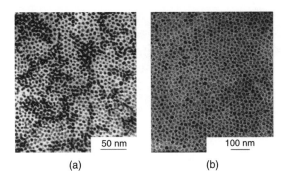

(a)   (b)

Figure 3.10 Typical TEM images of (a) synthesized $Co_{48}Pt_{52}$ nanoparticles and (b) synthesized $Fe_{50}Pd_{50}$ nanoparticles. (Chen and Nikles 2002a)

decomposition of iron pentacarbonyl at high temperature. The elemental composition can be adjusted by the molar ratio of precursor reagents added to the reaction. In a typical reaction process, a solution of 0.5 mmol platinum acetylacetonate and 0.1 mmol cobalt acetylacetonate in 10 ml octylether was heated to 100 °C under nitrogen protection. Then solution of 1.5 mmol 1,2-hexadecanediol in 10 ml dioctylether was added. Surfactants, including 0.5 mmol oleylamine, 0.5 mmol oleic acid, and 1 mmol iron pentacarbonyl, were added to the resulting purple solution by a syringe. After refluxing for 30 minutes, a black dispersion was obtained and 40 ml ethanol was used to precipitate the nanoparticles followed by centrifuge and washing processes. The as-prepared nanoparticles are easy to disperse in hexane and octane. To prepare the sample for TEM observation, the dispersion was spread onto a carbon coated copper grid followed by a slow evaporation of the solvent. TEM images of the thin and thick films consisting of $Fe_{49}Co_7Pt_{44}$ nanoparticles are shown in Figure 3.11a and 3.11b, respectively. Hexagonal arrays are displayed in both images. It seems the nanoparticles in the thin film formed a honeycomb array (AB stacking), while an ABC close-packed structure formed in the thick film. The electron diffraction pattern of the nanoparticles, shown as the inset of Figure 3.11b, clearly confirms the highly ordered crystalline phase for the nanoparticles. $Fe_{49}Co_7Pt_{44}$ nanoparticles have an average diameter of 3.5 nm and a very narrow size distribution. To measure

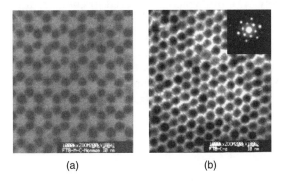

(a)   (b)

Figure 3.11 TEM images of (a) a thin film formed by $Fe_{49}Co_7Pt_{44}$ nanoparticles; (b) a thick film assembled from $Fe_{49}Co_7Pt_{44}$ nanoparticles. The inset shows the electron diffraction pattern confirming the single crystallinity. (Chen and Nikles 2002b)

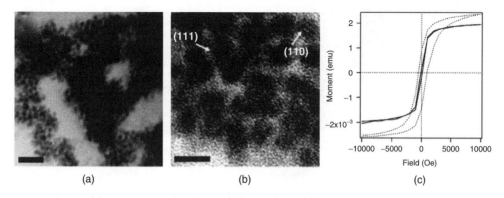

Figure 3.12   TEM images of FePt nanoparticles synthesized by using engineered viral templates; (a) low resolution image; (b) high-resolution image; (c) hysteresis loops of nanoparticles measured by a SQUID at 300 K (solid line) and 5 K (dashed line). (Reiss *et al.* 2004)

their magnetic hysteresis curves, a Princeton Micromag 2900 alternating gradient magnetometer using a 18 kOe saturating field was used. The as-prepared nanoparticles exhibit superparamagnetic characteristics. When treated at a high temperature ($\sim$550–700 °C), $Fe_{49}Co_7Pt_{44}$ nanoparticles transform to the tetragonal phase ($L1_0$) with higher coercivity, hence ferromagnetic nanoparticles can be obtained.

Another interesting synthetic strategy to prepare the technologically important magnetic material FePt is to use a type of peptides. This biological-related synthesis utilizes the biological interactions between peptides to control the nucleation of nanoparticles (Reiss *et al.* 2004). In a typical synthesis, a mixture was first prepared with 1 ml of engineered phage (gP8, 10^{12} phage/ml), 1 ml of 0.075 M $FeCl_2$ and 1 ml of 0.025 M H_2PtCl_6. The gP8 proteins were used because they can be modified to express fusion proteins, which have been shown in the synthesis of II-VI semiconducting nanoparticles. This mixture was then vortexed for 10 minutes followed by an addition of 1 ml 0.1 M $NaBH_4$ as a reducing reagent. Low-resolution and high-resolution transmission electron microscope (TEM) images of FePt nanoparticles prepared by using gP8 engineered phage as templates are shown in Figure 3.12a and 3.12b, respectively. An average diameter of 4.0 ± 0.6 nm can be estimated for the as-prepared FePt nanoparticles. The hysteresis curves for FePt nanoparticles were obtained from a superconducting quantum interference device. Figure 3.12c shows two curves measured at 5 K and 300 K, from which coercivities of 1350 and 200 Oe were estimated, respectively. The difference is believed to come from the existence of some high anisotropy FePt nanoparticles. The hysteresis loops also indicate particular $L1_0$ phase for as-prepared FePt nanoparticles. Compared with wet chemical methods, this biological strategy has several advantages. For instance, wet chemical methods usually require high temperature and inert atmosphere, whereas the biological synthesis may be conducted under ambient temperature, pressure and atmosphere. In addition, the biological synthesis of magnetic nanoparticles is generally easy due to the characteristics of proteins used as the template, hence it is cost-effective.

3.4.3 Synthesis of Magnetic Oxide Nanoparticles

Transition metal oxides are an important member of the family of inorganic solids. They exhibit many extraordinary physical properties, such as superconducting, ferroelectric, magnetic, ionic conducting and electrical conducting characteristics (Hyeon 2003). Meanwhile, these fascinating oxides have a variety of structures with attractive phenomena.

Thus, transition metal oxide nanomaterials have been explored and implemented in many technological fields, particularly for magnetic applications. Magnetic oxide nanoparticles, e.g. ferrites, have been widely used in ferrofluids and employed as magnetic storage media for many years. For most practical applications, monodisperse magnetic oxide nanoparticles are desirable. Prompted by this demand, synthesis of various monodisperse magnetic oxide nanoparticles has been achieved. Here synthesis of iron oxide and spinel ferrite will be reviewed.

γ-Fe$_2$O$_3$ micro-sized particles have been studied. It was documented that the magnetic behaviors of γ-Fe$_2$O$_3$ particles were affected by the degree of order in the distribution of cation vacancies. In the case of γ-Fe$_2$O$_3$ nanoparticles, a similar phenomenon was observed only for γ-Fe$_2$O$_3$ nanoparticles of diameter larger than 20 nm suggesting that the canting effects do not influence the atom moments in the interior of the ultrafine nanoparticle. It has been proposed by many researchers that, for nanoparticles around the frontier of 10 nm in diameter, the reduction in saturation magnetization may be attributed to the cationic disorder together with a magnetically disordered surface layer around the nanoparticles (Morales *et al.* 1999). The particle size distribution, shape distribution and magnetic interactions between nanoparticles should also be counted.

Alivisatos's group reported a general nonhydrolytic single-precursor approach to synthesize various dispersible transition metal oxide nanocrystals. In this technique, the molecular precursors were metal Cupferron complexes M^{x}Cup$_x$ (M: metal ion; Cup: *N*-nitrosophenylhydroxylamine, C$_6$H$_5$N(NO)O-), with the metal ion coordinated with the Cup ligand via the oxygen atoms in a bidentate manner (Rockenberger *et al.* 1999). In a typical synthesis, 0.3 M FeCup$_3$ in octylamine was heated to 60 °C and underwent evacuation and argon gas purging to get rid of oxygen and water. Meanwhile, 7 g of trioctylamine was also treated similarly at 100 °C, and then heated to 300 °C under vigorous magnetic stirring with argon protection. 4 ml of pretreated FeCup$_3$ stock solution was injected rapidly into the trioctylamine and reaction started and lasted for 30 minutes at 225 °C. During the reaction, the metal Cupferron complex decomposed and the solution's color changed from colorless to dark brown. After cooling down to room temperature, a precipitate consisting of nanocrystals was obtained. A clear and stable nanocrystal dispersion can be prepared by adding organic solvents (e.g. toluene, CHCl$_3$, etc.). Again this dispersion will undergo reprecipitation when methanol is added. Figure 3.13 is a low-resolution TEM image of an extended monolayer of as-prepared nanocrystals forming during the slow evaporation of solvent during the TEM sample preparation. The iron oxide nanocrystals have a narrow diameter distribution (6–7 nm) and they are separated from each other by the surfactant layers on their shells. A high-resolution TEM image (Figure 3.13, top left) reveals the crystalline phase of the nanoparticle. From the two observed lattice plane distances (4.77 and 4.11 Å) coupled with the angle between the crossed fringes ($\sim$50°) one can ascertain that the as-prepared nanoparticles are tetragonal γ-Fe$_2$O$_3$ with an ordered superlattice of cation vacancies.

Hyeon *et al.* reported two different approaches to synthesize highly crystalline and monodisperse γ-Fe$_2$O$_3$ nanocrystallites (2001). The first synthetic strategy was to prepare monodisperse iron nanoparticles followed by further oxidation to iron oxide. The iron nanocrystals were synthesized by the decomposition of iron pentacarbonyl (Fe(CO)$_5$) with a surfactant, oleic acid, at 100 °C. Then trimethylamine oxide was used as a mild oxidant to oxidize iron nanoparticles and monodisperse γ-Fe$_2$O$_3$ nanocrystallites were formed. In the second synthesis process, iron pentacarbonyl (Fe(CO)$_5$) was added into a solution of oleic acid and trimethylamine oxide, and iron oxide nanocrystals were obtained directly. Although experimental results confirmed that both synthetic approaches generated monodisperse oxide nanoparticles without a subsequent size-selection step, the

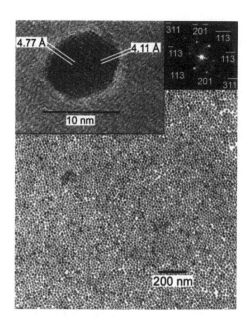

Figure 3.13 Low-resolution TEM image of a monolayer assembled from γ-Fe$_2$O$_3$ nanocrystals (10.0 $\pm$ 1.5 nm). Top left: high-resolution TEM image of an individual nanocrystal. Top right: electron diffraction pattern of the high-resolution TEM image. (Rockenberger *et al.* 1999)

first method provided a better control of the nanoparticle sizes and a better reproducibility. γ-Fe$_2$O$_3$ nanoparticles with different dimensions can be obtained by adjusting the reaction parameters. For example, γ-Fe$_2$O$_3$ nanocrystallites with an average diameter of 4 and 7 nm were obtained when the molar ratio of Fe(CO)$_5$ and oleic acid were set as 1:1 and 1:2 at the starting point, respectively.

Figure 3.14a is a TEM image showing the hexagonal arrangement of monodisperse γ-Fe$_2$O$_3$ nanoparticles with 11 nm diameter in a closed packed way. The high resolution transmission electron micrograph shown in the inset not only demonstrates the uniformity of the nanoparticle size but also reveals the highly crystalline nature of the γ-Fe$_2$O$_3$ nanoparticles. The temperature-dependent magnetization of γ-Fe$_2$O$_3$ nanoparticles was measured by using a superconducting quantum interference device following the zero-field cooling (ZFC) and field cooling (FC) procedures in an applied magnetic field of 100 Oe at 5–300 K. Figure 3.14b illustrates the relation between temperature and magnetization of 4, 13 and 16 nm γ-Fe$_2$O$_3$ nanocrystals. From the plots, it is obvious that the magnetizations reduce at certain temperatures for all samples. These temperatures, called blocking temperatures, were found to be around 23, 185 and 290 K for the γ-Fe$_2$O$_3$ nanocrystals of 4, 13 and 16 nm in diameters, respectively.

Besides γ-Fe$_2$O$_3$, hematite, α-Fe$_2$O$_3$, is another common iron oxide. Hematite is the most stable iron oxide under ambient conditions. Because of its magnetic properties and chemical stability, hematite is of scientific and technological importance. Regarding the hematite crystals, a rhombohedrally centered hexagonal structure of corundum-type is well documented. The crystalline structure of hematite is featured by a close-packed oxygen lattice where Fe^{3+} ions occupy two thirds of the octahedral sites. In α-Fe$_2$O$_3$ the magnetic moments of the two sublattices do not cancel each other completely hence a small magnetic moment in the direction of the basal plane is left. This characteristic is called parasitic or canted magnetism (Raming *et al.* 2002). For α-Fe$_2$O$_3$ nanoparticles with small dimension, they change from multidomain to single-domain. When the dimension is

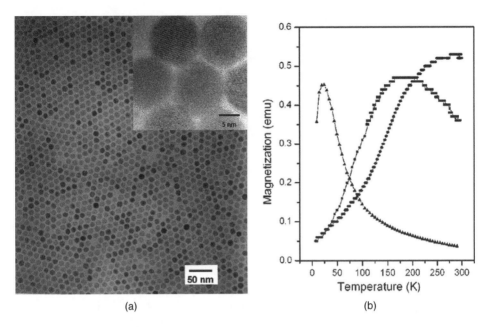

(a)   (b)

Figure 3.14 (a) TEM image of a two-dimensional assembly consisting of γ-Fe$_2$O$_3$ nanocrystals 11 nm in diameter arranged hexagonally. The inset shows a high-resolution image; (b) plot of the temperature-dependent magnetization of γ-Fe$_2$O$_3$ nanocrystallites with 4 (triangles), 13 (squares), and 16 nm (circles) in diameter measured with zero-field cooling at the applied magnetic field of 100 Oe. (Hyeon *et al.* 2001)

further decreased, superparamagnetic behavior is displayed. In this situation, the magnetic moment in the domain is fluctuated by thermal agitation randomly.

Since the unusual magnetic behavior of α-Fe$_2$O$_3$ is dependent upon the sizes and shapes of α-Fe$_2$O$_3$ particles, developing new routes for the synthesis of α-Fe$_2$O$_3$ nanocrystals with controlled sizes and shapes is sought for both investigation of their unusual magnetic behaviors and practical applications. However, so far only limited success on controllable synthesis has been achieved. Cao *et al.* reported a new reaction route to prepare rice- and cube-shaped single crystalline α-Fe$_2$O$_3$ nanostructures via a reaction between Fe(NO$_3$)$_3$ · 9H$_2$O and NH$_3$ · H$_2$O in ethylene glycol (EG) at 200 and 280 °C, respectively (2006). In their synthesis, two solutions were first made. One was Fe(NO$_3$)$_3$ · 9H$_2$O aqueous solution, and the other NH$_3$ · H$_2$O in ethylene glycol. Under stirring, the latter solution was added dropwise into the former solution. After 30 minutes, the mixture was transferred into a high-pressure vessel which was then heated to a high temperature. It was found that reactions occurring at 200 °C and 280 °C led to rice- and cube-shaped α-Fe$_2$O$_3$ nanoparticles, respectively. The final step was a carefully washing with deionized water and ethanol to remove the residual contaminants. A typical TEM image of monodisperse rice-shaped α-Fe$_2$O$_3$ with well-defined facets is shown in Figure 3.15(a). The rice-shaped α-Fe$_2$O$_3$ nanocrystals have average dimensions of 84.9 nm and 41.2 nm along their major and minor axes, respectively. Cube-shaped single-crystalline α-Fe$_2$O$_3$ nanoparticles were obtained when the reaction temperature was raised from 200 °C to 280 °C while other reaction conditions were kept identical. As shown in Figure 3.15b, the as-prepared α-Fe$_2$O$_3$ nanoparticles have an average dimension about 67 nm and have smooth surfaces. It was found that the dimension of cube-shaped α-Fe$_2$O$_3$ nanocrystals can be adjusted by changing the molecularity of the reaction and the reaction time. Figure 3.15c shows

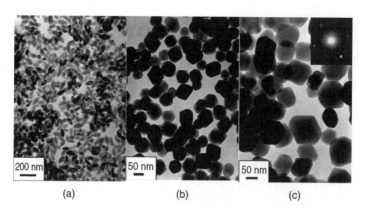

Figure 3.15 TEM images of hematite α-Fe_2O_3 nanoparticles with different shapes (a) rice-shape; (b) cube-shape ($\sim$67 nm); (c) cube-shape ($\sim$86 nm). (Cao *et al.* 2006)

α-Fe_2O_3 nanocrystals with an average 86 nm dimension which were prepared in a 10-hour reaction. Considering the reaction mechanism, α-Fe_2O_3 particles form via a two-step phase transformation, which can be described by: $Fe(NO_3)_3 + 3NH_3 \cdot H_2O \rightarrow Fe(OH)_3 + 3NH_4NO_3$, followed by $Fe(OH)_3 + \beta$-$FeOOH \rightarrow \alpha$-Fe_2O_3 (phase transformation).

Magnetic characterization of α-Fe_2O_3 nanoparticles was carried out on a vibrating sample magnetometer (VSM, LakeShore 7307) at room temperature. The hysteresis feature was observed indicating a weakly ferromagnetic property for as-prepared α-Fe_2O_3 nanoparticles at room temperature. In the measured hysteresis loops (M$-$H), no saturation was observed even at the maximum applied magnetic field. Further, cube-shaped α-Fe_2O_3 nanocrystals have higher values of remanent magnetization (M_r), squareness (M_r/M_s) and coercivity (H_c) than those of rice-shaped α-Fe_2O_3 nanoparticles.

Another type of iron oxide, magnetite (Fe_3O_4), is of great importance due to its unique structure and strong magnetic properties. A magnetite crystal is a cubic inverse spinel structure where oxygen atoms form an fcc closed packing and Fe cations occupy interstitial tetrahedral sites and octahedral sites. Magnetite is known as an important class of half-metallic materials because the electrons can hop between Fe^{2+} and Fe^{3+} ions in the octahedral sites at room temperature. Magnetite nanoparticles can be dispersed in either organic solvents or water, resulting in oil- or water-based suspensions, so-called ferrofluids. Ferrofluids have been widely used in industry including rotary shaft sealing, oscillation damping and position sensing, etc. Recently, magnetite nanoparticles have been applied in various biomedical applications such as targeted drug delivery, hypothermia, cancer therapy and magnetic resonance imaging, etc. One of the great advantages of using magnetite nanoparticles is that they can be manipulated by an external magnetic field so that a noninvasive technique for biomedical diagnostic or therapeutic purposes can be achieved. Detailed information on the biomedical applications of magnetic nanoparticles will be presented in Chapter 4.

For all technological and biomedical applications superparamagnetic nanoparticles with sizes smaller than 20 nm and a narrow size distribution are required. For this reason, the synthesis of magnetite nanoparticles with controlled size/shape and uniform physical/chemical properties is needed.

A widely used technique for magnetite nanoparticle synthesis is the high-temperature ($\sim$265 °C) reaction of Fe(acac)$_3$, iron (III) acetylacetonate, in phenyl ether with alcohol, oleic acid and oleylamine (Sun and Zeng 2002). This approach is able to produce monodisperse magnetite nanoparticles, as illustrated in Figure 3.16a. The dimension of

$$Fe(acac)_3 + ROH + RCOOH + RNH_2 + Ph_2O$$

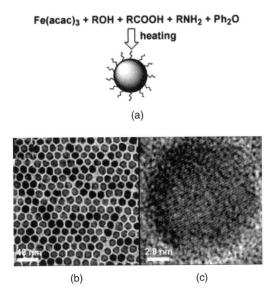

(a)

(b)   (c)

Figure 3.16 (a) Scheme of the synthesis process of magnetite nanocrystals; (b) low resolution TEM image of assembly of magnetite nanoparticles; (c) high-resolution TEM image of an individual nanoparticle. (Sun and Zeng 2002). Reprinted with permission from the *Journal of the American Chemical Society*, Sun. S and Zeng. H **124**, 8204-8205. Copyright 2002 American Chemical Society

nanoparticles can be controlled. Large monodisperse magnetite nanoparticles (up to 20 nm in diameter) can be synthesized by simply using the smaller nanoparticles as the seeds for further particle growth. Figure 3.16b is a typical TEM image showing an arrangement of 16 nm Fe_3O_4 nanoparticles. The as-prepared nanoparticles have a narrow size distribution, hence no subsequent size-selection process is required after the synthesis. The HRTEM image, shown in Figure 3.16c, reveals the atomic lattice fringes in an individual nanoparticle clearly indicating the single crystalline phase. The as-prepared magnetite nanoparticles can be dispersed into nonpolar solvent readily. And this seed-mediated growth method has the potential for mass production by scale up processes.

For large-scale synthesis of monodisperse nanocrystals, Park *et al.* reported a method using non-toxic and inexpensive metal salts as the reaction precursors (2004). This synthesis approach is very general and is capable of producing many transition metal oxides. Figure 3.17a depicts the overall procedure of synthesis. By using this method, monodisperse magnetite nanoparticles can be obtained. An obvious merit of this synthetic technique is its easy control of particle size by adjusting the synthesis conditions and there is no need for any additional size-selection process. Moreover, this technique offers other advantages over conventional methods such as the potential of ultra-large scale nanocrystal production, low cost of precursors, environmentally friendly synthesis due to the nontoxic reagents, etc. In a typical synthesis of monodisperse magnetite nanocrystals, 40 mmol iron-oleate complex was first prepared and a solution of oleic acid in 1-octadecene (5.7 g oleic acid in 200 g 1-octadecene) was added. The resulting mixture was heated to 320 °C at a ramping rate of 3.3 °C per minute and a severe reaction took place with a color change to brownish black. The reaction was kept for 30 minutes followed by a cooling process to room temperature. Next, 500 ml ethyl alcohol was added to initiate the precipitation of the magnetite nanoparticles. After centrifugation, washing and drying processes, about 40 g of magnetite nanocrystals (average 16 nm in diameter) could be obtained with a yield over 95 %. A number of organic solvents

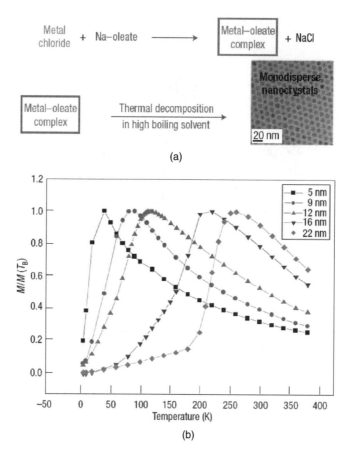

(a)

(b)

Figure 3.17 (a) Schematic diagram of the synthesis of monodisperse nanocrystals in ultra-large scale; (b) plot of the magnetization of magnetite nanocrystals *vs* temperatures measured after zero-field cooling (ZFC) using a field of 100 Oe. (Park *et al.* 2004)

such as toluene and hexane could be used to redisperse the as-prepared nanocrystals easily. To characterize the synthesized magnetite nanocrystals, a superconducting quantum interference device magnetometer (Quantum Design, MPMS5XL) was used and the temperature-dependent magnetization was measured with an applied magnetic field of 100 Oe at 5–380 K. The measurement results revealed a superparamagnetic behavior at high temperatures, as shown in Figure 3.17b. It was also found that, for 5 nm nanocrystals, the blocking temperature T_B is about 40 K, where the zero-field cooled magnetization starts to decrease and deviate from the field-cooled magnetization. Larger nanocrystals have higher block temperatures.

Another type of important magnetic oxide nanoparticles are so-called spinel ferrites, MFe_2O_4 (M = Mn, Co, Ni, Cu, Zn, Mg or Cd, etc.) which have been widely used for electronic applications over the past several decades. These materials, MFe_2O_4, or $MO \cdot Fe_2O_3$, have a cubic spinel structure where oxygen forms an fcc close packing, and M^{2+} occupy either tetrahedral or octahedral interstitial sites. Compared with iron oxide spinel ferrites provide flexibility to control both crystal structures and magnetic properties by choosing different non-iron metals in spinel ferrites and controlling their molar concentrations. Hence, it is possible to obtain great tenability in chemical composition and bonding through the variation of chemistry at the atomic level. Due in part to this versatility, magnetic spinel ferrite nanoparticles with excellent and controllable super-paramagnetic properties have attracted much interest and have the potential for many applications such as information storage, electronic devices, drug delivery, and medical diagnostics, etc. For example, $MnFe_2O_4$ is one of the most common spinel ferrites and its nanoparticles have found applications in MRI technology when they are used as the contrast enhancement agents.

Synthesis of $MnFe_2O_4$ nanoparticles has been developed recently by Liu *et al.* via a reverse micelles technique with a surfactant, sodium dodecylbenzenesulfonate (NaDBS) $[CH_3(CH_2)_{11}(C_6H_4)SO_3]$ Na (2000). In their synthesis, $Mn(NO_3)_2$ and $Fe(NO_3)_3$ were used as the precursor reagents. First 5 mmol of $Mn(NO_3)_2$ and 10 mmol of $Fe(NO_3)_3$ were first dissolved in 25 ml water completely, and 25 ml 0.4 M NaDBS aqueous solution was added into the metal salt solution. Then excess toluene was added to the resulting solution with rigorous stirring to form clear water-in-toluene reverse micelles. After stirring overnight, 40 ml 1.0 M sodium hydroxide aqueous solution was dropped into the reverse micelles and colloids were formed with stirring for an additional two hours. To separate the synthesized nanocrystals, the colloid was heated to evaporate most water and toluene followed by washing with water and ethanol to remove excess surfactant. The key reaction parameter is the volume ratio of water and toluene, which determines the dimension of spinel ferrite nanoparticles. Experimental results showed that a ratio of 5/100 (water/toluene) resulted in nanoparticles with an average diameter of 8 nm. TEM micrograph of the $MnFe_2O_4$ nanoparticles is shown in Figure 3.18a (Liu *et al.* 2000). The synthesized nanoparticles have a narrow dimension distribution with an average diameter of 7.7 ± 0.7 nm. Figure 3.18b plots the temperature dependence of magnetization and remanent magnetization decay (thermal remanent magnetization) of the $MnFe_2O_4$ nanoparticles, which was measured by a SQUID magnetometer (Quantum Design MPMS-5S). In the temperature-dependent magnetization measurement, the $MnFe_2O_4$ nanoparticles were cooled to 5 K under a zero magnetic field. When the temperature reached 5 K, a magnetic field of 100 G was applied. Then the temperature was raised with continuous magnetization measurements. As illustrated in the ZFC line in Figure 3.18b, the magnetization of nanoparticles increases with the temperature at the beginning. At 150 K, the magnetization reaches a maximum value and begins to drop as the temperature increases further.

To investigate the temperature dependence of remanent magnetization decay, the $MnFe_2O_4$ nanoparticles were first cooled to 5 K under a magnetic field of 100 G. Then, the magnetic field was switched off and the remanent magnetization was recorded as the temperature was raised up to 350 K. The measured result is shown in Figure 3.18c. It is obvious from the plot that the remanent magnetization decreases rapidly with increasing temperature at the initial period. The remanent value saturates and approaches zero at around 150 K. The hysteresis loops of nanoparticles were measured at different temperatures. Figure 3.18c displays the hysteresis feature of nanoparticles at 20 K with a coercivity about 190 G. It was noticed that the hysteresis feature of $MnFe_2O_4$ nanoparticles disappeared at measuring temperatures above 150 K. The hysteresis loop of nanoparticles measured at 400 K reveals direction changes in unison of the magnetization with the direction reversal of the applied magnetic field.

As mentioned above, magnetic MFe_2O_4 nanoparticles have been used widely not only as ferrofluids but also as promising candidates for biomedical applications such as molecular tagging, imaging and separation. Depending on specific M^{2+}, the densely packed nanocrystals of MFe_2O_4-based materials can exhibit high magnetic permeability and electrical resistivity or half-metallicity, and may be a potential candidate for future high-performance electromagnetic and spintronic devices (Sun *et al.* 2003a). Sun's

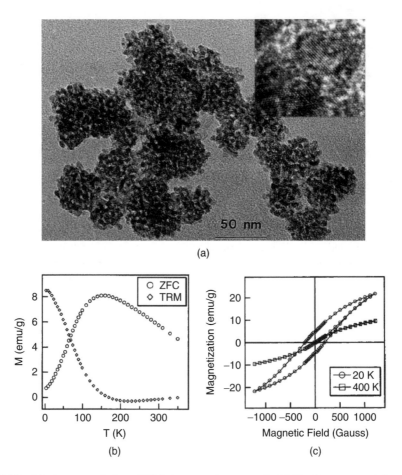

(a)

(b)

(c)

Figure 3.18 (a) Low resolution TEM image of prepared $MnFe_2O_4$ nanoparticles with a high-resolution image shown in the inset; (b) plots of temperature-dependent remanent magnetization decay for $MnFe_2O_4$ nanoparticles of ~8 nm in diameter cooled under a magnetic field of 100 G (TRM), and temperature-dependent magnetization under a magnetic field of 100 G for zero-field-cooled $MnFe_2O_4$ nanoparticles of ~8 nm in diameter (ZFC); (c) hysteresis loops of the $MnFe_2O_4$ nanoparticles with a diameter of ~8 nm measured at 20 K and 400 K. (Liu *et al.* 2000)

research group applied their organic phase process, which was used to prepare monodisperse Fe_3O_4 nanoparticles through the reaction of $Fe(acac)_3$ and a long-chain alcohol, to synthesize MFe_2O_4 nanoparticles (with M = Co, Ni, Mn, Mg, etc.). In this convenient organic phase process, different metal acetylacetonate precursors, $Co(acac)_2$ and $Mn(acac)_2$, were added to the mixture of $Fe(acac)_3$ and 1,2-hexadecanediol, and produced $CoFe_2O_4$ and $MnFe_2O_4$ nanoparticles, respectively. A typical process involves a reaction between metal acetylacetonate with 1,2-hexadecanediol, oleic acid and oleylamine at high temperature (up to 305 °C). The as-prepared monodisperse MFe_2O_4 nanoparticles have a tunable dimension ranging from 3 to 20 nm in diameter, which can be tailored by varying the reaction temperature or changing the metal precursors. Also, larger sized nanoparticles can be obtained by using the smaller nanocrystals as the seeds for further crystal growth. There are two advantages of this synthesis technique: there is no need for a low-yield fractionation procedure to achieve the desired size distribution and it is easy to scale up production for large quantities. The as-prepared nanoparticles are hydrophobic

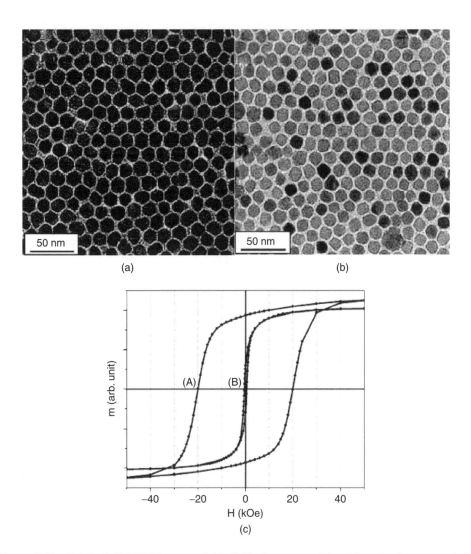

(a)   (b)

(c)

Figure 3.19   Bright field TEM images of (a) CoFe$_2$O$_4$ nanoparticles 14 nm in diameter and (b) MnFe$_2$O$_4$ nanoparticles 14 nm in diameter made from seed-mediated growth and deposited from their hexane dispersion on amorphous carbon-coated copper grid; (c) magnetization curves of the assembly made from CoFe$_2$O$_4$ nanoparticle 16 nm in diameter measured at (A) 10 K and (B) 300 K. (Sun *et al.* 2003a). Reprinted with permission from the *Journal of the American Chemical Society*, Sun. S and Zeng. H **124**, 8204-8205. Copyright 2002 American Chemical Society

inherently, hence they can be dispersed in nonpolar or weakly polar hydrocarbon solvents (e.g. hexane, toluene, etc). However, hydrophilic nanoparticles can be obtained by using a bipolar surfactant, such as tetramethylammonium 11-aminoundecanoate, thus aqueous nanoparticle dispersions can be easily prepared.

Figure 3.19a shows a TEM image of 14 nm CoFe$_2$O$_4$ nanoparticles made from the seed-mediated growth. It is obvious from the image that monodisperse nanoparticles have a very narrow diameter distribution. Similar monodisperse properties were observed for 14 nm MnFe$_2$O$_4$ nanoparticles, as shown in Figure 3.19b. Magnetic measurements of 16 nm CoFe$_2$O$_4$ nanoparticles revealed ferromagnetic characteristics at temperatures up to 300 K, illustrated in Figure 3.19c. The coercivity of the nanoparticle assembly is about

400 Oe. However, at 10 K, its coercivity changes to 20 kOe which is much larger than that of 16 nm Fe_3O_4 nanoparticles (around 450 Oe at 10 K). This coercivity difference indicates a great increase of the magnetic anisotropy for $CoFe_2O_4$ nanoparticles when Co cations are incorporated into the Fe-O matrix. On the contrary, the incorporation of Mn cations in the Fe-O matrix reduces the magnetic anisotropy of $MnFe_2O_4$ nanoparticles when compared with Fe_3O_4 nanoparticles of the same size. Measurements showed a coercivity of only 140 Oe for 14 nm $MnFe_2O_4$ nanoparticles at 10 K.

3.4.4 Synthesis of Magnetic Core–shell Nanoparticles

Interest in magnetic core–shell nanoparticles has been stimulated due to the drawbacks of uncovered magnetic nanoparticles, such as agglomeration, being highly reactive, etc. Nanostructure magnetic materials exhibit unique properties and have been expected to be used in data storage technology and biomedical systems. However, pure magnetic particles may not be very useful in practical applications. For instance, they are likely to form a large aggregation and their magnetic properties change readily. Most importantly, they can undergo rapid biodegradation when they are directly exposed to the biological system (Wang et al. 2005).

The ability to synthesize and assemble monodisperse core–shell nanoparticles is important for exploring the unique properties of nanoscale core, shell, or their combinations in technological applications. A good example is the assembly of gold nanoparticles on SiO_2 cores, which has been demonstrated to be a very effective approach to obtain core–shell SiO_2/Au nanoparticles (Westcott et al. 1998). Similarly, iron oxide/gold-core/shell nanoparticles have been synthesized by a deposition of gold onto Fe_2O_3 or partially oxidized Fe_3O_4 nanoparticles (9 nm) in an aqueous solution via hydroxylamine seeding to synthesize ~60 nm-sized particles (Lyon et al. 2004). It was reported that gold coated magnetic core–shell nanoparticles exhibited enhanced chemical stability because the gold layer protects the core from possible oxidation and corrosion. Moreover, with the help of the gold layer, core–shell nanoparticles possess better biocompatibility and affinity via amine/thiol functional groups. Synthesis of gold coated iron core–shell (Cho et al. 2005) and gold (silver)-coated magnetite (Mikhaylova et al. 2004) nanoparticles have also been reported by suing the reverse micelles method. Caruntu et al. prepared gold coated Fe_3O_4 nanoparticles by attaching 2–3 nm-sized gold nanoparticles via 3-aminopropyl triethylsilane onto 10 nm-sized Fe_3O_4 nanoparticles (Caruntu et al. 2005).

It is true that thin films of magnetic nanoparticles have been technologically interesting for a long time. For this reason, efforts have been made to prepare thin films of magnetic nanoparticles. Thin films of polymer-FePt nanoparticles and arrays have been deposited on Si substrates via a polymer-mediated assembly method (Sun et al. 2003b). Another example is thin film assemblies of magnetic nanoparticles prepared by an evaporation method together with the applied magnetic field parallel to a substrate (e.g. HOPG) (Pileni et al. 2004). To improve the stability and tenability of thin films consisting of magnetic nanoparticles, core–shell nanostructures are attractive. These magnetic core–shell nanoparticles are helpful for exploring the electronic, magnetic, catalytic, sensing and chemical or biological properties of the nanocomposite materials and are expected to find various applications (Wang et al. 2005a). In this section, syntheses of several representative core–shell magnetic nanoparticles are reviewed.

The first type is Fe_3O_4@Au nanoparticles. In Wang's synthesis, Fe_3O_4 nanoparticles were prepared by an initial synthesis and used as seeds for a subsequent reduction of $Au(OOCCH_3)_3$ on the surface of magnetic nanoparticles (2005a). First, 2 mmol $Fe(acac)_3$ was dissolved in a mixture of 20 ml phenyl ether, 2 ml oleic acid (~6 mmol) and 2 ml

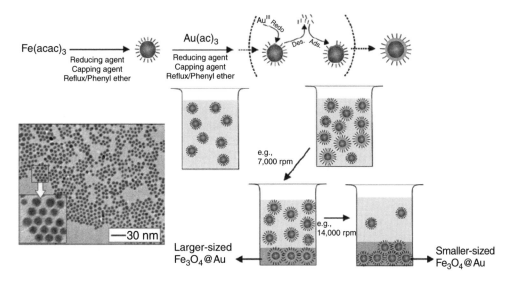

Figure 3.20 Schematic illustration of the synthesis processes and involved chemistry Fe_3O_4 and Fe_3O_4@Au nanoparticles preparation. TEM images are also shown. (Wang *et al.* 2005a)

oleylamine ($\sim$4 mmol) by vigorous stirring under a protection of argon. Then 10 mmol of 1,2-Hexadecanediol was added and the solution was heated to 210 °C followed by refluxing at the same temperature for 2 hours. After cooling down to room temperature, the synthesized 0.33 mmol Fe_3O_4 nanoparticles were dispersed in 10 ml phenyl ether and 2.2 mmol $Au(OOCCH_3)_3$, 12 mmol 1,2-hexadecanediol, 1.5 mmol oleic acid, and 6 mmol oleylamine were added. The mixture was heated to 180–190 °C under vigorous stirring and argon protection. After reaction for 1.5 hours, the mixture was cooled down and ethanol was added to precipitate a dark purple material which was separated by centrifuging. The synthesis processes of the Fe_3O_4 and Fe_3O_4@Au nanoparticles and involved chemistry are illustrated in Figure 3.20. As shown in the bottom left of Figure 3.20, the average size of the Fe_3O_4@Au was 6.6 $\pm$ 0.4 nm from the TEM image. The inset image reveals around 0.7 nm gold coating on Fe_3O_4 nanoparticles. The interparticle spacing is expected to be the capping monolayer, oleylamine and oleic acid, which helps nanoparticles to be isolated individually and uniformly. The magnetic properties were measured by SQUID. Compared with Fe_3O_4 nanoparticles, Fe_3O_4@Au core–shell nanoparticles display a decrease in magnetization and blocking temperature and an increase in coercivity. This can be explained by the decreased coupling of the magnetic moments induced by the increased interparticle spacing by shells (Wang *et al.* 2005).

Another promising shell material used in magnetic core–shell nanoparticles is silica (SiO_2). This type of core–shell system consists of anisotropic magnetic cores and isotropic silica spheres. The use of silica shell provides several unique characteristics such as hydrophilic behavior, biocompatibility in biological systems, and high suspension stability in various solvents. These properties make the magnetic core–shell nanoparticles an ideal candidate in targeted drug delivery systems. With the help of an external magnetic field, a large amount of drug loaded on microspheres can be transferred to the specific tissue with a small amount of magnetic core–shell colloid. The synthesis of crystallized spinel ferrite ($ZnMnFe_2O_4$) nanoparticles through the hydrothermal method has been reported. The synthesis reaction can be described as:

$$ZnCl_2 + MnCl_2 + FeCl_3 + NaOH \rightarrow ZnMnFe_2O_4 \qquad (3.1)$$

A typical synthesis starts from preparation of a 0.025 M ZnCl$_2$, 0.02 M MnCl$_2 \cdot$ 4H$_2$O, and 0.1 M FeCl$_3 \cdot$ 6H$_2$O aqueous solution (Wang *et al.* 2005). With stirring, the mixture was loaded into a Teflon-lined stainless autoclave. Then sodium hydroxide solution, prepared by dissolving 0.88 g NaOH in 5 ml water, was slowly dropped into the autoclave to initiate precipitation. The hydrothermal reaction was kept at 180 °C for 12 hours. The ZnMnFe$_2$O$_4$ nanoparticles were washed and collected by using a permanent magnet. TEM observation showed the as-prepared nanoparticles have a narrow size distribution with an average diameter of 10 nm. The second synthesis step involves silica coating on the as-prepared ZnMnFe$_2$O$_4$ nanoparticle surface by a hydrolysis of TEOS.

Figure 3.21a is an SEM micrograph of the SiO$_2$@ZnMnFe$_2$O$_4$ core–shell nanoparticles showing monodisperse spheres with a uniform dimension. From the TEM image, Figure 3.21b, the magnetic cores can be easily seen as dark spots inside spherical SiO$_2$ shells. The magnetic properties of synthesized SiO$_2$@ZnMnFe$_2$O$_4$ core–shell nanoparticles were characterized by using a vibrating sample magnetometer (VSM) at room temperature. Curve A and curve B in Figure 3.21c are hysteresis loops for ZnMnFe$_2$O$_4$ nanoparticles and SiO$_2$@ZnMnFe$_2$O$_4$ core–shell nanoparticles, respectively. It is obvious

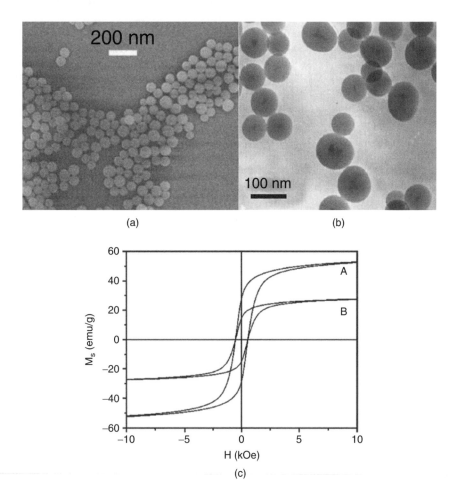

Figure 3.21 (a) SEM image of silica coated ZnMnFe$_2$O$_4$ nanoparticles; (b) TEM image showing the core–shell structure; (c) magnetic hysteresis loops of (A) ZnMnFe$_2$O$_4$ nanoparticles without silica coating and (B) silica coated ZnMnFe$_2$O$_4$ nanoparticles. (Wang *et al.* 2005)

that both uncoated nanoparticles and core–shell nanoparticles exhibit ferromagnetic properties, which are attributed to the well-crystallized spinel ferrite cores. Also, the measurement indicated that the core–shell nanoparticles have lower saturation magnetization (M_s), remnant magnetization (M_r) and saturation magnetic fields than those of uncoated $ZnMnFe_2O_4$ nanoparticles. However their coercivities (H_c) were nearly the same. Considering the possible reasons for the differences of magnetic parameters between coated and uncoated nanoparticles, it is believed that the core–shell nanoparticles have a lower magnetic content than uncoated nanoparticles. Since the magnetic behaviors of core–shell nanoparticles are completely dependent upon the core materials, the magnetization is reduced due to the existence of a nonmagnetic silica component in core–shell nanoparticles. Additionally, the uniformity or magnitude of magnetization of core–shell nanoparticles can be affected by the silica coating layer due to quenching of the surface moments.

Magnetite coated iron core–shell nanoparticles have been developed. Sun's research group reported an improved but simple, one-pot reaction that gives monodisperse Fe_3O_4@Fe nanoparticles (Peng *et al.* 2006). It was found that Fe nanoparticles in hexane dispersion were easily oxidized and agglomerated when exposed to air. Uncontrolled oxidation of iron nanoparticles usually gives amorphous iron oxide layers, which are not able to protect against further oxidation or agglomeration. On the contrary, well-crystalline iron oxide shells can provide effective and robust protection. With coated crystalline Fe_3O_4 shells, iron nanoparticles are more efficiently stable and functionalization can be achieved readily on the oxide surface. Consequently, stable Fe nanoparticle dispersion in phosphate-buffered saline (PBS) can be obtained. Sun's research group developed an efficient approach to prepare monodisperse core–shell Fe nanoparticles by decomposition of $Fe(CO)_5$ in octadecene (ODE) at 180 °C.

As shown in Figure 3.22a and 3.22b, the core–shell nanoparticles are monodisperse with a very narrow size distribution ($\sigma < 7\%$). The average dimension of cores and shells are 4 nm and 2.5 nm, respectively. The magnetic characterization results are shown in Figure 3.22(c), which was obtained from measurements on the core–shell nanoparticles precipitated from hexane dispersion. Line A in the figure indicates a superparamagnetic property with a saturation moment of about 66.7 emu/g, which corresponds to 102.6 emu/g [Fe] ([Fe] = Fe + Fe_3O_4). This value is slightly lower than that calculated for 2.5 nm@4 nm Fe_3O_4@Fe nanoparticles based on the relative weight percentage of Fe (7.86 g/cm^3, 218 emu/g) and Fe_3O_4 (5.18 g/cm^3, 80 emu/g), which is 123.5 emu/g [Fe]. Since the experimental value is close to the calculated value for 3.5 nm@3 nm Fe_3O_4@Fe nanoparticles (99.4 emu/g [Fe]), this discrepancy can be explained by a further oxidation occurring for the core–shell nanoparticles. To test the stability of the magnetic core–shell nanoparticles, as-synthesized 2.5 nm@4 nm Fe_3O_4@Fe nanoparticles dispersion was used to measure its magnetic moment drops as exposed to air at room temperature. As shown in line B, the magnetic moment drops quickly during the first 4 hours, stabilizes between 4 and 14 hours, and then drops to the value which is close to that of Fe_3O_4 nanoparticles. The plateau in both plots is probably caused by agglomeration occurring in the dispersions. To make the core–shell nanoparticles stable, a crystalline Fe_3O_4 shell was prepared by controlled oxidation using $(CH_3)_3NO$, an oxygen transferring agent. 5–nm@2.5–nm Fe_3O_4@Fe nanoparticles treated by this controlled oxidation show much better stability than untreated core–shell nanoparticles, as shown by line C. Slower moment dropping was observed while no agglomeration was detected over 14 hours. The results clearly show that the crystalline Fe_3O_4 shell provides a good protection to the Fe core and offers better nanoparticle dispersion.

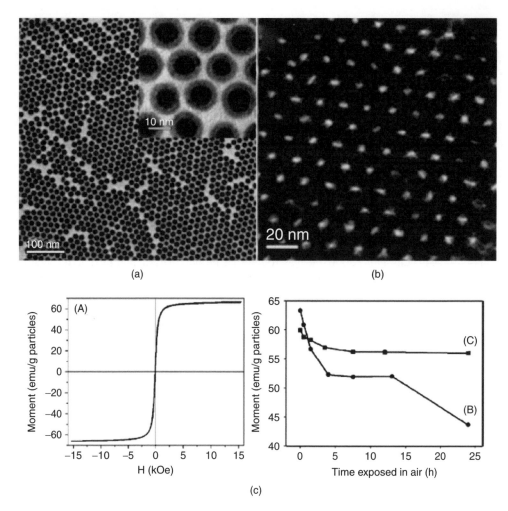

Figure 3.22 (a) TEM image of 4 nm/2.5 nm Fe/Fe$_3$O$_4$ nanoparticles with a high-resolution TEM image (shown in the inset); (b) TEM image of self-assembled Fe/Fe$_3$O$_4$ nanoparticle superlattice; (c) results from magnetic characterizations of Fe/Fe$_3$O$_4$ nanoparticles (A) hysteresis loop measured at room temperature; (B) magnetic moment drop of 4 nm/2.5 nm Fe/Fe$_3$O$_4$ nanoparticles; (C) magnetic moment drop of 2.5 nm/5 nm Fe/Fe$_3$O$_4$ nanoparticle dispersions versus exposure time to air at room temperature. (Peng *et al.* 2006)

Another representative example of magnetic core–shell nanoparticles is Co$_{core}$Pt$_{shell}$ nanoalloys (Park and Cheon 2001). The synthesis process includes a redox transmetalation reaction, which is depicted in Figure 3.23a. In the first step, Co nanoparticles were prepared from the thermolysis of Co$_2$(CO)$_8$ in toluene solution by refluxing for 8 hours. After adding ethanol, Co nanoparticles were precipitated and separated by centrifuging. Unlike Fe nanoparticles, Co nanoparticles are stable in air. To cover Co nanoparticles with Pt shell, Co nanoparticle colloids and Pt(hfac)$_2$ were refluxed in a nonane solution containing a stabilizer (e.g. C$_{12}$H$_{25}$NC). The obtained Co$_{core}$Pt$_{shell}$ nanoparticles are moderately monodisperse ($\sigma < 10\%$) without any additional size-selection process.

The TEM image reveals an average diameter of 6.27 nm ($\sigma = 0.58$ nm) for as-prepared nanoparticles, as shown in Figure 3.23b. High-resolution TEM observation indicated a

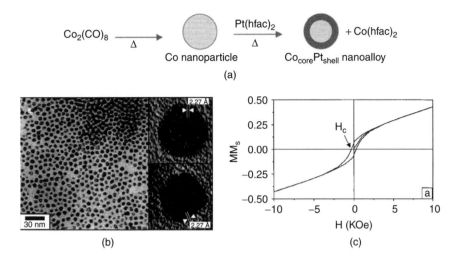

Figure 3.23 (a) Schematic depiction of the synthetic process of core–shell nanoalloys by transmetalation reaction; (b) low resolution and high-resolution TEM images of 6.27 nm ($\sigma = 0.58$) $Co_{core}Pt_{shell}$ nanoalloys; (c) measured hysteresis loop of as-prepared $Co_{core}Pt_{shell}$ nanoalloys. (Park and Cheon 2001)

smoother and homogeneous surface of the $Co_{core}Pt_{shell}$ particles, which is different from other types of core–shell nanoparticles with uneven outer particles attached to the surface. Figure 3.23b reveals the lattice image of a nanoparticle shell. The lattice distance was estimated to be 2.27 ± 0.03 Å, which corresponds to the Pt(111) lattice parameter (2.265 Å). EDAX spectrum also confirmed the existence of the Pt element on the nanoparticles with a stoichiometry of $Co_{0.45}Pt_{0.55}$. According to a simple close packing model simulation, a $Co_{core}Pt_{shell}$ nanoalloy is composed of a Co core with 4.75 nm diameter and a Pt shell with 1.82 nm thickness. It was expected that, by adjusting the amount of Pt precursor, the thickness of Pt shell could be controlled. Figure 3.23c is the hysteresis loop of $Co_{core}Pt_{shell}$ nanoalloys measured by a SQUID magnetometer. A coercivity of 330 Oe was obtained for nanoparticles at 5 K. The measurement results suggest that the $Co_{core}Pt_{shell}$ nanoalloys exhibit very similar magnetic properties to those of Co nanoparticles and Pt shells do not affect significantly the magnetic properties of $Co_{core}Pt_{shell}$ nanoalloys.

3.5　Bio-inspired Magnetic Nanoparticles

Biomimetic approaches, stimulated by a detailed understanding of biomineralization, have been envisioned for controlled synthesis of new materials with specific crystal morphology, phase and orientation. Generally, well-defined protein architectures or macromolecular templates are utilized in a bio-inspired approach to synthesize novel materials. The evolved molecular interactions are believed to play an important role for material formation.

Nanomaterials exhibit emergent magnetic, catalytic, electrical properties which could find versatile applications in the near future. Synthesis of nanomaterials in a controlled manner has been a challenge for several decades. So far, a number of physical and chemical techniques have been developed for nanomaterial synthesis. However, there is still an increasing demand for a novel synthesis technique which offers better control for nanomaterial formation. A promising approach is the bio-inspired synthesis, which uses biomacromolecules as templates for directed fabrication of nanomaterials. For instance, symmetric protein cage architectures can be used as constrained reaction environments to synthesize or encapsulate inorganic and organic nanomaterials. Phage-display techniques are able to identify peptides sequences specifically towards particular inorganic materials with unique crystal faces.

It is not surprising that the bio-inspired technique has been also utilized to synthesize magnetic nanoparticles. For this purpose, ferritin, a Fe-storage protein, can be used to grow CoPt nanoparticles which can convert to ordered $L1_0$ phase nanocrystals by annealing. Via the same technique, large 2D magnetic arrays can be obtained. This biomimetic approach makes protein engineering possible by controlling the directed evolution and screening of combinatorial peptide sequences with specific affinity toward inorganic materials.

CoPt nanoparticles with $L1_0$ phase have been synthesized successfully via a bio-inspired approach by Klem *et al.* (Klem *et al.* 2005). In their synthesis, the template was a protein-cage architecture of the small heat-shock protein, so-called MjHsp. It was obtained from the hyperthermophilic archeaon, *Methanococcus jannaschii*. Small peptide sequences (KTHEIHSPLLHK), obtained from screening a M13 bacteriophage library, have the capability of binding to the $L1_0$ phase of CoPt. The peptide sequence was incorporated into the template, resulting in a highly versatile cage-like structure with interior and exterior surfaces. Its exterior diameter is about 12 nm. Via chemical or generic modification, the surface properties can be tailored. Figure 3.24a depicts the protein assembly of MjHsp. For a 16.5 kDa MjHsp subunit, there are 147 amino acids which can be modified through site-directed mutagenesis or chemical modification. Importantly, there are large pores (around 3 nm) located at the three-fold axes. Molecules are able to access the interior of the protein cage through these pores. It is necessary to mention that this protein assembly has a relative high stability and it can stand up to 70 °C in a pH range of 5 11. In the bio-inspired synthesis of CoPt nanoparticles, the assembled protein cage served as size-constrained reaction vessels while the precursors, Co^{II} and Pt^{II}, entered the cage and were reduced by excess $NaBH_4$ at pH 8 and 65 °C for 10 minutes. A TEM image of as-prepared CoPt nanoparticles is shown in Figure 3.24b. The average diameter of nanoparticles is about 6.5 ± 1.3 nm. The inset image shows an electro diffraction pattern indicating a crystalline phase for the protein-encapsulated material.

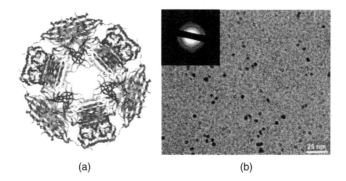

(a)   (b)

Figure 3.24 (a) Ribbon diagram of *Methanococcus jannaschii* MiHsp showing the threefold channel of the assembled protein cage; (b) TEM image of CoPt-mineralized Cp_Hsp with the electron diffraction pattern shown in the inset. (Klem *et al.* 2005)

Bacteria have also been used for nanoparticle synthesis. This bacteria-assisted biomineralization for magnetite synthesis can be classified into two categories, biologically controlled mineralization and biologically induced mineralization. The first technique can be traced back to 1975 when magnetic crystals formed from bacteria were discovered inside the cell (Blakemore 1975). This magnetite, termed magnetosomes, attaches to the cell membrane and allows the bacteria to orient along the magnetic field. However, the low ratio of magnetite nanocrystals to biomass is a major drawback of this approach. In the second synthesis approach, biologically induced mineralization, the necessary chemical situation can be created by the bacteria so that magnetite precipitation occurs on the outer surface of the cell membrane. A couple of thermophilic bacteria such as *thermoanerobacter* and *thermoanerobium* have shown their capability of reducing Fe^{3+} ions as part of their respiration processes.

Yeary *et al.* reported their successful microbial synthesis of the magnetite (2005). The bacteria was TOR-39 and a basal medium was used which was mixed from $NaHCO_3$, $CaCl \cdot 2H_2O$, NH_4Cl, $MgCl \cdot 6H_2O$, NaCl, HEPES (hydroxyethylpiperazine-N'-2-ethane-sulfonic acid), yeast extract, 0.1 % resazurin, trace mineral solution, and vitamin solution in proper concentration. The medium was prepared under nitrogen protection and the pH value was between 8.0 and 8.2. The precursor for magnetite production is amorphous Fe^{3+} oxyhydroxides (akaganeite), which was prepared by mixing 10 M NaOH solution and 0.4 M $FeCl_3 \cdot 6H_2O$ solution with rapid stirring. It was noticed that no extra reducing reagents were introduced in either medium or precursor solution. During the nanoparticle formation, the bacteria itself serves as a reducing agent. After reaction, amorphous iron hydroxide suspension was made from the precipitates. Then incubation was performed with the inoculation of sterile glucose, inoculums of bacteria, MOPS and nonsterile amorphous Fe^{3+} oxyhydroxide at 65 °C for 14 days. The final product was dried in a freeze dryer. To check the morphology of as-prepared nanoparticles, biomagnetite samples were prepared following Zhang's mounting procedure (Zhang 1998). Figure 3.25 is a typical TEM image showing a bacteria cell with dark spots of magnetite particles and needle-like Fe^{3+} oxyhydroxide. Nanoparticle agglomeration was also observed.

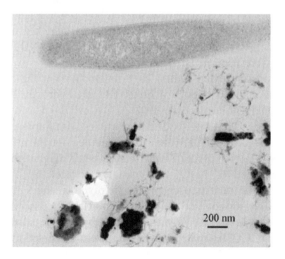

Figure 3.25 TEM image of a TOR bacterium with magnetite nanoparticles, shown as the black spots, and the needle-shaped material of Fe(III) oxyhydroxide. (Yeary *et al.* 2005)

3.6   Functionalization of Magnetic Nanoparticles

Functionalization of nanomaterials is an effective approach to modify their surface properties. For certain applications, it is essential to tune the surface characteristics of nanomaterials. Similarly, for magnetic nanoparticles, functionalization has been used, particularly for bioapplications. Surface modification of magnetic nanoparticles with organic molecules is an effective functionalization approach. Magnetic nanoparticles modified with organic molecules have been widely used for biomedical and biotechnological applications, such as cell separation, drug delivery, hyperthermia, automated DNA extraction, gene targeting, magnetic resonance imaging, etc. Biofunctionalization of magnetic nanoparticles with biological entities is an alternative functionalization approach which displays unique characteristics. Biofunctionalized magnetic nanoparticles have unique interaction between biological entities, and assisted by biological recognition elements attached to their surface magnetic nanoparticles can access the target specifically. For example, magnetic nanoparticles coated with antibodies can be used for sensitive immunoassays. Also, magnetic nanoparticles modified with single-stranded DNA can be used to identify organisms and analyze single-nucleotide polymorphism due to DNA hybridization capability. This type of specific bindings offers several advantages in biomolecule detection over conventional radioactive, electrochemical, optical and other techniques (Nidumolu *et al.* 2006). However, surface modification chemistries of magnetic nanoparticles are left far behind their synthesis achievements. Therefore, effective biofunctionalization of magnetic nanoparticles is required for their utility in various biological applications (Hong *et al.* 2005). In this section, several techniques of functionalizing magnetic nanoparticles with organic molecules and biological entities will be reviewed.

3.6.1   Functionalization with Organic Molecules

Organic molecules including chemically functional groups and polymer chains can be linked to magnetic nanoparticles by bindings.

Covalent bonding is a widely used functionalization technique. Wang and his team reported their synthesis of the molecularly-mediated assembly of Fe_3O_4 and $Fe_3O_4@Au$ nanoparticles via interparticle covalent or hydrogen bonding linkages (2005b). Their technique took advantage of the gold shells on Fe_3O_4 cores, which can bind to the thiolate groups specifically. In the synthesis process, 1,9-nonanedithiol (NDT) and mercaptoundecanoic acid (MUA) were used in an initial thiolate–oleylamine exchange reaction. The cross-linking was made by using dithiol as a linker (A) via the alkyl chain or using carboxylic acid functionalized thiol as a linker (B) via hydrogen bonding. Figure 3.26 illustrates the functionalization process.

Biofunctionalized core–shell iron oxide nanoparticles with fluorescent dyes can be used to enhance the contrast of magnetic resonance imaging, optical imaging and magnetic-related cancer therapy (Levy *et al.* 2002). These core–shell magnetic nanoparticles, named nanoclinics, are composed of Fe_2O_3 nanoparticles covered by a thin shell of silica and fluorescent dyes. The multistep chemistry involved in the synthesis is depicted in Figure 3.27. Luteinizing hormone-releasing hormone (LH-RH), serving as a biotargeting group, was attached to the surface of nanoparticles by functionalization. In a representative functionalization, COOH-coated nanoclinics were dispersed in anhydrous DMF by sonication. Then LH-RH and BOP (benzotriazolyl-N-oxytrisdimethylaminophos-phonium

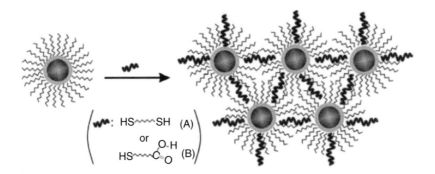

Figure 3.26 A schematic illustration of the functionalization of the $Fe_3O_4@Au$ nanoparticles. The functionalization involves a thiolate–oleylamine exchange reaction followed by cross-linking via linkers A or B for $Fe_3O_4@Au$ nanoparticles assembly (Fe_3O_4 core: dark circle, Au shell: gray circle, oleylamine monolayer: thin gray zigzag lines). (Wang *et al.* 2005b)

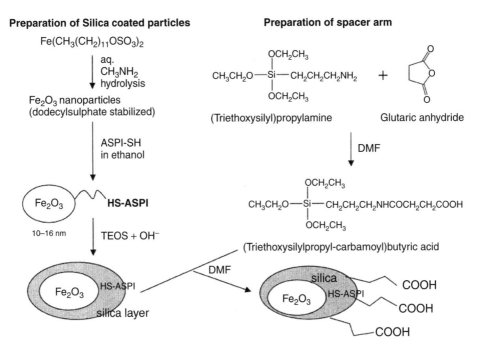

Figure 3.27 Schematic diagram of the synthetic protocol for Fe_2O_3 nanoparticle functionalization (Levy *et al.* 2002)

hexafluorophosphate) were added to the dispersion and sonicated for 5 hours for coupling of the peptide hormone analogue. In the next step, the proper amount of DIPEA (N, N-diisopropyl-ethylamine) was added dropwise to neutralize the reactive groups on the surface of the nanoclinics. After washing and centrifugation, the biofunctionalized nanoclinics were obtained which can be used for targeting of the receptor-specific cells and preferential destruction of cancer cells and tissues by an external DC magnetic field.

Functionalization of magnetic FePt nanoparticles by PEGylation has been demonstrated by Hong *et al.* coworkers (2005). The synthesized mixed-monolayer-functionalized FePt nanoparticles possessed enhanced stability and improved versatility. It is well documented that poly(ethylene glycol) (PEG) and PEGylated materials are biocompatible. Thus colloidal nanoparticles, including Au, CdSe, CdSe/ZnS and iron oxide, have reduced cytotoxicity and nonspecific protein binding after PEGylation. The PEGylation reagents are PEGylated thiol and dopamine ligands. After functionalization, thiol ligands are attached to the surface of FePt nanoparticles and they are exchangeable. Owing to this property, a large number of different functional ligands, ranging from small chemical functional groups to complicated biomacromolecules, are able to be incorporated into the nanoparticle surface. Assisted by the functional ligands, hybrid inorganic-organic materials could be realized for numerous bioapplications.

The PEGylation of FePt nanoparticles is depicted in Figure 3.28 (Hong *et al.* 2005). Use of PEG (**2**) did not lead to water-soluble nanoparticles, indicating that it is not an effective functionalization agent. This is probably because of the low coverage of the nanoparticle by the neural PEG thiol ligand (**2**). On the other hand, the combination of ammonium-terminated (**3**) and carboxylate-terminated (**4**) PEG-thiols is a better choice which generated well functionalized water-soluble FePt nanoparticles. It is believed that the much better water solubility of functionalized nanoparticles is attributed to the charge at the ligand chain end while the degree of functionalization does not change much. To further improve the stability of the nanoparticle solution, dopamine, a 1,2-enediol, was found to bind to the FePt nanoparticle strongly. Therefore, an ideal functionalization was performed by using a thiol-dopamine mixed-monolayer system. After functionalization, thiols cap the Pt atoms while dopamine ligands are on the surface of Fe sites. As the result, stable nanoparticle aqueous solutions and nanoparticles in biologically relevant media (e.g. PBS, cell culture medium, etc) were achieved.

As mentioned above, the most impressive application of magnetic nanoparticles is probably their targeted drug delivery. For this purpose, the magnetic labeling of drugs can be transported to the destination under the guidance of an external magnetic field

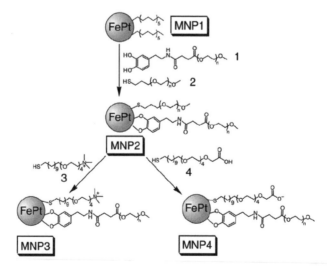

Figure 3.28 Schematic illustration of PEGylation of FePt nanoparticles. Thiol and dopamine ligands **1**–**4** were used for FePt nanoparticle functionalization and resulted in nanoparticles **MNP1**–**3**. The average molecular weight of mPEG was around 550 (i.e., $n \approx 11$). (Hong *et al.* 2005)

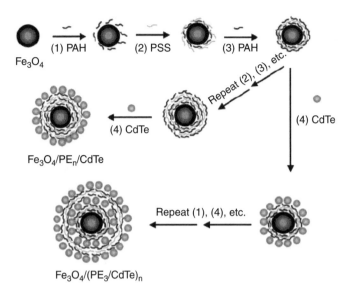

Figure 3.29 Schematic diagram of the Lay-by-Lay process to synthesize magnetic luminescent nanocomposites. (Hong *et al.* 2004)

followed by an effective targeted cell-specific drug release. This noninvasive technique is ideal for therapeutic purposes.

Magnetic luminescent nanocomposites have been developed by a layer-by-layer assembly synthesis. These magnetic nanomaterials, consisting of magnetic nanoparticles, semiconductor quantum dots and polyelectrolyte, could be used for luminescent tagging applications in biomedicine and biotechnology (Hong *et al.* 2004). The attachment of quantum dots and polyelectrolyte modifies the surface properties of magnetic nanoparticles, hence it is also a type of surface functionalization. Figure 3.29 is a schematic illustration of the layer-by-layer process to synthesize magnetic luminescent nanocomposites. In the first step, a primer of three polyelectrolyte layers was deposited by alternative adsorption of PAH (poly(allylamine hydrochloride)) and PSS (poly(sodium 4-styrenesulfonate)) on negatively charged Fe_3O_4 nanoparticles. Each polyelectrolyte adsorption was kept for 20 minutes. Between two adsorption processes, it is necessary to remove excess polyelectrolyte by separation/washing/redispersion cycles. The second step is the CdTe quantum dots coating by adding negatively charged CdTe quantum dots to treated Fe_3O_4 suspension obtained from step 1. This coating step lasted overnight and the excess quantum dots were removed. Step 1 and step 2 were repeated to prepare functionalized magnetic nanoparticles with the desired number of CdTe quantum dots/polyelectrolyte layers. The synthesized magnetic luminescent nanoparticles could be used in biolabeling, bioseparation, diagnostics and immunoassay, etc.

Bertorelle *et al.* reported their preparation and characterization of new magnetic fluorescent nanoparticles (γ-Fe_2O_3) and demonstrated their success in using them to label living cells (2006). The as-prepared functionalized magnetic nanoparticles are bifunctional and each particle consists of a magnetic oxide core (γ-Fe_2O_3) covered by a dimercaptosuccinic acid (DMSA) ligand and a covalently bonded fluorescent dye. As discussed above, the efficiency of cell internalization is determined substantially by the surface properties of the nanoparticles. Both fluorescence microscopy and magnetophoresis demonstrated the

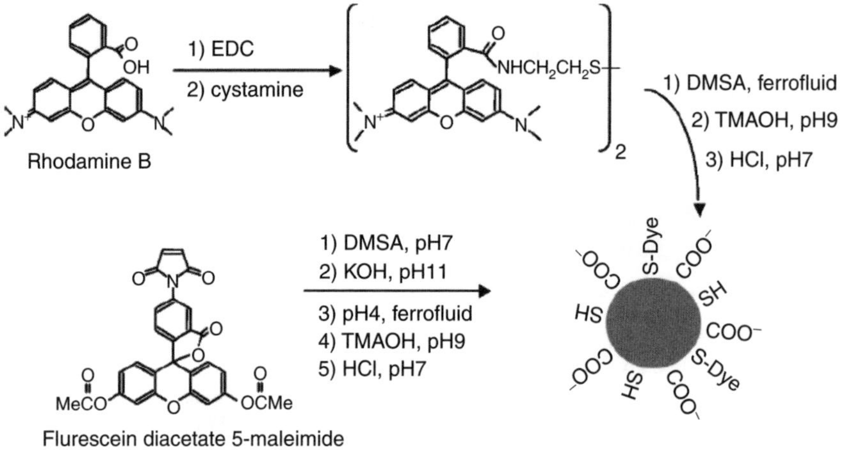

Figure 3.30 Schematic illustration of the functionalization process to prepare fluorescent magnetic nanoparticles. (Bertorelle *et al.* 2006)

bifunctional magnetic nanoparticles having a high affinity for cells. Figure 3.30 depicts the synthesis process, which is based on the strong interaction between dimercaptosuccinic acid (DMSA) and the positively charged surface of γ-Fe$_2$O$_3$ particles. DMSA has multiple carboxyl and thiol functional groups. It seems DMSA uses part of functional groups to attach the nanoparticles. The unbounded thiol and carboxyl groups are used for subsequent coupling and maintenance of surface charge, respectively. It was found that the coupling of DMSA/magnetic nanoparticles and DMAS/fluorescent dye could take place simultaneously. During the synthesis, the molar ratio of DMSA/Fe was kept at about 10 %. There are two fluorescent dyes, rhodamine B and a fluorescein derivative, which could be used as the coupling process to magnetic nanoparticles. Experiments indicated that mole ratios of 25 % and 20 % dye/DMSA resulted in maximum coupling rates for rhodamine B and fluorescein, respectively.

In a typical synthesis, 17 mg rhodamine B, 7 mg EDC (1-ethyl-3-(3-dimethylamino propyl)), and 4 mg cystamine were mixed to prepare aqueous solution with pH 7. Then, 1.25 ml acidic ferrofluid and 25 mg DMSA were added under stirring until a flocculate formed due to the growth of magnetic nanoparticles. The next step included a nanoparticle collection and a proper treatment. An external magnetite was applied on the supernatant to separate the magnetic material, which was washed and suspended in water under the alkaline medium. Hydrochloride acid was used to balance the pH value and the nanoparticles started to precipitate again with an addition of sodium chloride. Under the external magnetic field, nanoparticles precipitate completely resulting in a colorless supernatant. The final product was a dispersion of functionalized nanoparticles in Hepes buffer solution. The electrostatic interactions between anionic bifunctional nanoparticles and cell membranes provide a stable suspension which is an efficient candidate for magnetic fluorescent labeling for various cells, such as lymphocytes, stem cells, etc.

One of the common bioapplications of magnetic nanoparticles is bioseparation. Metal-chelate affinity chromatography (MCAC) is the most frequently used protocol for protein separations. Compared with conventional magnetic microparticles (e.g. resins, beads), magnetic nanoparticles offer several advantages in MCAC. Nanoparticles are able to move faster and enter into cells more easily due to their nanometer dimension. Also, they have a higher binding rate due to their high surface area and excellent solubilities. Additionally,

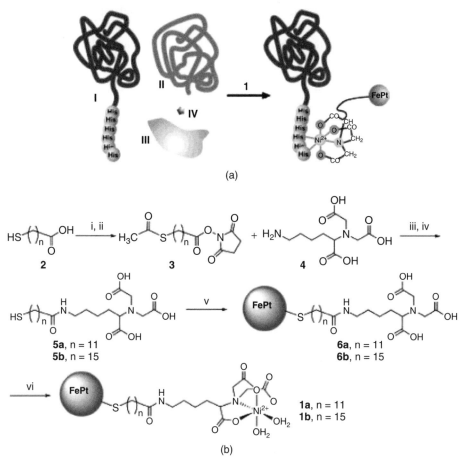

(a)

(b)

i) AcOCl, Zn; ii) NHS, DCC; iii) NaHCO₃; iv) H₂NNH₂, AcOH; v) FePt (3–4 nm); and vi) NiCl₂–6H₂O and buffer

Figure 3.31 (a) Schematic illustration of selective binding of the surface-modified magnetic nanoparticles to histidine-tagged proteins in a cell lysate (I: 6xHis tagged protein; II: other proteins; III: cell debris; IV: colloid contaminants); (b) synthesis process to prepare the FePt-NTA agents. (Xu *et al.* 2004)

their aggregation can be controlled by applying an external magnetic field so that they can be anchored onto a support for biological usage. The details of mechanism and technology of magnetic bioseparation will be addressed in the next chapter. Functionalization of magnetic nanoparticles can be used to enhance the bioseparation process. Xu's research group reported their functionalization of magnetic FePt nanoparticles with nitrilotriacetic acid (NTA) (2004). They also demonstrated these NTA-modified magnetic nanoparticles' potential of separating, transporting and anchoring proteins.

The magnetic nanoparticles used in the functionalization and bioseparation are less than 10 nm in diameter. By functionalization with NTA, a simple procedure can be achieved to separate pure proteins from a mixture of lysed cells within a short time (~10 mins). Figure 3.31a shows the surface-modified magnetic nanoparticles selectively bind to histidine-tagged proteins (6xHis tagged proteins). Extremely low concentration (about 0.5 pM) of proteins can be separated by this technique. Experiments also gave a binding capacity value which is about 2–3 mg protein per 1 mg NTA-modified FePt nanoparticles. Unlike microbead-based MCAC featured by nonspecific bindings, this

NTA-modified FePt nanoparticles-based MCAC provides completely specific bindings. Another merit of this approach is its capability of handling recombinant proteins because proteins can be anchored reversibly on a support under a magnetic field. A typical functionalization process to synthesize NTA-modified FePt nanoparticles is illustrated in Figure 3.31b. This synthesis included four steps. In the first step, the thiol groups of mercaptoalkanoic acids **2** were protected with acetyl chloride and their carboxylic groups were activated with N-hydroxysuccinimide. Intermediate agent **3** was created. The second step involved a reaction between **3** and N_α, N_α-bis(caboxymethyl) lysine **4**. And after deprotecting the thiol groups, **5** formed. In the third step, FePt nanoparticles were added to **5** under vigorous stirring, by which **5** was linked to FePt nanoparticles with Pt-S and Fe-S bonds. The resulting water soluble product was **6**. The final step was a reaction between **6** and excess $NiCl_2 \cdot 6H_2O$ followed by separation from the Ni^{2+} solution and washing by Tris buffer. The final product **1** is ready for specific protein binding in bioseparation, as illustrated in Figure 3.31a.

3.6.2 Functionalization with Biological Entities

Biofunctionalization can be defined as a surface modification technique which attaches biological entities to synthetic materials. To biofunctionalize magnetic nanoparticles, a number of biological entities can be used, such as proteins, ligands, antibodies, enzymes, etc.

Biofunctionalized magnetic nanoparticles consisting of a specific ligand coating on the outer surface can be used to isolate cells, cell organelles, nucleic acids, proteins, etc. Fluorescent-magnetic-biotargeting trifunctional nanoparticles, prepared by covalent coupling of hydrazide-containing biofunctional nanoparticles with IgG, avidin and biotin, can be used in visual sorting, apoptotic cell manipulation, and other biomedical applications (Wang *et al.* 2005). In the synthesis of trifunctional magnetic nanoparticles, sodium meta-periodate was first used to oxidize goat anti-rabbit IgG so that active aldehydes were created in the Fc fragment of IgG. The oxidation took place in an amber vial at room temperature under constant shaking. After the reaction was kept for half an hour, the mixture was introduced into a desalting column to remove unreacted sodium meta-periodate. Then, a suspension of hydrazide-containing bifunctional nanoparticles embedded with orange-red quantum dots was prepared by sonication and mixed with the oxidized antibody. The mixture was incubated for at least 6 hours at room temperature under constant shaking. Fluorescent-magnetic biotargeting trifunctional IgG-nanoparticles were obtained after washing with PBS. The synthesis process is schemed in Figure 3.32. For biosensing applications, magnetic nanoparticles possess several advantages over microparticles in detecting biological molecules. They have a higher magnetization density, and they are able to form much more stable suspensions. Also, they have less tendency to settle in microfluidics with faster velocities in solution. For the purpose of sensing, the recognition and specific capture of biological entities on the magnetic nanoparticle surface are critical for numerous bioanalytical applications in bio- and immunosensor diagnostic devices. The immobilization of biorecognition molecules (e.g. nucleic acids, proteins and other ligands) on magnetic nanoparticles offers modified surface properties and this biofunctionalization plays an important role in the development of such devices.

Magnetic nanoparticles can be functionalized with bioactive enzymes for versatile applications such as biosensors, biofuel cells, etc. Functionalization of magnetic nanoparticles with glucose oxidase has been demonstrated (Kousassi *et al.* 2005). This functionalization was realized by covalent bonding of glucose oxidase to magnetite nanoparticles by direct linkage via thiophene acetylation or carbodiimide activation. Figure 3.33 is a schematic of the functionalization procedure, which was adopted from the approach of immobilizing *Candida rugosa* lipase on the γ-Fe_2O_3. In a typical experiment, 5 g 11-bromoundecanoic

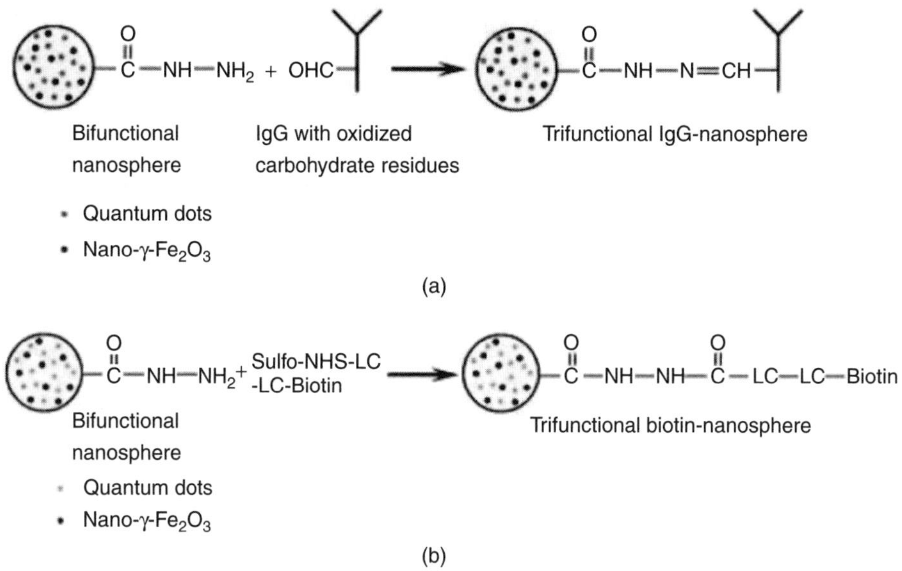

(a)

(b)

Figure 3.32 (A) Schematic diagram of trifunctional IgG-nanosphere synthesis. Oxidization was first conducted on IgG to create active aldehyde groups on its Fc fragment. Afterwards, the aldehydecontaining IgG was then covalently linked to the surface of a bifunctional nanoparticle via hydrazide groups. By using avidin, the same approach was also used to produce a trifunctional avidin-nanosphere by using avidin. (B) Schematic diagram of a trifunctional biotin-nanosphere synthesis in which hydrazide on the surface of the bifunctional nanosphere reacts with sulfo-NHS-LC-LCbiotin directly. (Wang *et al.* 2005)

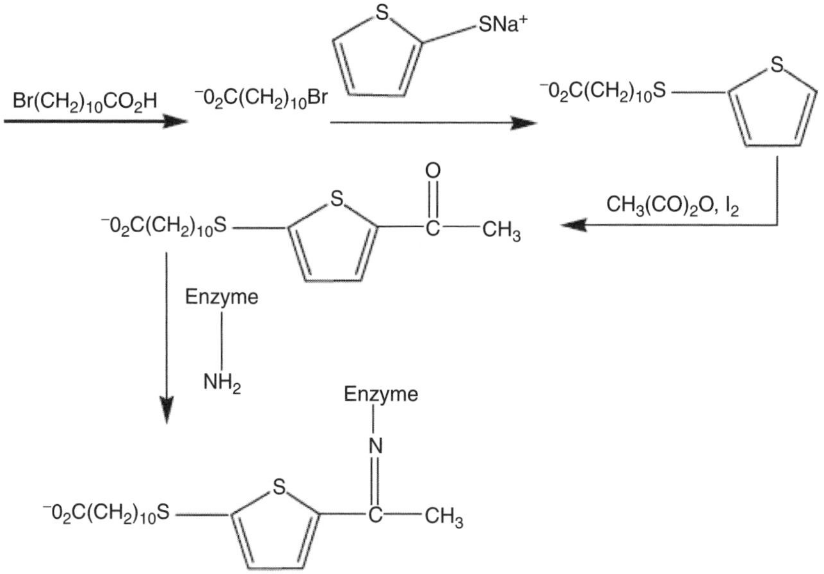

Figure 3.33 Schematic illustration of the process of glucose oxidase attachment including GOx-Fe$_3$O$_4$ I attachment, thiophene functionalization, and the acetylation of particles for GOx-Fe$_3$O$_4$ II attachment. (Kouassi *et al.* 2005)

acid was dissolved in 15 ml ethanol and 1.5 g Fe_3O_4 was added. By heating to a high temperature by microwave, reaction occurred in the mixture resulting in a covalent linkage between 11-bromoundecanoic acid and magnetite nanoparticles. In the subsequent functionalization step, 7 ml 2-thiophene thiolate was added to the resulting mixture containing covalently linked magnetite nanoparticles and again microwave heating was applied. In this step, functionalization of nanoparticles occurred through nucleophilic substitution with 2-thiophene thiolate. The next experimental step included washing with ethanol and refluxing with 4 ml acetic anhydride and 34.6 ml iodine for 1 hour. To attach glucose oxidase to nanoparticles, 2 ml GOx solution with 1000 units/ml in concentration was sonicated with 50 mg functionalized magnetic nanoparticles at 15 °C for 3 hours. The covalent bonding formed between the acetylated particles and the enzyme via C=N bond. The final step was magnetic separation of the supernatant which included unbounded enzymes and the bounded enzymes on magnetite nanoparticles exhibit bioactivities.

An interesting type of functionalization of magnetic nanoparticles with proteins involves a couple of biotin and stretavidin. It has been well documented that the biotin and streptavidin couple not only has a high binding affinity ($Ka = 10^{15} \, M^{-1}$) but also exhibit high specificity, hence it finds potential bioconjugation applications. The binding mechanism for the biotin and stretavidin couple lies in four biotin binding sites positioned in pairs on a stretavidin's opposite domains, which can serve as a bridge between the immobilized biotinylated moiety and target nanoparticles. In the case of magnetite nanoparticles, proteins can be functionalized through their cabodiimide-activated carboxylic groups reacting with the surface amine residues on nanoparticles. Various chemistries have been developed to functionalize surfaces for immobilization of specific biomolecules. Several forms of biotin that can be immobilized to surfaces have been commercially available.

Nidumolu and his team reported their synthesis of streptavidin-functionalized magnetic nanoparticles and investigation of their binding to biotinylated SAMs on gold and glass for biological recognition applications (2006). The schematic of magnetic nanoparticle-based functionalization of gold and glass surfaces is illustrated in Figure 3.34a. The whole process can be divided into three steps. In the first step, ferrous and ferric salts were coprecipitated in ammonium hydroxide to produce magnetite nanoparticles, which were then functionalized with the protein with FIFC-labeled streptavidin using a standard carboiimide activation. The second step involved functionalization of gold and glass surfaces with biotin SAMs. This was done by applying biotin-HPDP, reduced with butylphosphine, directly on gold coated glass slides. In the third step of specifically binding the functionalized nanoparticles to the biotinylated gold surfaces, 10 μL of the nanoparticle suspension was spotted onto the biotinylated surface for nanoparticle immobilization and kept in the dark for 1 hour followed by rinsing with HPLC-grade water and drying under nitrogen. To prevent nonspecific adsorption, the prepared glass slides were incubated in a 1 % solution of bovine serum albumin. After washing with deionized water and ethanol, the slides were checked by a transmission electron microscope. The TEM image of magnetite nanoparticles is shown in Figure 3.34(b). An average diameter of the nanoparticles can be estimated to be 9.8 ± 4.6 nm. TEM observation also revealed aggregation of nanoparticles which was most likely induced by the lack of surfactant in the nanoparticle synthesis and treatment. Sufficient amount of surfactant is believed to produce monodisperse nanoparticle suspensions. To confirm streptavidin attachment, FTIR analysis was performed. FTIR spectra of unfunctionalized and functionalized magnetite nanoparticles deposited onto gold slides were compared, as shown in Figure 3.34c. It can be seen from the spectra that characteristic peaks shifted or changed after the functionalization step.

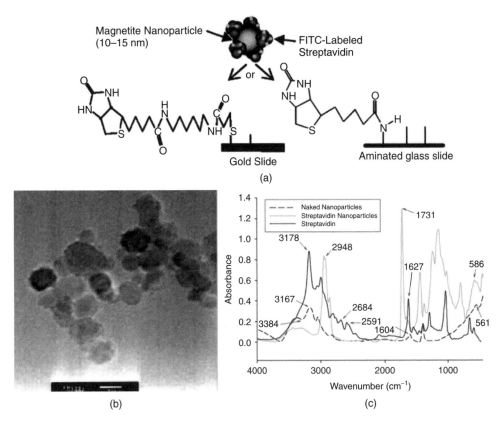

Figure 3.34 (a) Scheme description of functionalization of biotinylated gold or glass surfaces with streptavidin-modified magnetic nanoparticles; (b) TEM image of streptavidin-modified nanoparticles (scale bar: 10 nm); (c) FTIR spectra of streptavidin, unfunctionalized and functionalized magnetite nanoparticles on gold slides. (Nidumolu *et al.* 2006)

Before functionalization, magnetic nanoparticles show 3384 and 3167 cm^{-1} peaks indicating the presence of the N-H stretch of a primary amine, while the 1604 cm^{-1} peak is detected due to the N-H bend. For streptavidin-conjugated nanoparticles, a broad peak between 3300 and 3500 cm^{-1} for the N-H stretch and at 1448 cm^{-1} for the N-H bend were observed which is attributed to the secondary amines. For comparison purposes, the spectrum of streptavidin is shown in Figure 3.34c as well. The peak at 3178 cm^{-1} is a signature peak for asymmetric CH$_2$ stretching, which can be detected for streptavidin and streptavidin-functionalized nanoparticles. On the contrary, this peak did not show up in the unfunctionalized nanoparticle. Thus, the comparison of FTIR spectra confirms the magnetic nanoparticles functionalization with streptavidin.

Functionalized magnetic nanoparticles have been demonstrated to detect bimolecular interaction between biotin and streptavidin (Osaka *et al.* 2006). A schematic illustration of biotin-streptavidin reaction on magnetic nanoparticles is shown in Figure 3.35. The magnetic particles, about 200 nm in diameter, were modified by a silane-coupling agent, 3-aminopropyltriethoxysilane (APS) in toluene. It is the amine groups of APS that could be used for biological molecule bonding with magnetic nanoparticles. In the case of biotin, a cross-linking reaction of sulfo-LC-LC-biotin can occur and ensure a covalent bonding between biotin and the attached magnetic particle. The chemical structure of this covalent bonding is depicted in Figure 3.35. When an excess amount of streptavidin

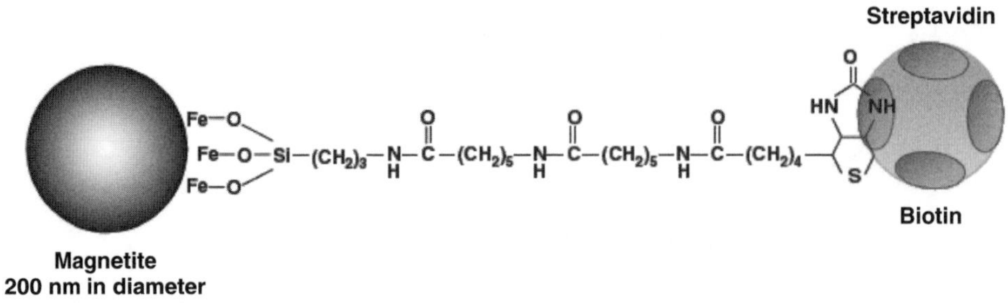

Figure 3.35 Schematic illustration of a streptavidin-modified magnetic nanoparticle. The particle is bound by a highly specific binding between biotin and streptavidin. (Osaka *et al.* 2006)

was added to the reaction mixture of biotin attached magnetic nanoparticles, the specific biotin-streptavidin bonding offered the capability of detection by biofunctionalized magnetic nanoparticles.

3.7 Future Prospects

Both the unique properties and potentially versatile applications have stimulated worldwide research efforts on magnetic nanoparticles for the past several decades. In addition to the existing applications in ultrahigh-density magnetic storage, magnetic nanoparticles have been demonstrated for attractive biomedical or biological applications, such as target drug delivery, hyperthermia, MRI contrast enhancement, bioseparation, etc. For these emerging bioapplications, there are two basic requisites for magnetic nanoparticles which are their controlled shape, size and crystallinity as well as their surface characteristics.

Considerable and impressive progress has been made in the controlled synthesis of various monodisperse magnetic nanoparticles with diameter less than 20 nm. However, there is still enough room left to either search for novel synthesis techniques or improve the current available synthesis with control. From a synthetic point of view, research efforts towards several directions are required. First, new synthesis approaches are needed for magnetic nanoparticles which cannot be produced or are extremely difficult to produce by known techniques. Secondly, a universal synthetic strategy is ideal to synthesize a number of monodisperse magnetic nanoparticles without additional size-selection processes. Importantly, large-scale production needs to be realized by a simple process at low cost. Thirdly, investigation on the magnetic nanoparticle formation and dispersion is necessary for better control synthesis on shape, size and crystallinity of nanoparticles.

Another important research aspect for magnetic nanoparticles is the scientific investigation on the relationship between their magnetisms and crystallinities and even individual atoms or small clusters. Correlation between nanoscale understanding and macroscopic measurements is essential.

Regarding the surface modification, functionalization of magnetic nanoparticles plays a crucial role for nanoparticle stabilization and various applications. For many practical usages, magnetic nanoparticles need to be dispersed in liquids (aqueous or organic solutions). To obtain better uniformity, preparation of monodisperse magnetic nanoparticles with controllable surface properties is a must. Thus, functionalization of magnetic nanoparticles with both organic molecules and biological entities has been well accepted as an effective approach to tailor their surfaces. Unfortunately, only a few functionalization techniques have been developed up to now. In order to precisely control the interaction between magnetic nanoparticles and biomolecules, more chemistry about either non-covalent bindings or covalent linkages needs to be developed. With such information, magnetic nanoparticles would be able to be modified specifically and utilized in desired bioapplications accordingly.

To achieve the above goals, multidisciplinary collaboration, in the field of chemistry, physics, materials science and engineering, bioengineering, etc, is undoubtedly necessary (Leslie-Pelecky and Rieke 1996). It is expected that the contributions from members of different research communities will greatly advance our ability to understand, control and fully take advantage of magnetic nanoparticles.

References

Bertorelle, F., Wilhelm, C., Roger, J., Gazeau, F., Ménager, C. and Cabuil, V. (2006). Fluorescence-modified superparamagnetic nanoparticles: intracellular uptake and use in cellular imaging, *Langmuir*, **22**, 5385–91.

Blakemore, R.P. (1975). Magnetotactic bacteria, *Science*, **190**, 377–90.

Cao, H., Wang, G., Zhang, L., Liang, Y., Zhang, S. and Zhang, X. (2006). Shape and magnetic properties of single-crystalline hematite (α-Fe$_2$O$_3$) nanocrystals, *ChemPhysChem*, **7**, 1897–901.

Caruntu, D., Cushing, B.L., Carunt, G. and O'Connor, C.J. (2005). Attachment of gold nanograins onto colloidal magnetite nanocrystals, *Chemistry of Materials*, **17**, 3398–401.

Chen, M. and Nikles, D.E. (2002a). Synthesis of spherical FePd and CoPt nanoparticles, *Journal of Applied Physics*, **91**, 8477–9.

Chen, M. and Nikles, D.E. (2002b). Synthesis, self-assembly, and magnetic properties of Fe$_x$Co$_y$Pt$_{100-x-y}$ nanoparticles, *Nano Letters*, **2**, 211–14.

Cho, S.-J., Idrobo, J.-C. Olamit, J., Liu, K., Browning, N.D. and Kauzlarich, S.M. (2005). Growth Mechanisms and Oxidation Resistance of Gold-Coated Iron Nanoparticles, *Chemistry of Materials*, **17**, 3181–6.

Donegá, C.deM., Liljeroth, P. and Vanmaekelbergh, D. (2005). Physicochemical evaluation of the hot-injection method, a synthesis route for monodisperse nanocrystals, *Small*, **1**, 1152–62.

Hong, R., Fischer, N.O., Emrick, T. and Rotello, V.M. (2005). 'Surface PEGylation and ligand exchange chemistry of FePt nanoparticles for biological applications', *Chemistry of Materials*, **17**, 4617–21.

Hong, X., Li, J., Wang, M., Xu, J., Guo, W., Li, J., Bai, Y. and Li, T. (2004). Fabrication of magnetic luminescent nanocomposites by a layer-by-layer self-assembly approach, *Chemistry of Materials*, **16**, 4022–7.

Hyeon, T., Lee, S.S., Park, J., Chung, Y. and Na, H.B. (2001). Synthesis of highly crystalline and monodisperse maghemite nanocrystallites without a size-selection process, *Journal of the American Chemical Society*, **123**, 12798–801.

Hyeon, T. (2003). Chemical synthesis of magnetic nanoparticles, *Chemical Communications*, 927–34.

Klem, M.T., Willits, D., Solis, D.J., Belcher, A.M., Young, M. and Douglas, T. (2005). Bio-inspired synthesis of protein-encapsulated CoPt nanoparticles, *Advanced Functional Materials*, **15**, 1489–94.

Kouassi, G.K., Irudayaraj, J. and McCarty, G. (2005), Activity of glucose oxidase functionalized onto magnetic nanoparticles, *Biomagnetic Research and Technology*, **3**, 1–10.

Leslie-Pelecky, D.L. and Rieke, R.D. (1996). Magnetic properties of nanostructured materials, *Chemistry of Materials*, **8**, 1770–83.

Levy, L., Sahoo, Y., Kim, K.-S., Bergey, E.J. and Prasad, P.N. (2002). Nanochemistry: synthesis and characterization of multifunctional nanoclinics for biological applications, *Chemistry of Materials*, **14**, 3715–21.

Liu, C., Zou, B., Rondinone, A.J. and Zhang, Z.J. (2000). Reverse micelle synthesis and characterization of superparamagnetic MnFe$_2$O$_4$ spinel ferrite nanocrystallites, *Journal of Physical Chemistry B*, **104**, 1141–5.

Lyon, J.L., Fleming, D.A., Stone, M.B., Schiffer, P. and Williams, M.E. (2004). Synthesis of Fe oxide core/Au shell nanoparticles by iterative hydroxylamine seeding, *Nano Letters*, **4**, 719–23.

Mayes, E., Bewick, A., Gleeson, D., Hoinville, J., Jones, R., Kasyutich, O., Nartowski, A., Warne, B. Wiggins, J. and Wong, K.K.W. (2003). Biologically derived nanomagnets in self-organized patterned media, *IEEE Transactions on Magnetics*, **39**, 624–7.

Mikhaylova, M., Kim, D.K., Bobrysheva, N., Osmolowsky, M., Semenov, V., Tsakalatos, T. and Muhammed, M. (2004). Superparamagnetism of magnetite nanoparticles: dependence on surface modification, *Langmuir*, **20**, 2472–7.

Morales, M.P., Veintemillas-Verdaguer, S., Montero, M.I., Serna, C.J., Roig, A., Casas, L., Martínez, B. and Sandiumenge, F. (1999). Surface and internal spin canting in γ–Fe$_2$O$_3$ nanoparticles, *Chemistry of Materials*, **11**, 3058–64.

Nidumolu, B.G., Urbina, M.C., Hormes, J., Kumar, C.S.S.R. and Monroe, W.T. (2006). Functionalization of gold and glass surfaces with magnetic nanoparticles using biomolecular interactions, *Biotechnology Progress*, **22**, 91–5.

Osaka, T., Matsunaga, T., Nakanishi, T., Arakaki, A., Niwa, D. and Iida, H. (2006). Synthesis of magnetic nanoparticles and their application to bioassays, *Analytical and Bioanalytical Chemistry*, **384**, 593–600.

Park, S.-J., Kim, S., Lee, S., Khim, Z.G., Char, K. and Kyeon, T. (2000). Synthesis and magnetic studies of uniform iron nanorods and nanospheres, *Journal of the American Chemical Society*, **122**, 8581–2.

Park, J. and Cheon, J. (2001). Synthesis of 'solid solution' and 'core–shell' type cobalt-platinum magnetic nanoparticles via transmetalation reactions, *Journal of the American Chemical Society*, **123**, 5743–6.

Park, J., An, K., Hwang, Y., Park, J.-G., Noh, H.-J., Kim, J.-Y., Park, J.-H., Hwang, N.-M. and Hyeon, T. (2004). Ultra-large-scale syntheses of monodisperse nanocrystals, *Nature Materials*, **3**, 891–5.

Peng, S., Wang, C., Xie, J. and Sun, S. (2006). Synthesis and stabilization of monodisperse Fe nanoparticles, *Journal of the American Chemical Society*, **128**, 10676–7.

Pileni, M.-P. (2003). The role of soft colloidal templates in controlling the size and shape of inorganic nanocrystals, *Nature Materials*, **2**, 145–50.

Pileni, M.P., Lalatonne, Y., Ingert, D., Lisiecki, I. and Courty, A. (2004). Self assemblies of nanocrystals: preparation, collective properties and uses, *Faraday Discussions*, **125**, 251–64.

Puntes, V.F., Krishnan, K.M. and Alivisatos, A.P. (2001). Colloidal nanocrystal shape and size control: the case of cobalt, *Science*, **291**, 2115–17.

Raming, T.P., Winnubst, A.J.A., van Kats, C.M. and Philipse, A.P. (2002). The synthesis and magnetic properties of nanosized hematite (α-Fe$_2$O$_3$) particles, *Journal of Colloid and Interface Science*, **249**, 346–50.

Reiss, B.D., Mao, C., Solis, D.J., Ryan, K.S., Thomson, T. and Belcher, A.M. (2004). Biological routes to metal alloy ferromagnetic nanostructures, *Nano Letters*, **4**, 1127–32.

Rockenberger, J., Scher, E.C. and Alivisatos, A.P. (1999). A new nonhydrolytic single-precursor approach to surfactant-capped nanocrystals of transition metal oxides, *Journal of the American Chemical Society*, **121**, 11595–6.

Sun, S. and Murray, C.B. (1999). Synthesis of monodisperse cobalt nanocrystals and their assembly into magnetic superlattices, *Journal of Applied Physics*, **85**, 4325–30.

Sun, S., Murray, C.B., Weller, D. Folks, L. and Moser, A. (2000). Monodisperse FePt nanoparticles and ferromagnetic FePt nanocrystal superlattices, *Science*, **287**, 1989–92.

Sun, S. and Zeng, H. (2002). Size-controlled synthesis of magnetite nanoparticles, *Journal of the American Chemical Society*, **124**, 8204–5.

Sun, S., Zeng, H., Robinson, D.B., Raoux, S., Rice, P.M., Wang, S.X. and Li, G. (2003a). Monodisperse MFe$_2$O$_4$ (M = Fe, Co, Mn) Nanoparticles, *Journal of the American Chemical Society*, **126**, 273–9.

Sun, S., Anders, S., Thomson, T., Baglin, J.E.E., Toney, M.F., Hamann, H., Murray, C.B. and Terris, B.D. (2003b). Controlled synthesis and assembly of FePt nanoparticles, *Journal of Physical Chemistry B*, **107**, 5419–25.

Wang, G.-P., Song, E.-Q., Xie, H.-Y., Zhang, Z.-L., Tian, Z.-Q., Zuo, C., Pang, D.-W., Wu, D.-C. and Shi, Y.-B. (2005). Biofunctionalization of fluorescent-magnetic-bifunctional nanospheres and their applications, *Chemical Communications*, 4276–8.

Wang, L., Luo, J., Fan, Q., Suzuki, M., Suzuki, I.S., Engelhard, M.H., Lin, Y., Kim, N., Wang, J.Q. and Zhong, C.-J. (2005a). Monodispersed core–shell Fe$_3$O$_4$@Au nanoparticles, *Journal of Physical Chemistry B*, **109**, 21593–601.

Wang, L., Luo, J., Maye, M.M., Fan, Q., Rendeng, Q., Engelhard, M.H., Wang, C., Lin, Y. and Zhong, C.-J. (2005b). Iron oxide-gold core–shell nanoparticles and thin film assembly, *Journal of Materials Chemistry*, **15**, 1821–32.

Wang, J., Zhang, K. and Zhu, Y. (2005). Synthesis of SiO$_2$-coated ZnMnFe$_2$O$_4$ nanospheres with improved magnetic properties, *Journal of Nanoscience and Nanotechnology*, **5**, 772–5.

Westcott, S.L., Oldenburg, S.J., Lee, T.R. and Hallas, N.J. (1998). Formation and Adsorption of Clusters of Gold Nanoparticles onto Functionalized Silica Nanoparticle Surfaces, *Langmuir*, **14**, 5396–401.

Xu, C., Xu, K., Gu, H., Zhong, X., Guo, Z., Zheng, R., Zhang, X. and Xu, B. (2004). Nitrilotriacetic acid-modified magnetic nanoparticles as a general agent to bind histidine-tagged proteins, *Journal of the American Chemical Society*, **126**, 3392–3.

Yeary, L.W., Moon, J.-W., Love, L.J., Thompson, J.R., Rawn, C. and Phelps, T.J. (2005). Magnetic properties of biosynthesized magnetite nanoparticles, *IEEE Transactions on Magnetics*, **41**, 4384–9.

Zhang, C., Vali, H., Romanek, C., Phelps, T.J. and Liu, S.V. (1998). Formation of single-domain magnetite by a thermophilic bacterium, *American mineralogist*, **83**, 1409–18.

4

Biomedical Applications of Magnetic Nanoparticles

4.1 Introduction

Nanotechnology, dealing with nanoscale objects, has been developed at three major levels: nanomaterials, nanodevices and nanosystems. At present, the nanomaterials level is the most advanced of the three. Nanomaterials are of great importance both in scientific investigations and commercial applications due to their size-dependent physical and chemical properties. Nanomaterials with various shapes have been developed successfully. Common morphologies are quantum dots, nanoparticles/nanocrystals, nanowires, nanorods, nanotubes, etc. It is desirable to have a full range within the nanomaterial family because many applications demand particular nanomaterials with special structures.

Magnetic nanoparticles, being a sub-family of nanomaterials, exhibit unique magnetic properties in addition to other specific characteristics. Their remarkable new phenomena include superparamagnetism, high saturation field, high field irreversibility, extra anisotropy, and temperature-depended hysteresis, etc. Research investigation has revealed that the finite size and surface effects of magnetic nanoparticles determine their magnetic behavior. For instance, a single magnetic domain forms when the size of a ferromagnetic nanoparticle is less than 15 nm. In other words, an ultrafine ferromagnetic nanoparticle displays a state of uniform magnetization under any field. Thus, at temperatures above the blocking temperature, these nanoparticles show identical magnetization behavior to atomic paramagnets (superparamagnetism) with an extremely large magnetic moment and large susceptibilities.

Magnetic nanoparticles have found many successful industrial applications, such as magnetic recording media for data storage, magnetic seals in motors, magnetic inks, etc. There are special requirements for each industrial application. For data storage applications, magnetic nanoparticles should possess a temperature-stable magnetic state which is also switchable to represent bits of information. Currently, magnetic nanoparticles are expected to lead to commercial exploration. Many potential applications in various fields have been expected by taking advantage of magnetic nanoparticles. Recently, tremendous research efforts have been stimulated on the usage of magnetic nanoparticles in the field of biomedical and biological applications.

Understanding of biological processes and hence developing biomedical means have been continuously pursued. These aims are one of strong driving forces behind the development of nanotechnology. The interests on magnetic nanoparticles for bioapplications

Nanomedicine: Design and Applications of Magnetic Nanomaterials, Nanosensors and Nanosystems
V. Varadan, L.F. Chen, J. Xie
© 2008 John Wiley & Sons, Ltd

comes from their comparable dimensions to biological entities coupled with their unique magnetic behaviors. Though common living organisms are composed of cells of about 10 μm size, the cell components are much smaller and generally in the nanosize dimension. Examples are viruses (20–450 nm), proteins (5–50 nm) and genes (2 nm wide and 10–100 nm long). Synthetic magnetic nanoparticles have controllable dimensions and just a few nanometer-diameter nanoparticles can be synthesized by carefully designing experimental procedures and controlling experimental conditions. With such a nanoscale dimension, it would be possible for magnetic nanoparticles to get close to a biological entity of interest. Moreover, the interaction between magnetic nanoparticles and biological entities can be adjusted by coating nanoparticles with biological molecules, called biofunctionalization. This offers a controllable means of 'tagging' or addressing the binding at nanoscale. The comparable dimensions and magnetic properties of magnetic nanoparticles have prompted the idea of using them as very small probes to spy on the biological processes at the cellular scale without introducing too much interference (Salata 2004). Actually, optical and magnetic effects have been treated as the most suitable approaches for biological applications owing to their non-invasive behavior.

In view of the magnetic properties of magnetic nanoparticles, they can be manipulated by an external magnetic field gradient, which is described by Coulomb's law. Magnetic nanoparticles are able to transport into human tissues due to the intrinsic penetrability of magnetic fields into human bodies. This 'action at a distance' opens up many potential bioapplications including transportation of magnetically tagged biological entities, targeted drug delivery, etc. Another important property of magnetic nanoparticles is their resonant response related to a time-varying magnetic field (Pankhurst *et al.* 2003). Hence energy transfer from the exciting field to the magnetic nanoparticles can be realized. In this way, toxic amounts of thermal energy are able to be delivered via magnetic nanoparticles to the targeted tumors resulting in malignant cell destruction. This process is named hyperthermia, which will be addressed in detail in this chapter. In addition to the site-specific drug delivery and hyperthermic treatment, magnetic nanoparticles have found other versatile bioapplications such as magnetic bioseparation, contrast enhancement of magnetic resonance imaging, gene therapy, enzyme immobilization, magnetic manipulation of cell membranes, immunoassays, magnetic biosensing, etc. (Sun *et al.* 2005). Each application depends upon the relationship between the external magnetic field and the biological system. Magnetic fields with proper field strength are not deleterious to either biological tissues or biotic environments. In a given bioapplication, magnetic nanoparticles are usually injected intravenously into the human body and are transported to the targeted region via blood circulation for biomedical diagnostic or treatment. An alternative means is using magnetic nanoparticle suspension for injection (Berry and Curtis 2003). It has been well accepted that a desirable magnetic medium should not contain nanoparticle aggregation, which will block its own spread. For this reason, a stable, uniform magnetic nanoparticle dispersion in either an aqueous or organic solvent at neutral pH and physiological salinity is required. The stability of this magnetic colloidal suspension depends on two parameters: an ultrasmall dimension and surface chemistry. The particle size should be sufficiently small to avoid precipitation due to gravitation forces while the charge and surface groups should create both steric and coulombic repulsions which stabilize the colloidal suspensions.

As discussed in Chapter 3, the magnetic properties of magnetic nanoparticles are determined by their elemental compositions, crystallinitys, shapes and dimensions. Various magnetic nanoparticles have been developed. Therefore, the selection of proper magnetic nanoparticles with the desired properties is the first but crucial step for certain bioapplications. For example, ferromagnetic nanoparticles (e.g. Fe nanoparticles) have a large magnetic moment and they can be the best material candidate in magnetic biosensors because they not only produce a better signal but respond to an applied magnetic field readily. On the other hand, iron oxide nanoparticles with superparamagnetic behavior do an excellent job when used to enhance the signals in magnetic resonance imaging examinations. With the help of iron oxide nanoparticles a sharpened image with detailed information can be achieved because of the change of behavior of nearby biomolecules by introduced nanoparticles (Bystrzejewski *et al.* 2005). For many biomedical applications, magnetic nanoparticles presenting superparamagnetic behavior (no remanence along with a rapidly changing magnetic state) at room temperature are desirable. Biomedical applications are commonly divided into two major categories: *in vivo* and *in vitro* applications. Consequently, additional restrictions apply on various magnetic nanoparticles for *in vivo* or *in vitro* biomedical applications.

It is rather simple for *in vitro* applications of magnetic nanoparticles. The size restriction as well as biocompatibility/toxicity are not so critical for *in vitro* applications, when compared with *in vivo* ones. Therefore, superparamagnetic composites containing submicron diamagnetic matrixes and superparamagnetic nanocrystals can be used. Composites with long sedimentation times in the absence of a magnetic field are also acceptable. It was noticed that functionalities may be provided readily for the superparamagnetic composites because of the diamagnetic matrixes (Tartaj *et al.* 2003).

On the other hand, severe restrictions must be applied for magnetic nanoparticles for *in vivo* biomedical applications. First of all, it is a requisite that the magnetic components should be biocompatible without any toxicity for the biosystems of interest. This is predominantly determined by the nature of the material (e.g. iron, nickel, cobalt, metal alloy, etc). For instance, cobalt and nickel are highly magnetic materials. However, both of them are rarely used due to their toxic properties and susceptibility to oxidation. Currently, the most commonly employed magnetic nanoparticles in biomedical applications are iron oxides including magnetite (Fe_3O_4), maghemite (γ-Fe_2O_3) and hematite (α-Fe_2O_3). The second requirement for magnetic nanoparticles is their particle sizes. Ultrafine nanoparticles (usually smaller than 100 nm in diameter) have high effective surface area, thus they can be attached to ligand easily. Also the lower sedimentation rate leads to a high stability for colloidal suspensions, and the tissular diffusion can be improved by using nanoparticles in nanometer dimensions. After injection, nanoparticles would be able to remain in the circulation and pass through the capillary systems to reach the targeted organs and tissues without any vessel embolism. Further, the magnetic dipole–dipole interaction among magnetic nanoparticles can be substantially reduced. The third requisite for magnetic nanoparticles is their biocompatible polymer coating which may be done during or after the nanoparticle synthesis. There are several functions of the coating layers: 1) they will prohibit agglomeration of nanoparticles; 2) they prevent structural or elemental changes; 3) unnecessary biodegradation can be stopped; 4) the polymer layer offers a covalent binding or adsorption attachment of drugs to the nanoparticle surface. In summary, for *in vivo* biomedical applications, magnetic nanoparticles must be made of a non-toxic and non-immunogenic material with ultrasmall particle sizes and high magnetization.

It is no doubt that interdisciplinary research collaboration is badly needed for clinical and biological applications of magnetic nanoparticles (Berry and Curtis 2003). Research fields involved include chemistry, materials science, cell engineering, clinical tests and other related scientific efforts. In this chapter, an overview of the biomedical applications of magnetic nanoparticles will be presented.

4.2   Diagnostic Applications

The diagnostic applications of magnetic nanoparticles are processes to detect malignant tissues or pathogenic bioaggregates by using magnetic nanoparticles. Common applications include enhancement of magnetic resonant imaging, magnetic labeling, magnetic separation and purification, spatially resolved magnetorelaxometry, biological assay system, magnetic nanosensors, etc.

4.2.1   Enhancement of Magnetic Resonance Imaging

In life science, an emerging technology, molecular imaging, has been playing a substantial role in both scientific investigation and practical bioapplications. Molecular imaging is the technology that helps us to understand diseases and search for the appropriate treatments. Since most disease processes have a molecular basis, molecular imaging with sufficient image resolution and image contrast at the molecular level is much needed for better diagnostic differentiations, early disease detection and objective therapy monitoring with imaging biomarkers (Cassidy and Radda 2005). There is a number of molecular imaging techniques available ranging from ultrasonic to gamma ray and X-ray frequencies in the electromagnetic spectrum. Molecular imaging operates at the interface between life science and physical science, and images are obtained by facilitating interaction with a biological entity at a molecular level. Thus, both the life and physical sciences are required to develop and implement a molecular imaging technique due to its diverse nature. Among them, optical imaging, nuclear imaging and magnetic resonance imaging (MRI) are the most widely used molecular imaging techniques. Magnetic nanoparticles are able to enhance the image contrast in MRI, which has attracted enormous attention. To better investigate the functions of magnetic nanoparticles as a contrast agent in MRI, it is necessary to first understand the working mechanism and physical principles in MRI.

As far as the imaging mechanism of MRI is concerned, the technology is based on the detection of nuclear spin in molecules. When a strong static magnetic field is applied, spinning nuclei precess at the Larmor frequency. Two different energy states form corresponding to spin-up and spin-down orientation. There are more spinning nuclei in the lower energy state than in the other one. Activated by the oscillating radio frequency (RF) magnetic field, some spinning nuclei jump from the lower energy state to the higher one. Eventually, spins at the higher energy state will return to their original state with a RF emission at the Larmor frequency. The RF signal can be detected by using a RF coil and it is related to the energy difference between the two energy states.

4.2.1.1   Physical Principles

Large numbers of protons exists in biological tissue, and an exceedingly small magnetic moment exhibits for a single proton. MRI relies on the resulting measurable effect in the presence of large external magnetic fields. In the case of a water molecule, about 6.6×10^{19} protons are available in every cubic millimeter of water. When a steady state field of B_0 of 1 T is applied, the magnetization effect is extremely small such that possibly only three of every million proton moments m can be aligned parallel to B_0. However, an effective signal of 2×10^{14} proton moments is observable for one cubic millimeter of water (Pankhurst et al. 2003).

Figure 4.1 illustrates a magnetic resonance for a large ensemble of protons with net magnetic moment m in the presence of an external magnetic field B_0. By applying a time-varying magnetic field perpendicular to the existing field B_0, the moment signal can be captured with resonant absorption and tuned to the Larmor precession of the protons

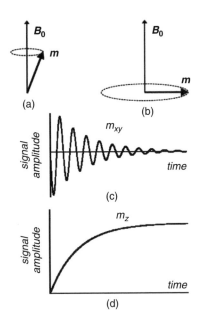

Figure 4.1   Schematic illustration of magnetic resonance for a large ensemble of protons with net magnetic moment *m* when an external magnetic field B_0 is applied; (*a*) net moment precesses around B_0 at the Larmor frequency, ω_0; (*b*) a second external field (oscillating at ω_0), with its direction perpendicular to B_0, is applied. Although this second field is much weaker than B_0, it induces the moment precession into the plane perpendicular to B_0; (*c*) and (*d*) the in-plane (*c*) and longitudinal (*d*) moment amplitudes relax back to their initial values as the oscillating field is removed at time zero. Pankhurst, Q.A., Connolly, J., Jones, S.K. and Dobson, J. (2003). Applications of magnetic nanoparticles in biomedicine, *Journal of Physics D: Applied Physics*, **36**, R167–R181. Reproduced with permission of the Institute of Physics

(frequency: f $= \omega_0/2\pi = \gamma B_0/2\pi$). For hydrogen protons, the gyromagnetic ratio $\gamma = 2.67 \times 10^8$ rad s^{-1} T^{-1}, and if the field $B_0= 1$ T, the Larmor precession frequency is calculated to be 42.57 MHz. To enhance the signal, resonantly tuned detection coils are usually applied in a pulsed sequence, which results in coherent response deriving and a relaxation process via induced currents in coils. The quality factor of signal can be enhanced up to 100 times by using these resonantly tuned detection coils.

If the steady state magnetic field B_0 is applied parallelly to the z-axis the relaxation signals can be described as:

$$m_z = m\left(1 - e^{-\frac{t}{T1}}\right) \tag{4.1}$$

and

$$m_{x,y} = m\sin(\omega_0 t + \varphi)e^{-\frac{t}{T2}} \tag{4.2}$$

where φ is a phase constant, and T_1, T_2 are the longitudinal (or spin–lattice) and transverse (or spin–spin) relaxation times, respectively. The longitudinal relaxation is a measure of the dipolar coupling between the proton moments and the surroundings because it reflects a loss of heat energy. The rapid transverse relaxation is driven by the loss of phase coherence caused by the magnetic interactions of the precessing protons. Meanwhile, T_2 in Equation (4.2) can be replaced by a shorter relaxation time, T_2^* when the local inhomogeneities in the applied longitudinal field are considered:

$$\frac{1}{T_2^*} = \frac{1}{T_2} + \gamma \frac{\Delta B_0}{2} \tag{4.3}$$

where ΔB_0 is the field variation induced by either the homogeneity distortions of the applied field, or variations in the magnetic susceptibility of the system.

4.2.1.2 MRI Contrast Enhancement Studies

The reason for using magnetic nanoparticles in MRI is to shorten both T_1 and T_2^* so that the image contrast can be enhanced. Iron oxide nanoparticles have been tested for contrast enhancement in MRI. SPM nanoparticles are commercially available products, and 'Feridex I. V.' was marked by Advanced Magnetics Inc. which is widely used for the organ-specific targeting of liver lesions. As shown in Figure 4.2, SPM particles are magnetically saturated in the common magnetic fields applied in MRI, and they are able to create a substantial locally perturbing dipolar field, and hence successfully decrease the value of T_1 and T_2^*. To further improve biocompatibility, iron oxide-based SPM nanoparticles are always coated by a polymer layer such as Dextran, which can be excreted via the liver after treatment. In MRI, biofunctionalized magnetic nanoparticles are selectively taken up by the reticuloendothelial system, which serves as a network to remove foreign substances from the blood circulation system. It was found that magnetic nanoparticles with diameter less than 10 nm had a longer half-life than those of 30 nm in diameter or larger in the blood stream. Larger nanoparticles are recognized by liver and spleen, while smaller ones are collected by the reticuloendothelial system in healthy tissues throughout the body. It is the differential uptake of magnetic nanoparticles by different tissues that determines the MRI contrast. For tumor cells, the relaxation times are not changed by introducing contrast agent (magnetic nanoparticles) due to the absence of an effective reticuloendothelial system in the tumor. Therefore, identification of malignant lymph nodes, liver tumors and brain tumors by comparing the image contrast can be readily achieved in the presence of magnetic nanoparticles in MRI. It has been also noticed that iron oxide nanoparticles have the capability of being encapsulated into target-specific agents. Branched polymer coating on the surface of nanoparticles can be used to carry DNA into the cell nucleus, which enables the intracellular interaction between magnetic

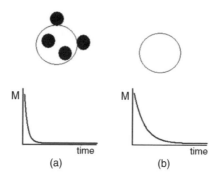

Figure 4.2 Comparison of the relaxation times for (a) magnetic particle-tagged cells having a T_2^* relaxation time (b) untagged cells having a longer T_2^* relaxation time. Pankhurst, Q.A., Connolly, J., Jones, S.K. and Dobson, J. (2003). Applications of magnetic nanoparticles in biomedicine, *Journal of Physics D: Applied Physics*, **36**, R167–R181. Reproduced with permission of the Institute of Physics

nanoparticles and stem cells. These bindings enhance the performance of magnetic nanoparticles in MRI contrast enhancement. Another type of common contrast agents used in MRI is PM gadolinium ion complexes.

As discussed above, magnetic resonance imaging is one of the most powerful medical tools for diagnostic purposes owing to its noninvasive process, high spatial resolution and multidimensional tomographic capabilities. However, this technique suffers from low-signal sensitivity. To overcome the weakness of current MRI techniques, signal enhancers need to be used. The signal-enhancing capabilities of magnetic nanoparticles/nanocrystals have been demonstrated. Magnetite (Fe_3O_4), as a member of clinically benign iron oxide-based nanoparticles, has been widely explored for signal-enhancing purposes. Jun *et al.* demonstrated the use of magnetite (Fe_3O_4) nanoparticles as the signal enhancer in MRI (2005).

In their work, magnetite nanoparticles were prepared by a well-developed chemical method, in which $Fe(acac)_3$ (acac = acetylacetonate) decomposes in a hot organic solvent followed by nanocrystal growth and precipitation. The as-prepared nanocrystals are highly crystalline and stoichiometric magnetite. By controlling the synthesis conditions, the size of nanocrystals can be controlled from 4 to 12 nm with high monodispersities ($\sigma < 5\%$). Figure 4.3a shows TEM images of magnetite nanocrystals with 4, 6, 9 and 12 nm, respectively. Experiments indicated that the magnetism and induced MRI signals were closed related with the dimension of nanocrystals used as the signal enhancer. Uniform magnetic suspensions for different average dimensions were prepared at an identical concentration of 1 μM. When a magnetic field of 1.5 T was applied, the spin–spin relaxation time (T_2) weighted spin-echo MRI of water soluble iron oxide (WSIO) nanocrystals showed the MRI images changing from white to black via gray colors as the dimension of magnetic nanocrystals increased from 4 to 12 nm. This grayscale color change can be clearly seen from Figure 4.3b. Color-coded images based on T_2 values are shown in Figure 4.3c. The result confirmed that the T_2-weighted MRI signal intensity continuously decreases as the size of nanocrystal increases. The general trend of size-dependent MRI signal is illustrated in Figure 4.3d. The magnetic properties of nanocrystals were measured by a superconducting quantum interference device (SQUID) magnetometer, as shown in Figure 4.3e. Larger mass magnetization values for bigger nanocrystals result in darker MRI images (low T_2 values) because the spin–spin relaxation process of protons in surrounding water molecules is facilitated by the magnetic spins of nanocrystals.

In addition to the *in vitro* MRI imaging with magnetic nanoparticles serving as the contrast enhancer, *in vivo* MRI is expected. The *in vivo* MRI offers active and extra functions for various cell trafficking, cancer diagnosis and gene expression, etc. However this biomedical application has been very limited until now. Researchers have been realizing the importance of magnetic nanoparticle-based *in vivo* MRI applications and research efforts have been made. Huh *et al.* utilized a well-defined magnetic nanoparticle-based probe system for *in vivo* diagnosis of cancer (Huh *et al.* 2005). The magnetic nanocrystals are characterized by small size, strong magnetism, biocompatibility and active functions for receptors. In their experiments, MRI probes were prepared by conjugation of a cancer-targeting antibody, Herceptin, to magnetic nanocrystals. By using this novel MRI probe, *in vivo* monitoring of human cancer cells which were implanted into live mice was demonstrated successful. By further conjugating fluorescent dye-labeled antibodies to these magnetic nanocrystal MRI probes, *in vitro*, *ex vivo* and *in vivo* MRI can be realized. And this new technology could find potential applications in an advanced multimodal detection system.

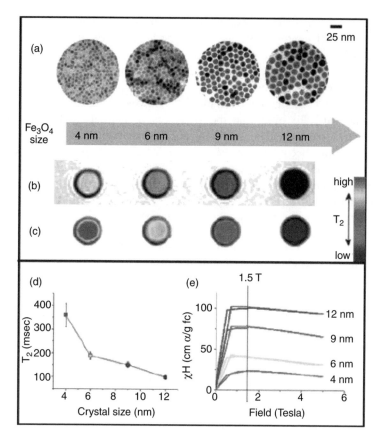

Figure 4.3 Effect of the nanosize of WSIO nanocrystals on magnetism and corresponding magnetic resonance signals; (a) TEM images of magnetite Fe_3O_4 nanocrystals of 4 to 6, 9 and 12 nm in diameter; (b) size-dependent T_2-weighted MR images of WSIO nanocrystals in aqueous solution at an applied field of 1.5 T; (c) size-dependent colors of WSIO nanocrystals in color-coded MR images based on T_2 values; (d) plot of T_2 value depending upon the size of WSIO nanocrystals; (e) comparison of the magnetization curves of WSIO nanocrystals with different sizes measured by a SQUID magnetometer. (Jun *et al.* 2005)

4.2.2 Magnetic Labeling

Generally, magnetic labeling is employed by using fluorescent-labeled magnetic nanoparticles in both *in vitro* and *in vivo* imaging. The combination of nanoscale dimensions, superparamagnetic properties and fluorescence for these bifunctional nanoparticles has prompted their use in medical imaging which combines fluorescent imaging techniques and the MRI approach. Both methods are of great importance in diagnostic and therapeutic applications because cellular monitoring provides valuable information for subsequent cell-based therapy. The advantage of the MRI technique is well known: it offers anatomically sensitive deep tissue imaging. Optical fluorescent techniques, both *ex vivo* by classical fluorescence microscopy and *in vivo* by fibered confocal fluorescent microscopy, have high spatial resolution in cellular imaging and molecular event quantification. There are several ways of utilizing the bifunctional nanoparticles. For instance, labeling cells with nanoparticles can be done *in vitro* followed by an *in vivo* transplantation. Then cell migration throughout the body can be monitored by *in vivo* imaging. Another approach is to use drug carrier-conjugated magnetic nanoparticles for target guidance to specific sites.

Magnetic fluorescent nanoparticles can be used to label living cells. In Bertorelle's work, bifunctional nanoparticles consist of magnetic oxide cores covered with a layer of dimercaptosuccinic acid (DMSA) ligand and a fluorescent dye covalently bonded on the surface (2006). Both fluorescence microscopy and magnetophoresis demonstrated that the bifunctional nanoparticles had a high cell affinity. When the association of magnetic nanoparticles with cells was investigated by fluorescence microscopy, both membranous fluorescence and intracellular vesicular fluorescence were observed indicating two types of magnetic labeling: nanoparticle adsorption on cell membranes and nanoparticle internalization inside cells. The localization patterns also confirmed that, in the internalization process, nanoparticles resided inside endosomes, submicrometric vesicles of the endocytotic pathway. More interestingly, it was found that the external magnetic field could be applied to manipulate the movement of magnetically labeled endosomes so that they could penetrate inside the cells and form in the cell cytoplasm along the direction of applied field. Additionally, the magnetophoresis measurements gave a critical cellular magnetic load which exhibited a potential of this new type of magneto-fluorescent nanoagent for medical use.

In their cell labeling process, the culture medium was a mixture of DMEM, 10 % inactivated fetal bovine serum, 50 units/ml penicillin, 40 mg/ml streptomycin, and 0.3 mg/ml L-glutamine. Human cervical cancer cells (HeLa) with a mean cell diameter of $14.6 \pm 0.8\,\mu m$ was selected as the model cell system. A Leica DMIRB microscope with a 63× oil-immersion lens and digital camera was used to obtain fluorescence images. For fluorescence microscopy, the cell growth took place on glass coverslips for two days followed by incubation with rhodamine iron nanoparticles suspension (5 mM in concentration) at $4\,^\circ C$ for one hour. Three samples were prepared and their cell labeling was compared. The first sample, obtained by immediate fixation with 3 % paraformaldehyde in phosphate-buffered saline (PBS) after incubation, was used to study membrane localization. The second sample underwent a 2-h chase at $37\,^\circ C$ in RPMI culture medium and three washes with RPMI followed by a fixation under a uniform field with 3 % paraformaldehyde in PBS. The purposes of this process were to restore the internalization pathway and access the magnetic properties. To prepare the third sample, incubation and chase processes were conducted on cells followed by the fixation submitted to a uniform magnetic field ($B = 80 \times 10^3$ A/m). Investigation was made on the interaction between hybrid nanoparticles and cells in culture to probe the magnetic labeling capacity of fluorescent magnetic nanoparticles. To prevent the dynamic internalization process, cells were cultured at $4\,^\circ C$, at which temperature only interactions between nanoparticles and cells need to be considered. Figure 4.4 shows fluorescent images of cells after labeling and culture. It is obvious from Figure 4.4a that the whole membrane was labeled as a result of binding of fluorescent nanoparticles on the cell membrane. Figure 4.4b shows the fluorescent pattern of the cells when heated to $37\,^\circ C$. Since the internalization pathway was restored, nanoparticles were chased to the cell interior via small vesicles and further confined inside endosomes. The fluorescent spots in the cell cytoplasm can be observed in Figure 4.4b. Thus, the fluorescent microscopy confirmed the capability of direct imaging and localization in living cells for magnetic nanoparticles attached with a fluorescent dye. As far as the cell affinity is concerned, DMSA coated nanoparticles exhibit a high electrostatic affinity due to their tiny volume and negative surface charge. Consequently, a massive capture of nanoparticles (up to 10^7 nanoparticles per cell) can be obtained on cell membranes. The mechanism for cells to ingest extracellular materials is the endocytotic pathway. When magnetic nanoparticles are involved, the first step of

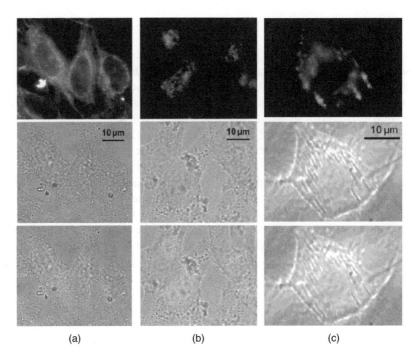

(a)   (b)   (c)

Figure 4.4 Fluorescence microscopy and transmission electron microscopy images showing the localization patterns of magnetic nanoparticles in living cells. The top images are fluorescence images and the center images are transmission electron microscopy images of the same cells while the bottom ones are overlays of both images, with the red color for fluorescence signal. (a) For the first sample, fluorescence appears on the cell membrane; (b) for the second sample, the fluorescence is visible in small, localized spots inside the cells; (c) for the third sample, chains of magnetic fluorescent endosomes inside the cell can be seen. (Bertorelle *et al.* 2006)

the endocytotic pathway is the internalization of nanoparticles within vesicles initiated by the membrane adsorption. And the next step is the delivery of nanoparticles to membrane organelles or endosomes with larger dimension (about 0.6 μm in diameter). Bertorelle's study visualized the membrane adsorption and nanoparticle residence within endosomes, which are the first and final steps of the endocytotic pathway, respectively. For the third sample, cell fixation was subjected to a uniform field to assess the magnetic properties of the fluorescent endosomes. As shown in Figure 4.4c, magnetic nanoparticles are enclosed in the endosomes. Under a magnetic field, a magnetic moment was conferred on these biological vesicles and the endosomes could act as small magnets. As a result, small chains formed in the cell cytoplasm caused by the dipole–dipole interactions. Obviously, the formation of chains demonstrated the association of both fluorescence and magnetic properties within the endosomes, and the bifunctional endosomes provided new biological tools to track the dynamics of endocytotic vesicles under magnetic fields.

Magnetically labeled cells have been demonstrated by iron oxide nanoparticles attachment, which can be imaged *in vivo* via MRI. Oca-Cossio and his team implemented this technology in insulin secreting cells (Oca-Cossio *et al.* 2004). It had to be established whether the existence of nanoparticles in the cytoplasm of the cells has any side effect on the insulin secretion. They used monocrystalline iron oxide nanoparticles to label mose insulinoma βTC3 and βTC-tet cells and investigated the effectiveness and consequence. After exposed to 0.02 mg/ml nanoparticle suspension for one day, internalization of magnetic nanoparticles in both cells was observed. At the same time, cell viability and apoptosis were maintained during the cell culture in the presence of magnetic nanoparticles. Furthermore, it was found that the magnetic labeling of insulin cells did not affect substantially the metabolic and secretory activities of the cells. With the help of magnetic nanoparticles, MRI images of labeled cells showed better contrast than sham-treated cells.

4.2.3 Spatially Resolved Magnetorelaxometry

Spatially resolved magnetorelaxometry is a novel technique to detect magnetic labeled biological entities by measuring the relaxation of their magnetization when the magnetizing field is removed. Similar to the applications discussed above, this technique also deals with magnetic labeling of biomolecules. Research investigation has demonstrated the feasibility of detecting and locating of immobilized magnetic nanoparticles *in vivo*. More and more interests in spatially resolved magnetorelaxometry have been attracted, especially in medical diagnostics, with respect to the developing technology in diseases at molecular level.

Magnetic nanoparticles are stable and biocompatible, hence they can be introduced into the living system. Magnetorelaxometry can be applied for opaque media or tissues, which is not possible for optical techniques. It is true that the relaxation of magnetic nanoparticles after alignment under a magnetic field pulse depends upon their dimensions as well as their immobilization. In magnetorelaxometry, magnetic nanoparticles bound to target biological entities have different relaxation times and relaxation behavior from unbound nanoparticles.

A system containing a single channel second order SQUID gradiometer as the magnetic sensor was developed to detect the relaxing magnetization of magnetic nanoparticles in objects (Romanus *et al.* 2002). Investigation by using this system aimed to demonstrate the potential use of the magnetic relaxation measurements in spatially-resolved determining the distribution of magnetic nanoparticles, which were injected into living beings. In order to determine the spatial distribution of the relaxation signals from magnetic nanoparticles at various locations, an adjustable sample holder which can be moved to positions of interest was designed and it was controlled by an automated measurement system. Figure 4.5a depicts the developed measurement system. In a typical measurement, biofunctionalized magnetic nanoparticles were first synthesized by a carboxydextran coating process and uniform aqueous dispersion of these particles (concentration $\sim$20 mmol/kg) was prepared. Then, this magnetic colloidal suspension was injected into the tail vein of C57Bl/6 mice followed by an *in vivo* measurement of the magnetic relaxation signal. A strong magnetic relaxation signal ($\sim$35 pT) was detected 15 minutes after injection. Figure 4.5b shows the strongest magnetic relaxation signal in the ventral region of the mice, suggestive of an enrichment of magnetic nanoparticles in that region. In addition to this strong signal, a weak relaxation signal (less than 5 pT) was also detected in the tail vein area. Further investigation even confirmed the strong relaxation signals coming from liver and spleen, which is in accordance with the well-accepted accumulation of magnetic nanoparticles by the phagocytes in both organs. For the purpose of comparison, control measurements were carried out without magnetic nanoparticles, and no magnetic relaxation signal was detected. It should be mentioned that the above measurements were obtained in a regular laboratory environment. Interference signals coming from the surroundings caused disturbance to the measurement. To overcome this problem, a magnetic shielding could be used. Another method is the implementation of different procedures in data acquisition and data evaluation.

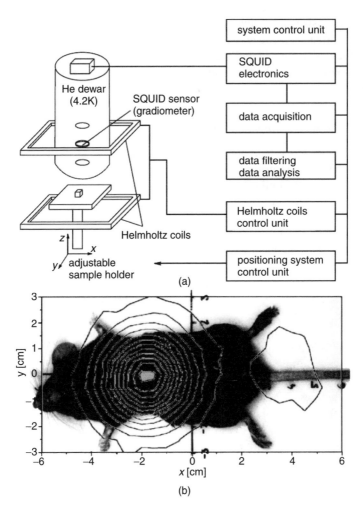

Figure 4.5   (a) Schematic illustration of the measurement set-up for *in vivo* measuring of the relaxation of magnetic nanoparticles; (b) photograph showing magnetic nanoparticle relaxation signals for an anesthetized mouse placed on the sample holder. (Romanus *et al.* 2002)

Magnetic Relaxation Immuno Assay (MARIA) is another example of magnetorelaxometry using magnetic nanoparticles as markers for the specific analysis of biomolecules (Ludwig *et al.* 2006). Superparamagnetic nanoparticles are used as markers in MARIA. Target biomolecules need to be labeled and then immobilized with magnetic nanoparticles. This immobilization is to suppress Brownian relaxation. Meanwhile, unimmobilized magnetic nanoparticles will undergo Brownian relaxation. When the magnetic field used to align their magnetic moments is removed, immobilized magnetic nanoparticles exhibit much faster Néel relaxation than unimmobilized ones. Thus, it is possible to separate the bound and unbound magnetic markers by their different relaxation behavior. Obvious merits of MARIA are its lack of need for washout of unbound markers and its capability of observing opaque media including blood. Research efforts have aimed to develop a new MARIA process to analyze viruses and bacteria in solution without immobilization (Ludwig *et al.* 2004).

4.2.4   Magnetic Separation and Purification

In modern biology or biomedicine, separation of particular biological entities (e.g. DNA, proteins, ions, molecules, etc.) from their native environment is often required for both scientific investigation and practical uses. Magnetically labeled biological entities exhibit magnetic properties so that they can be separated by applying an external magnetic field. Due to this reason, magnetic separation and purification have been widely used to prepare concentrated biosamples in biomedical applications.

A typical magnetic separation usually contains two steps: labeling desired biological entities with magnetic materials and separating magnetically tagged entities via a magnetic separation device. The first step is exactly the same as the magnetic labeling discussed in the previous subsection. To ensure an effective labeling, functionalization needs to be performed to deposit biocompatible molecules on magnetic nanoparticles. The organic coating not only provides a link between the particles and target biological entities, but improves the colloidal stability. In the case of iron oxide, common coating materials include polyvinyl alcohol (PVA), phospholipids, dextran, etc. A highly effective method of labelling cells is via antibody/antigen bonding due to its specific binding characteristics. By this technique, various biological entities (e.g. cancer cells, bacteria, red blood cells, Golgi vesicles, etc) have been demonstrated binding to immunospecific agents – coated magnetic nanoparticles. Similarly, magnetic microparticles can also be used for larger entities when incorporated in a polymer binder. In the second step of magnetic separation, when the mixture passes through a magnetic field, magnetic labeled targeted materials immobilize because of the force due to the magnetic field gradient and separate from the native solution.

The basis for the magnetic separation as well as drug delivery lies in the magnetic force applied on the magnetically labeled biomaterials. The relationship between the magnetic force $\boldsymbol{F}_m$ and the differential of the magnetostatic field energy can be described as:

$$\boldsymbol{F}_m = V_m \Delta \chi \nabla (1/2 \boldsymbol{B} \cdot \boldsymbol{H}) \qquad (4.4)$$

where $1/2 \boldsymbol{B} \cdot \boldsymbol{H}$ is the magnetostatic field energy density. It is the magnetic field gradient that causes the force to immobilize the tagged materials. When the fluid mixture passes through the magnetic region, the hydrodynamic drag force acting on the nanoparticles needs to be overcome by the magnetic force. The hydrodynamic force can be presented as:

$$F_d = 6\pi \eta R_m \Delta v \qquad (4.5)$$

where η is the medium viscosity (e.g. water), R_m is the radius of the magnetic particle, and Δv is the relative velocity of the cell to that of the medium. It should be pointed out that the buoyancy force, caused by the density difference of cells and medium can be neglected in biology and biomedical applications. From Equations (4.4) and (4.5), the velocity of the nanoparticles relative to the carrier fluid Δv can be expressed as:

$$\Delta v = \frac{R_m^2 \Delta \chi}{9 \mu_0 \eta} \nabla (B^2) \ \text{ or } \ \Delta v = \frac{\xi}{\mu_0} \nabla (B^2) \qquad (4.6)$$

where ξ is called 'magnetophoretic mobility', a parameter indicating how easy it is to manipulate the magnetic nanoparticle. For magnetic microparticles with high magnetophoretic mobility, the time requested for sufficient separation is short. On the contrary, magnetic nanoparticle-involved magnetic separation requires a longer time. However, using nanoparticles will reduce their potential interference with further tests on the separated biomolecules.

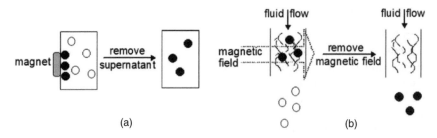

Figure 4.6 Schematic of two standard methods for magnetic separation. (a) A magnet is placed close to the container wall of a solution containing magnetically tagged (•) and unwanted (○) biomaterials. Due to the magnetic force exerted the tagged particles are attracted by the magnet and agglomerate while the unwanted biomaterials can be removed with the supernatant solution; (b) a solution containing tagged and unwanted biomaterials flows continuously through a region with strong magnetic field gradient which can be provided by packing the column with steel wool. Thus the tagged particles can be attracted. Afterwards, the tagged particles can be collected after removing the field followed by flushing through with water. Pankhurst, Q.A., Connolly, J., Jones, S.K. and Dobson, J. (2003). Applications of magnetic nanoparticles in biomedicine, *Journal of Physics D: Applied Physics*, **36**, R167–R181. Reproduced with permission of the Institute of Physics

Due to the magnetic properties of magnetic-labeled biomolecules, magnetic separation may be achieved by simply using a permanent magnet to aggregate magnetic entities on the wall of a test tube followed by removal of the supernatant. This simple process is illustrated in Figure 4.6a. However, this method is suffering from its low separation efficiency because of the slow accumulation rates. Thus better magnetic separation is needed. As discussed above, the magnetic force exerted on a magnetic object is proportional to the gradient of the applied magnetic field, as described by Equation 4.4. A common method of achieving high separation efficiency is to use a magnetizable matrix of wire or beads to apply a strong magnetic field on a flow column through which magnetically tagged fluid is pumped. The type of separator design is able to generate high magnetic field gradient and capture the magnetic biological entities efficiently as they flow with their carrier medium. Figure 4.6b depicts the schematic of this process. In spite of its faster separation, this method, however, has some problems caused by the settling and adsorption of magnetic materials on the matrix. To overcome this problem, an alternative separation design was developed, as shown in Figure 4.7. In this design, a magnetic gradient is created radially outwards from the center of the flow column by a quadrupolar arrangement. So obstruction will not take place during the separation. Additionally, biological entities with different magnetophoretic mobilities can be fractioned to achieve 'fluid flow fractionation' by adjusting the magnitude of the field gradient. In a typical separation process, the applied magnetic field is moved up the column while the fluid is kept static. Consequently, the particles can be moved up by the magnetic force and collected by a permanent magnet. The bottom section of fluid is then subjected to repeat separation processes with different magnetic strengths.

Magnetic separation can be used with optical sensing to perform 'magnetic enzyme linked immunosorbent assays'. Immuno-assays, combined with fluorescent enzymes, are able to detect cells labeled by the enzymes. In contrast to a conventional means in which the target cells need to bind to a solid matrix, magnetic-assisted immunosorbent assays utilize magnetic microspheres as the surface for cell immobilization. Importantly, magnetic separation is typically used to increase the concentration of the target materials. Recently, magnetic nanoparticles have been employed in this assay application. The use

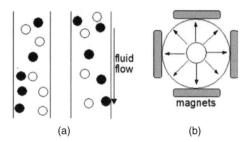

(a)   (b)

Figure 4.7 Schematic illustration of a rapid throughput method for magnetic separation. In this method, a solution containing magnetically tagged (•) and unwanted (○) biomaterials flows through an annular column placed within a set of magnets arranged in quadrature. (a) Longitudinal cross-section view of the annular column showing the flow of the solution; (b) transverse cross-section view of the four magnets with the resulting magnetic field lines. Under the magnetic force due to the magnetic field gradient, the tagged particles are attracted to the column walls, where they are held until the field is removed. The held tagged biomolecules can be recovered by flushing the column with water. To avoid complications coming from the near-zero field gradients the central core of the column is made of non-magnetic material. Pankhurst, Q.A., Connolly, J., Jones, S.K. and Dobson, J. (2003). Applications of magnetic nanoparticles in biomedicine, *Journal of Physics D: Applied Physics*, **36**, R167–R181. Reproduced with permission of the Institute of Physics

of nanoparticles improves the magnetic separation efficiency because their high mobility results in a short reaction time and more immobilized reagent. Furthermore, magnetic separation offers an effective way to localize labeled cells at certain locations, which is of great help for cell detection and number counting by optical scanning.

As an example, magnetic luminescent nanocomposites have been demonstrated in magnetic separation (Hong *et al.* 2004). These bifunctional nanocomposite nanoparticles consist of Fe_3O_4 magnetic cores with deposited CdTe quantum dots (QDs)/polyelectrolyte (PE) multilayers via a layer-by-layer (LbL) assembly approach. The magnetic cores have an average diameter of 8.5 nm. It was noticed that the combination of magnetic and luminescent properties is essential for magnetic separation or bioassays. Since each magnetic composite nanoparticle possesses a remnant magnetic field, it will act as a dipole magnet at nanoscale, thus aggregations and precipitation of the nanoparticles can be caused by applying an external magnetic field. The digital images in Figure 4.8 show a reversible process of magnetic separation and dispersion of these magnetic luminescent composite nanoparticles. As shown in Figure 4.8a, without an external magnetic field, the magnetic luminescent nanoparticles can be dispersed uniformly and the suspension exhibits an orange color under a UV radiation. In the presence of a magnetic field, magnetic luminescent nanoparticles aggregated on the wall of the test tube, where a permanent magnet was placed. Figure 4.8b clearly shows an orange agglomeration and a clear suspension due to the magnetic separation. This aggregation and suspension process was reversible and a site-specific transport of magnetic luminescent nanoparticles was expected in bioapplications.

Magnetic separation has also been used for DNA/RNA separation. In disease diagnostics, gene profiling and gene expression, separation and collection of target DNA/mRNA with base mismatches from a complex matrix is highly crucial. It has been well documented that genetic mutations coming from a single base mismatch in DNA sequences is the reason for most cancers. Separation and collection of the trace amount of mismatched gene products are desirable for both biomedical studies and biotechnology development. Zhao *et al.* reported their development of a genomagnetic nanocapturer (GMNC) using

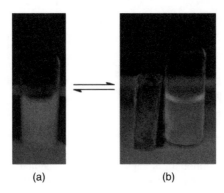

(a)   (b)

Figure 4.8 Digital pictures of the magnetic luminescent nanocomposites under UV irradiation (a) without an external magnetic field and (b) in presence of an external magnetic field. The strength of the applied magnetic field by a magnet was 1000 G. An aggregation process was observed indicated by a color change from orange to transparent when an external magnetic field was applied. The aggregations could disappear rapidly as the magnetic field was removed. (Hong *et al.* 2004)

magnetic nanoparticles functionalized with molecular beacons (2003). In their work, molecular beacons were used owing to their high sequence selectivity and excellent detection sensitivity, while magnetic nanoparticles provided an effective way for bioseparation. After functionalization of magnetic nanoparticles with molecular beacons, an effective GMNC was developed where magnetic nanoparticles act as carriers and molecular beacons serve as recognition probes for target gene sequences. Figure 4.9a illustrates the mechanism of GMNC discrimination, separation and base-mismatched DNA sequence collection. The structure of GMNC is illustrated in Figure 4.9b. There is a thin silica layer between magnetic cores and molecular beacons which serves as a biocompatible interface. With this silica layer, the magnetic nanoparticles can be readily functionalized with biomolecules. The layer was deposited by a base-catalyzed hydrolysis and polymerization reaction of TEOS in a well-developed sol-gel process. Figure 4.9c is a TEM image indicating the existence of a thin silica shell. The developed GMNC was demonstrated for gene separation. In the experiments, DNA/mRNA strands with a single base difference can be separated and collected from perfectly matched DNA strands by applying a magnetic field under controlled temperatures. Separation and collection of mRNA strands from both cultured cells and prepared complex matrix (cell lysates, tissue samples) was also achieved by using the same GMNC. Experiments suggested that the amounts of the recognition probes (molecular beacons) immobilized on the nanoparticle surface influenced the separation efficiency. And the separation capability of GMNC was completely determined by the magnetic properties of the magnetic cores. Therefore, in both synthesis and application of GMNC, care needs to be taken to tailor the immobilization of probes and the magnetic characteristics. With the help of the intrinsic fluorescent property of molecular beacons in GMNC, it is possible to observe and monitor the gene separation and collection processes *in situ*.

4.2.5 Biological Assay System

A biological assay system is widely used for biomaterial detection. Analyses of DNA, protein and other biotargets can be realized by using an array-based bioassay for advanced medical care and environmental measurements. Presently, biological immunoassay, taking advantage of specific antigen/antibody bonding, has dominated in both research investigation and practical diagnostics. For a regular biological assay system, the key aspect

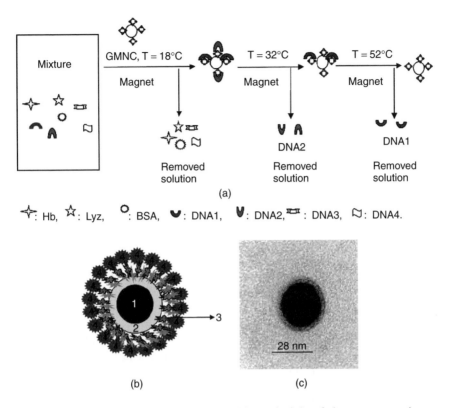

Figure 4.9 (a) Schematic illustration of the working principle of the genomagnetic nanocapturer for magnetic separation and collection of gene products with a one-base difference in DNA sequence; (b) schematic representation of a genomagnetic nanocapturer (1: magnetic nanoparticle, 2: silica layer, 3: biotin-avidin linkage, 4: molecular beacon DNA probe); (c) TEM image of a silica-coated magnetic nanoparticle showing the core–shell nanostructure. The diameter of the overall nanoparticle is about 28 nm. (Zhao *et al.* 2003)

is to obtain detectable signals from the antigen/antibody bonding. Currently, common microarray detection methods are fluorescence and chemiluminescence technologies due to their high sensitivities and versatile capabilities. However, both methods suffer from low signal intensity owing to photo-degradation. Among a number of approaches to improve sensitivity, molecular labeling is promising. Magnetic nanoparticles have exhibited extraordinary labeling capabilities in bioapplications. Also, magnetic detection could be an ideal means for bioassays. Recently, using a combination of magnetic nanoparticles and a patterned substrate covered with a self-assembled monolayer provides a promising technique for magnetic detection of biomolecular interactions (Osaka *et al.* 2006). In a magnetic immunoassay, magnetic nanoparticles attach to antibodies and serve as a biomarker to give magnetic signal after antibody/antigen coupling. Usually, the magnetic signals including remanence, relaxation and susceptibility can be detected via a superconducting quantum interference device (SQUID).

Again, superparamagnetic nanoparticles with diameters of a few tens of nanometers are ideal magnetic biomarkers in this application. The magnetic properties of superparamagnetic nanoparticles are degraded by thermal interference and their degradation is affected by their dimension significantly. For this reason, monodisperse magnetic nanoparticles are desirable. Practically, magnetic nanoparticles have a size distribution, thus both thermal

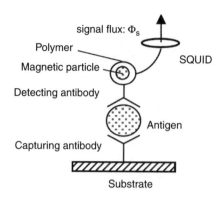

Figure 4.10 Schematic representation of magnetic immunoassay. The detecting antibody is labeled with a magnetic marker made of a magnetic nanoparticle a few tens of nanometers in diameter. The binding interaction between an antigen and its specific antibody can be detected by measuring the magnetic field of the marker by a SQUID. Reproduced from: Enpuku, K., Inoue, K. and Soejima, K. (2005). Properties of magnetic nanoparticles for magnetic immunoassays utilizing a superconducting quantum interference device, IEICE Transactions on Electronics, vol. E88-C, no. 2, Feb. 2005, pp. 158–167

noise and size distribution of particles need to be taken into account. Since the magnetic signal comes from the magnetic nanoparticles, their magnetic properties are expected to generate a large magnetic signal which is required for a high performance magnetic immunoassay with high sensitivities.

Figure 4.10 illustrates the basic configuration and mechanism for a magnetic immunoassay system (Enpuku *et al.* 2005). Detecting antibodies are first labeled by polymer protected superparamagnetic nanoparticles (Fe_2O_3 or Fe_3O_4 nanoparticles). Meanwhile, antigen binds specifically to capturing antibody attached on a substrate. Then reaction between antigens and detecting antibodies provides a robust bonding. The markers attached to the substrate generate a magnetic field, which can be detected by a SQUID system.

Detection of biomolecules on a self-assembled monolayer formed on magnetic nanoparticles and detection of biomolecular interaction between biotin and streptavidin using magnetic nanoparticle have been demonstrated (Osaka *et al.* 2006). In both experiments, biomolecules are detected via magnetic particles by measuring their magnetism. The utility of magnetic nanoparticles has several advantages: 1) magnetic nanoparticles are stable for storage and through the whole measurement; 2) apparatus needed is simple and inexpensive; 3) magnetic nanoparticles are low cost; 4) washing procedure can be done simply by using a permanent magnet.

Self-assembled monolayers (SAMs) can serve as a template for implementing biological recognition with high accuracy. Prompted by this idea, a novel method was developed for biological assay which combines magnetic nanoparticles and a silicon substrate with a self-assembly monolayer. Monomolecular layers are able to provide an ideal model to probe the molecular interaction at liquid–solid interfaces. A detection system using specific biomolecular interaction between biotin and streptavidin is shown in Figure 4.11a. Magnetite nanoparticles (Fe_3O_4) 200 nm in diameter were functionalized with streptavidin. To do so, magnetic nanoparticles were first modified by APS in toluene and the amine groups from APS were used to bind biological molecules. Biotin was able to attach to the magnetic nanoparticle surface via a cross-linking reaction between sulfo-LC-LC-biotin and the surface amine groups. Then, streptavidin can bind to magnetic labeled biotin

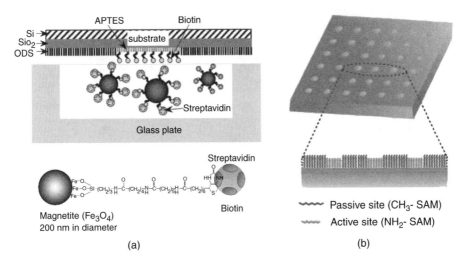

Figure 4.11 (a) Schematic illustration of biotin-streptavidin interaction on self-assembly monolayer-modified substrate. Biotin is attached to the substrate after APS patterning. Then streptavidin-attached magnetic nanoparticles are injected into a channel on a glass plate. The nanoparticles are bound by a specific binding between biotin and streptavidin; (b) schematic representation of the surface patterned with different monolayers. (Osaka *et al.* 2006)

specifically. When used for patterned monolayers in bioassays, substrates were patterned by APS followed by attachment with biotin. Streptavidin-modified magnetic nanoparticles were introduced and bound to patterned biotin on the substrate. Figure 4.11b illustrates the surface patterned with different monolayers.

Perez *et al.* reported the application of magnetic nanoparticles in a biological assay system capable of detecting low levels of viral particles in serum (2003). It was postulated that the multivalent interactions between virus and magnetic nanoparticles could lead to a highly sensitive assay. In their demonstration, herpes simplex virus (HSV) and adenovirus (ADV), which are relevant to human pathology as viral vectors in gene therapy, were used. Experiments started from dextran coating of superparamagnetic iron oxide nanoparticles. Then, anti-adenovirus 5 (ADV-5) or anti-herpes simplex virus 1 (HSV-1) antibodies were attached via Protein G coupling to the surface of dextran-modified magnetic nanoparticles, as shown in Figure 4.12a. Cross-link between the amino groups from the antibodies and the amino groups from the nanoparticles was initiated by a bifunctional linker, suberic acid bis(N-hydroxysuccinimide ester) (DSS). After this step, the overall diameter of nanoparticles was about 46 nm. Experiments showed that the detectable magnetic changes could be used to detect viral particles in biological media (cell lysates or serum) at low concentrations directly. Without any purification or amplification processes, the detectable lower limit could be 5 viral particles in 10 µl serum in biological samples. Figure 4.12b depicts the detectable range of this magnetic biological assay system. Using a higher magnetic field strength may further lower the detectable lower limit for sensing.

4.2.6 Biosensors

Tremendous research efforts have been stimulated by the rising demand for biosensors with high sensitivity, high reliability, fast response and excellent selectivity for decades. With the capability of detecting trace amounts of biomolecules in real time, biosensors have found versatile applications in the field of environmental control, hazard material

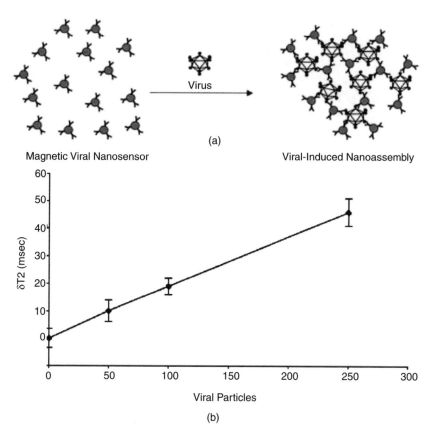

Magnetic Viral Nanosensor   Viral-Induced Nanoassembly

(a)

(b)

Figure 4.12 (a) Schematic diagram of viral-induced nanoassembly of magnetic nanoparticles; (b) plot of the dependence of water T2 in anti-HSV-1 magnetic nanosensors on the amount of viral particles. The magnetic field was 1.5 T at 60 MHz and the sample size was set as 100 µl. (Perez *et al.* 2003)

detection, pharmaceutics and clinical diagnostics. A common biosensor works as a biospecific surface interacting with a particular analyte, and generating detectable signals, such as electrochemical, optical, piezoelectrical and thermal responses. Currently, amperometric biosensors, detecting current signals coming from electrochemical redox reactions on electrodes, are extensively investigated in biosensor research, and widely used in commercial devices, mainly due to their superior properties, such as low cost, simple detection tools and procedures, broad detection range and high accuracy.

A regular amperometric biosensor consists of single enzyme hence it is able to detect a single specific analyte which undergoes a redox reaction giving an electrical signal. In many circumstances, it is desirable to detect multiple analytes (substrates) in biological media. For example, to detect two substrates in a mixture, two corresponding enzymes need to be immobilized to respective redox reactions. However, there is always a problem when the detecting substrates are oxidized at close oxidation potentials, since the detected current signal is the sum of two bioelectrocatalyzed transformations occurring concurrently. An effective approach to solve this problem is to use switching biocatalyst systems. In a dual biosensing system, when one enzyme is switched to 'off' state, the detected current signal comes completely from the redox reaction on the other enzyme which is 'on'. Therefore, searching controllable bioelectrocatalysis is required for dual or even multiple biosensing purposes.

The novel concept of using magnetic nanoparticles in biosensors, stemming from the magnetic control of bioelectrocatalysis, was developed by Hirsch *et al.* (2000). Dual biosensing of two analytes, glucose and lactate, by magneto-controlled bioelectrocatalysis has been demonstrated (Katz *et al.* 2002a). As enzymes are usually used as the redox center in amperometric biosensors, the first step in magneto-controlled bioelectrocatalysis is to bond redox modifier to magnetic nanoparticles (Perez *et al* 2004). For this purpose, both covalent bonding and physically adsorption can be used. A commonly used chemical process for chemical bonding is to immobilize 1,4-naphthoquinones on the electrode surface functionalized with amino groups. In this chemical reaction, amino-quinone derivatives, formed from the nucleophilic addition of amino groups to quinones, exhibit negatively shifted redox potentials due to the electron-donating characteristics from the amino groups. A similar approach applies for chemical binding of redox enzyme to magnetic nanoparticles. Figure 4.13 is the schematic of the synthesis of covalent linked redox-active units to functionalized magnetic nanoparticles (Katz *et al.* 2002b). In this synthesis, magnetic nanoparticles were first functionalized with [3-(2-aminoethyl)aminopropyl]-trimethoxysilane to attach amino groups. After treating the functionalized nanoparticles with 2,3-dichloro-1,4-naphthoquinone (**1**), magnetic nanoparticles were further attached with 2-amino-3-chloro-1,4-naphthoquinone groups (see Figure 4.13A). Similarly, reacted with carboxylic acids, carboxylic-acid-functionalized redox-active magnetic nanoparticles can be obtained via carbodiimide coupling. A number of carboxylic acids were tested, such as pyrroloquinoline quinine (**2**, PQQ), heme-containing undecapeptide, microperoxidase-11 (**3**, MP-11), *N*-(ferrocenylmethyl) amino-

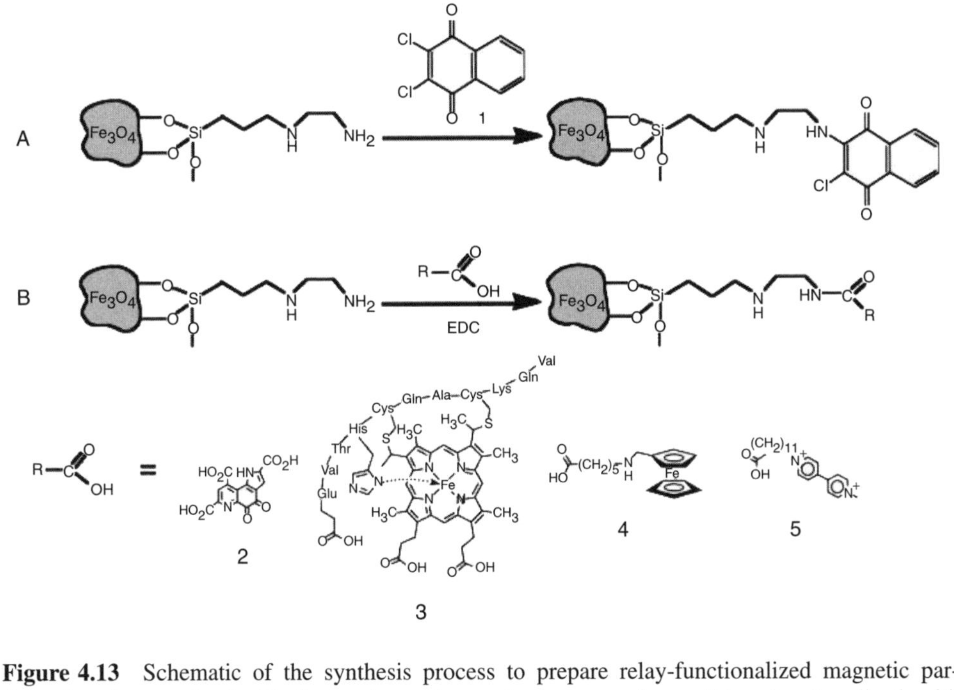

Figure 4.13   Schematic of the synthesis process to prepare relay-functionalized magnetic particles by the covalently binding redox-active units to magnetic particles functionalized with [3-(2-aminoethyl)aminopropyl]siloxane film. A) Binding of 2,3-dichloro-1,4-naphthoquinone to the functionalized magnetic nanoparticles; B) carbodiimide coupling of electron-relay carboxylic derivatives to the amino groups on the siloxane layer. (Katz *et al.* 2002b). Reproduced by permission of Wiley-VCH

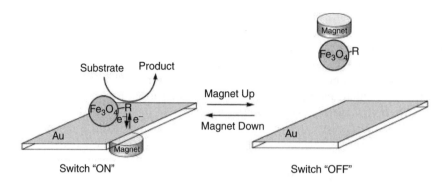

Figure 4.14 Schematic illustration of the processes of magneto-switched electron transfer and electrocatalyzed transformation assisted by redox-active units linked to magnetic nanoparticles. (Katz *et al.* 2002b). Reproduced by permission of Wiley-VCH

hexanoic acid (**4**), and *N*-methyl-*N*'-(dodecanoic acid)-4,4'-bipyridinium (**5**), as shown in Figure 4.13B. The obtained active redox unit-attached magnetic particles can be magneto-switching when used in amperometric biosensors. As depicted in Figure 4.14, by manipulating an external permanent magnet, the electron transfer for signal detection can be switched between 'on' and 'off'. When the magnet is placed under the gold electrode, the magnetic redox unit will be attracted towards the electrode, and their contact will ensure the transfer of electrons generated from the redox reaction. As the magnet is placed above the electrode, the magnetic redox unit will be raised up. Consequently, the electron transfer is blocked showing 'off' state. Besides the magnetic attraction of the magnetic redox units to the electrode, the rotation of the magnetic nanoparticles on the electrode by an external rotating magnet is also important. With this rotation, redox-functionalized magnetic nanoparticles can be treated as rotating micro-electrodes. In this case, rather than diffusion, convection of the substrate to the microelectrodes will dominate in the bioelectrocatalytic process resulting in an enhanced amperometric response.

Figure 4.15 shows the glucose sensing by rotating ferrocene functionalized magnetic particles. Here ferrocene is used as a mediator for electron relay purposes. Glucose reacts with glucose oxidase, the enzyme, and releases an electron to ferrocene. As the magnet is placed under the electrode, electrons can be collected by electrical contact of ferrocene

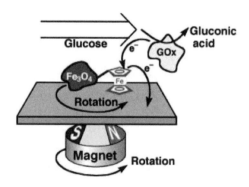

Figure 4.15 Schematic illustration of the bioelectrocatalytic oxidation of glucose in the presence of glucose oxidase (GOx) and ferrocene-functionalized magnetic particles. The oxidation process was enhanced with a circular rotation of the particles by rotating the external magnet beneath. (Katz *et al.* 2004)

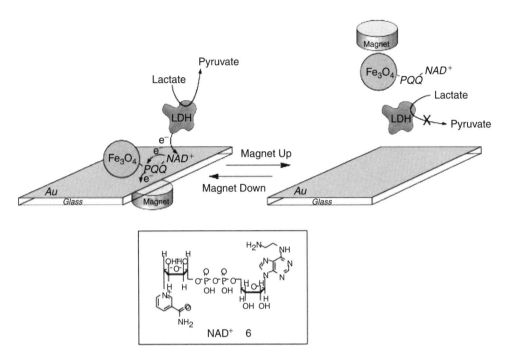

Figure 4.16 Schematic illustration of magneto-switched bioelectrocatalytic oxidation of lactate with lactate dehydrogenase (LDH) and PQQ-NAD$^+$-functionalized magnetic nanoparticles. The electron transfer can be controlled by an external magnet applied below the substrate. (Katz *et al.* 2002b). Reproduced by permission of Wiley-VCH

with the electrode. The detected current signal is proportional to the concentration of glucose in medium. By introducing the rotating magnet, the interaction between glucose, glucose oxidase, ferrocene and nanoparticles is convection-controlled. Therefore, amplified amperometric analysis is achieved. Further, the resulting detectable currents should be related to the rotation speed of the magnet as $I_{cat} \propto \omega^{1/2}$.

In lactate sensing, NAD$^+$-dependent redox enzymes are commonly used. An important step in redox reactions is the electrochemical regeneration of NAD(P)$^+$ by the oxidation of NAD(P)H. Among various available redox mediators, PQQ (**2**) is an effective catalyst for this process. Magneto-switchable bioelectrocatalyzed oxidation of lactate is schematically depicted in Figure 4.16. NAD$^+$-dependent enzyme lactate dehydrogenase (LDH) plays the same role as the glucose oxidase (GOx) in glucose sensing. Magnetite nanoparticles were functionalized by PQQ and amino-NAD$^+$ (**6**). The resulting functionalized magnetic nanoparticles serve as the mediator in electrochemical reactions occurring on the electrode. It is likely that the bioelectrocatalytic transformation of lactate can be regulated by adjusting the place of the magnet. For instance, 'on' state can be obtained when placing the magnet under the electrode. In this case, magnetic nanoparticles are attracted to the electrode, and the electrochemical regeneration of NAD$^+$ takes place via the electrochemical oxidation of PQQH$_2$ to PQQ followed by further oxidation of NADH to NAD$^+$. When functionalized magnetic nanoparticles are retracted from the electrode, both the bioelectrocatalyzed oxidation of lactate and electrochemical regeneration of NAD$^+$ are prohibited.

As the magnetic labeled enzyme systems have exhibited switchable biocatalysis, it is possible to develop a biosensing system capable of detecting multiple substrates by manipulating an external magnetic field. In a magneto-controlled dual biosensing system

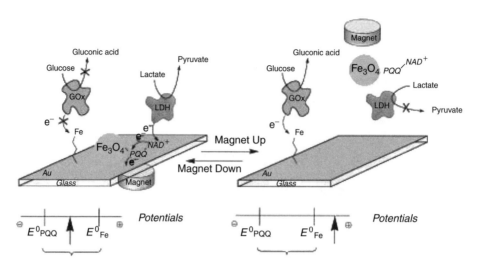

Figure 4.17 Schematic illustration of the working principle for magneto-controlled dual biosensing of glucose and lactate. The system consists of glucose oxidase (GOx), lactate dehydrogenase (LDH), PQQ-NAD$^+$ functionalized magnetic particles, and an Au-electrode coated with a monolayer of ferrocene units. (Katz *et al.* 2002b). Reproduced by permission of Wiley-VCH

for glucose and lactate, the magnetic attraction or retraction of magnetic-attached enzymes or mediators enables two redox tranformations occurring separately. Specific analysis of glucose and lactate in the presence of glucose oxidase (GOx) and lactate hydrogenase (LDH), respectively, is illustrated in Figure 4.17. A ferrocene monolayer was first coated on the gold electrode surface, which served as the mediator for the activation of GOx towards oxidation of glucose. Meanwhile, magnetite nanoparticles functionalized with PQQ-NAD$^+$ acted as the catalyst-cofactor and activated LDH for the bioelectrocatalyzed oxidation of lactate. Cyclic voltammogrametry analysis indicated that two quasi-reversible redox peaks occurred at $E^0 = -0.13$ V and 0.32 V, which corresponds to the potentials to active PQQ-NAD$^+$ and ferrocene, respectively. Interestingly, the potential applied on the system varied with the position of the magnet. When magnetic nanoparticles were attracted to the electrode, the applied potential ranged from -0.13 to 0.32 V, which is sufficiently positive to oxidize the PQQ units, but not positive enough for the redox potential of ferrocene. Thus selective oxidation of lactate was achieved. On the other hand, both oxidations of glucose and lactate occurred when the magnetic nanoparticles were retracted from the electrode because the applied potential was higher than 0.32 V. Subtracting the current due to the oxidation of lactate from the current derived from the latter case, the current signal coming from the glucose can be obtained. This work successfully demonstrated the feasibility of using magneto-switchable bioelectrocatalysis for multisensor devices for selectively detecting different substrates in a mixture. Potential future applications using this concept are expected.

4.3 Therapeutic Applications

4.3.1 Drug and Gene Target Delivery

Controlled drug delivery is of great importance for therapeutic applications, such as cancer treatment. Various materials have been investigated as the drug carrier to bring drugs,

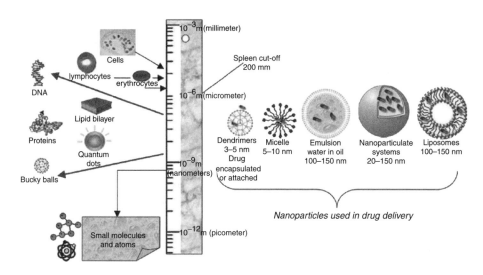

Figure 4.18 Nanoparticles for drug delivery applications. (Arruebo *et al.* 2007)

genes and proteins to target specific areas in the living body. Figure 4.18 provides a chart showing the candidate moieties at different dimension scales for this purpose (Arruebo *et al.* 2007). Common materials include dendrimers, micelles, emulsions, nanoparticles, liposomes, etc. In a typical drug delivery process, drugs are loaded in biocompatible carry materials, transferred to bodies and released in control for cancer therapy and the treatment of other ailments.

Nanomaterials have been considered as a candidate of nanosized drug delivery systems for many years mainly due to their excellent biocompatibility, subcellular size and targeting action. Various nanostructured materials have shown capability for drug delivery. The potential of magnetic nanoparticles for drug delivery applications stems from their inherent magnetic properties coupled with other common properties shared with other nanomaterials. Their unique magnetic properties include superparamagnetism, high saturation magnetization and high magnetic susceptibility. Also, chemical and biofunctionalization techniques have been developed to improve both the stability and the biocompatibility of magnetic nanoparticles. Researchers can select suitable coating materials on magnetic nanoparticles for particular drugs and certain targets. The most attractive characteristic of using magnetic nanoparticles in drug delivery is their capability of controlled delivery to a specific area by applying an external magnetic field. Selectively delivering drug molecules to the diseased site while not increasing its level in the healthy tissues of the organism is always desirable in drug delivery for cancer treatment (Zhou *et al.* 2005).

In cancer treatment, magnetically controlled drug targeting is based on binding anticancer drugs with ferrofluid and desorbing from the ferrofluid after reaching the area of interest by means of an external magnetic field (Lübbe *et al.* 2001). In this process, local effects occur including irradiation from immobilized therapeutic radionuclide or hyperthermia from the magnetic nanoparticles themselves. In regard to control release, the simplest method is the natural diffusion of immobilized drugs from the carriers (magnetic nanoparticles). Since this method cannot be well controlled, other techniques have been employed. Better control release can be triggered by changing the physiological conditions such as temperature, pH value, osmolality, etc.

The major limitation of magnetic drug delivery systems is that the external magnetic field should be lowered to a certain level, higher than which living bodies cannot withstand. However, a magnetic field at this level may not generate magnetic gradient high enough to control the targeted movement of nanoparticles or trigger the drug desorption because the magnetic gradient decreases with the distance. This problem can be alleviated by locating an internal magnet near the target by minimally invasive surgery. Another problem which may occur in magnetic drug delivery is the possible agglomeration of magnetic nanoparticles, especially after the removal of the applied magnetic field, due to their high surface energy (Arruebo *et al.* 2007). Further, smaller nanoparticles possess weaker magnetic force. Thus, the ultrasmall dimension of nanoparticles, which is required for superparamagnetism, may lead to difficult control of their movement or location in the presence of a relatively strong drag force from the blood flow. Consequently, magnetic drug delivery is more effectively controlled in blood flows with lower velocities.

To better understand the whole process of magnetic drug delivery, it would be helpful to discuss the basic physical principles and other related issues before giving several examples.

Chemotherapies have been conventionally used for cancer treatments. In these therapies, drugs are administered with a general systematic distribution. In addition to the effective treatment for the cancer tumor, deleterious side effects also take place when the drug attacks normal, healthy cells. Magnetic targeted drug delivery is then expected to reduce the unnecessary side effect and improve the efficiency by controlled localized delivery. Figure 4.19 depicts a hypothetical magnetic drug delivery system. In magnetically targeted therapy, a cytotoxic drug is immobilized with biocompatible magnetic nanoparticles forming a drug/carrier complex. Then this biocompatible ferrofluid is injected into the circulatory system of a patient. After the magnetic drug/carrier complex enters the bloodstream, a magnet is applied outside of the body generating a magnetic field gradient to capture magnetic carriers flowing in the circulatory system and concentrate the complex in the vicinity of the target area within the body. After the targeted tumor cells are encompassed with drug/carrier complexes, release of the drug is initiated in a controlled manner and the drug is taken up by the cells. After treatment, magnetic nanoparticles will be enriched and metabolized in organs, such as the spleen and liver.

Similar to magnetic separation, the physical principles underlying magnetic drug delivery involve the magnetic force exerted on magnetic nanoparticles by a magnetic field

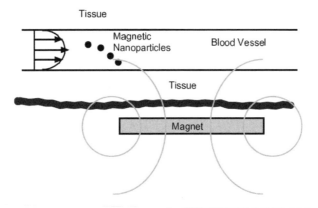

Figure 4.19 Schematic diagram of a hypothetical magnetic drug delivery system. A magnet is placed outside the body and the generated magnetic field gradient might capture magnetic carriers flowing in the circulatory system to the targeted area. Pankhurst, Q.A., Connolly, J., Jones, S.K. and Dobson, J. (2003). Applications of magnetic nanoparticles in biomedicine, *Journal of Physics D: Applied Physics*, **36**, R167–R181. Reproduced with permission of the Institute of Physics

gradient. Besides the physical parameters such as the magnetic field strength, gradient and volumetric and magnetic properties of nanoparticles, both hydrodynamic and physiological parameters influence the magnetic therapy. The former includes blood flow velocity, ferrofluid concentration, infusion route and circulation time, while the latter includes distance from the field source, strength and reversibility of the drug/carrier complexes, and tumor volume, etc. Investigation revealed an effective magnetic flux density for comer magnetic carriers of 0.2 T with field gradients of about 8 Tm^{-1} for femoral arteries and stronger than 100 Tm^{-1} for carotid arteries. This clearly demonstrated that more effective drug delivery can be obtained in slower blood flows and the area closer to the magnetic source. Mathematical models have been developed to simulate the 2D nanoparticle trajectories in various field/nanoparticle configurations. Modified hydrodynamic parameters for nanoparticles were suggested from this simulation since the movement of nanoparticles is no longer governed by Stoke's law near the walls.

So far, many magnetic materials have been demonstrated as the carrier in magnetic drug delivery. Optimizing the existing available magnetic materials and searching for novel magnetic materials are concurrently under research. In regard to the structural configurations, there are two common types, which are magnetic cores covered with biocompatible coatings and biocompatible porous polymers with embedded magnetic nanoparticles. The former has attracted more attention due to its simpler preparation and better controlled properties. A representative core–shell structure of a functionalized magnetic carrier is illustrated in Figure 4.20. Ferrite cores (magnetite, Fe$_3$O$_4$, or maghemite, γ-Fe$_2$O$_3$) are coated with biocompatible silica or polymer such as PVA or dextran. The coating plays a dual function: protection of magnetic cores from unnecessary chemical changes and providing functionalization potentials. Via these functional groups, cytotoxic drugs can be coupled to magnetic nanoparticles. Also, chemical functional groups, biotin, avidin, target antibodies and other biomolecules can be linked for multifunctionalization purposes. Recent research efforts have been continued into alternative magnetic cores (e.g. iron, cobalt, nickel, yttrium aluminum iron garnet, etc). For instance, cobalt/silica carriers have found applications in eye surgery to repair detached retinas.

It was Widder and and his team who demonstrated magnetic drug delivery for the first time (Widder *et al.* 1983). In their experiments, they injected magnetically carried target

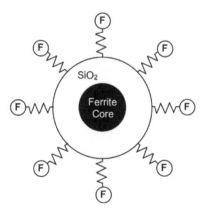

Figure 4.20 Schematic illustration of a functionalized core–shell magnetic nanoparticle for targeted drug delivery. This nanostructured material consists of a ferrite core, a shell of silica, SiO$_2$, and functional groups attached to the shell layer. Pankhurst, Q.A., Connolly, J., Jones, S.K. and Dobson, J. (2003). Applications of magnetic nanoparticles in biomedicine, *Journal of Physics D: Applied Physics*, **36**, R167–R181. Reproduced with permission of the Institute of Physics

cytotoxic drugs, doxorubicin, to sarcoma tumors implanted in rat tails and demonstrated the remission of the sarcomas. However, no remission was observed for drugs administered at ten times the dose but without the magnetic carrier. This pioneering work showed promise in the use of magnetic drug delivery for highly efficient cancer therapy. Since then, a number of research groups have reported their work using swine, rabbits, and rats for *in vivo* testing. Lübbe *et al.* conducted a Phase I clinical trial on ferrofluid for 14 patients and reported their success on directing ferrofluid to target sarcomas without obvious organ toxicity (Lübbe *et al.* 1996). Kubo *et al.* implanted magnets at solid osteosarcoma sites in hamsters and used magnetic liposomes to deliver cytotoxic drugs (Kubo *et al.* 2000). It was found that the magnetic drug delivery was able to deliver four times the level of drugs provided by non-magnetic delivery. Meanwhile, significantly improved anti-tumor activity and minimized side effects were observed. A hybrid vector using HVJ-E (hemagglutinating virus of Japan-envelop) and magnetic nanoparticles was developed to achieve enhanced transfection efficiency (Morishita *et al.* 2005). This novel vector was able to transfer plasmid DNA, oligonucleotide and proteins into cells rapidly by cell fusion. It is believed that the application of a magnetic force induces a rapid attachment of HVJ-E and cells, thus resulting in higher transfection efficiency. Both *in vitro* and *in vivo* studies were conducted to demonstrate the enhanced efficiency of the HVJ-E system with modified magnetic nanoparticles. A model of the functionalized graphite-encapsulated magnetic nanoparticles for drug delivery was proposed, as shown in Figure 4.21 (Bystrzejewski *et al.* 2005). The magnetic core material is FeNdB. The carboxyl groups on the nanoparticles' surface can directly linked to targeting ligands, drugs peptides, antibodies and receptors via covalent bonding. In a drug delivery process, the drug-functionalized magnetic nanoparticles can be directed along a magnetic field gradient to transfect the appropriate cells and recognize the target ligand.

Magnetic hollow silica nanospheres, composed of a coating of Fe_3O_4 magnetic nanoparticles and silica ($\sim$10 nm) on nanosized spherical calcium carbonate ($CaCO_3$) surfaces, have been demonstrated in magnetic drug delivery (Zhou *et al.* 2005). Electron microscope

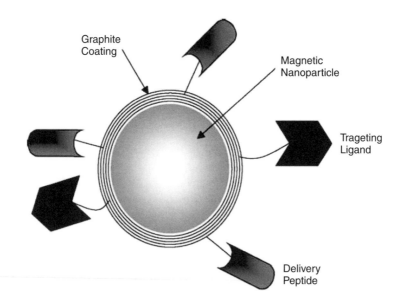

Figure 4.21 A schematic model of the multi-functionalized graphite-encapsulated FeNdB nanocrystal. (Bystrzejewski *et al.* 2005)

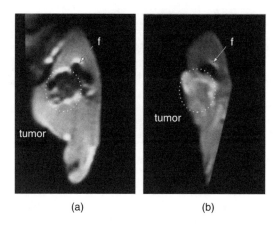

(a)   (b)

Figure 4.22 (a) MRI image of tumorous hind limb after i.a. application of ferrofluids; (b) MRI image of tumorous hind limb after i.v. application of ferrofluids. The area with a dotted circle indicates the tumor region. 'f' indicates the head of the femur. (Alexiou *et al.* 2002)

observations indicated that a calcium carbonate nanosphere was coated with a thin silica film embedded with magnetic nanoparticles. The superparamagnetic nanospheres were immersed in ibuprofen solution and UV measurements revealed a 12wt % of drug loading. When embedded inside the silica thin film, magnetic nanoparticles are stable and can maintain their magnetic function readily during biocycling. Therefore, these drug-loaded magnetic nanospheres are expected to be an ideal carrier candidate in nanomedicine.

Alexiou *et al.* reported their testing of drug bound ferrofluid in cancer treatment for rabbits (2002). Mitoxantrone bound ferrofluid (FF-MTX) was injected into tumor bearing rabbits (VX2 squamous cell carcinoma). After applying an external magnetic field of 1.7 T onto the tumor area for one hour, complete tumor remissions were observed without any obvious side effects such as leucocytopenia, alopecia and gastrointestinal disorders, etc. To study the distribution of the magnetic nanoparticles, they were labeled with iodine and histological investigations and MR imaging were performed. Figure 4.22a and 4.22b shows the left hind limb (implantation site) of two rabbits 6 hours after receiving intra-arterial and intravenous 50 % FF-MTX, respectively. In both images, the area 'f' is located at the head of the femur, and it appears to be hypodense. From Figure 4.22a, a tumor is shown at the medial portion of the hind limb, as indicated by the dotted circle. Conversely, intravenous FFMTX only shows a very discrete signal extinction, as can be seen from Figure 4.22b. The experimental results suggested this type of mitroxantrone bound ferrofluids to be a promising delivery system for anticancer agents (e.g. radionuclids, cancer-specific antibodies, anti-angiogenetic factors and genes, etc.) with strong and specific therapeutic efficacy.

The same research group also investigated the distribution of the chemotherapeutic agent bound magnetic carriers *in vivo* (Alexiou *et al.* 2005). Again, they used VX2 tumor-bearing rabbits for treatment with mitoxantrone bound magnetic nanoparticles. After incubating with ferrofluid-containing medium under an external magnetic field for one hour at room temperature, the structure of HeLa-cells and nanoparticle distribution were observed by transmission electron microscopy. Figure 4.23 presents TEM images showing ferrofluid (black spots) with a HeLa-cell. It was observed that the ferrofluids were disseminated throughout the intracellular space. Further, it was also found that ferrofluids were not taken up by the HeLa-cells after incubation at 4 °C.

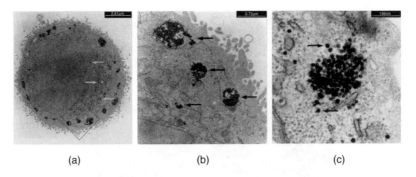

(a)   (b)   (c)

Figure 4.23 (a) Transmission electron microscopic image of a HeLa-cell after ferrofluid incubation under an applied external magnetic field (0.4 T). The visible black spots in the peripheral area (frame) and disseminated in the intracellular space (marked with arrows) are the administered ferrofluid; (b) magnified TEM image of the frame showing the presence of ferrofluids in the peripheral area of a HeLa-cell (marked with arrows); (c) TEM image showing the presence of magnetic nanoparticles in the intracellular space. Single particles (marked with arrows) might be surrounded by starch polymers which are not visible by electromicroscopic observation. (Alexiou *et al.* 2005)

A special magnetic drug delivery for cancer treatment, named radionuclide delivery, was developed and tested *in vitro* and *in vivo* in 1995 (Pankhurst *et al.* 2003). Different from cytotoxic drug/magnetic carrier complex systems, this drug delivery does not require drug release and the therapeutic agent remains attached to the magnetic carrier throughout the whole treatment. So there is no need for tumor cells to take up the agent. The effectiveness of the radionuclide occurs when it is targeted to a region of interest and the surrounding tumor tissues are affected by radiation. To test this delivery system *in vitro* and *in vivo*, a β-emitter (Y-90) was coupled to a magnetic carrier, and radiation concentrated to the desired tumor area. From the tests, the magnetic radionuclide delivery exhibited a significantly higher radioactivity at the tumor area than using the same complex without applying a magnetic field. Following this study, Häfeli and his team demonstrated the use of both yttrium-90 and rhenium-188 bounded magnetic carriers for cancer treatment in both animal and cell culture studies (Häfeli *et al.* 2001).

Magnetic nanogels labeled with radiation isotope can be used as targeted radiopharmaceutical carriers for cancer treatment. To prepare magnetic nanogels attached with amino groups, superparamagnetic Fe_3O_4 nanoparticles were added to an emulsion-free aqueous system and underwent further Hoffmann elimination. This photochemical method produced poly(acrylamide-vinyl amine) magnetic nanogels with a narrow size distribution (25–180 nm in center diameter). The as-prepared magnetic nanogels with core–shell structure exhibit high dispersibility and stability, which offer potential radiopharmaceutical carriers for cancer therapy. In radiotherapy of cancers, one of the most effective radioisotopes is L-histidine labeled with Re. During the loading process, magnetic nanogels with average diameter of 78 nm and a polydispersity index of 0.217 were linked with Re labeled L-histidine in the presence of glutaraldehyde in aqueous solution at room temperature (Sun *et al.* 2005). It was found that, under optimized reaction conditions, the Re labeling efficiency of histidine-immobilized magnetic nanogels was around 97 %. When the immobilization process occurred at 37 °C over 2 days in calf serum, over 83 % of radioactivity was retained. For comparison, magnetic nanogel labeling in the absence of histidine under the same conditions showed only 21 % of radioactivity. Therefore this type of radiation isotope-labeled magnetic nanogels with amino groups could be an ideal targeted radiopharmaceutical carrier against cancers.

Extensive investigation on magnetic drug delivery has prompted investigations on magnetic carriers for gene therapy. It is likely that magnetic carriers' surfaces are coated with a viral vector with therapeutic genes and transported to the target area by applying a magnetic field. Then the virus contacts with the tissue while gene transfection and expression start. Extracellular barriers cause short contact time of complexes and hinder the uptake of the complex. Gersting and his team evaluated the feasibility of magnetofection, a technique based on the principle of magnetic drug delivery (2004). A comparison was made of this technique with conventional nonviral gene transfer methods such as lipofection and polyfection. Their gene transfection was performed on permanent (16HBE14o-) and primary airway epithelial cells (porcine and human), and native porcine airway epithelium. Luciferase reporter gene expression, fluorescence and electron microscopy were used to examine the transfection efficiency and the morphology change, respectively. It was shown that the magnetofections sedimented and enriched at the cell surface quickly after applying a magnetic field. For both permanent and primary airway epithelial cells, their transfection efficiencies were enhanced. Based on the experimental results, magnetofection offers a potential method for gene transfection, and has the potential to be used for *in vivo* applications due to the very fast speed of the process.

As demonstrated in the examples discussed above, there are several major advantages of using magnetic nanoparticles in drug delivery. First, drug bound magnetic nanoparticles can penetrate through small capillaries and be taken up by cells. Secondly, the movement of drug loaded magnetic nanoparticles can be readily controlled by an external magnetic field. Hence, accumulation of efficient drugs at the target sites can be achieved. Thirdly, biocompatible magnetic core–shell nanoparticles allow sustained drug release over a period of time. Fourthly, after being injected into bodies, the drug uptake process can be visualized by MRI (Arruebo *et al.* 2007). Finally, unlike searching for new molecules, magnetic drug delivery involves low-cost research, which reduces the cost of drug loaded product. Despite many merits, there are some problems associated with magnetic drug delivery. For instance, magnetic carriers themselves are not biodegradable and may cause some toxic response over a long period. Additionally, the accumulation of the magnetic carriers may cause a tough embolization of the blood vessels in the treated area. Further, when applying this technique to human bodies, a magnetic field with a safe strength may not be effective due to the large distance between the target site and the magnet. Pre-clinical and experimental investigations have been carried out to overcome these limitations and improve the efficiency of magnetic drug delivery at a safe level.

4.3.2 Hyperthermia Treatment

Another major use of magnetic nanoparticles in therapeutic treatment is hyperthermia treatment for cancers. Gilchrist *et al.* did the experimental investigations for the first time when they heated various tissue samples with γ-Fe_2O_3 of 20–100 nm in diameter by a 1.2 MHz magnetic field (Gilchrist *et al.* 1957). Since then, studies have shown the feasibility of using the hyperthermic effect generated from magnetic nanoparticles by applying a high-frequency AC magnetic field as an alternate therapeutic approach for cancer treatment. Briefly speaking, the hyperthermic effect is generated from the relaxation of magnetic energy of the magnetic nanoparticles which is able to destroy tumor cells effectively (Levy *et al.* 2002).

Hyperthermia is a common cancer therapy in which certain organs or tissues are heated preferentially to temperatures between 41 °C and 46 °C. Artificially induced hyperthermia has been designed to heat malignant cells without destroying the surrounding healthy tissue. When heated to a higher temperature ($\sim$56 °C), coagulation or carbonization may occur. This 'thermo-ablation' induces a completely different biological response and hence is not considered as hyperthermia. A classical hyperthermia not only causes almost reversible damage to cells and tissues, but also enhances radiation injury of tumor cells (Jordan *et al.* 1999). For modern clinical hyperthermia trials, moderate temperatures (42–43 °C) are normally selected to optimize the thermal homogeneity in the target area. It is true that the heating effect will change the dose-dependent behavior of treated cells. However, the exact mechanism of thermal dose-response in hyperthermia is still unknown. There are great difficulties in identifying the individual cell as the target for hyperthermia. Instead, hyperthermia affects most biomolecules including proteins and receptor molecules. On the other hand, DNA damage by irradiation has been well understood due to the highly specific interaction.

As far as the underlying physics of the heating effect in hyperthermia is concerned, magnetic heating via magnetic nanoparticles essentially is determined by their sizes and magnetic properties (Mornet *et al.* 2004). Magnetic nanoparticles can be divided into two major categories: multi-domain and single-domain nanoparticles, which possess different heating effects. Multi-domain nanoparticles usually have larger dimensions and contain several sub-domains with definite magnetization direction for each. When they are exposed to a magnetic field, a phenomenon called 'domain wall displacements' occurs. This is featured by growth of the domain with magnetization direction along the magnetic field axis and shrinkage of the other. Figure 4.24a depicts this irreversible phenomenon. It can be seen that the magnetization curves for increasing and decreasing magnetic field do not coincide, and the area within the hysteresis loop represents the heating energy, named 'hysteresis loss', due to the AC magnetic field. For single-domain nanoparticles, since there is no domain wall, no hysteresis loss occurs leading to no heating. When exposed to an external AC magnetic field, rotation of magnetic moments from superparamagnetic

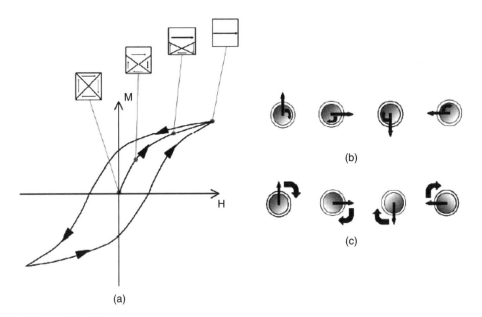

Figure 4.24   (a) Typical hysteresis loop of a multi-domain magnetic nanomaterial; (b) schematic of Neel rotation of magnetization in a static magnetic nanoparticle (the nanoparticle does not rotate); (c) Brown rotation of a magnetic nanoparticle (the nanoparticle rotates as a whole). (Mornet *et al.* 2004)

nanoparticles is assisted by the supplied energy which overcomes the energy barrier. Then these nanoparticles undergo Néel relaxation in which their moments relax to their equilibrium orientation. Simultaneously, heat is generated during this relaxation by thermal dissipation. Figure 4.24b illustrates the Néel relaxation. The Néel relaxation time t_N is related to the temperature, and can be described as:

$$t_N = t_0 e^{\frac{KV}{kT}} \tag{4.7}$$

where $t_0 \approx 10^{-9}$ s, T is the temperature and k is the Boltzmann constant. The frequency v_N for maximal heating via Néel relaxation corresponds to the frequency of the maximum imaginary component of the complex magnetic susceptibility $\chi''(v)$, and it can be obtained from the equation $2\pi v_N t_N = 1$. For both multi- and single-domain nanoparticles, rotational Brownian motion in a carrier also generates heat. This rotation is caused by the torque exerted on the magnetic moment by the AC magnetic field, as shown in Figure 4.24c. The Brown relaxation time t_B is described as:

$$t_B = \frac{3\eta v_B}{kT} \tag{4.8}$$

where η is the viscosity of the carrier. v_B is the frequency for maximal heating via Brown rotation, corresponding to the hydrodynamic volume of the particle, and it is given by the equation $2\pi v_B t_B = 1$.

As discussed above, ferrimagnetic or ferromagnetic nanoparticles (multi-domain and single-domain) generate magnetically induced heating under a AC magnetic field. To quantitate this heating, a formula is used:

$$P_{FM} = \mu_0 f \int H \, dm \tag{4.9}$$

which indicates the amount of heat generated per unit volume is related to the product of the frequency multiplied by the area of the hysteresis loop. In this case, eddy currents and resonance effects are ignored. For magnetic nanoparticles, P_{FM} can be determined from quasi-static measurements of the hysteresis loop by using a vibrating sample magnetometer (VSM) or superconducting quantum interference device (SQUID). Usually, specific absorption rate (SAR) in units of Wg^{-1} is used to quote the heat generation from magnetic nanoparticles. The heat generated per unit volume can be obtained by multiplying the SAR value by the density of the nanoparticles. It has been well documented that the orientation and magnetized domains of magnetic nanoparticles are dependent on their intrinsic features (elemental composition, crystallinity, magneto anisotropy, shape, dimension, etc.) and microstructural features (impurities, grain boundaries, vacancies, etc.). Generally, strongly anisotropic magnets such as Nd-Fe-B or Sm-Co exhibit substantial hysteresis heating. Since it is not possible to use fully saturated loops due to the constraints on the amplitude of H, maximum P_{FM} can only be obtained with a rectangular hysteresis loop. However, this is almost impossible to achieve *in vivo* because only an ensemble of uniaxial particles with perfect alignment with H could give the rectangular loop. For this reason, it is reasonable to estimate around 25 % of the maximum value for randomly aligned nanoparticles.

In magnetic hyperthermia treatment, after heat conducts into the area with diseased tissues, the surrounding temperature can be maintained above the therapeutic threshold of 42 °C for about half an hour to destroy the cancer. It is of great importance for hyperthermia to minimize the heat effect on healthy cells. Assisted by magnetic nanoparticles,

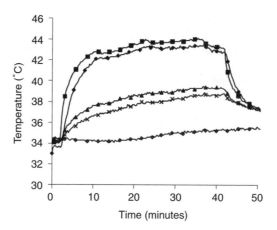

Figure 4.25   Animal trial data of hyperthermia treatments on rabbits. Preferential heating of tumor cells using intra-vascularly infused ferromagnetic microparticles was confirmed. (■) tumor edge, (♦) tumor centre, (▲) normal liver 1–2 cm from tumor, (×) alternative lobe, and (◊) core body temperature. Pankhurst, Q.A., Connolly, J., Jones, S.K. and Dobson, J. (2003). Applications of magnetic nanoparticles in biomedicine, *Journal of Physics D: Applied Physics*, **36**, R167–R181. Reproduced with permission of the Institute of Physics

it is possible to heat the specific area while unacceptable coincidental heating of healthy tissue is avoided. Figure 4.25 shows a preferential heating of tumor cells in hyperthermia treatments in rabbits using intra-vascularly infused ferromagnetic microspheres (Pankhurst *et al.* 2003). Likewise, ferromagnetic nanoparticles are able to ensure a similar localized heating as well.

Although the hyperthermia treatment for cancer has been demonstrated with therapeutic efficacy in animal models, however, there have been no report of successful hyperthermia treatment for human patients. The major reasons are the necessities of an adequate amount of magnetic nanoparticles and sufficiently high magnetic field which are not safe for human treatments. To date, laboratory research efforts on hyperthermia treatment for animals have all used magnetic field conditions which are not clinically acceptable. In most instances, hyperthermia treatments with a reduced amount of magnetic nanoparticles and reduced field strength or frequency cannot be effective due to the reduction of heat generated. Simulations suggest a sufficient level with heat deposition rate of $100\,\mathrm{m\,W\,cm^{-3}}$ to destroy cancer cells effectively in most circumstances. The practical frequency and strength of the external AC magnetic field are 0.05–$1.2\,\mathrm{MHz}$ and 0–$15\,\mathrm{k\,Am^{-1}}$, respectively. On the other hand, sufficient magnetic materials are needed to enrich around the cancer tissues to generate enough heat for hyperthermia treatment. Direct injection of ferrofluid into the tumor tissues is able to introduce a large amount of magnetic materials for heat generation. Antibody targeting and intravascular administration offer better preference heating, but the problem here is the small quantity. It is estimated that about 5–$10\,\mathrm{mg}$ of magnetic material concentrated in each $\mathrm{cm^3}$ of tumor tissues is able to generate enough heat for tumor cell destruction in human bodies. Magnetite (Fe_3O_4) and maghemite ($\gamma\text{-}Fe_2O_3$) nanoparticles are two common types used in hyperthermia treatments owing to their appropriate magnetic properties and their excellent biocompatibilities. Several examples will be given here.

The heat loss of maghemite nanoparticles covalently coated with polyethylene glycol was studied (Hergt *et al.* 2004). These maghemite nanoparticles have an average diameter of 15.3 nm according to the TEM observation. The magnetic properties of aqueous ferrofluid consisting of maghemite nanoparticles were characterized by DC-magnetometry. Measurement of susceptibility spectra in a frequency range between 20 Hz to 1 MHz and calorimetric measurements of specific loss power (SLP) at 330 and 410 kHz for field amplitudes up to 11.7 kA/m were performed, showing an extremely high SLP value in the order of 600 W/g. To investigate the role of Brownian relaxation, maghemite nanoparticles were suspended in gel. The susceptibility spectra obtained from experimental measurements were in good agreement with the simulated results, which were obtained by superposing Neel and Brown loss processes. Hergt's work demonstrated the very high specific heating power associated with maghemite nanoparticles in magnetic hyperthermia treatment.

Hilger and his team demonstrated the use of biocompatible magnetic nanoparticles in hyperthermia for breast cancer treatment (2005). Breast cancer is a very common disease for women and its treatment is mostly associated with medicine. At present, breast cancers can be diagnosed at an early stage due to the implementation of new techniques. Based on the diagnostic results, appropriate treatments will be performed. Generally, minimally invasive treatments, without any major deformation of the organ, are preferred for patients' emotional and physical welfare. The techniques developed include chemoembolisation, interstitial chemotherapy, cytotoxic agents, ultrasound treatments, intratumoral injection, etc. Magnetic nanoparticles with their capabilities of localized accumulation and heat generation were proposed for hyperthermia breast cancer treatment. In this treatment, there are two different situations: the treatment of *in situ* tumors and multi-focal tumors.

At an early stage, breast cancer cells remain in their original area and have not grown into the surrounding tissues. These *in situ* tumors are still small and confined within the borders of a duct or lobule. At such a stage, magnetically induced minimally invasive treatment may produce an ideal effect on the tumors. Thus, as the first implementation of the treatment, magnetic materials can be directly applied to the tumor tissues via radiological stereotactic methods for tumor puncture. Hilger *et al.* conducted systematic research on monitoring the localization of the magnetic materials, determining the temperature regime and experimental testing in mice and rats. The magnetic materials employed in their experiments were a mixture of core–shell nanoparticles consisting of magnetite and maghemite cores 10–20 nm in diameter and a dextran coating layer. Assisted by the dextran shell, antibodies and other biological entities can be coupled to the magnetic nanoparticles readily. Preference heating at the area of interest is desirable. Ideally, the nanoparticle dose which will be administrated into the bodies should be as low as possible. Applying an external magnetic field with proper frequency and amplitude is critical for specific heating. In the experiments, a field with amplitude of 10 kA/m and frequency of 410 kHz, which have been proved safe for animals, was used. Figure 4.26a and 4.26b are two radiographs showing the typical macroscopic observations of a tumor bearing mice before and after magnetic hyperthermia treatment, respectively. Compared with the tumor before treatment, a collapse of the subcutaneously implanted tumor resulting from the tumor cell destruction is obvious. This tumor destruction was definitely induced by the magnetic hyperthermia effect on subcutaneously implanted tumors in mice. From the radiographs, signs for the induction of coagulation necrosis including chromatin migration along the nuclear envelope and nuclear pyknosis and DNA damage can be differentiated, which provide additional evidence for the potential use of magnetic hyperthermia in breast cancer treatment. According to the long-term effect resulting from heating, it is conceived that the occurrence of hypertrophic granulation tissue is followed by healing through the formation of keloid.

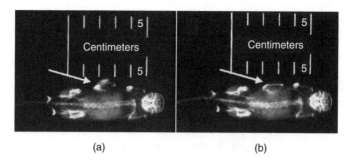

(a)   (b)

Figure 4.26 Radiographs of a tumor bearing mouse by typical macroscopic observation; (a) before magnetic tumor heating and (b) after magnetic tumor heating. Arrow indicates the tumor area. (Hilger *et al.* 2005)

Yan *et al.* demonstrated the use of Fe_2O_3 nanoparticles combined with magnetic fluid for hyperthermia treatment on human hepatocarcinoma SMMC-7721 cells *in vitro* and xenograft liver cancer in nude mice (2005). Their experiments verified the significantly inhibitory effect of magnetic ferrofluid in weight and volume on xenograft liver cancer. After infiltrating magnetic ferrofluid into the target tissues, a time-varied magnetic field was applied and its energy was transformed to heat energy by the magnetic nanoparticles resulting in a temperature rise to 42–45 °C. This generated heat is able to kill malignant tumor cells without injuring the normal cells nearby. The growth and apoptosis of SMMC-7721 cells treated with the ferrofluids containing Fe_2O_3 nanoparticles at various concentrations (2, 4, 6 and 8 mg/ml) were examined by MTT, flow cytometry (FCM) and transmission electron microscopy (TEM) after 30–60 minute treatments. It was observed that Fe_2O_3 nanoparticles-based ferrofluid could significantly inhibit the proliferation and increase the ratio of apoptosis of SMMC-7721 cells. These dose-dependent inhibitions were 26.5 %, 33.53 %, 54.4 %, 81.2 %, and 30.26 %, 38.65 %, 50.28 %, 69.33 %, for inhibitory rate and apoptosis rate, respectively. It was also observed from animal experiments that the tumors became smaller and smaller as the dosage of magnetic ferrofluid increased, as shown in Figure 4.27. The weight and volume inhibitory ratios were 42.10 %, 66.34 %, 78.5 %, 91.46 %, and 58.77 %, 80.44 %, 93.40 %, 98.30 %, respectively. In a comparison of the control and experimental groups, each group exhibited significant difference. According to histological examination, many brown uniform spots are located at the stroma in the margin of the tumors, which are identified as iron oxide nanoparticles.

4.3.3   Eye Surgery

Magnetic ferrofluids consisting of superparamagnetic nanoparticles have potential applications in eye surgery (Dailey *et al.* 1999). Concerning the structure of the eye, its anterior segment contains the aqueous humor and is bound by the cornea and lens-iris diaphragm, while its posterior segment includes the vitreous (gel/fluid), retina (neurosensory tissue), and choroid (heavily vascular) behind the lens-iris diaphragm. A main reason for vision loss is the detachment of the retina from the choroid, which serves to support retinal photoreceptors. In the normal way, the retina and choroid are bound with each other, and the subretinal space is maintained dry by a suction pump. When separation occurs, the retina dies and consequently leads to vision loss. The detaching of the retina from the underlying choroids is a normal part of aging. This process starts from liquefaction, collapse and separation of the vitreous gel from the retina, which may result in the formation of a tear at the site of vitreoretinal adhesion. It is the pathway provided by the retinal

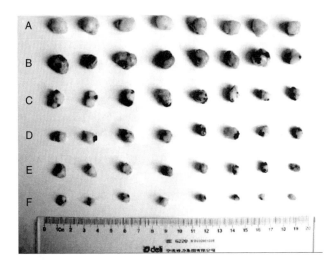

Figure 4.27 Digital picture showing the morphological changes of tumors from tumor-bearing nude mice treated by different methods. (A) Control experiment. The tumor was large with an average volume of $290.94 \pm 114.97\,mm^3$, and weight of $0.7070 \pm 0.0475\,g$; (B) nonhyperthermia experiment. Compared with the control experiment, the volume and weight of the tumors did not show significant difference; (C), (D), (E), and (F) hyperthermia experiments with Fe_2O_3 nanoparticles concentrations of 2, 4, 6 and 8 g/l, respectively. The tumors are smaller and smaller from C to E. Significant differences in both volume and weight of the tumors are observed, when compared with the control experiment. Many nanoparticles were observed in dark color under the envelope of tumors in nonhyperthermia and hyperthermia experiments. (Yan *et al.* 2005)

tear for vitreous fluid to pass through as well as underneath the retina, which results in the detaching of the retina.

In eye surgery for retina detaching, preventing additional fluid flow into the subretinal space thus allowing retinal reattachment is usually achieved by closing the holes in the retina. Currently, the available technique to treat retinal detachment is to use a scleral buckle, consisting of a silicone band, to compress the wall of the eye and hence close the holes in the retina. The strategy of the work of Dailey and his team was to develop an internal tamponade made from silicone fluid containing stabilized magnetic nanoparticles (4–10 nm in diameter). This tamponade was then held in place with an external magnetized scleral buckle. When applied on the eye, a ring of silicone coil in apposition to the retinal periphery was produced by encircling the magnetized scleral buckle and magnetic fluid. Figure 4.28 illustrates the internal tamponade using silicone magnetic fluids. One advantage of this technique is that no contact exists between the magnetic fluid and the lens, anterior chamber structures and macula because of the absence of magnetic fluid in the central vitreous cavity. Therefore the complication of the treatment modalities is reduced. Preparation of stable silicone magnetic fluids is key for this application. Nitrile containing triblock copolymers, (PDMS-b-PCPMS-b-PDMS)s, are effective stabilizers for this purpose. It can be prepared by living polymerization of D_3 using lithium silanolate ended poly (3-cyanopropyl) methylsiloxane as the macroinitiator. The triblock copolymers have microphase separated morphologies and are transparent liquids with high viscosities. The practical application of ferrofluid in eye surgery has not yet been realized. However, it is expected that magnetic nanoparticles could potentially be utilized to treat retinal detachment.

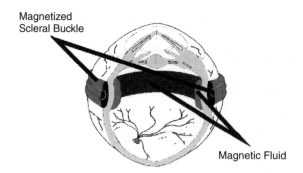

Magnetized
Scleral Buckle

Magnetic Fluid

Figure 4.28 Schematic diagram of the internal tamponade using silicone-based magnetic ferrofluids. (Dailey *et al.* 1999)

4.3.4 Antitumor Effects

We have discussed the applications of magnetic nanoparticles in cancer treatments. Their antitumor effects are either caused by the drugs carried by the magnetic nanoparticles or by hyperthermia effects induced by an external magnetic field. Interestingly, it was found that magnetic nanoparticles themselves exhibited antitumor effects.

Previous experiments on mammary adenocarcinoma revealed that injected biocompatible magnetic fluid determined the lysis of the tumor cells. An interesting phenomenon was observed regarding the significant difference of magnetic nanoparticles endocytosis by tumor cells and normal cells. Tumor cells are able to take up a large quantity of magnetic nanoparticles from an extracellular matrix while nanoparticles in normal cells are not present. In a typical process, a magnetic fluid is injected into the tumor area, where unspecific nanoparticles endocytosis by tumor cells initiates within from 20 minutes to 12 hours. After the tumor cells are overloaded with magnetic nanoparticles, the lysis process starts gradually. As to the reason for the antitumor effect of magnetic nanoparticles, supply depletion due to the endocytosis of magnetic nanoparticles could be attributed to the tumor lysis. In tumor tissue, food and oxygen supplies are needed for cell proliferation and vascular network alteration of tumor cells. When exposed to an excess amount of magnetic nanoparticles, the tumor cells may lose their specific endocytosis of necessary supplies and take up excessive foreign particles in competition to survive. Consequently, the proliferation of tumor cells is suppressed resulting in tumor lysis. Experiments confirmed that the nanoparticle endocytosis by tumor cells was not dependent upon the concentration of magnetic fluid used. Of course, it is desirable to achieve a suitable nanoparticle concentration in the tumor area for optimal antitumor effect. The magnetic properties of magnetic nanoparticles make it possible to use an external magnetic field to retain the magnetic nanoparticles in certain areas for efficient cell endocytosis and to avoid dispersion of magnetic nanoparticles.

In cancer treatment with magnetic nanoparticles only, the stability of ferrofluid is important for both target transportation and endocytosis of the tumor cells. It was observed that magnetic nanoparticles could be taken up by coated pit mediated endocytosis and phagocytosis. Therefore the first prerequisite step is to prepare a stable magnetic fluid. Magnetite nanoparticles Fe_3O_4 were selected to prepare the magnetic fluid mainly due to the fact that this colloidal magnetic iron oxide can undergo metabolism and can be excreted from the body readily (Sincai *et al.* 2005). The average size of magnetic nanoparticles ranged from 10 to 15 nm. An unsaturated fatty acid, laurel acid, was used as the stabilizer in magnetic fluid preparation. Compared with saturated fatty acids, unsaturated fatty acids are easier to

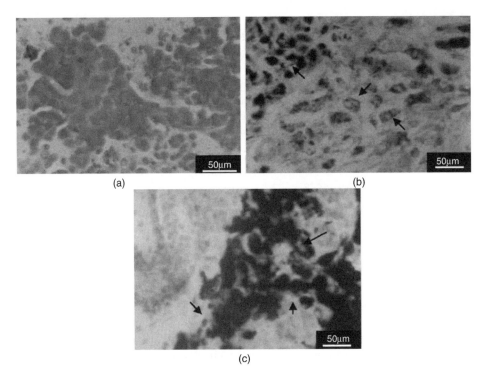

Figure 4.29 (a) Microscope image of smear from cat mammary tumor with tumor glandular cells after staining by Perls' method. This sample was taken before magnetic ferrofluid injection; (b) biopsy from cat mammary tumor at 1 hour after magnetic fluid injection. The arrows indicate tumor cells overloaded with magnetic nanoparticles which aggregated into solid clumps; (c) section through cat mammary tumor obtained two months after the magnetic fluid inoculation. The arrows indicate tumor cells overloaded with magnetic nanoparticles. (Sincai *et al.* 2005)

degrade in the body. The prepared magnetic fluid was biocompatible and had a saturation magnetization of about 80 Gs. Animal tests were performed on cats for treatment. Before treatment, fluid samples were taken from the tumors by needle puncture for cytohistological examination followed by adenocarcinoma diagnosis. For better microscopy imaging, the May Grunwald-Giemsa method was applied to stain the sample smears. Figure 4.29a is the optical image of a stained smear from a cat mammary tumor before magnetic fluid injection. Many large tumor-specific glandular cells are clearly shown. After magnetic fluid inoculation into cat mammary tumors directly, an external magnetic field of 0.1 T was applied close to the mammary tumor gland and maintained for 20–30 minutes. After one hour, a tumor sample was retrieved from one animal, and the sample smear was stained by Perls' method. As shown in Figure 4.29b, many magnetic nanoparticles were loaded with tumor cells.

Further analysis was performed after two months. Again, the sample obtained from the tumor was stained for microscope observation. It was found that some necrosis foci and black clusters of organic and ferrite residua existed after rejection in the tumor tissue of mammary glands. Figure 4.29c is a high magnification image showing solid clumps agglomerated from ferrite clusters. Microscope observations clearly revealed tumor cell destruction by magnetic fluid after two months. This experiment demonstrated that the overloading of magnetic nanoparticles induces the tumor cells' destruction. It seems that tumor cell lysis is determined by the magnetic properties of nanoparticles. When these small magnets enter into the cells, the intracellular metabolism can be affected by the change of plasmalemma's electric and chemical potentials. It was suggested that the existence of magnetic nanoparticles could stop the microtubule assembly and block the spindle division. As a result, the growth of the tumor cells was stopped and they were removed by macrophages after a lysis process. It is believed that the overloading of the magnetic nanoparticles is important for this antitumor effect. Tumor cells are able to take up up to ten times more magnetic nanoparticles than normal cells. Since the tumor cells are not able to get rid of the nanoparticles, they could disturb the metabolism of the cells.

4.4 Physiological Aspects

Since magnetic nanoparticles are always introduced into living bodies for both diagnostics and therapeutic treatments, it is of importance to consider the physiological aspects of magnetic nanoparticles/fluids for any biomedical application. First of all, magnetic nanoparticles/fluid should be safe and effective so that they can be used successfully in animals and human beings. Studies have been conducted to investigate the physiological effect of magnetic fluid in magnetically controlled drug delivery.

In magnetic drug delivery systems, ferrofluids consisting of magnetic nanoparticles serve as the drug carrier for delivery purposes. Investigation of the biological effects, bioavailability of the ferrofluids and their *in vivo* desorption time of the anticancer drug epirubicin is necessary to fully understand the underlying mechanism of action on the tissue at the microcirculatory level. Sometimes, magnetic drug delivery needs to be realized via a developed collateral circulation because of the possible tumor embolization. In addition to biological and technical details of ferrofluids, physiological parameters determine the efficacy of therapeutic treatments. The physiological parameters come from ferrofluid/drug complexes, magnetic fields and the targeted organisms. For ferrofluid/drug complexes, critical physiological parameters include particle size, surface characteristics of the particles, concentration of the fluid, volume of the fluid, binding strength of the drug/ferrofluid binding, desorption characteristics, accessibility to the organism and ferrofluid administration, etc. The physiological parameters relevant to the magnetic field are the geometry, strength and duration of the applied magnetic field. As regards patient organisms, important physiological parameters are size, weight, body surface, blood volume, cardiac output and systemic vascular resistance, circulation time, tumor volume and locations, vascular content of tumor and blood flow in tumor. For cancer treatment, it is desirable to introduce the minimum amount of magnetic fluid loaded with sufficient effective drug and apply the magnetic field at a safe strength and for the minimum duration. To achieve optimal treatment, physiological parameters need to be adjusted. For instance, adjusting of ferrofluid/drug and magnetic field characteristics is highly necessary according to the animal species to be treated and their sizes.

Specifically for *in vivo* drug delivery, precise quantification of the ferrofluid in the microcirculation of the target tumor tissues and drug desorption kinetics can be realized by microcirculatory techniques and fluorescent labeling. Magnetic nanoparticles not only are confined within microvessels of normal tissues, but could extravasate into the tumor interstitial space. Also, when labeled with fluorescent dyes, the drug desorption under the applied magnetic field can be visualized, which provides useful information for the distribution kinetics of the ferrofluid *in vivo* upon systemic intravenous applications. In this case, the reversible binding of drugs to the nanoparticles can be measured and determined.

In vivo experiments on animals have generated valuable information about the physiological data, both macro- and microcirculatory, for effective magnetic drug delivery.

Although these data from animals may arrive at approximations for complex new drug delivery forms in human beings, improvements are still required to make magnetic drug delivery a more effective therapy for human beings. It is apparent that further improvement can only be made through patient trials. From the physiological point of view, ferrofluid/drug complexes and their infusion scheme/route are more easily modified than magnetic field and other physiological variables. In particular, magnetic nanoparticle size, binding between particles and drugs, ferrofluid concentrations and volumes are primary parameters to be further optimized. Infusion routes for ferrofluids are arterial injection, intravenous injection and direct tumor injection. It seems the latter two are preferable infusion routes. Further, the amount of the injected ferrofluid and the duration time for magnetic field application need to be determined based on perfused tumor blood vessels for successful treatment. When treatments are switched from small to large animals and eventually to human beings, mathematical extrapolations are usually used to estimate the optimal physiological parameters for the first patient trials. But optimization of the system towards the best efficacy of the treatment is mandatory through further clinical studies. Only with optimized physiological parameters could the magnetic drug delivery be utilized to specifically treat tumors and infectious sites with drugs, cytokine-induced killer cells, biological response modifiers, gene products, etc. (Lübbe *et al.* 1999).

4.5 Toxic Effects

The health and safety issues related to synthetic nanostructured materials are concomitant with the developing nanotechnology. On one hand, nanotechnology aims to take advantage of nanomaterials and nanodevices with extraordinary performance and less consumption of energy. On the other hand, the development of nanotechnology has to face the nanomaterial-related safety problem with potential hazards and risks for humans and environments. To some extent, nanotechnology is only adaptable in areas where the potential advantages will exceed potential risks and where the potential risks can be well controlled.

For humans, exposure to nanoparticles (dimension less than 100 nm) has occurred throughout their evolutionary stages. However, such exposure has increased dramatically for the past several decades, mainly due to synthetic nanoparticles. Numerous nanomaterials can be prepared by modern chemical or physical methods. Most currently employed nanoparticles are made of transition metals, carbon, silicon and metal oxide, which may cause severe safety problems after inhalation, ingestion, skin uptake and injection by human bodies. It is well known that nanoparticles with large surface-to-volume ratio result in unfavorable biological responses. They can be absorbed via the lung when inhaled, and absorbed across the gastrointestinal tract when swallowed (Arruebo *et al.* 2007). However, it is not easy to generalize the toxicity of nanoparticles because their toxicity is determined by many material characteristics such as chemical composition, particle dimension, solubility, surface chemistry, shape and structure, crystallinity, as well as many physiological factors such as dose, method of administration, biodegradability, pharmacokinetics, biodistribution, etc.

Nanotoxicology is defined as safety evaluations of engineered nanostructured materials and nanomaterial-based nanodevices. Emerging concepts of nanotoxicology have been identified from the combination of biokinetic, epidemiologic and toxicologic studies with airborne ultrafine nanoparticles. It was found that nanoparticles were biologically active. In terms of toxicity, these biological activities may lead to a potential for inflammatory and pro-oxidant, but also antioxidant activities (Oberdörster *et al.* 2005). Studies have revealed some toxicological effects of nanoparticles for human beings. When contacting with skins, airborne nanoparticles are able to penetrate the skin and distribute in lymphatic channels. If nanoparticles are inhaled, they will deposit in all regions of the respiratory tract through diffusion. Because of their ultrafine dimension, they are able to be ingested by cells efficiently. Meanwhile, they could be transported across epithelial and endothelial cells into the blood and lymph circulation and reach sensitive target organs such as heart, spleen, etc. Neuronal experiments have shown nanoparticles' translocation along neurites to access the central nervous system and ganglia. It is necessary to point out that endocytosis and biokinetics are largely determined by the surface chemistry and surface modifications of nanoparticles. Hence toxicological effects of nanoparticles can be significantly reduced by coating biocompatible materials outside of the nanoparticles. Additionally, for further safety improvement, careful selections of appropriate doses/concentrations and the likelihood of increased effects in a compromised organism need to be considered. Information about detailed toxicological effects of nanomaterials is a must for further bioapplications of nanomaterials. Concerted effort is urgently required to develop methodologies and protocols concomitant with the implementation of toxicity studies (Warheit 2004). Thus an appropriate risk assessment, conducted by interdisciplinary teams in the fields of materials science, chemistry, toxicology, medicine, molecular biology and bioinformatics, needs to be arrived at for this nanotoxicology research.

For magnetic nanoparticles, studies on their toxic effects have been initiated owing to their versatile *in vivo* bioapplications. It was interesting to note that large magnetic nanoparticles exhibited higher cytotoxicity than smaller ones (Arruebo *et al.* 2007). This result was obtained by normalization of the surface area due to their different surface-to-volume ratios. In many cases, magnetic nanoparticles were injected into living bodies and allowed to stay for a period of time for either diagnostic or therapeutic purposes. Therefore, in addition to acute toxicity, long-term toxicity including degradation and stimulation of cells with subsequent release of inflammatory mediators should be seriously considered.

Apparently, studies on the toxicity of nanomaterials can start from *in vitro* testing, in which nanoparticles are incubated with tissues from cell cultures histologically. However, these *in vitro* toxicity tests have limited applications because of the different cytotoxicities of the same nanomaterial *in vitro* and *in vivo*. It is true that degradation products can be eliminated continuously from the application site *in vivo* (Neuberger *et al.* 2005). Thus the cytotoxicity is usually much lower *in vivo* than *in vitro*. There are two important criteria for nanoparticles in *in vivo* applications. First, the nanoparticle colloidal should be hydrophilic and its pH value should be close to 7.4. The second is the requirement that nanoparticles should be degraded and eliminated by the body system without residues in order to avoid nanoparticle accumulation in cell compartments (e.g. liposomes and tissues from the phagocytosis system).

Kim *et al.* conducted research to test the toxicity of core–shell magnetic nanoparticles (Kim *et al.* 2006). They administered silica covered magnetic nanoparticles intraperitoneally into mice and observed their distribution and toxic effects. After 4 weeks, it seems, nanoparticles penetrate the blood-brain barrier since nanoparticles were observed in the brain. Most importantly, apparent toxicity was not observed indicating that the existence of magnetic nanoparticles does not disturb the brain's function. Similar results were obtained with experiments done on diverse organs such as liver, lungs, kidneys, spleen, heart, testes and uterus. Their study demonstrated that magnetic nanoparticles with 50 nm in diameter did not have apparent toxicity, at least under the experimental conditions of that study.

Another feasible toxicity test *in vivo* was conducted to study the LD50 dosage, the mitotic index and effects on macrophages for radioactive magnetic nanoparticles with

80 nm in diameter (Neuberger *et al.* 2005). Nanoparticles were injected into mice and the studied parameters included the distribution of radioactivity within different tissues using a special indicator for Feisotopes, the relaxation time of liver and spleen in MRI, pathology of several organ systems by means of histology, the mutagenicity by means of a special test (Ames Salmonella Microsome Reverse Mutation Assay), the capability to treat a previously induced iron deficit anemia and chemscreen of blood and urine. After injection, nanoparticles were detected in both liver and spleen. Also, decrease of nanoparticle concentration was observed, indicating the incorporation of nanoparticles into the hemoglobin of erythrocytes. Even in the case of injections with large dosage (up to 3 mole Fe/kg), no acute or subacute toxic side effects were displayed.

Lacava *et al.* tested the biocompatibility of magnetic nanoparticles-based magnetoliposomes in potential thermal cancer therapy systems (Lacava *et al.* 2004). In their experiments, a bolus dose of magnetoliposomes were administered into adult female Swiss mice via endovenous treatment. Morphology and room temperature magnetic resonance studies were then conducted from 1 hour to 28 days after administration. The biocompatibility of magnetic nanoparticles was confirmed because of the absence of morphological alteration. According to the histological studies, magnetic nanoparticle clusters were detected without any concentration change in both liver and spleen tissues up to 28 days. This study also demonstrated that important pharmacokinetic parameters for liver and spleen, including the effective clearance and peak concentration, could be obtained from magnetic resonance studies.

To successfully take advantage of magnetic nanoparticles in potentially versatile bioapplications, both *in vitro* and *in vivo*, further studies on nanotoxicity are needed to understand the biological fate and potential toxicity of nanoparticles.

References

Alexiou, C., Schmid, R., Jurgons, R., Bergemann, Ch., Arnold, W. and Parak, F.G. (2002). Targeted tumor therapy with 'magnetic drug targeting': Therapeutic efficacy of ferrofluid bound mitoxantrone, *Lecture Notes in Physics*, **594**, 233–51.

Alexiou, C., Jurgons, R., Schmid, R., Hilpert, A., Bergemann, C., Parak, F. and Iro, H. (2005). *In Vitro* and *in vivo* investigations of targeted chemotherapy with magnetic nanoparticles, *Journal of Magnetism and Magnetic Materials*, **293**, 389–93.

Arruebo, M., Fernández-Pacheco, R., Ibarra, M.R. and Santamaría, J. (2007). Magnetic nanoparticles for drug delivery, *Nano Today*, **2**, 22–32.

Berry, C.C. and Curtis, A.S.G. (2003). Functionalisation of magnetic nanoparticles for applications in biomedicine, *Journal of Physics D: Applied Physics*, **36**, R198–R206.

Bertorelle, F., Wilhelm, C., Roger, J., Gazeau, F., Ménager, C. and Cabuil, V. (2006). Fluorescence-modified superparamagnetic nanoparticles: intracellular uptake and use in cellular imaging, *Langmuir*, **22**, 5385–91.

Bystrejewski, M., Huczko, A. and Lange, H. (2005). Arc plasma route to carbon-encapsulated magnetic nanoparticles for biomedical applications, *Sensors and Actuators B*, **109**, 81–5.

Cassidy, P.J. and Radda, G.K. (2005). Molecular imaging perspectives, *Journal of the Royal Society Interface*, 1–12.

Dailey, J.P., Phillips, J.P., Li, C. and Riffle, J.S. (1999). Synthesis of silicone magnetic fluid for use in eye surgery, *Journal of Magnetism and Magnetic Materials*, **194**, 140–8.

Enpuku, K., Inoue, K. and Soejima, K. (2005). Properties of magnetic nanoparticles for magnetic immunoassays utilizing a superconducting quantum interference device, *Japanese Journal of Applied Physics*, **44**, 149–55.

Gersting, S.W., Schillinger, U., Lausier, J., Nicklaus, P., Rudolph, C., Plank, C., Reinhardt, D. and Rosenecker, J. (2004). Gene delivery to respiratory epithelial cells by magnetofection, *Journal of Gene Medicine*, **6**, 913–22.

Gilchrist, R.K., Medal, R., Shorey, W.D., Hanselman, R.C., Parrott, J.C. and Taylor, C.B. (1957). Selective inductive heating of lymph nodes, *Annals of Surgery*, **146**, 596–606.

Häfeli, U., Pauer, G., Failing, S. and Tapolsky, G. (2001). Radiolabeling of magnetic particles with rhenium-188 for cancer therapy, *Journal of Magnetism and Magnetic Materials*, **225**, 73–8.

Hergt, R., Hiergeist, R., Hilger, I., Kaiser, W.A., Lapatnikov, Y., Margel, S. and Richter, U. (2004). Maghemite nanoparticles with very high AC-losses for application in RF-magnetic hyperthermia, *Journal of Magnetism and Magnetic Materials*, **270**, 345–57.

Hilger, I., Hergt, R. and Kaiser, W.A. (2005). Towards breast cancer treatment by magnetic heating, *Journal of Magnetism and Magnetic Materials*, **293**, 314–19.

Hirsch, R., Katz, E. and Willner, I. (2000). Magneto-switchable bioelectrocatalysis, *Journal of American Chemical Society*, **122**, 12053–4.

Hong, X., Li, J., Wang, M., Xu, J., Guo, W., Li, J., Bai, Y. and Li, T. (2004). Fabrication of magnetic luminescent nanocomposites by a layer-by-layer self-assembly approach, *Chemistry of Materials*, **16**, 4022–7.

Huh, Y., Jun, Y., Song, H., Kim, S., Choi, J., Lee, J., Yoon, S., Kim, K., Shin, J., Suh, J. and Cheon, J. (2005). *In vivo* magnetic resonance detection of cancer by using multifunctional magnetic nanocrystals, *Journal of the American Chemical Society*, **127**, 12387–91.

Jordan, A., Scholz, R., Wust, P., Fhling, H. and Felix, R. (1999). Magnetic fluid hyperthermia (MFH): cancer treatment with AC magnetic filed induced excitation of biocompatible superparamagnetic nanoparticles, *Journal of Magnetism and Magnetic Materials*, **201**, 413–19.

Jun, Y., Huh, Y., Choi, J., Lee, J., Song, H., Kim, S., Yoon, S., Kim, K., Shin, J., Suh, J. and Cheon, J. (2005). Nanoscale size effect of magnetic nanocrystals and their utilization for cancer diagnosis via magnetic resonance imaging, *Journal of the American Chemical Society*, **127**, 5732–3.

Katz, E., Sheeney-Haj-Ichia, L., Bückmann, A.F. and Willner, I. (2002a). Dual Biosensing by magneto-controlled bioelectrocatalysis, *Angewandte Chemie International Edition*, **41**, 1343–6.

Katz, E., Sheeney-Haj-Ichia, L. and Willner, I. (2002b). Magneto-switchable electrocatalytic and bioelectrocatalytic transformations, *Chemistry: A European Journal*, **8**, 4138–8.

Katz, E., Willner, I. and Wang, J. (2004). Electroanalytical and bioelectroanalytical systems based on metal and semiconductor nanoparticles, *Electroanalysis*, **16**, 19–44.

Kim, J.S., Yoon, T., Yu, K.N., Kim, B.G., Park, S.J., Kim, H.W., Lee, K.H., Park, S.B., Lee, J. and Cho, M.H. (2006). Toxicity and tissue distribution of magnetic nanoparticles in mice, *Toxicological sciences*, **89**, 338–47.

Kubo, T., Sugita, T., Shimose, S., Nitta, Y., Ikuta, Y. and Murakami, T. (2000). Targeted delivery of anticancer drugs with intravenously administered magnetic liposomes in osteosarcoma-bearing hamsters, *International Journal of Oncology*, **17**, 309–16.

Lacava, Z.G.M., Garcia, V.A.P., Lacava, L.M., Azevedo, R.B., Silva, O., Pelegrini, F., De Cuyper, M. and Morais, P.C. (2004). Biodistribution and biocompatibility investigation in magnetoliposome treated mice, *Spectroscopy*, **18**, 597–603.

Levy, L., Sahoo, Y., Kim, K.-S., Bergey, E.J. and Prasad, P.N. (2002). Nanochemistry: synthesis and characterization of multifunctional nanoclinics for biological applications, *Chemistry of Materials*, **14**, 3715–21.

Lübbe, A.S., Bergemann, C., Huhnt, W., Fricke, T., Riess, H., Brock, J.W. and Huhn, D. (1996). Preclinical experiences with magnetic drug targeting: tolerance and efficacy, *Cancer Research*, **56**, 4694–701.

Lübbe, A.S., Bergemann, C., Brock, J. and McClure, D.G. (1999). Physiological aspects in magnetic drug-targeting, *Journal of Magnetism and Magnetic Materials*, **194**, 149–55.

Lübbe, A.S., Alexiou, C. and Bergemann, C. (2001). Clinical applications of magnetic drug targeting, *Journal of Surgical Research*, **95**, 200–6.

Ludwig, F., Heim, E. and Schilling, M. (2004). Magnetorelaxometry of magnetic nanoparticles – a new method for the quantitative and specific analysis of biomolecules, 2004 *4th IEEE Conference on Nanotechnology*, 245–8.

Ludwig, F., Heim, E., Menzel, D. and Schilling, M. (2006). Investigation of superparamagnetic Fe_3O_4 nanoparticles by fluxgate magnetorelaxometry for use in magnetic relaxation immunoassays, *Journal of Applied Physics*, **99**, 08P106–1-3.

Mornet, S., Vasseur, S., Grasset, F. and Duguet, E. (2004). Magnetic nanoparticle design for medical diagnosis and therapy, *Journal of Materials Chemistry*, **14**, 2161–75.

Morishita, N., Nakagami, H., Morishita, R., Takeda, S., Mishima, F., Terazono, B., Nishijima, S., Kaneda, Y. and Tanaka, N. (2005). Magnetic nanoparticles with surface modification enhanced gene delivery of HVJ-E vector, *Biochemical and Biophysical Research Communications*, **334**, 1121–6.

Neuberger, T., Schöpf, B., Hofmann, H., Hofmann, M. and von Rechenberg, B. (2005). Superparamagnetic nanoparticles for biomedical applications: Possibilities and limitations of a new drug delivery system, *Journal of Magnetism and Magnetic Materials*, **293**, 483–96.

Oberdörster, G., Oberdörster, E. and Oberdörster, J. (2005). Nanotoxicology: An emerging discipline evolving from studies of ultrafine particles, *Environmental Health Perspectives*, **113**, 823–39.

Oca-Cossio, J., Mao, H., Khokhlova, N., Kennedy, C.McE., Kennedy, J.W., Stabler, C.L., Hao, E., Sambanis, A., Simpson, N.E. and Constantinidis, I. (2004). Magnetically labeled insulin-secreting cells, *Biochemical and Biophysical Research Communications*, **319**, 569–75.

Osaka, T., Matsunaga, T., Nakanishi, T., Arakaki, A., Niwa, D. and Iida, H. (2006). Synthesis of magnetic nanoparticles and their application to bioassays, *Analytical and Bioanalytical Chemistry*, **384**, 593–600.

Pankhurst, Q.A., Connolly, J., Jones, S.K. and Dobson, J. (2003). Applications of magnetic nanoparticles in biomedicine, *Journal of Physics D: Applied Physics*, **36**, R167–R181.

Perez, J.M., Simeone, F.J., Saeki, Y., Josephson, L. and Weissleder, R. (2003). Viral-induced self-assembly of magnetic nanoparticles allows the detection of viral particles in biological media, *Journal of the American Chemical Society*, **125**, 10192–3.

Perez, J.M., Josephson, L. and Weissleder, R. (2004). Use of magnetic nanoparticles as nanosensors to probe for molecular interactions, *Chembiochem*, **5**, 261–4.

Romanus, E., Hückel, M., Groβ, C., Prass, S., Weitschies, W., Bräuer, R. and Weber, P. (2002). Magnetic nanoparticle relaxation measurement as a novel tool for *in vivo* diagnostics, *Journal of Magnetism and Magnetic Materials*, **252**, 387–9.

Salata, O.V. (2004). Applications of nanoparticles in biology and medicine, *Journal of Nanobiotechnology*, **2**, 1–6.

Sincai, M., Ganga, D., Ganga, M., Argherie, D. and Bica, D. (2005). Antitumor effect of magnetite nanoparticles in cat mammary adenocarcinoma, *Journal of Magnetism and Magnetic Materials*, **293**, 438–41.

Sun, H., Yu, J., Gong, P., Xu, D., Zhang, C. and Yao, S. (2005). Novel core–shell magnetic nanogels synthesized in an emulsion-free aqueous system under UV irradiation for targeted radiopharmaceutical applications, *Journal of Magnetism and Magnetic Materials*, **294**, 273–80.

Tartaj, P., Morales, M. del P., Veintemillas-Verdaguer, S., González-Carreño, T. and Serna, C.J. (2003). The preparation of magnetic nanoparticles for applications in biomedicine, *Journal of Physics D: Applied Physics*, **36**, R182–R197.

Warheit, D.B. (2004). Nanoparticles: health impacts?, *Materials Today*, 32–5.

Widder, K.J., Morris, R.M., Poore, G.A., Howard, D.P. and Senyei, A.E. (1983). Selective targeting of magnetic albumin microspheres containing low-dose doxorubicin-total remission in Yoshida sarcoma-bearing rats, *European journal of cancer and clinical oncology*, **19**, 135–9.

Yan, S., Zhang, D., Gu, N., Zheng, J., Ding, A., Wang, Z., Xing, B., Ma, M. and Zhang, Y. (2005). Therapeutic effect of Fe_2O_3 nanoparticles combined with magnetic fluid hyperthermia on cultured liver cancer cells and xenograft liver cancers, *Journal of Nanoscience and Nanotechnology*, **5**, 1185–92.

Zhao, X., Tapec-Dytioco, R., Wang, K. and Tan, W. (2003). Collection of trace amount of DNA/mRNA molecules using genomagnetic nanocapturers, *Analytical Chemistry*, **75**, 3476–83.

Zhou, W., Gao, P., Shao, L., Caruntu, D., Yu, M., Chen, J. and O'Connor, C.J. (2005). Drug-loaded, magnetic, hollow silica nanocomposites for nanomedicine, *Nanomedicine: Nanotechnology, Biology, and Medicine*, **1**, 233–7.

5

Magnetosomes and their Biomedical Applications

5.1 Introduction

5.1.1 Magnetotactic Bacteria and Magnetosomes

Magnetotactic bacteria (MTB) have the ability to align in and navigate along geomagnetic field lines (Komeili 2007; Lang and Schuler 2006). This ability is mainly due to the existence of magnetosomes, which are intracellular magnetic crystals enclosed by membranes. MTB and magnetosomes are at the center of a variety of interdisciplinary efforts, and various applications of magnetotactic bacteria and magnetosomes have been developed in diverse disciplines from geobiology to biotechnology. Magnetosomes can also be used as a model system for the study of biomineralization and cell biology in bacteria.

The basic mechanism for the ability of magnetotactic bacteria to use geomagnetic fields for orientation and navigation can be analyzed using transmission electron microscopy (Komeili 2007). As shown in Figure 5.1, in a magnetotactic bacterium, there are one or more chains of magnetosomes, and these chains are fixed within the bacterium making it passively aligned in external magnetic fields. In most magnetotactic bacteria, magnetosomes are either magnetite (Fe_3O_4), or greigite (Fe_3S_4). However, it is reported that there is one type of marine bacteria that contains both magnetite magnetosomes and greigite magnetosomes (Bazylinski *et al.* 1995). Usually, magnetosomes have a permanent magnetic dipole in the order of 10^{-12} emu (ergs/G). In a natural environment, a magnetic bacterium is oriented as it swims by the torque exerted by the geomagnetic field on its magnetic dipole (De Waard *et al.* 2001).

5.1.1.1 Distribution of Magnetotactic Bacteria

Usually, magnetotactic bacteria can be found in chemically stratified water columns or sediments, and they exist mostly in or below the microaerobic redox transition zone, between the aerobic zone of upper waters or sediments and the anaerobic regions. Different kinds of magnetotactic bacteria have been isolated from sediments in diverse aquatic environments, such as marine, river, lake, pond, beach, estuary, drains and wet soil. The magnetotactic bacteria with magnetite usually exist in the transition zones between the oxygen-rich and the oxygen-starved water and sediment. Most magnetotactic bacteria

Nanomedicine: Design and Applications of Magnetic Nanomaterials, Nanosensors and Nanosystems
V. Varadan, L.F. Chen, J. Xie
© 2008 John Wiley & Sons, Ltd

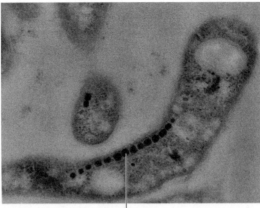

Magnetosome

Figure 5.1   Transmission electron micrograph of magnetotactic bacteria, *magnetospirillum magneticum* AMB-1. (Komeili 2007)

can only live in environments with very limited oxygen, and some types of magneto-tactic bacteria can only survive in environments that are completely anaerobic. Usually, magnetotactic bacteria with greigite exist in strictly anaerobic environments.

It is very interesting that magnetosome-like structures and magnetic minerals can also be found in eukaryotic organisms. Algae, fish, termites, pigeons, honeybees and even humans have been shown to have magnetic particles, and sometimes it appears that these magnetic particles participate in the direction-sensing behavior of the organism. As magnetosome-like chains of magnetite have survived in sediments for quite a long time, bacterial magnetite can be used as 'magnetofossils' to survey the history of life in ancient rocks and even to search for life signatures in extraterrestrial samples. Therefore, magnetotactic bacteria have the potential to be a relatively simple and genetically tractable model system for understanding fundamental evolutionary and mechanistic aspects of biomineralization (Komeili 2007).

5.1.1.2   Basic Structures of Magnetotactic Bacteria

Magnetotactic bacteria are motile prokaryotes, which can swim following the lines of the geomagnetic field, with their flagella. Different types of bacteria may have different arrangements of flagella, which may be polar, bipolar or in tufts. Moreover, the arrange-ment of magnetosomes inside the bacterial cell also shows considerable diversity. In most of the magnetotactic bacteria, the magnetosome chains with various lengths and numbers are aligned parallel to the axis of the bacteria, as this direction has the highest magnetic efficiency. But in some magnetotactic bacteria, dispersed aggregates or clusters of mag-netosomes can be found at one side of the bacterium, usually corresponding to the site of flagellar insertion (Frankel and Bazylinski 2004).

All magnetotactic bacteria contain magnetosomes enveloped by a membrane vesicle, adjacent to the cytoplasmic membrane. The magnetosome membrane, usually about 3 to 4 nm thick, attaches the magnetosome crystals to particular positions inside the cell, and it also provides biological control on the nucleation and growth of the magnetosome crystals. The magnetosome membrane in the genus *Magnetospirillum* is a lipid bilayer consisting of neutral lipids, free fatty acids, glycolipids, sulfolipids and phospholipids (Frankel and Bazylinski 2004). It is generally accepted that magnetosome vesicles are

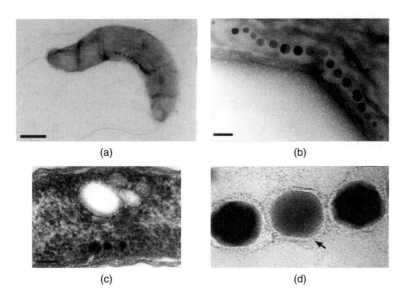

(a) (b)

(c) (d)

Figure 5.2 Transmission electron micrographs of *M. gryphiswaldense* cells showing the magneto-some chain and isolated magnetosomes. (a) Single *M. gryphiswaldense* cell with the magnetosome chain localized at mid-cell; (b) enlarged view of the magnetosome chain; (c) ultrathin section of a *M. gryphiswaldense* cell with magnetosomes; (d) isolated magnetosome particles with intact magnetosome membranes. The magnetosome membrane is indicated by arrows. The bar in (a) is 0.5 μm, and the bars in (b) and (c) are 0.1 μm. Lang, C. and Schuler, D.(2006). Biogenic nanoparticles: production, characterization, and application of bacterial magnetosomes, *Journal of Physics: Condensed Matter*, **18**, S2815–S2828. Reproduced by permission of the Institute of Physics

formed before the deposition of the mineral crystals, because empty and partially empty membrane vesicles can be found in iron-starved bacteria.

Generally speaking, one magnetotactic species exclusively produces one type of magnetosomes, either magnetite or greigite, and the only exception found is one marine organism that produces both magnetite and greigite magnetosomes. As shown in Figure 5.2, the magnetosomes in a magnetotactic bacterium are organized into chains along the axis of the bacterium. The magnetosomes in each species have a narrow size distribution and consistent morphologies, and the magnetosomes in chains have a consistent crystallographic orientation. The formation and arrangement of magnetosomes are realized intracellularly in magnetotactic bacteria, under the strict control of magnetotactic bacteria (Lang and Schuler 2006; Frankel and Bazylinski 2004).

Though magnetotactic bacteria exhibit great diversity, they possess important common features (Bazylinski and Frankel 2004). They are all Gram-negative, and all of them can move based on self-propulsion. They all contain magnetosomes, and a negative tactic and/or growth responses to atmospheric concentrations of oxygen.

5.1.2 Basic Properties of Magnetosomes

In a magnetotactic bacterium, magnetosomes are formed under strict biological and chemical controls (Bazylinski and Frankel 2004). In the mineralization procedure, the mineral composition of the magnetosomes is under strict chemical control. For example, some cultured magnetotactic bacteria synthesize magnetite instead of greigite, even in the presence of hydrogen sulphide in the growth medium. Furthermore, the magnetite crystals in magnetosomes have high chemical purity, and the impurity levels of other metal ions in the crystals are extremely low. In addition, no proteins can be found in magnetite magnetosome crystals. In the following, we discuss the morphologies, magnetic properties and crystalline structures of magnetosomes.

5.1.2.1 Morphologies and Magnetic Properties of Magnetosomes

Magnetosomes exhibit a wide range of morphologies. Figure 5.3 shows the typical morphologies and intra-cellular organizations of magnetosomes in different kinds of magnetotactic bacteria. The magnetic crystals may have various morphologies, and can be arranged in one, two or multiple chains; they may also be irregularly arranged. It should

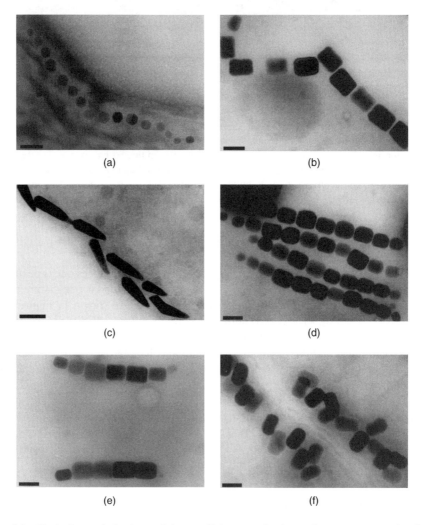

(a)

(b)

(c)

(d)

(e)

(f)

Figure 5.3 Typical morphologies and intra-cellular organizations of magnetosomes in different kinds of magnetotactic bacteria. The magnetic crystals may have the following shapes: cuboctahedral (a), elongated hexagonal prismatic (b, d, e, f) and bullet-shaped morphologies (c). The particles are arranged in one chain (a, b, c), two chains (e) or multiple chains (d), and they may be irregularly distributed (f). The bars in the pictures represent 100 nm. (Schuler and Frankel 1999)

be noted that though magnetite and greigite magnetosome crystals exhibit various particle morphologies, they are consistent in a single magnetotactic bacterial species.

The morphology, size and intracellular organization of magnetic crystals are under a species-specific biological control, and such control is genetically regulated by a specific set of genes within the genome of magnetotactic bacteria (Lang and Schuler 2006). The magnetosome membrane plays an important role in the control of crystal growth by providing spatial constraints for shaping of species-specific crystal morphologies. To realize biomineralization of magnetic crystals, a precise regulation of the redox potential, pH and the prevalence of a supersaturating iron concentration within the vesicle is required, and the magnetosome membrane plays important roles in the transport and accumulation of iron, the nucleation of magnetic crystals and the control of redox and pH value.

The size of magnetosomes in a magnetotactic bacterium is crucial for the magnetic properties of the bacterium. Typically magnetosome magnetite and greigite crystals have a length in the range of 30 to 130 nm. In this size range, both magnetite and greigite crystals have permanent, single-magnetic-domain (PSD) structures. They are uniformly magnetized, and have the maximum magnetic moment per unit volume. Magnetic crystals whose size is larger than the PSD size are non-uniformly magnetized due to their multi-magnetic domain structures, and this greatly decreases their magnetic dipole moments. However, magnetic crystals smaller than PSD size exhibit superparamagnetism, and thus have zero remanent magnetization. Therefore, the magnetosomes produced by magnetotactic bacteria have the optimum particle size so that they possess the maximum permanent magnetic dipole moment (Frankel and Bazylinski 2004). Usually the magnetic moment of a magnetite crystal is about three times of that of a greigite crystal with the same shape and size.

The arrangement of the PSD magnetosomes also plays an important role in determining the magnetic properties of a magnetotactic bacterium. For magnetosomes arranged in chains, the magnetic interactions between the magnetosomes make each magnetosome moment orient spontaneously along the chain axis and in parallel with others. So the whole magnetic moment of a magnetosome chain can be obtained by algebraically summing the magnetic moments of the individual magnetosomes in the chain. It should be noted that chain arrangements in magnetotactic bacteria are achieved because the magnetosomes are physically constrained by the magnetosome membranes in the chain configuration (Frankel and Bazylinski 2004). If the magnetosome crystals are free to float in the cytoplasm, they may tend to aggregate, and such an aggregation has a much smaller net dipole moment than a magnetosome chain with the same number of magnetosome crystals.

Finally, it should be noted that while most of the mature magnetosome crystals are in the PSD size range, the crystal sizes can be genetically modified. As shown in Figure 5.4, magnetic particles isolated from a mutant strain exhibit a narrower size distribution and smaller diameters than those from the wild-type strain. Due to their small sizes, the magnetic particles isolated from a mutant strain are usually superparamagnetic (Hoell *et al.* 2004).

5.1.2.2   Crystalline Structures of Magnetosomes

The magnetic properties of magnetosome crystals are related to their crystalline structures, and knowledge of the crystalline structures of magnetosomes is helpful for the study of the formation mechanism of magnetosomes and for the development of biomedical applications of magnetosomes. We discuss below the crystalline structures of magnetite and greigite crystals in magnetosomes.

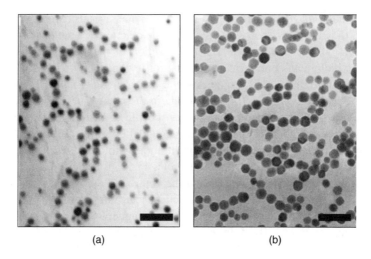

(a)   (b)

Figure 5.4 Electron micrographs of magnetosomes isolated from: (a) *M. gryphiswaldense* mutant strain (MSR1K) with reduced particle size, and (b) from the wild-type strain (MSR-1). The bars in the pictures represent 50 nm. Lang, C. and Schuler, D.(2006). Biogenic nanoparticles: production, characterization, and application of bacterial magnetosomes, *Journal of Physics: Condensed Matter*, **18**, S2815–S2828. Reproduced by permission of the Institute of Physics

(1) Magnetosome Magnetite Crystals

Magnetite forms from a precursor, an amorphous iron oxide phase. Generally speaking, the magnetite crystals in magnetotactic bacteria have high structural perfection. It is thought that, as shown in Figure 5.5, all the morphologies of magnetosome magnetite crystals can be obtained by combining the {111}, {110} and {100} forms with some distortions (Frankel and Bazylinski 2004). In the cuboctahedral crystal morphology, the symmetry of the face-centered cubic spinel structure is preserved, and all the symmetry-related crystal faces are developed equally. While in the elongated and prismatic morphologies, some of the symmetry-related faces are unequally developed, and this may be due to the anisotropic growth conditions, for example, an anisotropic ion flux into the magnetosome membrane.

(2) Magnetosome Greigite Crystals

Some marine, estuarine and salt-marsh magnetotactic bacteria contain iron sulfide magnetosomes, which mainly consist of greigite (Fe_3S_4). Greigite-producing magnetotactic bacteria mainly include a multicellular, magnetotactic prokaryote and various relatively large, rod-shaped bacteria, though none of them is currently available in pure culture (Frankel and Bazylinski 2004). The greigite crystals in the magnetosomes are developed from their nonmagnetic precursors, such as tetragonal FeS and sphalerite-type cubic FeS. Typical crystalline structures of magnetosome greigite crystals are shown in Figure 5.5. Similar to magnetosome magnetite crystals, magnetosome greigite crystals have species-specific morphologies.

5.1.3 Magnetotaxis and Magneto-aerotaxis

Magnetotaxis refers to the orientation and navigation of magnetotactic bacteria along magnetic field lines, while aerotaxis refers to the movement of magnetotactic bacteria

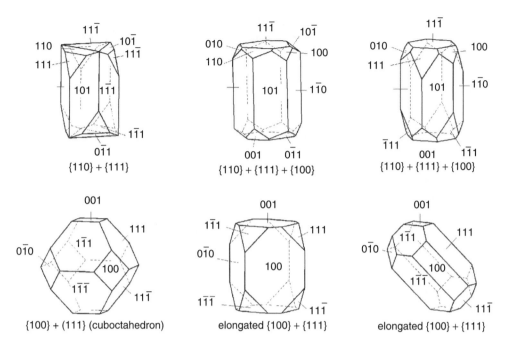

Figure 5.5 Morphologies of Fd3m magnetite and greigite crystals comprising the low index forms {100}, {111} and {110}. Anisotropic growth causes symmetry breaking in all but the cuboctahedron (lower left). (Frankel and Bazylinski 2004)

along or opposite to the gradient of oxygen concentrations. It has been demonstrated that in magnetotactic bacteria, magnetotaxis and aerotaxis work together, and this behavior is usually called magneto-aerotaxis.

Magnetotaxis of a magnetotactic bacterium is its passive orientation along the local magnetic field due to the torque exerted by the magnetic field on the magnetic dipole moment of the bacterium (Bazylinski and Frankel 2004). As the magnetosomes in magnetotactic bacteria are usually arranged in chains, and the chain of the magnetosome particles is fixed in the magnetotactic bacterium, the torque exerted on the magnetic dipoles by the geomagnetic field can orientate the whole magnetotactic bacterium. Usually a chain of 10 to 20 magnetosomes with dimension 50 nm would have sufficient magnetic moment for orientating the magnetotactic bacterium.

Similar to most of the free-swimming bacteria, magnetotactic bacteria are self-driven, and their movements are based on the rotation of their helical flagella. Magnetotactic bacteria are passively oriented because of their magnetosomes, and actively migrate along the geomagnetic field. It should be noted that the word magnetotaxis, which is used for describing the behavior of magnetotactic bacteria, is actually a misnomer (Bazylinski and Frankel 2004). The behavior of a magnetotactic cell is contrary to an actual tactic response. A magnetotactic cell does not swim up or down a magnetic field gradient, and it is not attracted or pulled to either geomagnetic pole. A live magnetotactic bacterium behaves like a small, self-propelled magnetic compass needle. A dead magnetotactic bacterium cannot move, though it aligns along the geomagnetic field lines.

It is more accurate to describe the tactic behavior of magnetotactic bacteria as magneto-aerotaxis, rather than magnetotaxis, because the alignment of magnetotactic bacteria in applied magnetic fields is passive and the sensing of oxygen gradients determines their swimming direction (Frankel *et al.* 1997). Though the magnetic dipole of a magnetotactic bacterium remains oriented along the local magnetic field, the migration direction along the magnetic field lines is determined by the sense of flagellar rotation, which in turn is controlled by aerotactic receptors. Therefore a magnetotactic bacterium is essentially a self-propelled magnetic dipole with an oxygen sensor. Experiments have quantitatively demonstrated that magnetotactic bacteria swim away from advancing oxygen gradients faster in the presence of a magnetic field than in its absence (Komeili 2007).

Magnetotactic bacteria exhibit two distinct types of responses, axial and polar, to applied magnetic fields (Komeili 2007). In axial magneto-aerotaxis, the bacteria use the magnetic field as an axis of swimming with no preference for either pole. Most magnetotactic bacteria are polar magneto-aerotactic bacteria, and they persistently swim toward only one pole of the magnet. Usually, north-seeking magnetotactic bacteria exist in the Northern Hemisphere, while south-seeking ones exist in the Southern Hemisphere. The swimming pattern of polar magnetotactic bacteria makes them move toward the bottom of their aquatic habitats where the oxygen levels are low. So magneto-aerotaxis simplifies the search for low-oxygen environments to a one-dimensional search rather than the three-dimensional search normally associated with other aerotactic systems. However, the existence of a distinct population of south-seeking bacteria in the Northern Hemisphere casts serious doubts on the above model for the function of magneto-aerotaxis (Simmons *et al.* 2006). It is yet to be determined whether these Northern Hemisphere-residing south seekers are an exception, or whether they are hinting at the true function of this remarkable behavior.

5.2   Magnetosome Formation

Biomineralization of magnetosomes is a process including the accumulation of iron, the deposition of the magnetic crystal within a specific compartment and the assembly, alignment and intracellular organization of particle chains. Biomineralization provides a way of producing magnetite crystals with very narrow size distributions, and it does not require the extreme conditions of temperature, pH value and pressure, which are commonly required in the industrial fabrication of these materials. The formation of magnetosomes is an interesting and quite complicated process. Several mechanisms have been proposed for the formation of magnetosomes and further improvements are needed for these mechanisms. In the following discussion, we consider several generally accepted mechanisms for magnetosome formation.

5.2.1   Biochemistry and Gene Expression

Knowledge about the biochemical and genetic controls on magnetite formation is important for understanding the process of how magnetosomes are produced and organized in magnetotactic bacteria. The convergence of genetic, proteomic and genomic approaches may provide important information about magnetosome formation.

As indicated above, magnetosome membranes (mam) play an important role in the formation of magnetosome minerals, and much effort has been made on the research of mam proteins. Mam proteins exist in magnetosome membranes but could not be found in the soluble fraction, the cytoplasmic or outer membranes. The approaches for the study of mam proteins generally fall into two categories (Frankel and Bazylinski 2004). One is N-terminal amino acid sequencing of the mam proteins, followed by reverse genetics to obtain the gene sequences for these proteins; and the other is performing biochemical protein comparison of mutants that do not produce magnetosomes with wild-type strains, then again using reverse genetics to determine gene sequences.

Mam genes may be involved in magnetite biomineralization. Grunberg *et al.* (2001) cloned and sequenced some of the mam genes in *M. gryphiswaldense*. The proteins resulting from these gene sequences exhibited the following homologies: MamA to tetratricopeptide repeat proteins; MamB to cation diffusion facilitators and MamE to HtrA-like serine proteases. However, the gene sequences of MamC and MamD showed no homology to existing proteins (Frankel and Bazylinski 2004).

Three other genes encoding mam specific proteins, mms6, mms16 and mpsA, can be obtained from *Magnetospirillum* strain AMB-1, and *M. magnetotacticum*. The mms6 gene is the most abundant of the three genes, and it is bound to magnetite and may function in regulation of crystal growth. The mpsA gene exhibits homology to an acyl-CoA transferase, while the mms16 gene shows GTPase activity and is possibly involved in mam vesicle formation by invagination and budding from the cytoplasmic membrane (Frankel and Bazylinski 2004).

5.2.2 Formation Procedure

The formation of bacterial magnetosome is an intricate procedure involving several steps, including the formation of the magnetosome vesicle, the uptake of iron by the cell, the transport of iron into the magnetosome vesicle and the biomineralization of magnetite (or greigite) inside the magnetosome vesicle. As shown in Figure 5.6, in the procedure of magnetite formation, Fe ions are transported into the cell and deposited within the magnetosome membrane vesicles to form a saturated Fe ion solution. The redox conditions within the magnetosome membrane vesicles are controlled so that the ratio between two kinds of Fe ions [Fe(III)]/[Fe(II)] is about 2, and in this condition, magnetite is the most stable Fe-oxide phase. The magnetosome membranes also provide sites for the nucleation and growth of magnetite crystals, and the crystal morphology is affected by the interactions between the magnetosome membranes and the faces of the growing crystal (Frankel and Bazylinski 2004).

We discuss below the important steps in the formation of magnetite magnetosomes, and greigite magnetosomes are formed in a similar way. It should be noted that some questions about magnetosome formation are yet to be answered. For example, though it is generally accepted that the steps of uptake, transport and mineralization are sequentially ordered, it is not clear whether the uptake of iron occurs before or after the formation of vesicle, or whether theses two steps happen concurrently (Bazylinski and Frankel 2004).

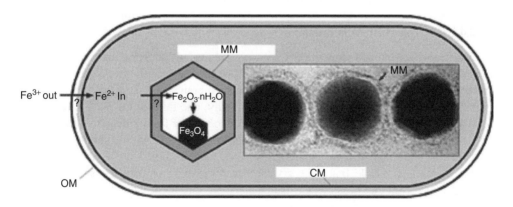

Figure 5.6   Model for magnetosome mineral formation in magnetotactic bacteria. CM, cytoplasmic membrane; MM, magnetosome membrane; OM, outer membrane. (Frankel and Bazylinski 2004)

5.2.2.1   Iron Uptake in Magnetotactic Bacteria

Magnetotactic bacteria contain up to about three per cent of iron by dry weight, which is several orders higher than non-magnetotactic species, but they do not have unique iron-uptake systems (Bazylinski and Frankel 2004). Fe(II) is soluble (up to 100 mM at neutral pH), and bacteria can take up Fe(II) by nonspecific mechanisms. But Fe(III) is insoluble. To take up Fe(III), microorganisms usually produce siderophores, which are specific ligands with low-molecular weight (<1 kDa), and can bind and solubilize Fe(III) for uptake.

5.2.2.2   Magnetosome Vesicle Formation

In several *Magnetospirillum* species, magnetosome membranes consist of lipid bilayers about 3–4 nm thick. As both the magnetosome membranes and the cytoplasmic membranes comprise phospholipids, fatty acids and some proteins, and for nearly all the magnetotactic bacteria, the magnetosomes are anchored to the cytoplasmic membrane, it is usually assumed that the magnetosome membrane vesicles originate from the cytoplasmic membranes.

As to the formation of magnetosome vesicles, it is not quite certain whether the magnetosome vesicles are formed before the magnetite nucleation, or whether the magnetite nucleation occurs in the periplasm and the growing magnetite crystals are invaginated by the cytoplasmic membrane (Bazylinski and Frankel 2004). In iron-starved cells of *M. magnetotacticum* and strain MV-1, empty and partially filled magnetosome vesicles can be found, suggesting that for these bacteria, the vesicle is produced before magnetite nucleation and precipitation.

5.2.2.3   Iron Transport into the Magnetosome Membrane Vesicle

Whenever the magnetosome membrane vesicles are formed, it is necessary to transport additional iron into the vesicles for the growth of the crystals. Though it is not quite certain, for most magnetotactic bacteria, which type of redox is transported into the magnetosome vesicle, it is generally accepted that Fe(II) is transported into magnetosome vesicles of *M. magneticum* strain AMB-1. Moreover, it is found that the transport of iron into magnetosome vesicles may be assisted by two types of proteins, MamB and MamM, which are abundant in the magnetosome membrane.

5.2.2.4   Controlled Biomineralization

The biomineralization of magnetite involves a complicated oxidation and reduction procedure. Generally speaking, the cell takes up Fe(III), and reduces it to Fe(II), which is subsequently transported into the magnetosome membrane vesicle. After that, Fe(II) is reoxidized to form hydrous Fe(III) oxides. Eventually, one third of the Fe(III) ions in the hydrous oxides are reduced and magnetite is produced after further dehydration. Though the exact mechanism is unclear, usually the size and shape of the magnetite phase are controlled by the magnetosome membrane vesicle. One possible way is that the magnetosome membrane vesicle limits the crystal size by placing physical constraints on the growing crystal. Another possible way is that some proteins are asymmetrically distributed in the magnetosome membrane, and thus in certain directions the growth of crystals is facilitated, while in other directions the growth of crystals is retarded (Bazylinski and Frankel 2004).

5.2.3 Cell Biology of Magnetosomes (Komeili 2007)

Research on bacterial cell biology can provide broader insights into the evolution of eukaryotic cell structure. Membranous organelles, a hallmark for eukaryotic cells, have been studied for various types of bacteria, and magnetosomes are among the most studied bacterial organelles. Here we discuss some conclusions from the research on the molecular mechanisms of magnetosome formation (Komeili 2007).

5.2.3.1 Magnetosome Membrane

Based on electron microscopic studies and lipid composition analyses, it is confirmed that the magnetosome membrane is a lipid bilayer vesicle originating from the cell membrane. In addition the electron cryotomographic (ECT) imaging of magnetosomes can provide unprecedented three dimensional views of the magnetotactic bacteria. Figure 5.7 shows an ECT imaging of *Magnetospirillum magneticum* sp. AMB-1 (AMB-1) in a near native state without disruptive staining, fixation and sectioning procedures. Such figures confirmed many known features of magnetosomes, for example, the presence of a membrane prior to magnetite biomineralization and the juxtaposition of the chain against the membrane. Furthermore, all the ECT images indicate that magnetosome membranes are invaginations of the cell membrane regardless of their places in the chain and the sizes of the magnetite crystals inside, and this suggests that magnetosome membranes are not vesicles (Komeili *et al.* 2006). But it is not certain whether this conclusion is applicable to all magnetotactic bacteria. The ECT imaging of *Magnetospirillum gryphiswaldense sp.* MSR-1 (MSR-1) by Scheffel *et al.* (2006) confirmed that the magnetosome chain is close to the cell membrane, but the presence of membrane invaginations is yet to be further studied.

Further information about the magnetosome membrane can be derived from the research on the biology of magnetosomes (Komeili 2007). As a magnetosome is a cell membrane

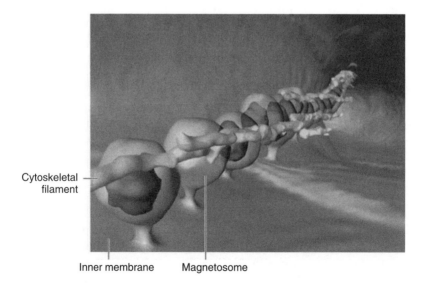

Cytoskeletal filament

Inner membrane   Magnetosome

Figure 5.7 An ECT three-dimensional reconstruction of *Magnetospirillum magneticum* sp. AMB-1. The background is the inner membrane, and magnetosomes can be seen as invaginations of the inner membrane. A network of cytoskeletal filaments surrounds the magnetosome chain. (Komeili 2007)

invagination, the magnetosome chains are naturally connected to the cell, ensuring the alignment of magnetotactic bacteria in magnetic fields. Moreover, there is an open channel between the magnetosome and the periplasm, so that ferrihydrite, a precursor to magnetite, formed in the periplasm, can be transported into the magnetosome. As the open channel between the magnetosome and the periplasm may also provide an opportunity for the contents of the two compartments to mix, there may be a barrier which exists between the two compartments, or specific proteins for the purification of the magnetosome lumen.

5.2.3.2 Magnetosome Chain Formation

In magnetotactic bacteria, magnetosomes are organized end to end into chains. However, purified magnetosomes would collapse into aggregates if their magnetosome membranes and proteins were removed through a detergent treatment. So there exist specific mechanisms for the stabilization of magnetosome chains in magnetotactic bacteria. ECT imaging studies of AMB-1 and MSR-1 have found the sub-cellular structures that ensure the stability of the magnetosome chain. As shown in Figure 5.7, in both of these organisms, a network of filaments surrounds the magnetosome chain. These filaments may be formed by a magnetosome-specific cytoskeletal protein, and their dimensions are similar to those of bacterial actin-like proteins.

Different magnetotactic bacteria may have different mechanisms for chain formation. For AMB-1, under iron-starvation conditions, cells can grow, but magnetite cannot be formed. Therefore chains of empty magnetosome membranes are formed. Once iron is added, magnetite crystals are formed simultaneously within multiple adjacent magnetosomes until full-sized crystals are achieved (Komeili *et al.* 2004). The magnetic chains in cells of MSR-1 are formed by a different mechanism. Before magnetite formation, the magnetosomes in these cells are dispersed throughout the cells. When biomineralization proceeds, magnetic interactions seem to be required for aligning magnetosomes into a chain (Scheffel *et al.* 2006). The differences in the mechanisms of chain formation may be due to the differences in growth conditions, or due to the fundamental differences in the molecular mechanisms of chain organization in the cell (Komeili 2007).

It should be noted that the chain formation in the cell growth under normal conditions is not yet fully understood. Individual magnetotactic bacterium cells in a population have a relatively constant number of magnetosomes per chain, and the chains are separated down the middle during cell division. Therefore, during each cell cycle, all the cells need to duplicate their magnetosomes. Though preliminary research on magnetosome dynamics during synchronized growth of AMB-1 cells has been undertaken, it is still not clear whether the spatial or temporal production of magnetosomes and their assembly into the chain are regulated by cell cycle (Yang *et al.* 2001).

5.2.3.3 Model for Magnetosome Formation

Figure 5.8 shows a simple model for magnetosome formation. As the magnetosome membrane is the site where biomineralization takes place, and it pre-exists magnetite formation, magnetosome formation starts with the biogenesis of the magnetosome membrane and the targeting of magnetosome proteins to this compartment. Though the proteins responsible for producing the magnetosome membrane have not yet been identified, it is likely that some of the genes involved in this step are present in the magnetosome island (MAI) because a deletion of this region leads to a failure in magnetosome membrane biogenesis (Komeili 2007). Protein sorting to the magnetosome membrane may occur concurrently with the invagination of the cell membrane. However, as magnetosomes are continuous

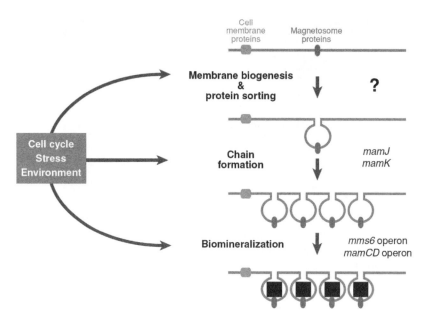

Figure 5.8   Model for magnetosome formation. Magnetosomes are formed in three steps. First, a membrane invagination is derived from the inner membrane and magnetosome proteins are sorted away from cell membrane proteins. Second, individual invaginations are assembled into a chain with the help of MamJ and MamK proteins. Third, iron is transformed into highly ordered magnetite crystals within the magnetosome membrane with the possible involvement of genes from the mamCD and mms6 operons. Cell cycle, environmental conditions and cellular stress may feed in at any of the three steps to modulate the formation of magnetosomes. (Komeili 2007)

with the cell membrane, magnetosome proteins might be transported to the membranes after magnetosomes have been formed, possibly by a diffusion and capture mechanism.

Once magnetosome membrane invaginations are formed, they will be assembled into a chain under the effects of the MAI-encoded proteins, MamK and MamJ. Though both of these proteins interact with the magnetosome membrane, they do not have transmembrane domains. So other proteins may participate in the formation of magnetosome chains. In the final step, the biomineralization of magnetite is realized. This step encompasses all the reactions involving iron, including its uptake from outside of the cell, oxidation to ferrihydrite in the periplasm, transport into the magnetosome and partial reduction to form magnetite, and it also includes the control of shape and size, possibly through the gene products of the mms6 and mamCD operons. Other processes, such as environmental influences on magnetite synthesis and magnetosome chain length, as well as the cell cycle regulation of magnetosome formation, can be included in different steps of this model (Komeili 2007).

5.3   Cultivation of Magnetotactic Bacteria

Magnetotactic bacteria are fastidious organisms and it is very difficult to isolate and grow magnetotactic bacteria in pure culture. The research and applications of magnetosomes are hindered by the limited availability of natural or cultured magnetotactic bacteria. Much effort has been made to develop methods for the cultivation of magnetotactic bacteria. We discuss below two typical methods for the cultivation of magnetotactic bacteria.

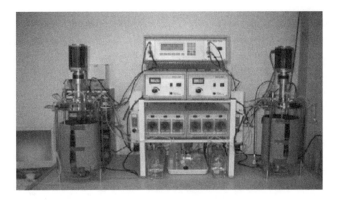

Figure 5.9 Dual-vessel (2 × 10*l*) laboratory fermenter system for mass culture of *M. gryphiswaldense*. Lang, C. and Schuler, D.(2006). "Biogenic nanoparticles: production, characterization, and application of bacterial magnetosomes", Journal of Physics: Condensed Matter, 18, S2815–S2828. Reproduced by permission of the Institute of Physics

5.3.1 Mass Cultivation of Magnetotactic Bacteria

Biochemical and biophysical characterization of magnetosomes requires large quantities of magnetosomes from magnetotactic bacteria cultivated under special conditions. In the cultivation of magnetotactic bacteria, *M. gryphiswaldense*, shown in Figure 5.2(a), is often selected as a model organism as it can be cultivated in simple liquid media containing short organic acids as a carbon source (Lang and Schuler 2006). This organism is often used in genetic analysis, and its genome sequence has been almost determined.

5.3.1.1 Synthesis of Magnetosomes

Many magnetotactic strains, such as *M. magnetotacticum, M. magneticum and M. gryphiswaldense*, produce magnetite only under microaerobic conditions, and higher oxygen concentrations, for example normal atmospheric conditions, repress the formation of magnetite crystals and inhibit their growth. Therefore, the control of a microoxic environment is important for the culture of magnetotactic bacteria.

Figure 5.9 shows an experimental system for mass cultivation of magnetotactic bacteria in an automated oxygen-controlled fermenter, so low oxygen partial pressures can be continuously maintained in this system (Lang and Schuler 2006). Installed with a highly susceptible oxygen amplifier and accessory equipment for the gas supply, a twin bioreactor is used for the microaerobic cultivation of microaerophilic bacteria under required conditions. The defined low oxygen partial pressures are regulated by a cascade control via separate and independent gassing with nitrogen and air. Nitrogen supply is controlled by a flow-meter installed in line with a pulsed solenoid gas valve, and the air supply is regulated by a thermal mass-flow controller and an additional pulsed solenoid control valve. The switch between nitrogen and oxygen gassing depends on the actual oxygen partial pressure in the medium. This system can produce magnetosomes in amounts sufficient for their characterization and applications.

This oxystat fermenter is used to determine the optimal oxygen partial pressures for the formation of magnetite crystals and the cultivation of *M. gryphiswaldense*. Magnetite crystals can only be formed below a threshold value of 10 mbar, and the formation procedure will be inhibited at higher oxygen concentrations. Figure 5.10 shows the relationship between the amount of magnetite formed and pO_2 which exists. The most favorable

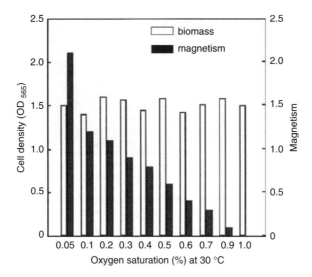

Figure 5.10 Biomass production and cell magnetism of oxystat grown *M. gryphiswaldense* at different constant oxygen concentrations. Lang, C. and Schuler, D.(2006). Biogenic nanoparticles: production, characterization, and application of bacterial magnetosomes, *Journal of Physics: Condensed Matter*, **18**, S2815–S2828. Reproduced by permission of the Institute of Physics

condition for magnetite biomineralization is 0.25 mbar. It is found that particles grown at 10 mbar displayed smaller sizes (about 20 nm), while the particles produced under optimal conditions (0.25 mbar) have a diameter around 42 nm. Therefore, growth conditions affect the morphology and size of particles (Lang and Schuler 2006). Similarly, iron limitation may reduce the sizes of the magnetic particles produced, and small magnetic particles can also be obtained by the addition of iron to iron-starved cells shortly before cell harvest.

5.3.1.2 Isolation and Purification of Magnetosomes

As magnetosome particles have higher density and stronger magnetism than organic cell constituents, they can be purified from cells by a straightforward isolation protocol. After cell disruption by French press and removal of cell debris by centrifugation, the magnetosomes can be separated from the crude extracts by magnetic separation columns (Grunberg *et al.* 2004). After magnetic separation, ultracentrifugation is performed, which results in suspensions of purified magnetosome particles with intact enveloping membrane structures. To solubilize the magnetosome membranes, strong detergents or organic solvents should be used (Lang and Schuler 2006). After this treatment, the agglomeration of membrane-free magnetite particles can be obtained, as shown in Figure 5.11.

5.3.2 Continuous Cultivation of Magnetotactic Bacteria

The synthesis of magnetic crystals within magnetotactic bacteria is a very slow process. To obtain a significant fraction of magnetotactic bacteria in the culture medium, an incubation period of several days is usually required. Typically, in the culture of magnetotactic bacteria, only part of the bacteria is both magnetic field susceptible and motile. Therefore batch production of magnetotactic bacteria is not only a gradual process but also has low yield.

Bahaj *et al.* (1997) proposed a continuous removal process, in which the culture medium is circulated from the culture reservoir through an orientation magnetic separator to recover

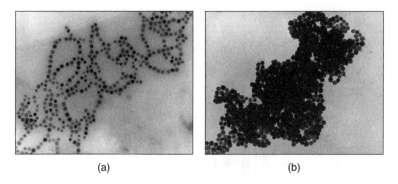

(a)   (b)

Figure 5.11 (a) Isolated magnetosome particles with intact magnetosome membranes have a strong tendency to chain formation; (b) removal of the magnetosome membranes by SDS treatment results in the agglomeration of membrane-free particles. Lang, C. and Schuler, D.(2006). Biogenic nanoparticles: production, characterization, and application of bacterial magnetosomes, *Journal of Physics: Condensed Matter*, **18**, S2815–S2828. Reproduced by permission of the Institute of Physics

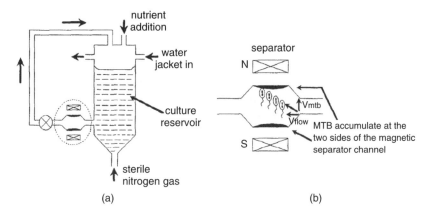

(a)   (b)

Figure 5.12 (a) Continuous culture and recovery system for magnetotactic bacteria (MTB). The orientation magnetic separator is enclosed in the dashed circle; (b) enlargement of the orientation magnetic separator. (Bahaj *et al.* 1997)

the motile, magnetic field susceptible fraction of magnetotactic bacteria, and the medium is then retuned to the top of the culture reservoir, as shown in Figure 5.12(a). Therefore, magnetotactic bacteria in which magnetosomes are not fully developed are not discarded. Instead, they continue their growth cycle. With periodic addition of nutrients, the culture process is self-sustaining.

In the operation of this experimental system, the fluid flow and bacterial flagellar velocities are taken into consideration. Motile, magnetic field susceptible MTB are captured on the separator chamber walls adjacent to the magnetic poles. In the use of this orientation magnetic separator for the recovery of bacteria, the limiting case should be considered. As shown in Figure 5.12(b), if a bacterium swimming towards the north-pole magnet enters the separator adjacent to the south-pole magnet, it must swim across the entire width of the separator to achieve separation. Under the limiting case condition, this crossing is achieved over the entire length of the separator chamber. For a given velocity of fluid flow, a threshold bacterium velocity exists below which complete separation of bacteria cannot be achieved (Bahaj *et al.* 1997).

The separation of magnetotactic bacteria is affected by many factors (Bahaj *et al.* 1997). If the bacteria are non-motile, or non-field susceptible, they travel through the orientation magnetic separator and return to the culture reservoir. Some magnetotactic bacteria with a low migration velocity (V_{MTB}) cannot satisfy the limiting condition, so usually they will be re-circulated. However due to their random positions where they enter the separator, they may be captured in subsequent passes. Moreover, the physical entrapment of bacteria on separator walls also affects accumulation efficiency.

Control experiments can be done by using a pair of magnets mounted on a holder which is rotated around the separator at frequency v_{rotate} (Bahaj *et al.* 1997). If the fluid flow rate is zero, a bacterium will swim in a circle of diameter, d, which is determined by the frequency of the rotating field v_{rotate} and the migration velocity of the magnetotactic bacterium V_{MTB}:

$$v_{rotate} = \frac{V_{MTB}}{\pi d} \tag{5.1}$$

For example, if $v_{rotate} = 0.75\,\text{Hz}$ and $V_{MTB} = 100\,\mu\text{ms}^{-1}$, the diameter of the circle d should be 24 μm. If the fluid flow rate is larger than zero, under the effects of the rotating magnets, when the bacteria pass through the separator, they will move in spiral tracks, avoiding contact with the walls of the separator.

5.4   Characterization of Magnetosomes

Though bacterial magnetic nanoparticles have shown great biomedical and nanotechnological potential, many unsolved fundamental questions prevent the applications of magnetosomes at technical scale (Lang and Schuler 2006). Characterization of magnetosomes is crucial for the study and application of this special type of magnetic nanomaterials. Magnetic measurements (Penninga *et al.* 1995), magnetic force microscopy (Proksch *et al.* 1995) and electron holography (Dunin-Borkowski *et al.* 1998) are typical methods in the characterization of the microstructure and magnetic properties of magnetotactic bacteria. We discuss below the typical methods for the biochemical characterization, microstructure characterization, magnetization characterization and relaxation characterization of magnetotactic bacteria.

5.4.1   Biochemical Characterization

Detailed knowledge about the biochemical composition and protein content of isolated magnetosomes is required for the functionalization and subsequent applications of magnetosomes. Usually the membranes of isolated magnetosomes are analyzed for the identification of magnetosome-associated proteins (Lang and Schuler 2006). First, the magnetosome membrane (MM) to be analyzed is solubilized by boiling in a buffer containing SDS and 2-mercaptoethanol. The samples are subsequently subjected to one-dimensional SDS–polyacrylamide gel electrophoresis (SDS–PAGE) (Figure 5.13(A)) or to Tricine–SDS–PAGE (Figure 5.13(B) and(C)). Improved protein separation can be obtained by performing two-dimensional gel electrophoresis (Lang and schuler 2006). After electrophoresis the proteins are blotted onto a membrane and the N-termini of separated proteins are sequenced by Edman degradation.

Mass spectroscopy can also be used to identify the magnetosome-associated proteins, either after size separation from single spots or bands, or from total tryptic digests of the entire magnetosome preparations (Lang and Schuler 2006). Magnetosomes are usually reduced with dithiotreitol, alkylated with iodacetamide and digested with

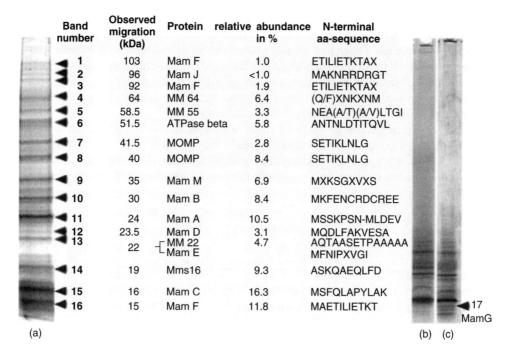

Band number	Observed migration (kDa)	Protein	relative abundance in %	N-terminal aa-sequence
1	103	Mam F	1.0	ETILIETKTAX
2	96	Mam J	<1.0	MAKNRRDRGT
3	92	Mam F	1.9	ETILIETKTAX
4	64	MM 64	6.4	(Q/F)XNKXNM
5	58.5	MM 55	3.3	NEA(A/T)(A/V)LTGI
6	51.5	ATPase beta	5.8	ANTNLDTITQVL
7	41.5	MOMP	2.8	SETIKLNLG
8	40	MOMP	8.4	SETIKLNLG
9	35	Mam M	6.9	MXKSGXVXS
10	30	Mam B	8.4	MKFENCRDCREE
11	24	Mam A	10.5	MSSKPSN-MLDEV
12	23.5	Mam D	3.1	MQDLFAKVESA
13	22	MM 22	4.7	AQTAASETPAAAAA
		Mam E		MFNIPXVGI
14	19	Mms16	9.3	ASKQAEQLFD
15	16	Mam C	16.3	MSFQLAPYLAK
16	15	Mam F	11.8	MAETILIETKT

17
MamG

(a)   (b)   (c)

Figure 5.13 MM-associated proteins separated by one-dimensional PAGE. (A) Summary of MM proteins detected by Coomassie stain in 1D SDS–PAGE (16 %). Proteins from indicated bands were identified by N-terminal amino acid analysis (Edman degradation); (B) Coomassie and (C) silver-stained SDS–Tricine gels (16.5 %) of MM proteins. Proteins from marked spots are identified by mass spectrometric sequencing. Lang, C. and Schuler, D.(2006). Biogenic nanoparticles: production, characterization, and application of bacterial magnetosomes, *Journal of Physics: Condensed Matter*, **18**, S2815–S2828. Reproduced by permission of the Institute of Physics

trypsin to completion. Then the magnetic moiety is removed, and the supernatant is chromatographically separated using a capillary liquid chromatography system. After that, the eluted peptides are analyzed by a mass spectrometer. Figure 5.13 also shows the proteins identified by mass spectrometry.

5.4.2 Microstructure Characterization

The magnetic crystals in each type of magnetotactic bacteria have specific morphologies, and usually their easy magnetization axes are parallel to the chain axis. Many efforts have been made to study the magnetic microstructures of magnetosome chains, which may be helpful for understanding the function of biogenic magnetic crystals in magnetic field sensing in higher organisms, and for developing their biomedical applications (Dunin-Borkowski *et al.* 1998).

The microstructures of magnetotactic bacteria can be characterized by scanning force microscopy, including atomic force microscopy, magnetic force microscopy and transmission electron microscopy.

5.4.2.1 Scanning Force Microscopy Study

Magnetic nanoparticles in diluted aqueous suspensions can be used in both diagnostic applications, for example MRI contrast enhancement, and therapeutic treatments, for

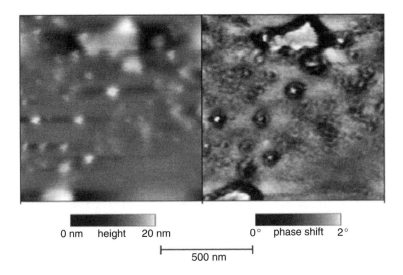

0 nm   height   20 nm   0°   phase shift   2°

500 nm

Figure 5.14 AFM (left) and MFM (right) images of isolated magnetosomes and magnetosome clusters. Before the measurements, the sample is magnetized vertically to the plane. (Albrecht *et al.* 2005)

example targeted drug delivery and hyperthermia. Isolated magnetosomes are preferred for these applications, and in some cases, it is advantageous to use single magnetosomes. Therefore, it is necessary to unambiguously clarify whether single, isolated magnetosomes exist in ensembles. This information can be obtained by using atomic force microscopy (AFM) and magnetic force microscopy (MFM) which have nanometer spatial resolutions. AFM can be used to unambiguously demonstrate the existence of single magnetosomes, and MFM can be used to particularly address the magnetic properties of single magnetosomes. AFM and MFM can often be used in combination to study bacterial magnetosomes (Albrecht *et al.* 2005).

During the image acquisition procedure, the magnetization of the sample under study may be affected by the magnetic tip of MFM. To minimize this effect, before image acquisition, the sample is magnetized by applying an external magnetic field, which is perpendicular to the main direction of the magnetic field generated by the MFM tip (Albrecht *et al.* 2005). The AFM image in Figure 5.14 shows isolated magnetosomes along with magnetosome clusters with different sizes. It can observed that the size of a single magnetosome is typically around 40–45 nm. As shown in Figure 5.14, the MFM image of an isolated magnetosome consists of a dark ring around a white center. The dark ring represents the attraction between the MFM tip and the isolated magnetosome, and the white center represents the repulsion between them.

To get more detailed information, a line cut is made across an isolated magnetosome, as indicated in the MFM image shown in Figure 5.15(a). The measurement results along line cut, as shown in Figure 5.15(b) agree well with the simulated dB_z/dz results of a sphere which is homogeneously magnetized in the direction normal to the sample plane, i.e., in the z direction, as shown in Figure 5.15(c). Therefore, an isolated magnetosome exhibits a magnetic mono-domain structure whose dipole moment is normal to the substrate plane.

5.4.2.2 Electron Holography Study

The relationship between the physical and the magnetic microstructures of magnetite nanocrystals in magnetotactic bacteria can be observed by using the off-axis electron

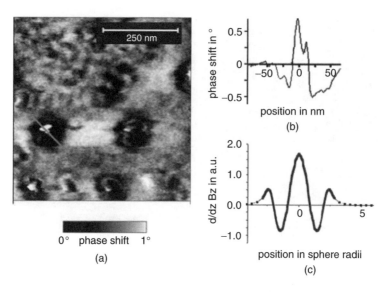

Figure 5.15 (a) Enlargement of a part of Figure 5.14. A line cut across a single magnetosome is indicated; (b) cross-section of the MFM image along the cut line indicated in (a); (c) calculated dBz/dz results of a sphere that is homogeneously magnetized vertically to the sample plane. The small tilt of the measured profile may be caused by a slight tilt of the magnetization axis. (Albrecht *et al.* 2005)

holography technique in a transmission electron microscope (TEM) (Dunin-Borkowski *et al.* 1998). Using this technique, high-resolution information about the magnetotactic bacteria can be obtained, and magnetic and electric microfields in the magnetotactic bacteria can be analyzed. As shown in Figure 5.16, in a TEM system for the generation of off-axis electron holograms, the field emission electron gun coherently illuminates the

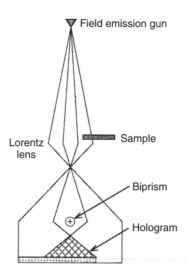

Figure 5.16 Schematic illustration of the system for the generation of off-axis electron holograms (Dunin-Borkowski *et al.* 1998)

sample under test, which occupies approximately half of the view field. The electrostatic biprism is a thin quartz fiber with diameter less than 0.7 μm, and it is coated with gold. The presence of the positively charged biprism causes overlap of the object wave and the reference wave, resulting in the holographic interference pattern.

Figure 5.17(a) shows a TEM image of a single cell of *M. magnetotacticum*. The magnetosome chain is not symmetrical, with larger crystals at its left end, and there is an obvious defect in the chain, as indicated by the arrow in the figure. To obtain the electron holograms as shown in Figure 5.17(b), an electrical voltage (120 V) is applied to the electrostatic biprism. The intensity and position of the resulted interference fringes record the amplitude and phase of the sample wave, respectively. The phase is closely related to the mean inner electric potential, and the in-plane component of the magnetic induction integrated in the incident beam direction. If we neglect the dynamical diffraction effects, the phase can be expressed by (Dunin-Borkowski *et al.* 1998):

$$\phi(x) = \left(\frac{2\pi}{\lambda}\right)\left(\frac{E + E_0}{E(E + 2E_0)}\right)\int V(x, z)\, dz - \left(\frac{e}{\hbar}\right)\iint B_\perp(x, z)dxdz \qquad (5.2)$$

where the x direction is in the plane of the sample, the incident beam is in the z direction, the magnetic induction component $B_\perp$ is perpendicular to both the x direction and the z direction, V is the mean inner potential of the sample, λ is the wavelength and E and E_0 are the kinetic and rest mass energies of the incident electrons, respectively. Usually, the phase is dominated by the mean inner potential, which should be removed to quantify the magnetization. For this purpose, the magnetic field of the microscope objective lens is used to reverse the magnetization of the magnetite chain, and two holograms corresponding to these two states can be obtained. From the phases of the two holograms, the magnetic contribution and the mean potential contribution to the phase can then be derived.

Figure 5.17(c) shows the contours formed from the magnetic contribution to the holographic phase of the *M. magnetotacticum* cell in field-free conditions. The contour density is proportional to the component of the magnetic induction in the sample plane integrated in the direction of the incident beam. It can be seen that the field gradients are low at the ends of the chain and at gaps between individual magnetosomes. In some magnetite crystals, the magnetic field lines bend to achieve their minimum magnetostatic energy, while in some other magnetite crystals, the directions of the magnetic field lines differ slightly from the direction of the chain axis. It seems that the small deviations of the crystal positions from the chain axis may severely affect the field direction in the crystals (Dunin-Borkowski *et al.* 1998).

5.4.3 Magnetization Characterization

The characterization of magnetization properties of magnetosomes and magnetotactic bacteria is important for the research and applications of magnetosomes and magnetotactic bacteria. Different magnetometers have been used for the measurement of the hysteresis of isolated magnetosome particles (Hergt *et al.* 2005; Eberbeck *et al.* 2005). In the following subsection, we first discuss typical magnetization properties of magnetosomes measured by conventional magnetometers, and then introduce an instrument developed for measuring the magnetization hysteresis loops of magnetotactic bacteria.

5.4.3.1 Magnetization Loops of Bacterial Magnetosomes

Vibrating sample magnetometers can be used to measure the magnetization loops of aqueous magnetosome suspensions. The percentages of magnetosomes and magnetic crystals

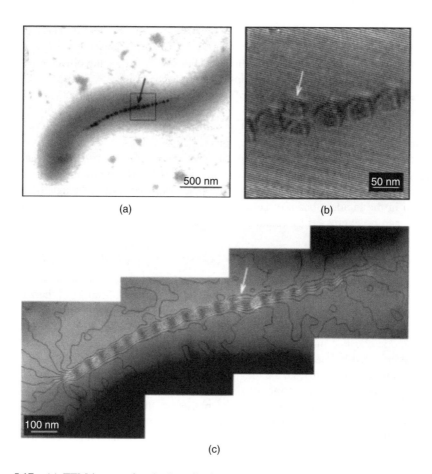

(a)   (b)

(c)

Figure 5.17   (a) TEM image of a single cell of *Magnetospirillum magnetotacticum* strain MS-1; (b) off-axis electron hologram of the region marked in (a) obtained in field-free conditions; (c) contours formed from the magnetic contribution to the holographic phase. In each figure, the arrow indicates the same double crystal defect in the chain. (Dunin-Borkowski *et al.* 1998)

in the suspensions can be derived by comparing the values of the saturation magnetization obtained theoretically and experimentally (Hergt *et al.* 2005). To separate the effects due to the Neel relaxation and the Brownian relaxation, magnetosomes can be immobilized by dissolving the magnetosomes in a liquid gelatine sol, and subsequently solidifying cooling down the magnetosome suspension below the sol-gel transition temperature. To fabricate textured samples in which all the magnetosomes are magnetically aligned in one direction, in the solidification procedure, a constant magnetic field should be applied. The required strength of the constant magnetic field is related to many factors, such as the viscosity of the sol, and the size and magnetization of the magnetosomes.

Similar to other magnetic materials, the magnetic hysteresis loss of magnetosomes is dependent on the amplitude of the applied magnetic field. Such dependence can be characterized by measuring minor hysteresis loops of magnetosome samples with different maximum amplitude of the magnetic field (Hergt *et al.* 2005). Figure 5.18(a) shows several minor loops for a magnetosome suspension in gel measured by vibrating sample magnetometers. The magnetic hysteresis loss of the sample can be calculated by

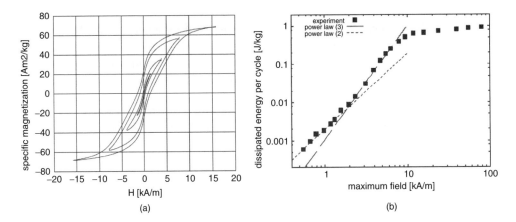

Figure 5.18 (a) Minor hysteresis loops of magnetosomes immobilized in gel. Measurements are made at room temperature, and the field loop amplitudes are 1.6, 4.0, 8.0 and 16.0 kA/m, respectively; (b) the relationship between the magnetic hysteresis loss of magnetosomes immobilized in gel and the field amplitude of minor loops. (Hergt *et al.* 2005)

integrating these minor hysteresis loops, and thus its dependence on field amplitude can be obtained. Figure 5.18(b) shows that, for a weak magnetic field, the field dependence of hysteresis loss follows a square law, while for a strong magnetic field, the field dependence of hysteresis loss follows a third-order power law. In this example, the transition field is about 2 kA/m.

5.4.3.2 Measurement of Hysteresis Loops of Magnetotactic Bacteria

De Waard *et al.* (2001) developed a magnetometer for measuring the magnetization curves of magnetic structures that can be microscopically observed, such as magnetosome chains in magnetotactic bacteria which can be suspended or swimming in water. In such a magnetometer, four coils generate continuous magnetic fields for the guidance and orientation of the magnetotactic bacteria, and two additional coils generate pulsed magnetic fields to change magnetization levels in small steps. This system can also be used to distinguish the magnetic bacteria and the non-magnetic bacteria in natural samples. This system may be useful in the study of production of nonmagnetic mutants from magnetic cells for genetic analysis, and the genetic engineering related to magnetic bacteria (Penninga *et al.* 1995). In the following subsection, we discuss the working principle and the structures of this type of magnetometer.

(1) Working Principle

This magnetometer can measure the hysteresis loops of the magnetosome chains in individual magnetotactic bacteria. In the hysteresis measurements, individual bacteria under test are suspended in water inside a capillary, and are observed or recorded by a microscope or a video camera. Usually, the capillaries used in the magnetometer have a rectangular cross-section.

Using the magnetometer developed by De Waard *et al.* (2001), the hysteresis loops of magnetosomes can be measured in following four steps. In the first step, an external DC guiding magnetic field $\mathbf{B}_G$ (up to about 10 G) is applied to orientate the magnetic bacterium under study, and the direction of this guiding field is denoted positive direction. By applying a strong magnetic pulse $\mathbf{B}_p$ with peak amplitude up to 800 G and duration

1–3 ms in the positive direction, the maximum (saturation) remanent magnetization of the bacterium's magnetosome chain can be achieved. In the second step, the bacterium is subjected to magnetic pulses $\mathbf{B}_p$ with increasing amplitude in the negative direction. To obtain sufficient details in the loop, the amplitude of the pulse field is increased in small steps. Once the pulse field exceeds the coercive force of the magnetosome chain, the cell will immediately rotate 180° after the pulse.

In the third step, the guiding field after each pulse is reversed, and the time required for the bacterium to rotate over a defined angle perpendicular to the rotation axis is determined. An alternative way is to rotate the DC guiding magnetic field in the horizontal plane at fixed frequency. During the rotation, the amplitude of the guiding magnetic field is decreased until the bacterium does not follow the rotating magnetic field. Each of the above steps can provide results for the determination of the remanent magnetization following the pulse, relative to the saturation remanent magnetization. In the fourth step, the above process for increasing pulse-field amplitude is repeated until the saturation remanent magnetization in the negative direction is achieved, and the first half of the hysteresis loop can be obtained based on the measured points. The other half of the hysteresis loop can be obtained by rotating the bacterium 180° with the guiding field and repeating the measurement process discussed above.

(2)   Structure of the Magnetometer

Figure 5.19(a) shows a photo of the magnetometer system. In the microscope shown at the left side of the photo, the normal microscope stage is replaced by a lucite stage, which holds a multiple magnetic coil system and a capillary containing the magnetic bacteria. To ensure that a specific part of the specimen can be chosen for investigation, the capillary is movable over several millimeters in the north (N)–south (S) and east (E)–west (W) directions. During measurement, the sample under test can be simultaneously observed through the standard binocular eyepieces of the microscope and a mono-ocular which is connected to a TV camera. The control system shown on the right hand side of the photo provides suitable electrical currents for the coils.

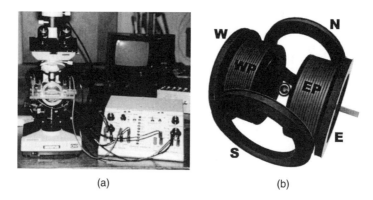

(a)   (b)

Figure 5.19   Structure of the magnetometer. (a) Photograph of the magnetometer system. At the left side is a microscope with video camera and coil system on the microscope stage. At the right is the control system for guiding magnetic bacteria and changing their magnetization; (b) schematic view of the coil system. The four coils N, W, E and S, placed in square, provide the guiding magnetic field. The two coils EP and WP inside the square generate a pulse field influencing the magnetization level of the bacteria held at the center. (De Waard *et al.* 2001)

The coil system which is schematically shown in Figure 5.19(b) generates horizontal magnetic fields, which are parallel to the plane of the microscope stage. The magnetic fields consist of two types. One is the guiding magnetic field that makes the bacteria move along defined tracks through the liquid, and the other is the pulsed magnetic field that changes the magnetization levels of the bacteria under study step by step. The guiding magnetic field is generated by two pairs of flat coils, designated N, S and E, W, respectively. The planes of the coils are vertical and placed in a square, so the axes of the magnetic fields in the N–S and E–W are perpendicular to each other, and cross at the middle point (C) of the horizontal plane. The thin, flat capillary containing the bacteria under study is placed at the middle point C. The coils producing the N–S and E–W components of the guiding magnetic field are serially connected so that the magnetic field components generated by them add in the center.

Different experiments may require the bacteria under study to move in different paths. To make the bacteria swim in a horizontal square path, the two pairs of coils are powered sequentially and, in this case, the non-motile bacteria rotate in steps of $\pi/2$. If the magnetic field is continuously rotating, the motile bacteria will swim in a circular path, while the nonmotile bacteria will continuously rotate about the centers of their permanent magnetic dipoles. To generate the rotating magnetic field, the N–S and E–W coil pairs are simultaneously powered by sinusoidal currents with a $\pi/2$ phase difference between them.

As shown in Figure 5.19(b), another pair of coils, consisting of the east pulsed (EP) coil and the west pulsed (WP) coil, generates the horizontal pulsed magnetic field. The EP and WP coils are arranged parallel to each other, and are placed between the E and W guiding-field coils. The capillary tube containing the bacteria under study is usually placed at point C, which is midway between the pulse-field coils (EP and WP) and is also the point where the axes of two pairs of guiding-field coils (N–S and E–W) cross. As the sample is usually small compared to the distance between the coils, the guiding field and the pulsed field are homogeneous at point C where the sample is placed. The pulse width of the pulsed field is adjustable in the range of 1 to 5 ms in five steps, while the amplitude of the pulsed field is adjustable from zero up to about 6800 Gauss in small steps. For most magnetic bacteria, saturate magnetization can be achieved if the amplitude of a pulsed field is greater than 600 Gauss. In addition to the arrangement of magnetic fields discussed above, the pulsed magnetic field and a guiding magnetic field can be used in combination. In this case, usually the E–W coil pair is connected to a DC electric current source, while the N–S coil pair is left open.

5.4.4 Susceptibility Characterization

Characterization of complex susceptibility is important for the study of the relaxation mechanisms in magnetosome suspensions (Harasko *et al.* 1993). Many methods have been developed for characterizing the susceptibility of magnetosome suspensions, as discussed in Chapter 2 and Chapter 8. Figure 5.20 shows spectra of the complex susceptibility of a magnetosome aqueous suspension, and measurements are made in the frequency range 100 Hz–1 MHz with amplitude of the magnetic field less 0.1 kA/m. The maximum imaginary part of the complex susceptibility at 1.5 kHz is due to Brownian relaxation of the magnetosomes in aqueous suspension. Further experiments indicate that the immobilization of magnetosomes in aqueous suspension can greatly decrease the imaginary part of the complex susceptibility at low frequencies (<10 kHz) mainly due to elimination of the Brownian relaxation path, while it hardly affects the losses at frequencies above about 100 kHz (Hergt *et al.* 2005).

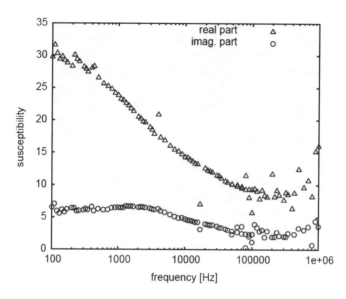

Figure 5.20 Spectra of the real and imaginary parts of the complex susceptibility of a magneto-some aqueous suspension. (Hergt *et al.* 2005)

Magnetosomes may be used in hyperthermia for cancer treatments. For ferrofluids with superparamagnetic particles whose mean core diameter is less than about 20 nm, the specific power p due to magnetic loss can be derived from the imaginary part of susceptibility χ'' based on the linear small amplitude assumption (Hergt *et al.* 2005):

$$p(f) = \frac{\mu_0 H^2 2\pi f}{2} \cdot \frac{\tilde{\chi}''(f)}{\rho} \qquad (5.3)$$

where f is the frequency of the magnetic field, H is the amplitude of the magnetic field, and ρ is the mass density of magnetic particles.

5.4.5 Trajectory Characterization

To use magnetotactic bacteria for practical applications, such as industrial and environmental clean-up of metal pollution, the related properties of magnetotactic bacteria, such as growth conditions, morphology, magnetic properties and movement speed, should be investigated. The magnetic moment of magnetotactic bacteria could be measured by projecting the video images of movement of a magnetotactic bacterium onto a screen. The properties of the magnetotactic bacterium could be derived from the trace of the bacterium trajectory. This method does not have high accuracy, and is labor intensive. Bahaj and James (1993) proposed a method to characterize the magnetic properties of magnetotactic bacteria, based on image processing techniques. This method can also quickly and accurately determine the speed, size and morphology of bacteria.

Under application of an external magnetic field, magnetotactic bacteria will align along the field lines due to torque produced by the interactions between the magnetic moments of the magnetosomes and the external magnetic field. The bacteria then swim along the directions of the alignments. The magnetic moment and the movement velocity of an individual bacterium can be derived from the trajectories due to various field reversals (Bahaj and James 1993).

Under the application of an external magnetic field, the movement of a bacterium in suspension can be described by:

$$m B_0 \sin \theta - 8 \pi \eta R^3 \frac{d\theta}{dt} = 0 \qquad (5.4)$$

where m is the magnetic moment (Am^2) of the bacterium, R is the radius of the bacterium with a spherical shape, η is the viscosity of the suspension (for water, η is 10^3 Pa·s), k is the Boltzmann constant, T is the absolute temperature (K), v(m/s) is the mean velocity of the bacterium and B_0 is the inductance of the applied magnetic field. The first term in Equation (5.4) represents the torque on the bacterium induced by the external magnetic field, while the second term represents the viscous drag torque on the bacterium. Following a magnetic field reversal, a bacterium executes a U-turn, and the magnetic moment, m, can be obtained from the time τ for the bacterium to execute the U-turn and the diameter L of the U-turn (Bahaj and James 1993):

$$\tau = 8 \pi \eta \frac{R^3}{m B_0} \ln \frac{2 m B_0}{kT} \qquad (5.5)$$

$$L = \frac{8 \pi^2 R^3 v \eta}{m B_0} \qquad (5.6)$$

The above analysis can be extended to the non-spherical case of the spirillum magnetotactic bacteria (Bahaj *et al.* 1993). In the modified model, the non-spherical bacterium is taken as an ensemble of spheres of diameter $2a$ with a length l. It is assumed that the viscous torque Ω is a combination of a shear and a rotational term with no first order wake effects, the couplings between the spheres are not strong and the Reynolds number is much smaller than one. Based on Stoke's Law, a shape factor S_I for each sphere in the ensemble can be estimated, and thus the drag torque in Equation (5.4) can be written as:

$$\Omega_I = S_I \pi \eta a^3 \frac{d\theta}{dt} \qquad (5.7)$$

where $I = l/2a$.

The trajectory characterization system mainly consists of a microcomputer, a video recorder and a frame grabber which can produce pseudo colors running under a suite of software. Two pairs of Helmholtz coils are used. One is to cancel the earth's magnetic field, and the other generates the magnetic field manipulating the bacterium. When the field direction is reversed, the U-turn trajectories of the magnetotactic bacteria are observed using a microscope, and video recorded as shown in Figure 5.21. The shallow depth of microscope field indicates that only the magnetotactic bacteria executing perfectly planar movements are recorded. According to Equations (5.4)–(5.7), the magnetic moment and other properties of the magnetotactic bacteria can be derived from their U-turn trajectories.

5.5 Biomedical Applications of Magnetosomes

Magnetosomes possess unique magnetic, chemical and biological properties, which could hardly be matched by those of synthetic particles, and they can be used in a variety of biomedical applications (Lang and Schuler 2006). Magnetosome magnetic crystals have a consistent shape and a narrow size distribution in the single magnetic domain range. Due

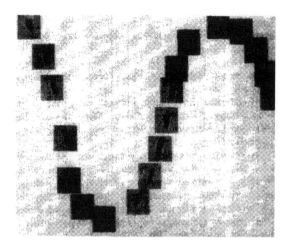

Figure 5.21 U-turn generated by magnetotactic spirillum sample at the reversals of magnetic field. (Bahaj and James 1993)

to the existence of magnetosome membranes, consisting of lipids and proteins, bioactive substances can be easily coupled to the surfaces of the magnetosomes, and this property is very important for their biomedical applications.

Both *in vivo* and *in vitro* biomedical applications of magnetosomes have been developed (Lang and Schuler 2006; Amemiya *et al.* 2005; Nakayama *et al.* 2003). For *in vivo* applications, magnetosomes can be used in hyperthermia and as contrast agents in magnetic resonance imaging. For *in vitro* applications, magnetosomes can be used in labeling, immobilization and separation of specific biomolecules, such as proteins and nucleic acids. Besides, magnetosomes can be used as an ideal model for the study of organelle development in bacteria (Komeili *et al.* 2006; Komeili *et al.* 2004; Scheffel *et al.* 2006).

Generally speaking, the applications of magnetotactic bacteria fall into two categories. In one category, the whole living cells and their magnetotactic behaviors are utilized, while in the other category, only isolated magnetosome particles are used (Schuler and Frankel 1999). In the following subsections, we give several examples for these two categories.

5.5.1 Applications of Magnetic Cells

Many applications based on the movements of magnetotactic bacteria under application of external magnetic fields have been developed. We will discuss two examples: Bacterial actuation and manipulation, and treatment of heavy metal contaminants.

5.5.1.1 Bacterial Actuation and Manipulation

Generally speaking, the motility of magnetotactic bacteria can be used in two approaches. In one approach, magnetotactic bacteria themselves are manipulated, while in the other approach, magnetotactic bacteria are used to manipulate other objects. There are many examples for the manipulation of magnetotactic bacteria. Darnton *et al.* (2004) demonstrated the application of bacteria as functional components, by attaching *Serratia marcescens* flagellated bacteria to polydimethylsiloxane or polystyrene, forming a bacterial carpet to move the fluid. Bahaj *et al.* (1998) explored the motility of magnetotactic bacteria by separating magnetotactic bacteria in low magnetic field. Lee *et al.* (2004) used microelectromagnet arrays for micromanipulation of magnetotactic bacteria.

Martel *et al.* (2006) demonstrated the application of magnetotactic bacteria for manipulation of other objects. In his work, a directional magnetic field generated by a programmed electrical current is used to modify the torques on the magnetosome chains in *Magnetospirillum gryphiswaldense* bacteria, and the magnetotactic bacteria are then used to push 3 μm beads along defined paths. The magnetotactic bacteria used have a length of about 1–3 μm with a swimming speed of about 40–80 μm/s. As each magnetotactic bacterium has a magnetosome chain, the swimming direction of magnetotactic bacteria is mainly based on magnetotaxis. The magnetic field lines generated by an electrical conductor network only modify the swimming direction of the magnetotactic bacteria, while they do not induce any propulsion force on the magnetotactic bacteria or the beads. Figure 5.22 shows the swimming speed distribution of magnetotactic bacteria

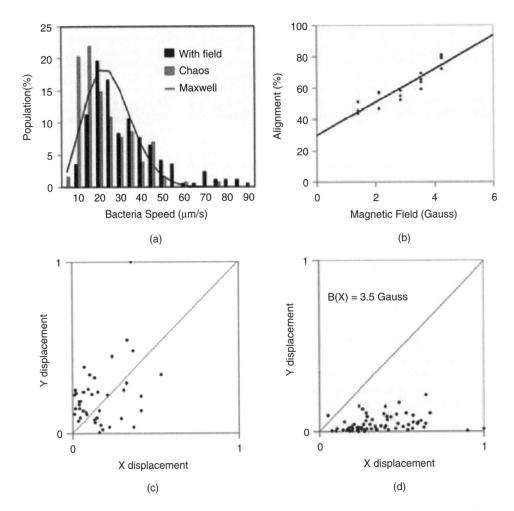

Figure 5.22 Motility and response to magnetic fields of magnetotactic bacteria. (a) Speed distributions of the sample with 295 MTB, with and without the presence of a magnetic field, showing a Maxwell distribution; (b) extrapolation line showing the percentage of the magnetotactic bacteria oriented within ±30° along the line of the magnetic field at various intensities; (c) swimming direction of the bacteria without magnetic field; (d) swimming direction with a magnetic field of 3.5 Gauss along the x axes. The data in (c) and (d) are normalized to unity. (Martel *et al.* 2006)

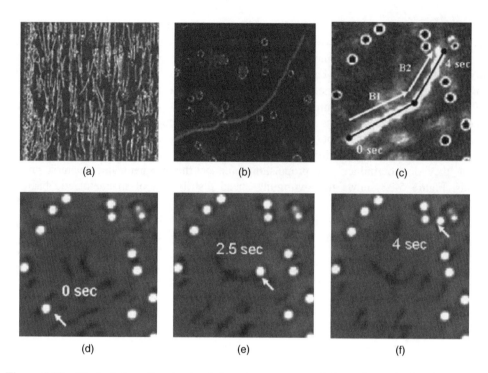

Figure 5.23 Manipulation of a microbead through a magnetotactic bacterium. (a) Direction control of magnetotactic bacteria by an external magnetic field; (b) movement path of a magnetotactic bacterium which is not attached to any microbeads; (c) displacement path of a microbead which is pushed by a magnetotactic bacterium with a direction change due to the orientation change of the magnetic field represented by B1 and B2; (d) initial position of the microbead tagged by a magnetotactic bacterium (at $t = 0$ seconds); (e) position of the microbead when the direction of magnetic field is changed (at $t = 2.5$ seconds); (f) final position of the manipulated microbead (at t = 4.0 seconds). The length of the images edges is 36.0 µm. (Martel *et al.* 2006)

and the effect of the magnetic field on the swimming direction of the magnetotactic bacteria.

Figure 5.23 shows an example of manipulating a microbead through a magnetotactic bacterium. As shown in Figure 5.23(a), when an external magnetic field is applied, magnetotactic bacteria move in a controlled direction. Figure 5.23(b) shows the movement of a magnetotactic bacterium which is not attached to any microbeads. Figure 5.23(c) shows an example of controlled manipulation of a microbead by a magnetotactic bacterium. The positions of the manipulated microbead at 0 seconds, 2.5 seconds and 4 seconds are shown in Figure 5.23(d), (e), and (f) respectively. It can be found that, after about 2.5 seconds, the direction of the microbead was intentionally shifted about 30° anticlockwise (Martel *et al.* 2006).

5.5.1.2 Treatment of Heavy Metal Contaminants

Some metallic ions can be immobilized and accumulated using microorganisms, and such bioaccumulations are beneficial for both industry and the environment (Bahaj *et al.* 1994). For example, using bioaccumulation techniques, effluents carrying heavy metals or radionuclides can be detoxicated, and the valuable elements such as the platinum or gold can be removed from solutions.

Magnetotactic bacteria can be restricted to move in a desired direction by applying an external magnetic field. If a low intensity, focusing magnetic field is applied across a chamber with solution containing magnetotactic bacteria, the motile bacteria can be separated from the solution. This procedure is different with traditional high gradient magnetic separation. In this case, the external magnetic field only orientates the magnetotactic bacteria, but does not magnetically attract these bacteria. For target metals that can be effectively removed by magnetotactic bacteria, and have no adverse effects on the motility of magnetotactic bacteria, orientation separation by magnetotactic bacteria offers a simple and efficient treatment method. The absolute separation efficiency of orientation separation by magnetotactic bacteria is related to the uptake percentage of the target metal, the chamber size and the separation time. A basic requirement is that the time for a magnetotactic bacterium to swim across the chamber is equal to the separation time (Bahaj *et al.* 1994).

In spite of the success of the applications of magnetotactic bacteria in magnetic orientation separation, several problems are yet to be solved. Some common metals, such as copper and zinc inhibit bacterial motility even at 1 ppm concentrations (Bahaj *et al.* 1994). Therefore great care must be taken in choosing suitable magnetotactic bacteria for magnetic orientation separation.

5.5.2 Applications of Isolated Magnetosome Particles

Similar to artificially synthesized magnetic nanoparticles, the biomedical applications of magnetosome particles can be generally classified into diagnosis and therapy. One typical diagnosis application of magnetosome particles is magnetic separation. Due to their small size and thus large surface-to-volume ratio, isolated magnetosome particles can be used to immobilize large quantities of bioactive substances. After immobilization, these bioactive substances can be separated using external magnetic fields. Another diagnosis application of isolated magnetosome particles is the contrast agent for magnetic resonance imaging (Hartung *et al.* 2007; Lang and Schuler 2006; Bulte and Brooks 1997). Magnetosome particles can be used to enhance the contrast between normal and diseased tissues, and to indicate the status of an organ.

Hyperthermia treatment is a promising therapy application for isolated magnetosome particles. In a hyperthermia treatment, magnetic particles promote cell necrosis in tumors by controlled heating, which is related to the loss processes resulting from the reorientation of the magnetic moments of the magnetic particles. The method requires magnetic particles with high specific loss powers. It has been found that magnetosomes isolated from *M. gryphiswaldense* have exceptionally high specific power losses ($960 \, g \, W^{-1}$ at $10 \, kA \, m^{-1}$ and $410 \, kHz$), which is substantially higher than artificial particles (Hergt *et al.* 2005). In addition, isolated magnetosome particles can be used as tumor-specific drug carriers, and they can also be used for introducing DNA into target cells (Takeyama *et al.* 1995).

In most of the biomedical applications, isolated magnetosome particles are functionalized to perform the desired functions. Therefore, functionalization of magnetosome particles is crucial for their biomedical applications. We will now discuss the functionalization of magnetosomes, and then the applications of magnetosome particles in the detection of biomolecules and magneto Immuno-PCR.

5.5.2.1 Functionalization of Magnetosomes

Isolated magnetosome particles possess obvious advantages for biomedical applications. They have very high magnetizations. Depending on the bacterial species from which they are produced, magnetosome particles exhibit unique shapes and sizes. Furthermore, the

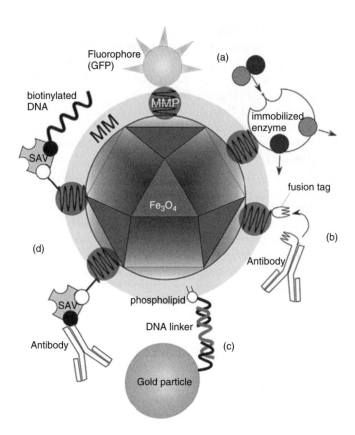

Figure 5.24 Schematic view of a hybrid bacterial magnetic nanoparticle. The magnetosome is modified by (a) magnetosome-specific expression of enzyme and fluorophore proteins by genetic fusion to the MMP, (b) fusion tags that serve as anchor groups for subsequent conjugation with various biomolecules, (c) conjugation with gold particles or quantum dots through a DNA linker and (d) biotinylation of membrane lipids and proteins, which would facilitate the subsequent streptavidin-mediated conjugation to various molecules such as nucleic acids or antibodies. In the figure, MM stands for magnetosome membrane, MMP for magnetosome protein and SAV for strep-tavidin. Lang, C. and Schuler, D.(2006). Biogenic nanoparticles: production, characterization, and application of bacterial magnetosomes, *Journal of Physics: Condensed Matter*, **18**, S2815–S2828. Reproduced by permission of the Institute of Physics

magnetosome membrane encapsulating the magnetic crystal is a natural coating of the magnetosome. It mainly consists of phospholipids and a specific type of proteins, and it is accessible for the attachment of artificial functional moieties (Ceyhan *et al.* 2006). Using genetic techniques, the biochemical composition of the magnetosome membrane can be altered for a special purpose. Figure 5.24 schematically shows different functionalization approaches for isolated magnetosome particles. These functionalizations could be realized by the generation of chimeric proteins specifically displayed on the magnetosome surface (Lang and Schuler 2006). Besides, it has been experimentally demonstrated that magne-tosome proteins can be used for realizing functional genetic fusions (Komeili *et al.* 2004; Schultheiss *et al.* 2005).

Figure 5.25 shows an example of modification of magnetosomes by a modular syn-thetic chemical approach (Ceyhan *et al.* 2006). In this example, the biogenic magnetosome particles (MPs) 1 are produced by the magnetotactic bacterium *Magnetospirillum*

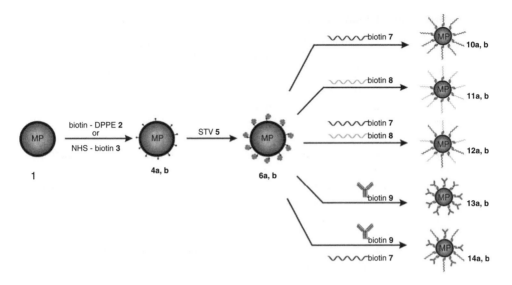

Figure 5.25 Schematic drawing of the preparation of oligofunctional DNA- and/or protein-modified magnetosome particles. Biotin groups were attached to the magnetosome membrane of MPs 1 either by incorparation of [1, 2-dipalmitoyl-sn-glycero-3-phosphoethanolamine-N-(biotinyl) (sodium salt)] (biotin-DPPE, 2) or by the covalent modification of the proteins within the magnetosome membrane by using sulfo-N-hydroxysuccinimide ester sodium salt (NHS-biotin, 3). Magnetosomes modified with 2 are labeled a, while magnetosomes modified with 3 are labeled b. The surface-bound biotin groups of the resulting MPs 4a and 4b were used to bind streptavidin (STV, 5), and the resulting STV-functionalized MPs 6a and 6b were functionalized with biotinylated DNA oligomers 7 and 8 and/or antibody 9. (Ceyhan *et al.* 2006)

gryphiswaldense. These magnetosome particles are 38 nm in size, and have cubocta-hedral shapes. The magnetosome has a monocrystalline magnetite core. First, the biotin groups are coupled to the magnetosome membrane of the magnetosome particles, and this could be realized in two different approaches: the incorporation of the biotinylated lipid biotin-DPPE, 2, or the covalent modification of the proteins within the magnetosome membrane by using NHS-biotin, 3. Then the protein streptavidin (STV, 5) is attached to the membrane-bound biotin groups of MPs 4a and 4b, resulting in STV-functionalized MPs 6a and 6b, respectively. The biotin-binding capacity of particles 6 is then used for the attachment of functional biomolecular entities, such as biotinylated DNA oligonu-cleotides (7, 8) whose DNA sequences are listed in Table 5.1, and biotinylated antibodies (9). The resulting MPs, 10–14, can be used for the detection of complementary targets, nucleic acids and proteins which can be identified by the MP-bound DNA oligomers or antibodies, respectively.

Table 5.1 Oligonucleotide sequences used

Compound	Sequence
7	5'-biotin-TCC TGT GTG AAA TTG TTA TCC GCT-3'
8	5'-GCA CTT GAG AGC (dT$_{12}$)-biotin-3'

5.5.2.2 Detection of Biomolecules

Magnetic particles are often used as magnetic markers for quantitative detection of molecular interactions, for example, between DNA–DNA, antigen–antibody and ligand–receptor. The magnetic measurements are usually made using a superconducting quantum interference device (SQUID) magnetometer (Enpuku *et al.* 2001), giant magneto-resistive (GMR) sensors (Edelstein *et al.* 2000) or by the measurement of magnetic permeability (Kriz *et al.* 1998). Theoretically, these methods are sensitive enough to detect single magnetic particles (Edelstein *et al.* 2000).

Amemiya *et al.* (2005) developed a method for the detection of streptavidin using biotin conjugated to single domain bacterial magnetosome particles (MPs). In this method, the magnetosome particles are used as magnetic markers, and the magnetic signals from MPs are measured using magnetic force microscopy (MFM). This method is sensitive enough for immunoassay and DNA detection.

Figure 5.26 shows the process of biotin-MPs immobilization for the detection of streptavidin. The magnetosome particles are isolated from *Magnetospirillum magneticum* AMB-1. The Sulfo-NHS-LC-LC-Biotin or Biotin-PEG-NHS containing N-hydroxysuccinimidyl esters are immobilized on magnetosome particles, resulting in biotin-MPs (Amemiya *et al.* 2005).

Figure 5.27 shows typical AFM and MFM images of magnetosome particles. Figure 5.27(b) is obtained using a standard MFM cantilever with coercivity force H_c = 300–400 Oe. The resolution of the AFM image is sufficient for identifying single magnetosome particles. The bright color represents repulsive interactions between the tip and sample, while the dark color represents the attractive interactions between them. It should be noted that in MFM imaging, the bright and dark contrasts depend on the magnetic polarity. Depending on which magnetic pole is detected, usually the MFM image from magnetosome particles should generate both bright and dark images. However, as shown in Figure 5.27(b), only dark images can be observed using the 300–400 Oe cantilever. But, as shown in Figure 5.27(d), the magnetic polarity of MPs can be indicated by the use of cantilever with higher coercive force (H_c = 650 Oe) under vacuum. The attractive interactions detected between the standard cantilever tip and magnetosome particles may be due to the relatively low coercive force (H_c = 300–400 Oe) of the cantilever and the subsequent magnetization of the cantilever by the magnetosome particles (Amemiya *et al.* 2005).

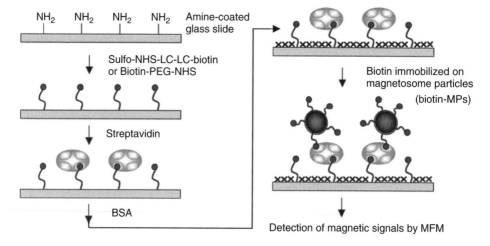

Figure 5.26   Magnetic detection for streptavidin by using biotin-MPs. (Amemiya *et al.* 2005)

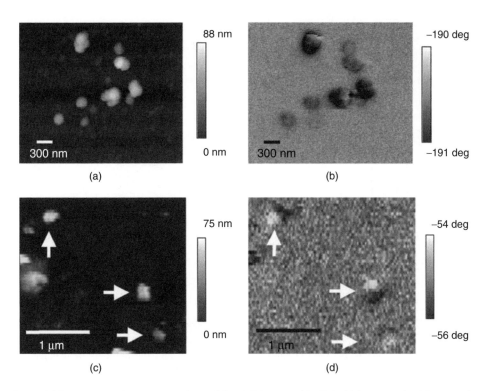

Figure 5.27 Images of MPs on a glass slide. AFM (a) and MFM (b) images of BMPs at a low magnification. The MFM imaging was performed by using the cantilever with $H_c = 300–400\,$Oe in air. AFM (c) and MFM (d) images of BMPs at high magnification. The MFM imaging was performed by using the cantilever with high coercivity $H_c = 650\,$Oe) in vacuum. (Amemiya *et al.* 2005)

Using MFM, the number of streptavidin on modified glass slides can be obtained by directly counting the magnetosome particles. Figure 5.28 shows the relationship between streptavidin concentrations and the numbers of magnetosome particles at an area of $20\,\mu m \times 20\,\mu m$. In the experiment, after the glass slides are treated with various concentrations of streptavidin ($0.001–100\,$pg/ml), the biotin-MPs are applied to the biotin immobilized on the glass slides (Amemiya *et al.* 2005).

5.5.2.3 Magneto Immuno-PCR

Nowadays, the Immuno-polymerase chain reaction (Immuno-PCR), first described by Sano *et al.* (1992), is a well-developed technique for ultrasensitive analysis of biomarkers. Theoretically, the Immuno-PCR technique can be adapted to detect any antigen. Compared to the standard Enzyme-Linked ImmunoSorbent Assay (ELISA) technique, the Immuno-PCR technique has up to around a ten thousand-fold gain in sensitivity. Meanwhile, the Immuno-PCR technique has a linear dynamic range up to five orders of magnitude. Moreover, the Immuno-PCR technique has many additional advantages, such as small sample volume and high tolerances against drug and matrix effects (Wacker *et al.* 2007).

The use of magnetic particles may make Immuno-PCR an automated one-step immunoassay, and the magnetic particles used should have uniform size and morphology,

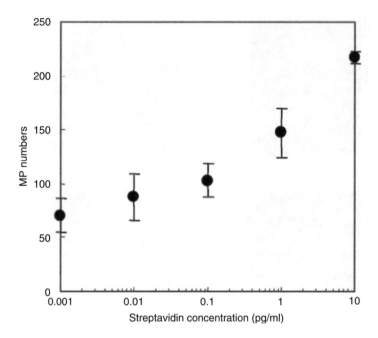

Figure 5.28 Relationship between streptavidin concentration and the numbers of magnetosome particles in an area of $20\,\mu m \times 20\,\mu m$. Before the biotin-MPs are applied, the glass slides are treated with various concentrations of streptavidin ($0.001-10\,pg/ml$). (Amemiya *et al.* 2005)

biocompatibility and high magnetization. Due to the attractive advantages of magneto-somes over synthetic magnetic particles, Ceyhan *et al.* (2006) used magnetosomes isolated from the magnetotactic bacterium *Magnetospirillum gryphiswaldense*, and modified with oligonucleotides and antibodies. The Immuno-PCR using antibody-functionalized magneto-some particles is often called Magneto Immuno-PCR (M-IPCR), which is often used for automatable high sensitive antigen detection.

The working principle of the M-IPCR technique is similar to that of a two-sided (sandwich) immunoassay. Here we discuss the M-IPCR technique using the chemically modified magnetosome particles, which bear streptavidin molecules at their magneto-some membranes. To elucidate its performance, the M-IPCR technique is used to detect the recombinant Hepatitis B surface Antigen (HBsAg) in human serum. As shown in Figure 5.29, this assay is based on a two-sided (sandwich) immunoassay. It should be noted that, in a standard immunoassay, the capture antibodies are typically bound to a solid phase; however, for M-IPCR, the capture phase is the antibody-functionalized biogenic magnetosome particles. In the experiment, these magnetosome particles are mixed with a DNA–anti-HBsAg antibody conjugate directly within the biological matrix, for example, standardized human serum. After the detection complex is formed, the magnetosomes are collected using an external magnetic field (Wacker *et al.* 2007).

The use of isolated magnetosome particles has obvious advantages over the use of synthetic magnetic microbeads. As shown in Figure 5.30, the limit of detection (LOD) using magnetosome particles is $320\,pg/ml$ HBsAg in standardized human serum, while the LOD using the commercial microbeads is $8\,ng/ml$. Therefore the M-IPCR using magneto-some particles has a 25-fold higher sensitivity. This enhanced performance may be due to the smaller size, monodispersity and higher magnetization of isolated magnetosome particles.

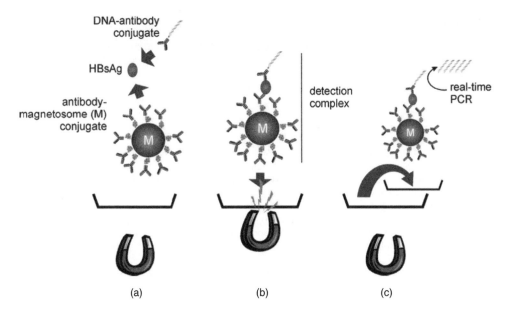

Figure 5.29 Procedure of the Magneto Immuno-PCR (M-IPCR). (a) HBsAg specific magnetosome–antibody conjugate and DNA–antibody conjugate are incubated simultaneously with the serum sample containing HBsAg resulting in a signal-generating detection complex; (b) a magnet is used to concentrate the complex to be detected, and the unbound materials are removed by subsequent washing steps; (c) after resuspending, a defined volume of the detection complex solution is transferred to a microplate containing the PCR mastermix to enable real-time PCR detection of the immobilized antigen. (Wacker *et al.* 2007)

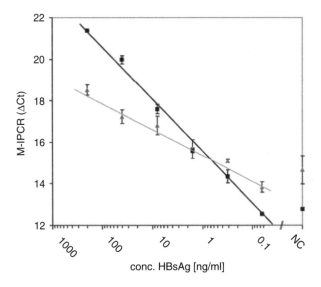

Figure 5.30 Comparison of Magneto Immuno-PCR (M-IPCR) using the magnetosome particles (black squares) and commercial magnetic microbeads (gray triangles). Both techniques showed linear regressions. The use of magnetosomes resulted in an improved LOD of 320 pg/ml and decreased standard deviations. This led to a 25-fold higher sensitivity, compared to the magnetic microbeads, which showed an LOD of 8 ng/ml. (Wacker *et al.* 2007)

References

Albrecht, M., Janke, V., Sievers, S., Siegner, U., Schuler, D. and Heyen, U. (2005). Scanning force microspy study of biogenic nanoparticles for medical applications, *Journal of Magnetism and Magnetic Materials*, **290–1**, 269–71.

Amemiya, Y., Tanaka, T., Yoza, B. and Matsunaga, T. (2005). Novel detection system for biomolecules using nano-sized bacterial magnetic particles and magnetic force microscopy, *Journal of Biotechnology*, **120**, 308–14.

Bahaj, A.S., James, P.A.B. and Moeschler, F.D. (1998b). *Low magnetic-field separation system for metal-loaded magnetotactic bacteria*, **177**, 1453–4.

Bahaj, A.S., James, P.A.B. and Moeschler, F.D. (1997). Continuous cultivation and recovery of magnetotactic bacteria, *IEEE Transactions on Magnetics*, **33**(5), 4263–5.

Bahaj, A.S., Croudace, I.W. and James, P.A.B. (1994). Metal uptake and separation using magnetotactic bacteria, *IEEE Transaction on Magnetics*, **30**, 4707–9.

Bahaj, A.S. and James, P.A.B. (1993). Characterization of magnetotactic bacteria using image processing techniques, *IEEE Transactions on Magnetics*, **29**(6), 3358–60.

Bahaj, A.S., James, P.A.B., Ellwood, D.C. and Watson, J.H.P. (1993). Characterisation and growth of magnetotactic bacteria: implication to clean up of environmental pollution, *Journal of Applied Physics*, **73**(10), 5394–6.

Bazylinski, D.A. and Frankel, R.B. (2004). Magnetosome formation in prokaryotes, *Nature Reviews Microbiology*, **2**(3), 217–30.

Bazylinski, D.A., Frankel, R.B. Heywood, B.R., Mann, S., King, J.W., Donaghay, P.L. and Hanson, A.K. (1995). Controlled biomineralization of magnetite (Fe_3O_4) and greigite (Fe_3S_4) in a magnetotactic bacterium, *Applied and Environmental Microbiology*, **61**, 3232–9.

Bulte, J.W.M. and Brooks, R.A. (1997). Magnetic nanoparticles as contrast agents for imaging, in *Scientific and Clinical Applications of Magnetic Carriers* (eds Hafeli, U., Schutt, W., Teller, J. and Zborowski, M.), Plenum, New York, London, pp. 527–43.

Ceyhan, B., Alhorn, P., Lang, C., Schuler, D. and Niemeyer, C.M. (2006). Semisynthetic biogenic magnetosome nanoparticles for the detection of proteins and nucleic acids, *Small*, **2**(11), 1251–5.

Darnton, N., Turner, L., Breuer, K. and Berg, H.C. (2004). Moving fluid with bacterial carpets, *Biophysical Journal*, **86**, 1863–70.

De Waard, H., Hilsinger, J. and Frankela, R.B. (2001). Instrument for the measurement of hysteresis loops of magnetotactic bacteria and other systems containing submicron magnetic particles, *Review of Scientific Instruments*, **72**(6), 2724–30.

Dunin-Borkowski, R.E., McCartney, M.R., Frankel, R.B., Bazylinski, D.A., Posfai, M. and Buseck, P.R. (1998). Magnetic microstructure of magnetotactic bacteria by electron holography, *Science*, **282**, 1868–70.

Eberbeck, D., Janke, V., Hartwig, S., Heyen, U., Schuler, D., Albrecht, M. and Trahms, L. (2005). Blocking of magnetic moments of magnetosomes measured by magneto relaxometry and direct observation by magnetic force microscopy, *Journal of Magnetism and Magnetic Materials*, **289**, 70–3.

Edelstein, R.L., Tamanaha, C.R., Sheehan, P.E., Miller, M.M., Baselt, D.R., Whitman, L.J. and Colton, R.J. (2000). The BARC biosensor applied to the detection of biological warfare agents, *Biosensors & Bioelectronics*, **14**, 805–13.

Enpuku, K., Minotani, T., Hotta, M. and Nakahodo, A. (2001). Application of High Tc SQUID magnetometer to biological immunoassays, *IEEE Transactions on Applied Superconductivity*, **11**, 661–4.

Frankel, R.B. and Bazylinski, D.A. (2004). Magnetosomes: nanoscale magnetic iron minerals in bacteria, in *Nanobiotechnology*, (eds Niemeyer, C. and Mirkin, C.), WILEY-VCH, Weinheim.

Frankel, R.B., Bazylinski, D.A., Johnson, M.S. and Taylor, B.L. (1997). Magneto-aerotaxis in marine coccoid bacteria, *Biophysical Journal*, **73**, 994–1000.

Grunberg, K., Muller, E.C., Otto, A., Reszka, R., Linder, D., Kube, M., Reinhardt, R. and Schuler, D. (2004). Biochemical and proteomic analysis of the magnetosome membrane in Magnetospitillum gryphiswaldense, *Applied and Environmental Microbiology*, **70**, 1040–50.

Grunberg, K., Wawer, C., Tebo, B.M. and Schuler, D. (2001). A large gene cluster encoding several magnetosome proteins is conserved in different species of magnetotactic bacteria, *Applied and Environmental Microbiology*, **67**, 4573–82.

Harasko, G., Pfutzner, H., Rapp, E., Futschik, K. and Schuler, D. (1993). Determination of the concentration of magnetotactic bacteria by means of susceptibility measurements, *Japanese Journal of Applied Physics*, **32**, 252–60.

Hartung, A., Lisy, M.R., Herrmann, K.H., Hilger, I., Schuler, D., Lang, C., Bellemann, M.E., Kaiser, W.A. and Reichenbach, J.R. (2007). Labeling of macrophages using bacterial magnetosomes and their characterization by magnetic resonance imaging, *Journal of Magnetism and Magnetic Materials*, **311**, 454–9.

Hergt, R., Hiergeist, R., Zeisberger, M., Schuler, D., Heyen, U., Hilger, I. and Kaiser, W.A. (2005), Magnetic properties of bacterial magnetosomes as potential diagnostic and therapeutic tools, *Journal of Magnetism and Magnetic Materials*, **293**, 80–6.

Hoell, A., Wiedenmann, A., Heyen, U. and Schuler, D. (2004). Nanostructure and field-induced arrangement of magnetosomes studied by SANSPOL, *Physica B - Condensed Matter*, **350**, E309–E313.

Komeili, A. (2007). Molecular mechanisms of magnetosome formation, *Annual Review of Biochemistry*, **76**, 351–66.

Komeili, A., Li, Z., Newman, D.K. and Jensen, G.J. (2006). Magnetosomes are cell membrane invaginations organized by the actin-like protein mamK, *Science* **311**, 242–5.

Komeili, A., Vali, H., Beveridge, T.J. and Newman, D.K. (2004). Magnetosome vesicles are present before magnetite formation, and MamA is required for their activation, *Proceedings of the National Academy of Sciences of the United States of America*, **101**, 3839–44.

Kriz, K., Gehrke, J. and Kriz, D. (1998). Advancements toward magneto immunoassays, *Biosensing and Bioelectroics*, **13**, 817–23.

Lang, C. and Schuler, D. (2006). Biogenic nanoparticles: production, characterization, and application of bacterial magnetosomes, *Journal of Physics: Condensed Matter*, **18**, S2815–S2828.

Lee, H., Purdon, A.M., Chu, V. and Westervelt, R.M. (2004). Controlled assembly of magnetic nanoparticles from magnetotactic bacteria using microelectromagnets arrays, *Nano Letters*, **4**, 995–8.

Martel, S., Tremblay, C.C., Ngakeng, S. and Langlois, G. (2006). Controlled manipulation and actuation of micro-objects with magnetotactic bacteria, *Applied Physics Letters*, **89**, 233904.

Nakayama, H., Arakaki, A., Maruyama, K., Takeyama, H. and Matsunaga, T. (2003). Single-nucleotide polymorphism analysis using fluorescence resonance energy transfer between DNA-labeling fluorophore, fluorescein isothiocyanate, and DNA intercalator, POPO-3, on bacterial magnetic particles, *Biotechnology and Bioengineering*, **84**(1), 96–102.

Penninga, I., de Waard, H., Moskowitz, B.M., Bazylinski, D.A. and Frankel, R.B. (1995). Remanence curves for individual magnetotactic bacteria using a pulsed magnetic field, *Journal of Magnetism and Magnetic Materials*, **149**, 279–86.

Proksch, R.B. *et al.* (1995). Magnetic force microscopy of the submicron magnetic assembly in a magnetotactic bacterium, *Applied Physics Letters*, **66**, 2582–4.

Sano, T., Smith, C.L. and Cantor, C.R. (1992). Immuno-PCR: very sensitive antigen detection by means of specific antibody–DNA conjugates, *Science*, **258**, 120–2.

Scheffel, A., Gruska, M., Faivre D., Linaroudis, A., Plitzko, J.M. and Schuler, D. 2006. An acidic protein aligns magnetosomes along a filamentous structure in magnetotactic bacteria, *Nature*, **440**, 110–14.

Schuler, D. and Frankel, R.B. (1999). Bacterial magnetosomes: microbiology, biomineralization and biotechnological applications, *Applied Microbiology and Biotechnology*, **52**, 464–73.

Schultheiss, D., Handrick, R., Jendrossek, D., Hanzlik, M. and Schuler, D. (2005). The presumptive magnetosome protein Mms16 is a poly(3-hydroxybutyrate) granule-bound protein (phasin) in Magnetospirillum gyphiswaldense, *Journal of Bacteriology*, **187**, 2416–25.

Simmons, S.L., Bazylinski, D.A. and Edwards, K.J. (2006). South-seeking magnetotactic bacteria in the northern hemisphere, *Science*, **311**, 371–4.

Takeyama, H., Yamazawa, A., Nakamura, C. and Matsunaga, T. (1995). Application of bacterial magnetic particles as novel DNA carriers for ballistic transformation of a marine cyanobacterium, *Biotechnology Techniques*, **9**, 355–60.

Yang, C.D., Takeyama, H., Tanaka, T., Hasegawa, A. and Matsunaga, T. (2001). Synthesis of bacterial magnetic particles during cell cycle of Magnetospirillum magneticum AMB-1, *Applied Biochemistry and Biotechnology*, **91–3**, 155–60.

Wacker, R., Ceyhan, B., Alhorn, P., Schueler, D., Lang, C. and Niemeyer, C.M. (2007). Magneto Immuno-PCR: a novel immunoassay based on biogenic magnetosome nanoparticles, *Biochemical and Biophysical Research Communications*, **357**, 391–6.

6

Magnetic Nanowires and their Biomedical Applications

6.1 Introduction

Many areas of science and engineering have been impacted by the integration of the biological sciences and the physical sciences at the micro and nano scales. Among various areas, biomagnetics is extremely interesting and promising. For example, nanofabricated magnetic particles have been used to selectively probe and manipulate biological systems. This is a rapidly growing field, and a broad range of applications have been developed, such as cell separation, biosensing, studies of cellular function, as well as a variety of potential medical and therapeutic applications (Reich *et al.* 2003). Most of the magnetic particles used have a spherical shape, usually consisting of a magnetic core and a shell that allows the functionalization of bioactive ligands to realize desired biomedical purposes. As the applications of magnetic particles are becoming more and more popular in the research of medicine and biotechnology, it would be advantageous if the magnetic particles could perform a variety of functions. Magnetic nanowires are a type of magnetic particle with significant potential for this purpose. Nanowires, also called nanorods in some literatures, have quasi-one-dimensional anisotropic structures with extremely high aspect ratios.

In the biomedical applications of magnetic particles, to optimize the properties of magnetic particles for desired biomedical applications, the composition, shape, size and surface chemistry of the particles should be controlled. In this respect, nanowires have attractive advantages. Magnetic nanowires possess unique properties that are quite different from those of bulk ferromagnetic materials, spherical particles and thin films. As schematically shown in Figure 6.1, the architecture and composition of a nanowire along its axis can be modulated precisely, and this could be used to precisely control the magnetic properties of nanowires for special biomedical applications.

Most of the magnetic nanowires used in biomedicine are metal cylinders electrodeposited in nanoporous templates. Their radius can be controlled in the range of 5 to 500 nm, and their length can be controlled up to about 60 μm. The multiple segment structures, as shown in Figure 6.1(b) and (c), provide extraordinary flexibilities for the control of their magnetic properties, as the shape and composition of each segment and the coupling between the layers can be precisely controlled (Sun *et al.* 2005; Chen *et al.* 2003a). Furthermore, as shown in Figure 6.1(d), by using ligands that selectively bind to different segments of a multi-segment wire, spatially modulated multiple functionalization

Nanomedicine: Design and Applications of Magnetic Nanomaterials, Nanosensors and Nanosystems
V. Varadan, L.F. Chen, J. Xie
© 2008 John Wiley & Sons, Ltd

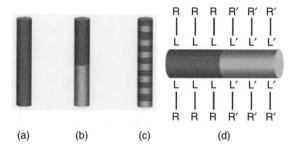

(a)   (b)   (c)   (d)

Figure 6.1 Inherent shape anisotropy of and functionalization of nanowires. (a) Single-component nanowire; (b) two-component nanowire in which the aspect ratio of each segment is greater than one; (c) two-component multilayer nanowire, in which the aspect ratio of each segment is less than one; (d) functionalization of a two-component nanowire. In the figure, ligands L and L' selectively bind to the two components, and thus the functional groups R and R', corresponding to L and L' repectively, are spatially separated. (Sun *et al.* 2005). Reproduced by permission of IBM

can be realized in these wires, and this feature can be used to improve the performances of magnetic nanowires in biomagnetic applications (Reich *et al.* 2003).

It should be noted that both single-component magnetic nanowires and multiple-segment magnetic nanowires are widely used for both scientific research and practical applications. Many important magnetic properties, for example, Curie temperature, coercivity, saturation field, saturate magnetization, remanent magnetization and the orientation of easy magnetization axis, can be modified by changing the diameter, thickness and composition of the magnetic/non-magnetic segments in multiple-segment nanowires (Sun *et al.* 2005). Many efforts have been made to develop methods for synthesis and manipulation of magnetic nanowires, and to explore their biomedical applications, and encouraging progress has been made.

In this chapter, before discussing the synthesis methods, characterization techniques and typical applications of magnetic nanowires, the basic knowledge about the magnetism of nanowires will be discussed. We start by introducing various nanowire structures.

6.1.1 Arrayed and Dispersed Nanowires

In most of their applications, nanowires are either arrayed or dispersed. Figure 6.2(a) shows an example of arrayed Ni nanowires with diameter about 200 nm. It should be noted that the Ni nanowires shown in Figure 6.2(a) are randomly arrayed. As will be discussed later, nanotubes can be arrayed in defined patterns for special purposes. Figure 6.2(b) shows dispersed Co nanowires with diameter about 70 nm. In biomedical applications, dispersed nanowires are usually suspended in solutions.

6.1.2 Single-segment, Multi-segment and Multilayer Nanowires

As it is desirable for a single nanomaterial to perform multiple functions simultaneously, multi-segment nanostructures are intensively investigated, and their inherent multiple functionalities are explored. Figure 6.1 shows that nanowires with multiple chemical components aligned along their long axis can be developed (Hurst *et al.* 2006). Figure 6.3(a) shows a piece of single-segment nickel nanowire. It should be noted that a single segment nanowire could be made of single element metal, alloy, or oxide. Figure 3 (b) shows a

(a)   (b)

Figure 6.2   (a) Arrayed Ni nanowires with diameter about 200 nm (Liu *et al.* 2007); (b) dispersed
Co nanowires with diameter about 70 nm (Maurice *et al.* 1998)

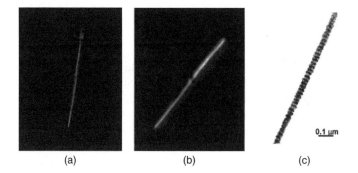

(a)   (b)   (c)

Figure 6.3   (a) Single-segment nickel nanowire (Bauer *et al.* 2003); (b) two-segment nickel-gold
nanowire (Bauer *et al.* 2003); (c) cobalt-copper multilayer nanowire. (Piraux *et al.* 1994)

piece of two-segment nickel-gold nanowire. Figure 6.3(c) shows a piece of cobalt-copper
multilayer nanowire.

6.1.3   Other Nanowire Structures

Besides the typical types discussed above, there are other types of nanowires, such as
encapsulated nanowires and core–shell nanowires.

6.1.3.1   Encapsulated Nanowires

Due to the significant electronic, magnetic and nonlinear optical properties of carbon
nanotubes, the encapsulation of foreign materials inside the hollow cavities of carbon
nanotubes has been extensively investigated. Many materials, such as iron, iron oxide
and cobalt, are encapsulated in carbon nanotubes, and their properties have been studied
(Bao *et al.* 2002). However, as their diameters are usually about several nanometers, the
applications of foreign materials encapsulated in carbon nanotubes in device development
are drastically limited. It has been found that materials encapsulated in the hollow cavities
of carbon micrometer tubes have more application value than materials encapsulated in
carbon nanotubes, and researchers are investigating the various physical and chemical

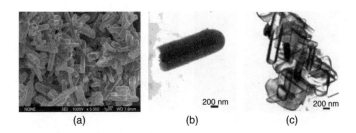

Figure 6.4 SEM and TEM images of magnetite nanowires encapsulated in carbon microtubes. (a) SEM image of magnetite nanowires encapsulated in carbon microtubes; (b) TEM image of a single magnetite nanowire encapsulated in carbon microtube; (c) TEM image of a single magnetite nanowire coexisted with the CMTs. (Xu *et al.* 2004)

phenomena of this types of materials (Xu *et al.* 2004). The materials encapsulated in carbon microtubes usually exist in the form of nanowires.

By pyrolyzing an ethanol/ferrocene mixture in an autoclave at 600 °C, magnetite (Fe$_3$O$_4$) nanowires encapsulated in carbon microtubes can be synthesized (Xu *et al.* 2004). The carbon microtubes have diameters in the range of 150 to 1500 nm and lengths up to 6 um. The inner magnetite nanowires are single crystalline with diameters in the range of 55 to 750 nm, and they exhibit ferromagnetic properties at room temperature. Figure 6.4(a) shows an SEM image of the as-prepared magnetite nanowires encapsulated in carbon microtubes. They are straight with a diameter range of 150–1500 nm and lengths up to 6 μm with at least one closed end. Figure 6.4(b) shows a TEM image of a single magnetite nanowire partly coated by carbon. Figure 6.4(c) shows a TEM image of the carbon microtubes obtained after HCl acid treatment. In these carbon microtubes, only one piece of single magnetite nanowire encapsulated in carbon microtube can be found.

6.1.3.2 Core–shell Nanowires

Quasi one-dimensional magnetic structures, where the magnetic domains, single-crystalline grains or nanocrystals are aligned in series, are desired for the study of magnetoresistance. Zhang *et al.* (2004) produced MgO/Fe$_3$O$_4$ core–shell nanowires by epitaxially depositing a Fe$_3$O$_4$ shell layer onto MgO nanowires serving as the supporting cores. The synthesized core–shell nanowires can be used to investigate the transport behavior of Fe$_3$O$_4$ in its quasi one-dimensional systems (Zhang *et al.* 2004). It should be noted that such core–shell nanowires can also be regarded as magnetic nanotubes filled with nonmagnetic materials.

Figure 6.5 shows the morphology of the MgO/Fe$_3$O$_4$ core–shell nanowires. Figure 6.5(a) shows that the MgO core is covered by a uniform layer of Fe$_3$O$_4$. The MgO core has a diameter of about 25 nm and the coated shell has a thickness of about 7 nm. Figure 6.5(b) shows the high-resolution transmission electron microscopy (HRTEM) image of the interface between the core and shell. It indicates that the MgO core and epitaxial magnetite shell have a relatively sharp interface. The continuous lattice fringes from the core to the shell confirm the perfect single-crystal epitaxial growth of the magnetite layer. As shown in Figure 6.5(c), the lattice spacing between two planes is about 2.947 Å, which corresponds to the distance of two (220) planes of magnetite. The angle between the interface of core–shell and lattice fringe of (220) plane is 45 °, indicating that the axial growth direction core is along [100]. Due to the tunneling of spin-polarized electrons across the anti-phase boundaries, at room temperature, the as-synthesized nanowires have a magnetoresistance of about 1.2 % under a magnetic field of B = 1.8 T (Zhang *et al.* 2004).

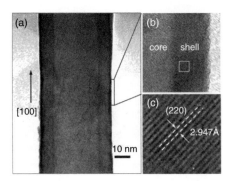

Figure 6.5 TEM image of a core–shell nanowire showing the uniform magnetite coating on the MgO core. To inspect the core–shell interface, the boxed region in (a) is enlarged in (b), and the boxed area in (b) is further enlarged in (c) to determine the lattice spacing in the shell region. (Zhang *et al.* 2004)

6.2 Magnetism of Magnetic Nanowires

Due to their quasi one-dimensional structure, magnetic nanowires exhibit unique magnetic properties (Heydon *et al.* 1997; Fert and Piraux 1999). The magnetic properties of a nanowire are related to many parameters of the nanowire, such as composition, length and diameter. For a multi-segment nanowire, its magnetic properties are also related to the layer thickness and the spacing between layers. The low dimensionality of nanowires brings about fundamental magnetic anisotropy. Some magnetic properties of magnetic nanowires, such as coercivity, remanence, saturation magnetic field and saturate magnetization, are dependent on the direction of the externally applied magnetic field. The giant magnetoresistance of a multilayer nanowire is caused by the segmented structure of the nanowire (Hurst *et al.* 2006).

In the following, we discuss the basic magnetism of magnetic nanowires, including shape anisotropy, switching in single-domain particles, magnetization hysteresis loop and the giant magnetoresistance of multi-segmented structures.

6.2.1 Shape Anisotropy

When a magnetic field is applied to a spherical object, the orientation of the magnetic field does not affect the magnetization of the spherical object. However, the magnetization of a nonspherical object is dependent on the orientation of the magnetic field. It is easier to magnetize a nonspherical object when the magnetic field is applied along the long axis of the object than along its short axis. For an object under an external magnetic field, the magnetic field inside the object is usually called the demagnetizing field, as this field tends to demagnetize the material. The demagnetizing field, H_d, is proportional to the magnetization $\mathbf{M}$ that creates it, but in an opposite direction, as given by:

$$\mathbf{H}_d = -N_d\mathbf{M} \tag{6.1}$$

where the demagnetizing factor N_d is related to the shape of the object. The calculation is quite complicated, and the exact value of N_d can be calculated only for an ellipsoidal object with uniform magnetization all over the object. To an ellipsoidal object with semi-axes a, b and c ($c \geq b \geq a$), the sum of demagnetization factors along the three semi-axes (N_a, N_b and N_c) equals to 4π, i. e. $N_a + N_b + N_c = 4\pi$.

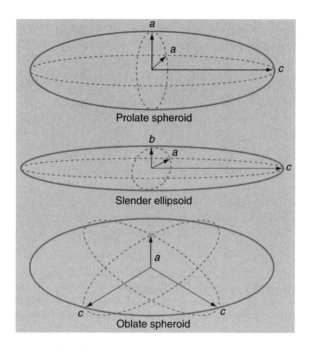

Figure 6.6 Three typical ellipsoids: prolate spheroid (c > a = b), slender ellipsoid (c ≫ a > b), and oblate spheroid (c = b > a). (Sun *et al*. 2005). Reproduced by permission of IBM

For a given magnetization direction, the magnetostatic energy E_D (erg/cm^3) is given by:

$$E_D = \frac{1}{2} N_d M_s^2 \tag{6.2}$$

where M_s (emu/cm^3) is the saturate magnetization of the object, and N_d is the demagnetization factor for the magnetization direction.

Figure 6.6 shows three typical ellipsoids, often used in the study of magnetic nanowires: prolate spheroid, slender ellipsoid and oblate spheroid. In the following, we discuss the demagnetization factors of these three types of ellipsoids.

6.2.1.1 Prolate Spheroid (c ⟩ a = b)

A single-component nanowire with circular cross-section can be approximated as a prolate ellipsoid. The demagnetization factors of a prolate ellipsoidal object are given by (Sun *et al*. 2005):

$$N_a = N_b = 4\pi \frac{m}{2(m^2 - 1)} \times \left[m - \frac{1}{2(m^2 - 1)^{1/2}} \times \ln \left(\frac{m + (m^2 - 1)^{1/2}}{m - (m^2 - 1)^{1/2}} \right) \right] \tag{6.3}$$

$$N_c = 4\pi \frac{1}{m^2 - 1} \times \left[\frac{m}{2(m^2 - 1)^{1/2}} \times \ln \left(\frac{m + (m^2 - 1)^{1/2}}{m - (m^2 - 1)^{1/2}} \right) - 1 \right] \tag{6.4}$$

where m is the aspect ratio: m = c/a.

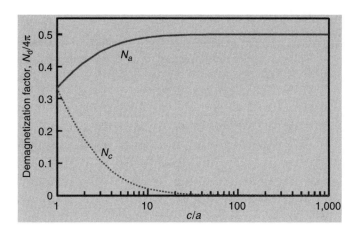

Figure 6.7 The relationship between the normalized demagnetization factors ($N_a/4\pi$ and $N_c/4\pi$) of a prolate spheroidal object and its aspect ratio (m = c/a). It can be found that if the aspect ratio is larger than 10, $N_a/4\pi \approx 0.5$ and $N_c/4\pi \approx 0$. (Sun *et al.* 2005). Reproduced by permission of IBM

For a prolate spheroid, the relationship between the normalized demagnetization factors ($N_a/4\pi$ and $N_c/4\pi$) and its aspect ratio (m = c/a) can be calculated based on Equations (6.3) and (6.4), and the results are shown in Figure 6.7. A high aspect ratio nanowire can be approximated as a high aspect ratio (large m) prolate spheroid. Under this approximation, the hard magnetization axis is perpendicular to the axis of the nanowire, and demagnetization factor of the nanowire in this direction is 2π; while the easy magnetization axis is parallel to the axis of the nanowire, and the demagnetization factor in this direction equals to zero. Therefore, the difference of the shape anisotropy energy of the nanowire along the two axes is given by (Sun *et al.* 2005):

$$K_u = \Delta E_D = E_{Da} - E_{Dc} = \pi M_s^2 \tag{6.5}$$

It should be noted that, to validate the long cylinder approximation, usually the aspect ratios of the nanowires should be larger than 10.

6.2.1.2 Slender Ellipsoid (c ≫ a ⟩ b)

A nanowire deposited into a template pore with noncircular cross-section can be approximated as a slender ellipsoid. The demagnetization factors along the three semi-axes a, b and c of a slender ellipsoid are given by (Sun *et al.* 2005):

$$N_a = 4\pi \frac{b}{a+b} - \frac{1}{2}\frac{ab}{c^2}\ln\left(\frac{4c}{a+b}\right) + \frac{ab(3a+b)}{4c^2(a+b)} \tag{6.6}$$

$$N_b = 4\pi \frac{a}{a+b} - \frac{1}{2}\frac{ab}{c^2}\ln\left(\frac{4c}{a+b}\right) + \frac{ab(3a+b)}{4c^2(a+b)} \tag{6.7}$$

$$N_c = 4\pi \frac{ab}{c^2}\left[\ln\left(\frac{4c}{a+b}\right) - 1\right] \tag{6.8}$$

6.2.1.3 Oblate Spheroid (c = b 〉 a)

A disk-shaped magnetic segment in a multiple-segment nanowire can be approximated as an oblate spheroid. The demagnetization factors of an oblate spheroid are given by (Sun et al. 2005):

$$N_a = 4\pi \frac{m^2}{m^2 - 1} \left[1 - \frac{1}{(m^2 - 1)^{1/2}} \times \arcsin \frac{(m^2 - 1)^{1/2}}{m} \right] \tag{6.9}$$

$$N_b = N_c = 4\pi \frac{m^2}{2(m^2 - 1)} \left[\frac{m}{(m^2 - 1)^{1/2}} \times \arcsin \frac{(m^2 - 1)^{1/2}}{m} - 1 \right] \tag{6.10}$$

where the aspect ratio m is given by m = c/a.

A magnetic segment in a multi-segment nanowire usually has a disk shape. Such a magnetic segment can be approximated as an oblate spheroid with a very low aspect ratio. The hard magnetization axis is parallel to the axis of the multi-segment nanowire, and the demagnetization factor N_a along this direction equals to 4π; while the easy magnetization axis is perpendicular to the axis of the multi-segment nanowire, and demagnetization factor N_c along this direction is zero. Therefore, the difference of the shape anisotropy energies in these two axes is given by (Sun et al. 2005):

$$K_u = \Delta E_D = E_{Da} - E_{Dc} = -2\pi M_s^2 \tag{6.11}$$

6.2.2 Switching in Single-domain Particles (Sun et al. 2005)

Based on the Brown equation, the magnetization configuration in a magnetic nanowire can be obtained by minimizing the total free energy (Frei et al. 1957). In the determination of magnetization configurations, usually many effects are ignored, such as the crystalline anisotropy energy (E_{ca}), the magnetoelastic energy (E_{EA}) and the effects related to the microstructure and the surface (Sun et al. 2005).

The formation of multiple magnetic domains inside a bulk ferromagnetic material can minimize the energy of the material. However, a magnetic particle will be in a single-domain state if its size is smaller than a critical value. For a prolate spheroidal particle as shown in Figure 6.6, the critical radius r_{sd}, which is in the short axis direction, is given by (Frei et al. 1957):

$$r_{sd} = \sqrt{\frac{6A}{N_c M_s^2} \ln\left(\frac{2r_{sd}}{a_1} - 1\right)} \tag{6.12}$$

with:

$$N_c = 4\pi \frac{1}{m^2 - 1} \times \left[\frac{m}{2(m^2 - 1)^{1/2}} \times \ln\left(\frac{m + (m^2 - 1)^{1/2}}{m - (m^2 - 1)^{1/2}}\right) - 1 \right] \tag{6.13}$$

where A is the stiffness constant (erg/cm), N_c is the demagnetization factor, M_s is the saturate magnetization (emu/cm^3) and a_1 is the near-neighbor spacing (cm). According to Equations (6.12) and (6.13), the critical radius of a nickel nanowire with an aspect ratio of 10 is about 300 nm, so the diameter of a single-domain nickel nanowire with an aspect ratio of 10 should be less than about 600 nm. However, for cobalt and iron nanowires with the same aspect ratio, to have single-domain structures, their diameters

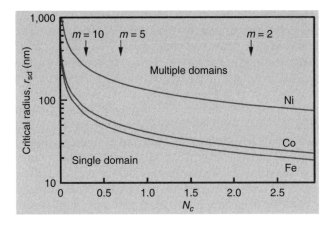

Figure 6.8 The relationship between the critical radius for a single-domain prolate spheroid and the demagnetization factor along the c-axis N_c. In the calculation, it is assumed that $A = 1 \times 10^{-6}$ erg/cm, $M_s(Ni) = 485$, $M_s(Co) = 1,440$, $M_s(Fe) = 1,710$, $a_1(Ni) = 0.2942$ nm, $a_1(Fe) = 0.2482$ nm, and $a_1(Co) = 0.2507$ nm. (Sun *et al.* 2005). Reproduced by permission of IBM

should be less than about 140 nm, because their critical radius is about 70 nm. Figure 6.8 shows the relationship between the critical radius and the demagnetization factor N_c. Generally speaking, when the aspect ratio (m = c/a) increases, demagnetizing factor N_c will decrease, and thus the critical radius will increase (Sun *et al.* 2005).

Figure 6.9 shows two models often used for analyzing the magnetization reversal modes of a single-domain particle: the coherent rotation model and the curling model (Sun *et al.* 2005). When no magnetic field is applied, the magnetic moments are aligned along its

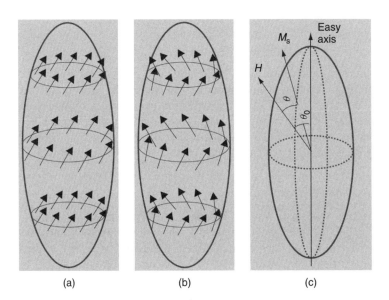

Figure 6.9 Two magnetization reversal models for in a single-domain prolate spheroid. (a) Coherent rotation model; (b) curling model; (c) coordinate system for analysis of coherent rotation. (Sun *et al.* 2005)

easy axis, which is the longest major axis. In the coherent rotation model as shown in Figure 6.9(a), during the magnetization reversal process, all the magnetic moments in the particle remain parallel to each other, and rotate away from the easy magnetization axis to minimize the exchange energy in the system. However, the demagnetization energy will be increased due to the existence of the magnetization component along the hard axis. In the curling model as shown in Figure 6.9(b), as the neighboring magnetic moments are not constrained to be parallel, there is no net magnetization along the hard axis, and thus the demagnetization energy can be minimized. However, the exchange energy may increase because the magnetic moments are not parallel to each other.

The switching mode of a magnetic particle is determined by the competition between the exchange energy and the demagnetization energy, and the particle size plays an important role in this competition. In the curling model, the size decrease will cause the increase in the relative angle between neighboring moments, and thus the increase in the exchange energy density, favoring the coherent rotation. Meanwhile, the increase in aspect ratio causes the increase in demagnetization energy density, which makes the curling process more favorable. There exists a critical size for the transition between the two reversal processes. If the external magnetic field is applied along the easy magnetization axis, the critical radius r_c (cm) is given by:

$$r_c = q \left(\frac{2}{N_a}\right)^{1/2} \frac{A^{1/2}}{M_s} \tag{6.14}$$

where q is the smallest solution of the Bessel functions, N_a is the demagnetizing factor along the minor axis, A is the constant of exchange stiffness (erg/cm) and M_s is the saturate magnetization (emu/cm^3). The value of q is related to the aspect ratio of the prolate spheroidal particle, and it is in the range of 1.8412 to 2.0816. The lower limit corresponds to an infinitely long cylinder, while the upper limit corresponds to a sphere. For an infinitely long cylinder, as $N_a = 2\pi$, we have (Sun *et al.* 2005):

$$r_c = \frac{q}{\pi^{1/2}} \times \frac{A^{1/2}}{M_s} \tag{6.15}$$

Three parameters, nucleation field, switching field and coercivity, are often used in describing the magnetization reversal process. The nucleation field H_n is a theoretical concept. At the nucleation field H_n, the magnetization just starts to change from a saturated single-domain state. A rapid change in magnetization occurs at the switching field H_s and the magnetization changes sign at the coercivity H_c. The switching mechanism can be experimentally determined from the angular dependence of the coercivity (Sun *et al.* 2005).

6.2.3 Magnetization Hysteresis Loops

The magnetization hysteresis loop of a sample illustrates how this sample responds to an external magnetic field, and theoretically, the magnetization hysteresis loop of an arbitrary sample can be obtained by minimizing the total free energy of the object in an external magnetic field (Sun *et al.* 2005). The hysteresis loop of an object is affected by many factors, such as the material and microstructure of the object, the shape and size of the object, the orientation of the magnetizing field, and the magnetization history of the sample. For arrays of nanoparticles, the interactions between individual particles may also affect the hysteresis loop. Figure 6.10 schematically shows two typical magnetization hysteresis loops for an array of single-segment ferromagnetic nanowires.

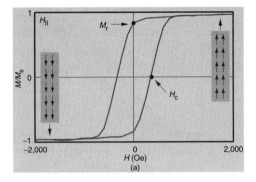
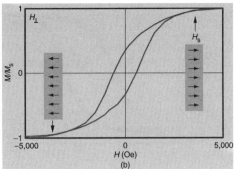

Figure 6.10   Hysteresis loops for a nickel nanowire array. The diameter of the nanowires is 100 nm, and their length is 1 μm. (a) The applied magnetic field H is parallel to the axis of the nanowires; (b) the applied field H is perpendicular to the axis of the nanowires. (Sun *et al.* 2005). Reproduced by permission of IBM

The parameters often used in describing the characteristics of a sample include the saturate magnetization M_s, the remanent magnetization M_r, the saturation field H_{sat} and the coercivity H_c. As shown in Figure 6.10, the saturation field H_{sat} is the field required for the sample to achieve the saturate magnetization M_s; the remanent magnetization M_r is the magnetization of the sample when the external magnetic field is moved away; the coercivity H_c is the magnetic field corresponding to the zero magnetization. There is another important parameter, switching field H_s, which is often used in analyzing magnetic nanomaterials. It is defined as the field at which the slope of the M–H loop reaches its maximum value, i.e., $d^2M/dH^2 = 0$. Actually, it is the field required to switch the magnetization from one direction to the opposite direction (Sun *et al.* 2005). Usually, the switching field H_s is equal to the coercivity H_c.

The saturate magnetization M_s of an object is achieved when all the magnetic moments in the object are aligned in the same direction. Therefore, the saturate magnetization M_s is an intrinsic property of a magnetic material, which is not related to the size and shape of the sample (Sun *et al.* 2005). Table 6.1 gives the M_s values of several typical magnetic materials.

The magnetic behaviors of a nanowire array are mainly determined by two factors. The first one is the magnetic properties of the individual nanowires. The second is the interactions among the individual magnetic nanowires, which are related to the geometry parameters of the nanowire array. In the following discussion, we concentrate on the influences of the geometrical characteristics on the magnetic behaviors of magnetic nanowire arrays.

Table 6.1   Saturation magnetization of typical magnetic materials. (Sun *et al.* 2005). Reproduced by permission of IBM

Material	M_s (emu/cm^3) at 290 K
Fe	1,710
Ni	485
Co	1,440
bcc $Fe_{0.5}Co_{0.5}$ (permedur)	1,950
Fcc $Ni_{0.78}Fe_{0.22}$	800

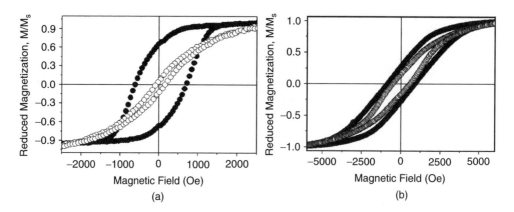

Figure 6.11   Axial (●) and in-plane (o) hysteresis loops for two arrays of Ni (a) and Co (b) nanowires. (Vazquez *et al.* 2005). Reprinted from Vazquez, M. Pirota, K. Torrejon J. Navas, D. Hernandez-Velez, M. (2005) "Magnetic behaviour of densely packed hexagonal arrays of Ni nanowires: Influence of geometric characteristics", *Journal of Magnetism and Magnetic Materials*, **294**, 174–181 with permission from Elsevier

6.2.3.1   Easy Magnetization Direction

To analyze the magnetic behavior of a nanowire array, it is necessary to find out the intrinsic easy magnetization axis of individual nanowires. One approach is to analyze the magnetization process when the sample is magnetized along the nanowire axis or in the plane of the membrane (Vazquez *et al.* 2005). Figure 6.11(a) shows the axial and in-plane magnetization hysteresis loops of a nickel nanowire array. The Ni nanowires have a diameter of 23 nm and a length of 3000 nm, and the array has a lattice parameter of 65 nm. It is clear that the axial loop has much larger values of coercivity and remanence, so the individual Ni nanowires have an effective longitudinal uniaxial anisotropy, originating from the shape anisotropy that would overcome other components such as magnetocrystalline and magnetoelastic anisotropies. Therefore, it is appropriate to consider as a first approximation individual nanowires to be single-domain with axial magnetization.

Figure 6.11(b) shows the magnetic hysteresis loops of an array of Co nanowires in a hexagonal lattice with 105 nm interwire distance. The Co nanowires have a diameter of 35 nm, and length of 800 nm. In this case, both the axial and the in-plane loops exhibit similar values of coercivity and remanence, so there is a balance between the axial shape anisotropy and the magnetocrystalline anisotropy.

6.2.3.2   Crystalline Long-range Ordering of an Array

It has been found that if each nanowire individually acts as a magnetic dipole, its magnetization process will contribute to the whole hysteresis loop of the array with a small square shaped loop. For an array of identical non-interacting nanowires, a simple macroscopic hysteresis loop would be observed with a single Barkhausen jump. However, as the distance among nanowires is usually smaller or comparable to their diameter and length, there are magnetostatic interactions among them (Vazquez *et al.* 2005). In this case, the neighboring wires give rise to an effective stray field that adds to the applied field. The strength of the stray field depends on the geometrical characteristics of the array. Here we discuss the influence of the size of the long-range crystalline ordering degree.

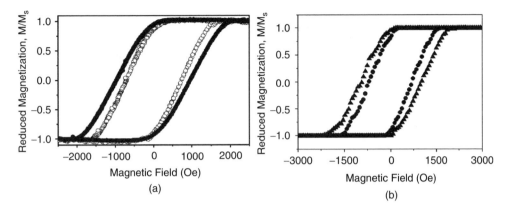

Figure 6.12 (a) Experimental hysteresis loops of nickel nanowire arrays 35 nm diameter and 105 nm lattice parameter after 3 h (o) and 72 h (●) first anodization; (b) simulated hysteresis loops of nanowire arrays of (a): 3 h (●) and 72 h (△) first anodization time. (Vazquez *et al.* 2004). Reprinted from Vazquez, M. Pirota, K. Torrejon, J. Navas, D. Hernandez-Velez, M. (2005) "Magnetic behaviour of densely packed hexagonal arrays of Ni nanowires: Influence of geometric characteristics", *Journal of Magnetism and Magnetic Materials*, **294**, 174–181 with permission from Elsevier

The nanopores on a template, and consequently nanowires, get ordered into domains with hexagonal symmetry by self-organization during first anodization process. The sizes of those domains increase with time, and this can be interpreted as the increase of the long-range ordering of the array. Figure 6.12(a) shows the difference in the hysteresis loops for different ordering degrees. After 72 hours of first anodization, the average size of crystalline domains of hexagonal symmetry increases up to around 3 μm², but with a reduction in homogeneity of diameter of about 6 % in comparison with the array obtained after only 3 hours of first anodization (Vazquez *et al.* 2005). It can be seen that the coercivity, remanence and field to reach saturation increase with the time of first anodization.

The influence of this spatial ordering on the magnetostatic interactions among nanowires and consequently on the magnetization process has been analyzed using Monte Carlo and iterative models (Vazquez *et al.* 2004), and as shown in Figure 6.12(b), the simulation results agree well with the experimental results. This simulation indicates that the magnetostatic interactions among the nanowires should be taken as multi-polar rather than dipolar. This is reasonable because the magnetic charges at the ends of single-domain nanowires are usually closer to those in adjacent nanowires instead of to those of the same nanowire, as the interwire distance is typically smaller than the nanowire length (Vazquez *et al.* 2005).

6.2.3.3 Filling Factor

The filling factor of a nanowire array is mainly determined by the ratio between the nanowire diameter and the lattice parameter or interwire distance. Figure 6.13 shows the hysteresis loops of nanowire arrays with different filling factors. It can be found that the higher filling factor results in a decrease in axial coercivity and remanence. This indicates that the uniaxial anisotropy of the nanowire array along the nanowire axis decreases with the increase of the filling factor, or in an alternative view, the effective easy magnetization axis of a nanowire array is somehow intermediate between the nanowire axis and the membrane plane (Vazquez *et al.* 2005).

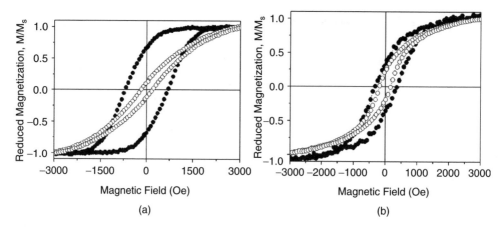

Figure 6.13   Axial (•) and in-plane (o) hysteresis loops of nickel nanowire arrays after filling the nanopores in membranes. (a) 35 nm and (b) 53 nm pores diameter. The lattice parameter is 105 nm. (Vazquez *et al.* 2005). Reprinted from Vazquez, M. Pirota, K. Torrenjon J. Navas, D. Hernandez-Velez, M. (2005) "Magnetic behaviour of densly packed hexagonal arrays of Ni nanowires: Influence of geometric characteristics", *Journal of Magnetism and Magnetic Materials*, **294**, 174–181 with permission from Elsevier

6.2.3.4 Nanowire Length

Figure 6.14(a) shows the hysteresis loops for different arrays of Ni nanowires with nanowire length ranging from 500 to 2000 nm. Figure 6.14(b) shows a nearly linear dependence of the coercivity and remanence of the nanowire arrays on the nanowire length. These conclusions suggest that different lengths of nanowires may result in the different strengths of magnetostatic interaction among nanowires (Vazquez *et al.* 2005).

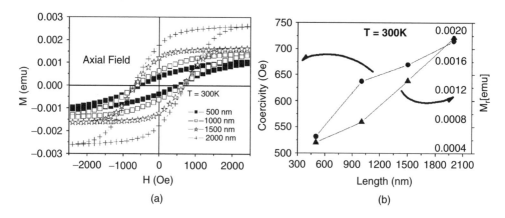

Figure 6.14   (a) Axial hysteresis loops for arrays of nanowires with different lengths; (b) coercivity and remanence versus the length of the nanowires. (Vazquez *et al.* 2005). Reprinted from Vazquez, M. Pirota, K. Torrenjon, J. Navas, D. Hernandez-Velez, M. (2005) "Magnetic behaviour of densly packed hexagonal arrays of Ni nanowires: Influence of geometric characteristics", *Journal of Magnetism and Magnetic Materials*, **294**, 174–181 with permission from Elsevier

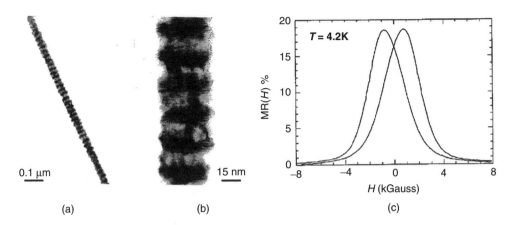

Figure 6.15 (a), (b) TEM images of a single nanowire with 10 nm thick Co (light bands) and Cu (dark bands) layers; c) plot of MR versus applied magnetic field (parallel to the nanowire film) at T = 4.2 K. The forward and reverse curves display little overlap, indicating a large coercive, or switching, field. (Piraux *et al.* 1994)

6.2.4 Multiple-segment Nanowires

Multi-segment ferromagnetic/nonmagnetic nanowires may exhibit giant magnetoresistance (GMR) (Hurst *et al.* 2006). Figures 6.15(a) and (b) show the transmission electron microscopy (TEM) images of multilayer Co–Cu nanowires produced within the pores of track-etched polycarbonate membranes (Piraux *et al.* 1994). As shown in Figure 6.15(c), the shape of the magnetoresistance (MR) curve for a Co–Cu nanowire is similar to those of the conventional Co–Cu multilayered films. Many other nanowires synthesized from magnetic non-magnetic couples, such as Ni–Cu, Fe–Cu and CoNiCu–Cu, also exhibit GMR effects.

The MGR effect of multilayered nanowires is related to the special electronic structures of the magnetic and non-magnetic layers, and this effect can be analyzed using the Mott model (Tsymbal and Pettifor 2001; Hurst *et al.* 2006). The Mott model assumes that the electrical conductivity in ferromagnetic metals can be described in terms of two independent conduction channels formed by the spin-up electrons and the spin-down electrons, and that the resistance of the spin-up and spin-down electron conduction channels in the ferromagnetic metals can be different. In a Co–Cu multi-component nanowire system, the non-magnetic Cu layers function as spacer layers. Due to the presence of these spacer layers, each magnetic Co layer can be taken as a single magnetic domain, in which the individual electron spins are aligned, so the magnetic moment of the layer is in one direction. Without an external magnetic field, the orientations of the magnetic moments in each of the Co layers are random with respect to one another. In this case, the resistance is high for both the spin-up and spin-down electron conduction channels, and thus the nanowire as a whole has high resistance. When an external magnetic field is applied, the magnetic moments of each Co layer will align in the same direction as that of the applied magnetic field. In this case, one electron conduction channel will have high resistance, while the other electron conduction channel will have little or no resistance. Therefore, with the application of the magnetic field, the resistance of the multi-component nanowire as a whole is decreased.

6.3 Template-based Synthesis of Magnetic Nanowires

Many efforts have been made on the synthesis of nanowires, and many important methods have been developed, such as ion-beam and electron-beam nanolithography, evaporation condensation, vapour–liquid–solid (VLS) growth, hydrothermal synthesis, and chemical synthesis (Sarkar *et al.* 2007). Among the various methods developed for nanowire synthesis, the templating method is the most attractive, as it can be used for synthesizing nanowire arrays with desired composition, size and aspect ratio. In this method, nanowire arrays are synthesized inside the pores of a template, through chemical or electrochemical techniques. Therefore, in selecting a template for synthesizing nanowires, several factors should be taken into consideration, especially the size, shape and density of the pores on the template (Piraux *et al.* 2005).

Electrochemical template synthesis allows the fabrication of single-segment and multi-segment nanowires. Using this technique, different segments can be introduced along the axis of a nanowire, and it is particularly attractive for the realization of multi-functionality. Furthermore, the materials for individual segments may be metals, alloys, metal oxides or electronically conducting polymers, and so specific magnetic, optical or electrical properties can be achieved (Wildt *et al.* 2006).

Magnetic nanowires possess unique and controllable magnetic properties due to their inherent shape anisotropy and the small wire dimensions. Using template-based synthesis methods, we can exploit the magnetic shape anisotropy and the magnetocrystalline anisotropy associated with single component, alloy or multi-segment nanowires (Mallet *et al.* 2004). In the following subsections, we discuss the important issues related to the template-based synthesis methods, including the fabrication of templates, electrochemical deposition method and electroprecipitation method.

6.3.1 Fabrication of Nanoporous Templates

In a template-based synthesis method, usually a membrane with uniform cylindrical pores with nanometer diameters is used as a template. When material is deposited into the pores of the membrane, it adopts their shape. If the template is dissolved, the material can retain the high aspect ratio of the pores, resulting in wires with nanometer diameters (Bauer *et al.* 2004). Therefore, the properties of the synthesized nanowire arrays or single nanowires are related to the template properties, for example, the roughness of the pore surfaces, and the spatial and size distributions of the pores (Sun *et al.* 2005).

Though a wide range of nanoporous materials have been used in nanowire synthesis, the most popular templates used are track-etched polymer membranes and anodic porous alumina membranes (Fert and Piraux 1999; Sarkar *et al.* 2007). Under optimized fabrication conditions, the nanoporosity of these two types of templates can be well defined, and the geometrical parameters of the pores can be controlled over a wide range. Depending on the thickness of the nanoporous materials, these templates may be self-supported or supported on a substrate (Piraux *et al.* 2005). We discuss below the fabrication of ion-track-etched membranes and anodic aluminum oxide templates.

6.3.1.1 Ion-track-etched Membranes

Polymer membranes, typically with thickness 5–50 μm, are usually fabricated by the nuclear track-etch technique. In this technique, high-energy heavy ion beams are used to bombard polymer films, such as polycarbonate, PET, polyimide and PVDF. The bombarded films are subsequently irradiated using UV light to increase the track-etching selectivity. The tracks are then revealed by etching the films with a suitable solution, and the etching time determines the diameter of the resulting pores. Usually, the diameter of the pores can range from 10 to 2000 nm, and the pore densities can be as high as 10^9 per square centimeter (Hurst *et al.* 2006; Piraux *et al.* 2005). Figure 6.16 shows the scanning

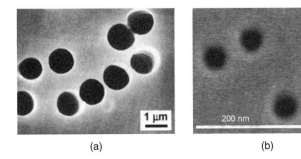

(a)　　　　　　　　　　　　(b)

Figure 6.16 SEM images of polycarbonate membranes. (a) Membrane with about 1 μm diameter pores (Hurst *et al*. 2006); (b) membrane with about 35 nm diameter pores. (Piraux *et al*. 2005)

electron microscopy (SEM) pictures of two polycarbonate membranes, with pore diameter about 1 μm and 35 nm respectively.

Iron-track-etched polymer templates have proven to be excellent templates for the preparation of nanowire arrays with uniform dimensions and small roughness at the surface (Piraux *et al*. 2005; Fert and Piraux 1999). However, the random nature of pore formation during the fabrication process may cause a large number of intersecting pores that will detrimentally affect the number, homogeneity and dimensions of the multi-segment nanowires during their synthesis (Hurst *et al*. 2006).

6.3.1.2 Anodic aluminum oxide templates

Sarkar *et al*. (2007) made an extensive review on the preparation of porous anodic aluminum oxide template (AAO) templates. Usually, AAO templates are fabricated using the two-step anodization method developed by Masuda and Fukuda (1995). In this method, a thin sheet of high purity aluminum is first annealed and then subjected to an electropolishing solution, which removes the top layers of aluminum oxide on the surface leaving it with a mirror-like shine. Subsequently, the aluminum is subjected to an anodization step in an acidic electrolyte solution. The sheet is then placed in a chromate solution to remove the barrier oxide layer, and a second anodization is performed in the same acidic solution. Finally, the aluminum is removed. As shown in Figure 6.17, the membranes fabricated using this method contain cylindrical pores of uniform diameter arranged in a hexagonal array, and these pores usually do not intersect each other (Hurst *et al*. 2006).

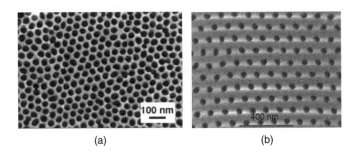

(a)　　　　　　　　　　　　(b)

Figure 6.17 SEM images of AAO membranes with different pore densities. (a) High density (Hurst *et al*. 2006); (b) low density. (Piraux *et al*. 2005)

To obtain templates with desired diameter, length and density of pores, the anodization parameters, such as the anodization voltage, current, electrolyte bath temperature and composition, should be carefully controlled during the fabrication procedure (Sarkar *et al.* 2007). The template thickness, and subsequently the pore length, is dependent on the anodization time, with longer anodizations resulting in thicker templates. Moreover, pores with different diameters can be achieved by varying the composition and concentration of the acidic electrolyte solution, and the temperature and voltage of the anodization. Pore diameters of 5–33 nm, 30–70 nm, and 150–267 nm can be achieved by using H_2SO_4, oxalic acid, and H_3PO_4, respectively, under varying temperatures and voltages (Hurst *et al.* 2006). By controlling the anodization parameters, pore densities as high as 10^{11} to 10^{12} per square centimeter can be achieved. Typically, the templates fabricated in this method have porosities are in the range 15–65 % (Piraux *et al.* 2005).

Generally speaking, there are three types of alumina membranes: self-supported alumina membrane, nanoporous alumina on an aluminum sheet and nanoporous alumina supported by a silicon (or quartz) substrate (Piraux *et al.* 2005). An advantage of using alumina templates is that nanowires can be synthesized at high temperatures. However, thermal stability of alumina templates on aluminum is limited by the melting point of aluminum, and self-supported alumina membranes tend to warp under thermal treatment. In contrast, alumina membranes supported on silicon (or quartz) substrates exhibit better thermal stability since the silicon (or quartz) wafer prevents the alumina membranes from warping and cracking under thermal treatment.

6.3.2 Electrochemical Deposition

Electrochemical deposition offers distinct advantages over other methods for the synthesis of one-dimensional nanostructures, and is widely used in the fabrication of magnetic nanowires. This method does not require expensive instrumentation, high temperatures or low-vacuum pressures. As nanowires have a high growth rate, this method is also not time-consuming. Using this method, multi-segment nanowires can be easily synthesized by changing the plating solution and accordingly varying the potential of the deposition. Furthermore, by varying the shape of the electrical pulse bringing about the deposition, the interface between multiple electrodeposited components can be controlled (Hurst *et al.* 2006). We will first discuss the fabrication procedure, followed by the synthesis of multi-segmented nanowires and alloy nanowires.

6.3.2.1 Fabrication Procedure

Electrodeposition is a process in which an electrical current passes through an electrolyte of metallic ions, and a reduction takes place when the ion encounters the cathode (working electrode). In the electrodeposition using a nanoporous membrane as a structure to create nanowire arrays, electrodeposition takes place in the channels of the membrane (Whitney *et al.* 1993; Bauer *et al.* 2003). As shown in Figure 6.18, electrodeposition of nanowires is usually done in a three-electrode arrangement, consisting of a reference electrode, a specially designed cathode and an anode or counter electrode (Bera *et al.* 2004).

Figure 6.19 shows the time dependence of the electrodeposition current in the fabrication of nanowires. As the material is electrodeposited, the nanowires grow from the bottom. At Region I, materials are deposited in the pores of the membrane until the pores are fully filled. At Region II, the material grows out of the pores, forming hemisphere caps on the ends of the nanowires. At Region III, the hemisphere caps formed at Region II

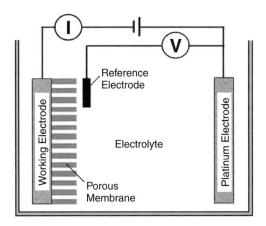

Figure 6.18 Three-electrode arrangement for electrodeposition of nanowires. (Bera *et al.* 2004)

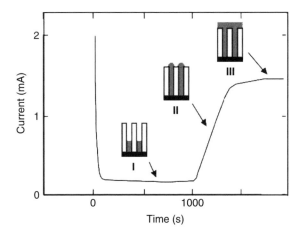

Figure 6.19 Time dependence of the deposition current in the synthesis of nickel nanowires. The nanowires have a diameter of 60 nm, and thickness of the polycarbonate template used is 6 μm. (Bauer *et al.* 2004)

form a contiguous surface over the surface of the membrane. To obtain single nanowires, the growth should be stopped somewhere within Region I.

Figure 6.20 shows the general process for the synthesis of nanowires through electrochemical deposition. In the first step, a thin metal film is evaporated onto one face of the template (Hurst *et al.* 2006). This metal film is used as a working electrode that is responsible for electrodepositing materials in the pores. Typically, before the desired components are deposited, a layer of sacrificial metal is deposited into the pores to prevent a puddling effect, which causes one end of the rod to have a deformed mushroom shape. After the sacrificial layer is deposited, the desired components are deposited sequentially. After the deposition of the desired components, the nanowires are released by chemically dissolving the thin film electrode, sacrificial metal layer and template.

For some sensing and nanoelectrode applications, the magnetic nanowires remain in the template and function as an array. In the production of nanowire arrays, the final quality control of nanowire arrays should be carefully performed (Hernandez-Velez *et al.* 2005).

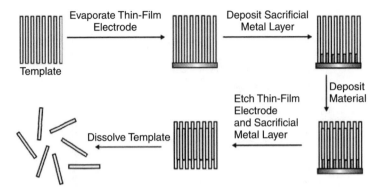

Figure 6.20 General procedure for the synthesis of nanowires by the deposition of materials into nanoporous templates. (Hurst *et al.* 2006)

The final quality control usually includes checking the nanopores filling by the Rutherford Back-Scattering technique, smoothing the upper surface of the membrane by suitable polishing and etching the bottom surface to eliminate the dendrites at the lower part of nanowires that may deteriorate the magnetic behavior of the array (Vazquez *et al.* 2005).

6.3.2.2 Synthesis of Multi-segment and Multilayer Nanowires

An advantage of electrodeposition is that different materials can be sequentially deposited in the templates, yielding nanowires comprised of segments of different materials. This is usually achieved by altering the applied potential of a solution with more than one precursor, or by changing the deposition solution (Bauer *et al.* 2003; Bauer *et al.* 2004; Blondel *et al.* 1997). Similar methods have been used for the fabrication of magnetic multilayer nanowires in membrane, as shown in Figure 6.21. Fert and Piraux (1999) developed a pulse-plating method in which two metals are deposited from a single solution by switching between the deposition potentials of the two constituents. Such a single-bath method can be used to make various multilayer nanowires, such as Co/Cu, NiFe/Cu, Ni/Cu, and CoFe/Cu (Piraux *et al.* 2005; Chen *et al.* 2003; Blondel *et al.* 1994). We will now briefly discuss the single-bath method for the growth of Co/Cu multilayer nanowires.

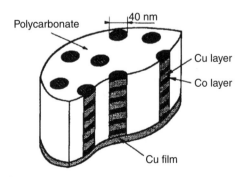

Figure 6.21 Array of multilayered nanowires in nanoporous track-etched polymer membrane. (Fert and Piraux 1999). Reprinted from Fert, A. Piraux, L. (1999). Magnetic nanowires. *Journal of Magnetism and Magnetic Materials* **200**, 338–358, with permission from Elsevier

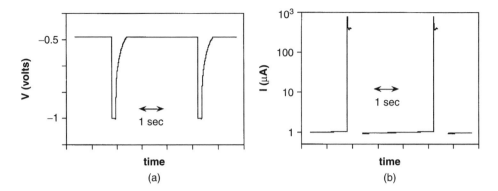

Figure 6.22 Time traces of (a) the potential, and (b) plating current of pulse-plated Co(10 nm)/Cu(2 nm) multilayer nanowires. (Fert and Piraux 1999). Reprinted from Fert, A. Piraux, L. (1999). Magnetic nanowires. *Journal of Magnetism and Magnetic Materials* **200**, 338–358, with permission from Elsevier

Fert and Piraux (1999) synthesized multilayer Co/Cu nanowires from a single sulfate bath using potentiostatic control and a pulsed deposition technique. The nobler element (Cu) is kept in a dilute concentration so that its reduction rate is slow and limited by diffusion. The concentration of the Cu ions is less than 1 % of the concentration of Co ions. Pure Cu is deposited at potentials between −0.1 and −0.5 V versus Ag/AgCl electrode, while almost pure Co is deposited at −1 V. The deposition rates of Cu and Co are 0.5 and 30 nm/s respectively. The potential is switched when the deposition charges for the nonmagnetic and the magnetic layers, Q_{NM} and Q_M respectively, reach the set values. Such a procedure is required to give uniform layer thicknesses all along the filament. Upon switching to the Cu deposition potential, the current is slightly anodic, indicating small redissolution of the magnetic metal. To limit redissolution effects as the potential is set to more positive value, sufficient time is required to allow the potential to rise to the Cu potential deposition (Fert and Piraux 1999). Then, the potential is reset for the deposition of the next copper layer. Figure 6.22 shows typical cathode current transients resulting from potentiostatic pulses.

Figure 6.23 shows the mean plating current recorded during each pulse for Cu deposition at −0.5 V and Co deposition at −1 V, plotted as a function of the number of electrodeposited bilayers. The electrodeposition process is stopped when the wires emerge from the surface, as indicated by the sudden increase of the plating current (Fert and Piraux 1999).

6.3.2.3 Synthesis of Alloy Nanowires

Using different electrolytic solutions and depositing at different potentials, a variety of alloys with controlled composition can be fabricated, such as NiCu, NiFe, CoNiFe, CoFe, CoPd and CoPt (Encinas *et al.* 2002; Mallet *et al.* 2004). As an example, in what follows, we discuss the synthesis of Co_XPt_{1-X} alloy nanowires.

Co_XPt_{1-X} alloy nanowires can be fabricated by electrochemical template synthesis from a solution containing both Co(II) and Pt(II) ions. Single-phase, fcc Co_XPt_{1-X} alloy nanowires can be obtained over a wide range of deposition conditions. Due to their shape and magnetocrystalline anisotropy, Co_XPt_{1-X} alloy nanowires exhibit large coercivity and high remanence along the axis of the nanowire. In the method developed by Mallet *et al.*

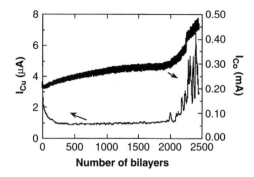

Figure 6.23 Mean plating current recorded during pulses for Cu deposition at -0.5 and Co deposition at -1 V as a function of the number of Co(10 nm)/Cu(2 nm) bilayers. The sudden increase of the plating current for both Co and Cu is due to the emergence of the first nanowires from the surface of the membrane. Reprinted from Fert, A. Piraux, L. (1999). Magnetic nanowires. *Journal of Magnetism and Magnetic Materials* **200**, 338–358, with permission from Elsevier

(2004), Co_xPt_{1-x} alloys are electrodeposited into a nanoporous template from a solution containing 0.01 M $CoSO_4$, 0.01 M $Pt(NO_2)_2(NH_3)_2$, 0.085 M $NaCH_3COO$, 0.052 M $N(CH_2\text{-}CH_2\text{-}OH)_3$ and 0.094 M Na_2CO_3. Sulfuric acid (H_2SO_4) is used to adjust the pH to 5.7. Deposition is carried out in a conventional three-electrode cell with a counter electrode (platinum) and a reference electrode (Ag/AgCl).

Figure 6.24 shows a typical current transient for the deposition of Co_xPt_{1-x} nanowires. After the initial decrease due to depletion of metal ions within the pores, the deposition current does not change much during the filling of the pores. When the pores are filled, the deposition current increases dramatically because the deposition area is no longer constrained. The inset of the figure is an SEM image of a Co_xPt_{1-x} nanowire after removal from the template. The nanowire has a diameter of about 48 nm diameter, and it is clear that it is uniform and free from voids.

Figure 6.25 shows X-ray diffraction patterns for the synthesized Co_xPt_{1-x} nanowires. The nanowires exhibit a disordered fcc structure with a strong peak at 42.3 $°$ corresponding

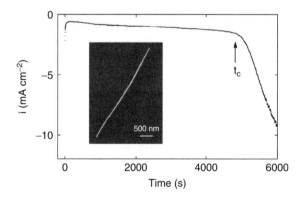

Figure 6.24 Current transient for deposition for Co_xPt_{1-x} nanowires in a 6 μm-thick polycarbonate template at -0.9 V (Ag/AgCl). The current is normalized to the area of the template. The inset shows an SEM image of a 4 μm long, 48 nm diameter $Co_{0.65}Pt_{0.35}$ nanowire after removal from the template. (Mallet *et al.* 2004)

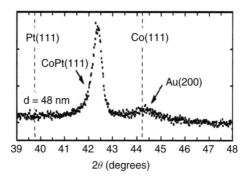

Figure 6.25 X-ray diffraction patterns for the $Co_{0.65}Pt_{0.35}$ nanowires electrodeposited. (Mallet *et al.* 2004)

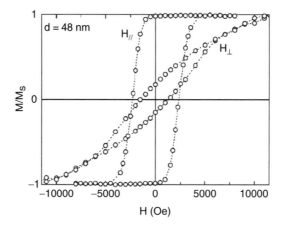

Figure 6.26 Hysteresis loops of a $Co_{0.65}Pt_{0.35}$ nanowire array in a polycarbonate template. (Mallet *et al.* 2004)

to fcc CoPt(111). The absence of peaks for Pt(111) at $39.8\,°$ and for Co(111) at $44.2\,°$ indicates that the nanowires are single phase (Mallet *et al.* 2004).

Figure 6.26 shows the magnetization hysteresis loops of the nanowire array when the external magnetic field is parallelly and perpendicularly applied to the axis of the nanowires, respectively. When the external magnetic field is parallelly applied to the axis of the nanowires, the hysteresis loop is relatively square. The hysteresis loop has a squareness of 0.98 and a coercivity of 2320 Oe. When the magnetic field is applied perpendicular to the axis of the nanowire, the hysteresis loop has a squareness of 0.16 and a coercivity 1390 Oe. Therefore, the synthesized nanowires have a strong magnetic anisotropy, and the easy magnetization axis is along the axis of the nanowires.

6.3.3 Electroprecipitation

Terrier *et al.* (2005) synthesized Fe_3O_4 nanowires using an electroprecipitation method. The templates used are polycarbonate track-etched membranes, and the pores on the membranes are $6\,\mu m$ long and have an average diameter of 350 nm. Both sides of a membrane are covered with Au. The bottom Au layer has a sufficient thickness to close the pores, and serves as a working electrode. The upper layer of Au is thin (about 20–100 nm), and the holes are open. The membrane is encapsulated in such a way that the electrolyte reaches the working electrode only through the pores. In such a configuration, the iron oxide is deposited from the bottom of the pores until it establishes an electrical contact while reaching the top Au layer.

In this method, an aqueous solution is used, containing 3 g/l of ammonium ferrous sulfate $(NH_4)_2SO_4 \cdot FeSO_4 \cdot 6H_2O$, 3 g/l of potassium nitrate KNO_3, hydrochloric acid HCl and hydrazine hydrate $N_2H_4 \cdot H_2O$. The quantity of hydrazine and HCl is adjusted so that the starting pH value of the solution is close to 5.5. The process is taking place in an Ar atmosphere to avoid unwanted reactions with oxygen. According to Equation (6.16), hydroxide ions are produced on the surface of the working electrode, inside the pores, when applying a voltage of typically −0.8 V versus Ag/AgCl. This chemical reaction increases the alkalinity of the solution at the surface of the electrode, and it is a necessary condition to produce ferrous hydroxides $Fe(OH)_2$ as indicated in Equation (6.17), which takes place at a pH of 11–13. Around the pH in the range of 11–13, ferrous hydroxides are partially transformed into ferric hydroxides, as indicated in Equation (6.18). The hydroxide compounds are then combined in a cascade process to form magnetite, which is a 2:1 mixture of Fe^{3+} and Fe^{2+} ions, as indicated in Equation (6.19).

$$NO_3^- + 7H_2O + 8e^- \rightarrow NH_4^+ + 10OH^- \tag{6.16}$$

$$Fe^{2+} + 2OH^- \rightarrow Fe(OH)_2 \tag{6.17}$$

$$2Fe(OH)_2 + N_2H_4 + 4H_2O \rightarrow 2Fe(OH)_3 + 2NH_4OH \tag{6.18}$$

$$2Fe(OH)_3 + Fe(OH)_2 \rightarrow Fe_3O_4 + 4H_2O \tag{6.19}$$

For temperatures between 70 and 90 °C, the synthesized magnetite is expected to have high purity, since iron oxides such as akaganeite β-FeOOH, hematite α-Fe_2O_3, or other ferrous/ferric compounds undergo a dissolution-recrystallization process involving hydrazine that yields magnetite (Terrier et al. 2005).

The XRD pattern shown in Figure 6.27(a) indicates a polycrystalline magnetite structure. There is no peak corresponding to other iron oxides/hydroxides such as hematite or goethite, in the XRD pattern. Figure 6.27(b) shows a TEM image of a fragment of a nanowire after dissolution of the membrane. The nanoparticles surrounding the wire are due to the growth that occurs on the top Au layer when the growing nanowires reach the top Au layer. Figures 6.27(c) and (d) are high-resolution TEM (HRTEM) pictures of the same sample. The nanowires have a diameter of about 350 nm, and the average grain size in the nanowires is in the range of 50 to 70 nm.

Figure 6.28 shows the temperature dependence of magnetizations under the field-cooled (FC) and zero-field-cooled (ZFC) conditions, and measurements are performed under magnetic fields of 30 and 75 mT. The difference between FC and ZFC curves is typical of the relaxation of granular systems. The ZFC curves exhibit a clear break in the slope at 110 K. Because the temperature of this break is independent of the applied external magnetic field, this break is intrinsic to the system. It appears as a magnetic manifestation of the Verwey transition (Terrier et al. 2005).

6.3.4 Self-assembly of Nanowires

Self-assembly is a process by which small-scale components spontaneously assemble into functional systems without human intervention, and this is a promising technique

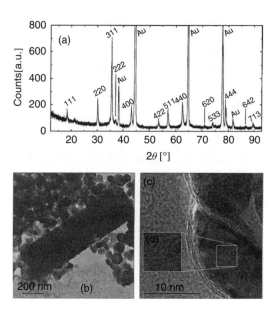

Figure 6.27 (a) XRD spectrum of a Fe_3O_4 nanowires in membrane; (b) TEM image of a Fe_3O_4 nanowire of 350 nm in diameter; (c) HRTEM picture of the granular structure of the nanowire, with (d) insight into atomic planes. (Terrier *et al*. 2005)

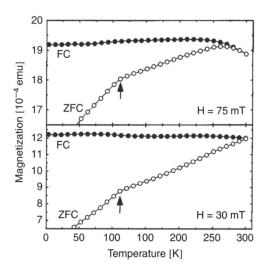

Figure 6.28 Temperature dependence of magnetizations under the field-cooled (FC) and zero-field-cooled (ZFC) conditions. (Terrier *et al*. 2005)

for fabricating nanometer-scale functional devices in the bottom-up approach (Liu *et al*. 2007; Tanase *et al*. 2001). Different driving forces have been used in controlling the self-assembly process, for example, magnetostatic forces, capillary forces, electrostatic forces, van der Waals forces, biospecific interactions. The assemblies of both single-segment nanowires and multi-segment nanowires have been studied. Compared to single-segment nanowires (Martin *et al*. 2002), multi-segment structure provides more flexibilities for self-assembly. By taking advantage of the different properties and

Figure 6.29   SEM image of a bundle of Au-Ni nanowires. (Love *et al.* 2003)

functions of the individual segments, multiple avenues for self-assembly are possible for a single system (Hurst *et al.* 2006). Here we discuss the magnetic assembly of multi-segment nanowires. The self-assembly of magnetic nanowires in suspension is discussed in Section 6.6.3.

Different types of bundle structures can be formed based on magnetic interactions among multi-segment nanowires (Hurst *et al.* 2006). Figure 6.29 shows a bundle structure of Au–Ni nanowires (Love *et al.* 2003). Each multi-segment nanowire consists of ferromagnetic Ni segments separated by diamagnetic Au spacer segments. The magnetic interactions between the Ni segments of the nanowires are responsible for the assembly, and the Au spacer segments increase the stability of the assembly by producing multiple, simultaneous magnetic interactions along the length of the nanowire. In these multi-segment nanowires, the Ni segments have a disk shape, with thickness smaller than diameter. This type of structure causes the magnetic moments of the Ni segments to align perpendicular to the long axis of the nanowire. Therefore, individual nanowires bundle side-to-side, forming the bundle structure shown in Figure 6.29.

6.4   Characterization of Magnetic Nanowires

Nanowires have two quantum-confined dimensions and one unconfined dimension. They possess unique electrical, electronic, thermoelectrical, optical, magnetic and chemical properties, which are different from those of their parent counterpart. The physical properties of nanowires are influenced by the morphology of the nanowires, diameter-dependent band gap, carrier density of states etc. (Sarkar *et al.* 2007; Doudin *et al.* 1997). In the following subsections, we discuss the methods for the characterization of the electrical properties, magnetization properties and magnetic anisotropy of magnetic nanowire arrays and single magnetic nanowires.

6.4.1   Electrical Properties

In the study of the electrical properties of magnetic nanowires, usually the anisotropic magnetoresistance (AMR) of magnetic nanowires is characterized. The electrical resistivity of a nanowire is related to the relative orientations of the magnetization and the electric current, as given by:

$$\rho = \rho_\perp + (\rho_\parallel - \rho_\perp)\cos^2\theta \qquad (6.20)$$

where $\rho_\parallel$ and $\rho_\perp$ are the resistivities with the magnetization parallel and perpendicular to the electric current respectively, and θ is the angle between the magnetization and the current. In measurements, the electrical current is along the axis of the nanowire. Usually the AMR ratio is defined as $(\rho_\parallel - \rho_\perp)/\rho_0$, with ρ_0 the resistivity of the nanowire without external magnetization. Though the AMR ratio is small in Ni and Co (typically of the order of 1%), it can be used as a tool to investigate the reversal of magnetization in nanowires (Wegrowe *et al.* 1998) as the resistance measurements can be performed with an accuracy of one part in $10^4 - 10^5$.

6.4.1.1 Nanowire Arrays

The conventional two-probe method can be used for the measurement of magnetoresistance of nanowire arrays. In this method, the nanowires in an array are connected in parallel to each other, and the two measurement probes contact the top and bottom faces of the template. Usually, silver epoxy and silver paint are used as the connecting and contacting materials.

Figure 6.30 shows the measurement results of magnetoresistance of parallel arrays of Ni and Co nanowires at T = 4.2 K. In the experiment, the current is along the axis of the nanowires (Piraux *et al.* 1997; Fert and Piraux 1999). For the arrays of cobalt nanowires with $\phi = 90$ nm, if a magnetic field is applied parallel to the current, the resistance decreases; while if the magnetic field is perpendicular to the current, the resistance increases. This is due to the dependence of the resistance on the angle between magnetization and current, and shows the coexistence of transversally and axially magnetized domains in the zero-field state. However, for the arrays of Ni nanowires with $\phi = 90$ nm, the resistance hardly changes when an external magnetic field is parallel to the nanowire axis. This is consistent with a squareness ratio M_r/M_s close to one, and confirms that the Ni nanowires keep their magnetization along the axis of the nanowire throughout the reversal process.

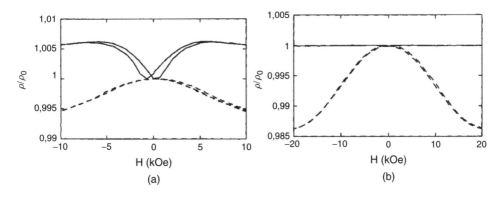

Figure 6.30 Anisotropic magnetoresistance curves obtained for arrays of Co nanowires with $\phi = 90$ nm (a) and for Ni nanowires with $\phi = 90$ nm (b). The field is applied perpendicular (dashed line) and parallel (full line) to the wire axis. (Piraux *et al.* 1997)

6.4.1.2 Single Nanowires

The measurement of single nanowires is more complicated that the measurement of nanowire arrays, and one challenge is how to make electrical contacts to single nanowires. Two methods have been developed for the measurement of the electrical properties of single nanowires: the self-contacting method and the lithography method (Piraux *et al.* 2005).

(1) Self-contacting Method

This method is applicable for nanowire arrays in templates with relatively low pore density (Piraux *et al.* 2005). Figure 6.31 shows the working principle of this method. As shown in Figure 6.31(a), in addition to the thick metal cathode from which the nanowires start to grow inside the pores, another thin metallic layer is deposited on the other side exposed to the electrolyte prior to electrodeposition. This film is very thin so that it does not close the pores, and thus the pores can still be filled by the solution. As the nanowires do not grow at the same velocity, the first emerging nanowire slows down the growth of the others by favoring the formation of a film on the surface metallic layer. As shown in Figure 6.31(b), the emerging of the first nanowire can be detected by a sharp increase of the plating current, and electroplating is interrupted immediately to avoid the emergence of other nanowires once such a sharp increase is observed. Using this procedure, electrical contacts are established on the two extremities of a nanowire, so two-probe measurements of electric resistance can be conducted. As a nanowire has an extremely large aspect ratio, the contact resistance is usually much smaller than the electrical resistance of the single nanowire sample.

(2) Lithography Method

In this method, electron beam lithography is used to connect a single nanowire (Piraux *et al.* 2005). Before the lithography process, several pre-treatment steps are needed, including the dissolution of the membrane, the dispersion of an appropriate density of nanowires

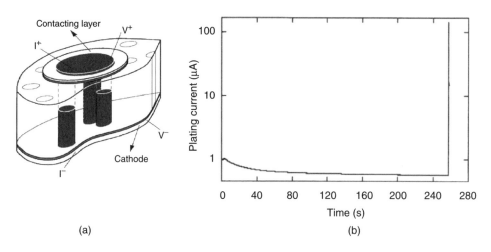

(a)   (b)

Figure 6.31 (a) Schematic drawing of the self-contacting method for single-wire measurements (Piraux *et al.* 2005); (b) sharp increase of the plating current when the first nanowire is contacted to the metal strip. (Fert and Piraux 1999). Reprinted from Fert, A. Piraux, L. (1999). Magnetic nanowires. *Journal of Magnetism and Magnetic Materials* **200**, 338–358, with permission from Elsevier

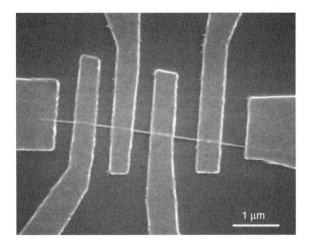

Figure 6.32 SEM picture of the multi-probe connection of a 35 nm Co nanowire. (Vila *et al.* 2002)

on a substrate and the cleaning of the nanowires for removing the polycarbonate residues from the surface of the nanowires and ensuring low contact resistance. Due to the lithography process, a multiprobe connection on an isolated nanowire can be realized, as shown in Figure 6.32. Using a multiprobe connection, electrical measurements can be performed without any contact resistance contribution on different segments of the same wire, and the length of a segment can be as small as 500 nm. Therefore, this method can provide the information of different segments of a nanowire, and can be used to investigate the inhomogeneous character of a nanowire.

Figure 6.33 shows the magnetoresistance of a 0.5 μm-long segment of a Co nanowire at room temperature. The resistances, when the external magnetic field is parallel and perpendicular to the axis of the nanowire, respectively, are measured. It is clear that, for the Co nanowire, the magnetoresistance signal in the parallel configuration is much larger that in the perpendicular configuration. It is interesting that, after the saturation in either the perpendicular or the parallel direction, similar low resistance state is obtained, with a value close to the one obtained for the saturation in the perpendicular direction. It seems that the magnetization at zero magnetic field is in a direction close to perpendicular to the axis of the nanowire (Piraux *et al.* 2005).

6.4.2 Magnetization Properties

The characterization of the magnetization properties of nanowire arrays and single nanowires is important for the study of nanomagnetism and the applications of magnetic nanowires. The magnetization properties of nanowire arrays can be characterized using the conventional vibrating sample magnetometers (VSM). For the characterization of single nanowires, usually superconducting quantum interference devices (SQUID) and magnetic force microscopy (MFM) are used.

6.4.2.1 Arrays of Nanowires

Magnetic nanowire arrays are usually electrodeposited in anodic aluminum oxide (AAO) porous templates or polymer membranes. For the nanowire arrays in AAO templates, due to the extremely high pore densities of the template, the dipolar interactions greatly

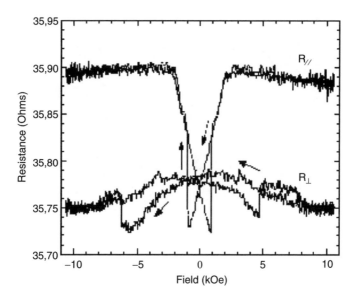

Figure 6.33 AMR characteristics of a 0.5 μm long segment of a single 60 nm Co nanowire at room temperature. $R_\parallel$ and $R_\perp$ represent the resistances when the applied magnetic field is parallel and perpendicular to the axis of the nanowire, respectively. (Vila *et al.* 2002)

affect the magnetic properties of the filled templates. Therefore, the magnetic properties of nanowire arrays in AAO templates are closely related to the diameter, packing density and aspect ratio of the magnetic nanowires.

However, nanoporous polycarbonate membranes have a low pore density, and can be used to investigate the magnetic properties of almost magnetically isolated magnetic nanowires. The dipolar interactions among the nanowires are expected to be small if the average spacing between the nanowires is much larger than 1 μm. Figure 6.34 shows the coercive field and remanent ratio M_r/M_s for Co and Ni-nanowire arrays as a function of nanowire diameter (Fert and Piraux 1999; Piraux *et al.* 1997). The nanowire diameter is in the range 30–500 nm, and the external magnetic field is in the axial direction. As shown in Figure 6.34(a) and (b), when the diameter of the nanowires increases, the remanent magnetization of the Co nanowire array decreases more quickly than that of the Ni nanowire array. However, as shown in Figure 6.34(c) and (d), the coercivity values for both Co and Ni nanowires increase quickly along the decrease of the diameter of the nanowires, indicating a crossover towards a single-domain structure.

6.4.2.2 Single Nanowires

In the following subsection, we discuss SQUID method and MFM method for the measurement of the magnetization of single nanowires.

(1) SQUID Method

Wernsdorfer *et al.* (1996) measured the magnetization of nickel single nanowires using a microbridge DC SQUID, as shown in Figure 6.35. The nanowires are fabricated by electrodeposition method in polycarbonate membranes, and their diameters are in the range of 40 to 100 nm. The flux induced by a single nanowire can be detected by a SQUID magnetometer.

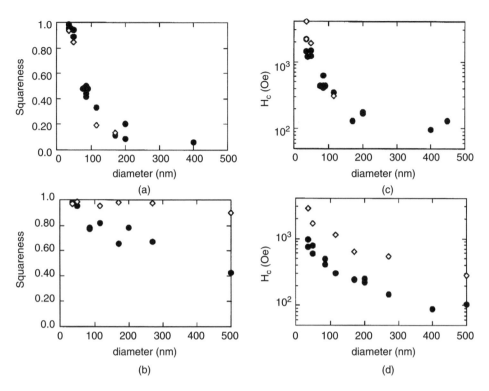

Figure 6.34 The coercivity and squareness coefficient M_r/M_s of arrays of magnetically isolated Co (a, c) and Ni (b, d) nanowires embedded in membranes with relatively small pore densities. The external magnetic field is parallel to the nanowire axis. The filled circles represent the data obtained at room temperature, while the open lozanges represent the data obtained at T = 35 K. (Fert and Piraux 1999). Reprinted from Fert, A. Piraux, L. (1999). Magnetic nanowires. *Journal of Magnetism and Magnetic Materials* **200**, 338–358, with permission from Elsevier

Figure 6.35 SEM image of a microbridge DC SQUID loaded with a nickel nanowire with diameter 65 ± 4 nm. (Wernsdorfer *et al.* 1996)

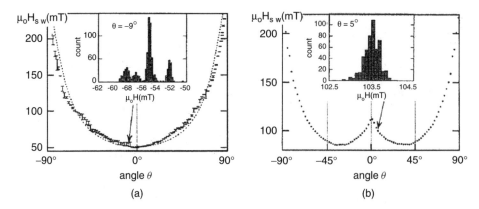

Figure 6.36 (a) Angular dependence of the switching field of a nickel nanowire with diameter 92 ± 4 nm, and length $5\,\mu$m. The bars indicate the width of switching field distribution, and the dotted line is based on the curling model of magnetization reversal. The inset is the histogram of the switching field for $\theta = -9°$; (b) angular dependence of the switching field of a nickel nanowire with diameter 50 ± 5 nm, and length $3.5\,\mu$m. In this case, the width of the switching field distribution is smaller than the dot size. The inset is the histogram of the switching field for $\theta = 5°$. (Wernsdorfer *et al.* 1996)

Using microbridge DC SQUID as shown in Figure 6.35, the switching field between the two opposite polarities can be studied as a function of the angle θ between the applied magnetic field and the axis of the nanowires (Wernsdorfer *et al.* 1996; Fert and Piraux 1999). As shown in Figure 6.36(a), for a nickel nanowire with diameter larger than about 75 nm, the angular dependence of the switching field has maxima at $\pm 90°$. The inset of Figure 6.36(a) blows up the complex histogram of the switching fields measured by repeated experiments at the same angle. The several maxima in this complex histogram suggest that the magnetization reversal may be caused by a nucleation process with several sites competing for the nucleation. As shown in Figure 6.36(b), for a nickel nanowire with diameters smaller than 75 nm, another maximum at $\theta = 0$ appears in the angular dependence, suggesting the angular dependence of the Stoner-Wohlfarth (SW) model.

(2) MFM Method

The magnetization properties of single nanowires can also be observed using magnetic force microscopy (MFM) (Piraux *et al.* 2005; Ebels *et al.* 2000; Henry *et al.* 2001; O'Barr and Schultz 1997). Figure 6.37(a) shows the MFM image of a Co nanowire with diameter 35 nm after the application of a strong magnetic field parallel to the axis of the nanowire ($H_\parallel$). The dark and bright contrasts at the ends of the nanowire correspond to a charge distribution which arises when the magnetization is in a single-domain state and aligned parallel to (or close to) the axis of the nanowire. In contrast to the saturation in the parallel magnetic field $H_\parallel$, a multi-domain state with head-to-head domain walls can be induced by saturation in a magnetic field perpendicular to axis the nanowire ($H_\perp$), as shown in Figure 6.37(b). This formation of the multi-domain structure is due to the fact that, when the magnetizing field decreases from the perpendicular saturation field to zero, the magnetization can rotate clockwise or counterclockwise toward the wire axis. The dark and bright contrasts visible along the axis of the nanowire arise from the magnetic volume and surface charges located at the domain walls. The domain walls shown in

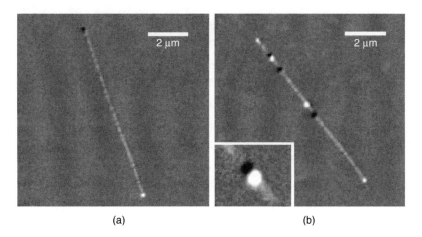

(a)   (b)

Figure 6.37   MFM images of a 35 nm Co wire after application and removal of (a) a longitudinal saturating field and (b) a transverse saturating field. (Piraux *et al.* 2005)

Figure 6.37(b) are stabilized at the pinning sites and can be moved along the wire by applying a field $H_\parallel$ which is larger than the local depinning field (Ebels *et al.* 2000).

6.4.3 Ferromagnetic Resonance

The effective magnetic anisotropy in a nanowire ensemble is mainly determined by shape anisotropy, dipolar interwire coupling and magnetocrystalline anisotropy. The effective anisotropy field (H_{EF}) with a uniaxial symmetry parallel to the axis of nanowires is given by (Piraux *et al.* 2005):

$$H_{EF} = H_S - H_{DIP} + H_K \tag{6.21}$$

where $H_S = 2\pi M_S$ is the shape anisotropy for an infinite long cylinder, H_{DIP} is the dipolar coupling between the nanowires which lowers the magnetostatic energy and H_K is the magneto-crystalline anisotropy. In a mean-field approximation, H_{DIP} can be written as $H_{DIP} = 6\pi M_S P$, with P indicating the packing of the array. The magneto-crystalline anisotropy H_K may add ($H_K > 0$ along the wire axis) or compete ($H_K < 0$ perpendicular) with the shape anisotropy.

Among the different approaches for the investigation of the static and dynamic properties of magnetic nanowire arrays, ferromagnetic resonance (FMR) is a very powerful technique. Assuming $H_{EF} > 0$, for a given constant frequency f, if the external magnetic field is parallel to the wires, the ferromagnetic resonance conditions are (Piraux *et al.* 2005)

$$f/\gamma = H_R + 2\pi M_s(1 - 3P) + H_K \qquad (H_K > 0) \tag{6.22}$$

$$(f/\gamma)^2 = (H_R + 2\pi M_s(1 - 3P) + H_K) \times (H_R + 2\pi M_s(1 - 3P)) \quad (H_K < 0) \tag{6.23}$$

where H_R is the resonance field, and γ is the gyromagnetic ratio. For Co and Ni, the γ value is about 3.05 and 3.09 GHz/kOe, respectively. According to Equations (6.22) and (6.23), the effective field H_{EF} can be derived from FMR measurements.

Due to the small diameter of the nanowires (compared to the skin depth) and the insulating nature of the nanoporous templates, the magnetically filled porous membranes

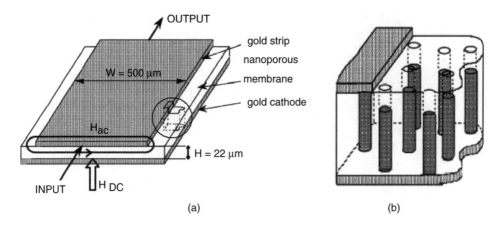

(a)   (b)

Figure 6.38 Schematic description of a microstrip transmission line fabricated using a dielectric substrate filled with magnetic nanowires. (a) Overall structure of the microstrip; (b) enlargement of the circled part in (a). (Encinas *et al.* 2001a)

are particularly suitable for the fabrication of the microstrip transmission line that can be used to investigate the ferromagnetic resonance properties at gigahertz frequencies (Goglio *et al.* 1999). Figure 6.38 shows a microstrip transmission line deposited on the free surface of the membrane. The microwave signal propagating along the microstrip transmission line produces a microwave pumping field that is perpendicular to the nanowires and induces a precession of the magnetization around the static equilibrium position. At ferromagnetic resonance, the incident microwave signal is absorbed, resulting in a minimum of the transmitted power.

Figure 6.39 shows typical microwave absorption curves of a nickel nanowire array as a function of the intensity of the magnetic field parallel to the wires (Encinas *et al.* 2001b). The nickel nanowires have a diameter 100 nm, and the membrane porosity P of

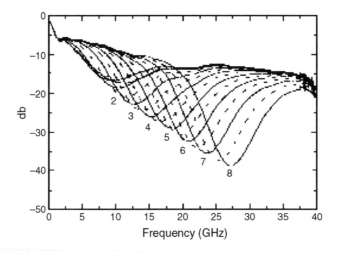

Figure 6.39 Microwave absorption spectra as a function of the applied magnetic field intensity. In the measurements, the static magnetic field is parallel to the nanowires. Continuous lines correspond to measurements made with the applied field value indicated by the numbers. (Encinas *et al.* 2001b)

the nanowire array is about 27 %. It can be seen that the nanowire array has an intense absorption to the microwave signal, and the position and intensity of the absorption peak change along with the change of the applied magnetic field.

Using iron-track-etched polymer membranes, the spacing between the nanowires can be controlled over a wide range, and effects of dipolar interactions between the nanowires can be investigated from the limit of isolated nanowires to the limit of almost touching nanowires (Piraux *et al.* 2005). Based on FMR measurements, Encinas *et al.* (2001b) studied the dipolar interactions between nanowires grown in alumina membranes of porosity ranging from 4–5 % up to 35–38 %. Figure 6.40 shows the dispersion relation and the corresponding hysteresis loops of nickel nanowire arrays with different diameters and packaging densities. As shown in Figure 6.40(a), for low packing densities, the ferromagnetic resonance frequencies in the saturated states agree well with the values expected for an array of isolated wires. At higher packaging densities, the ferromagnetic

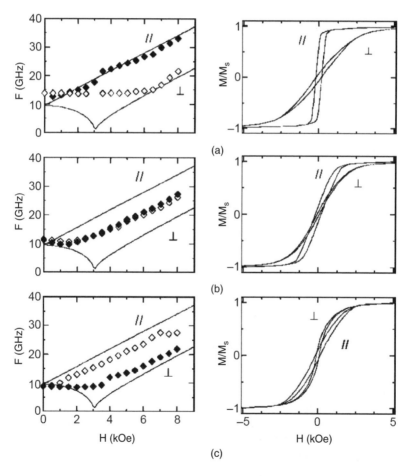

Figure 6.40 Dispersion relations (left) and the corresponding hysteresis loops (right) of nickel nanowire arrays. The filled symbols represent the results obtained when the magnetic field is parallel to the nanowire axis, while the open symbols represent the results obtained when the magnetic field is perpendicular to the nanowire axis. The solid lines show the theoretical results for isolated nanowires. The nanowire diameter and packing density are (a) 180 nm, 4 %, (b) 100 nm, 27 %, and (c) 115 nm, > 35 %. (Encinas *et al.* 2001b)

resonances for both magnetic field directions have almost identical frequencies, as shown in Figure 6.40(b). When the packaging density is higher than a critical value which is around 33 % for a regular array, a switching of the effective easy magnetization axis can be observed, from the direction parallel to the nanowire axis to the direction perpendicular to the nanowire axis, as shown in Figure 6.40(c). Therefore, by changing the packaging densities, nanowire arrays with different effective anisotropies can be obtained, from the isolated nanowires with the easy axis parallel to the nanowire axis to the arrays whose the easy magnetization axis is perpendicular to the nanowire axis, and it is possible to obtain quasi-isotropic arrays. The deviations from the calculated frequency-field dispersion curves at low magnetic field region are due to the fact that the nanowire arrays are not in the single-domain state at low magnetic fields.

6.5   Functionalization of Magnetic Nanowires

Surface functionalization of nanowires is necessary for realizing their biocompatibility and functionality. It is often used to tailor surface properties (such as hydrophilicity, hydrophobicity, surface charge), impart other properties (such as fluorescence), and introduce molecular recognition for small molecules (such as drugs), biopolymers (such as peptides, proteins and DNA), protein assemblies (such as viruses) and other nanoparticles (such as particle-DNA conjugates) (Wildt *et al.* 2006). Multi-segment nanowires represent a unique platform for engineering multifunctional nanoparticles for varieties of biomedical applications (Nicewarner-Pena *et al.* 2001) and bring exciting new perspectives for surface functionalization which can modify the chemical and biological properties of nanowires (Martin *et al.* 1999).

Generally speaking, there are two approaches for surface functionalization: chemical functionalization and biological functionalization. Functionalizing surfaces with chemical or biological groups has become an important tool in nanobiotechnology. In the following, we discuss the chemical functionalization and biological functionalization, and we will discuss the assembly by surface chemistry.

6.5.1   Chemical Functionalization

The most direct approach to functionalizing multi-segment nanowires is to transfer the well-developed surface chemistry at planar metal interfaces to the nanowire geometry. For example, Martin *et al.* (1999) have demonstrated selective chemistry with non-magnetic gold–platinum nanowires. For magnetic nanowires, it has been shown that porphyrins with terminal carboxylic acid groups bind selectively to the native oxide on nickel in single-component nickel nanowires (Tanase *et al.* 2001), and in two-segment gold–nickel nanowires (Bauer *et al.* 2003).

Based on the knowledge of surface coordination chemistry, multi-segment nanowires can be selectively functionalized with receptors of interest for applications in sensing, molecular recognition, drug delivery, separations and imaging. As shown in Figure 6.41, due to their selective headgroups, ligands selectively bind to the desired segments of a two-segment Au–Ni nanowire. The tail groups (R and R$'$) of these ligands can be used to target different biological entities. This coupling chemistry can be used to functionalize a specific segment of a multi-segment magnetic nanowire.

The hard–soft acid–base concepts originally developed in fluid solution are often used in surface coordination chemistry (Bauer *et al.* 2003). Soft bases, such as thiols, coordinate to soft metals such as gold, while carboxylic acids bind to harder metal oxides such as NiO.

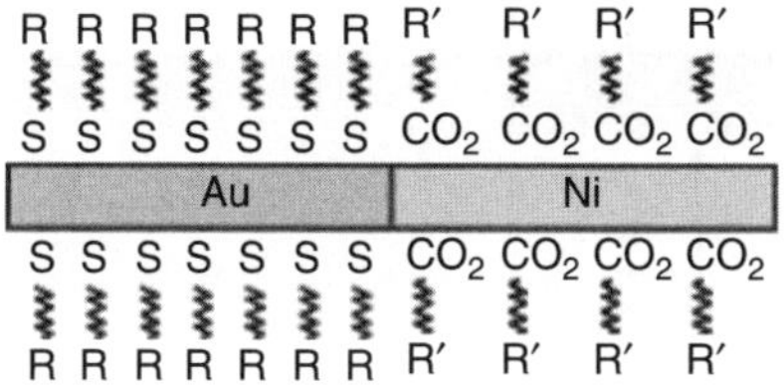

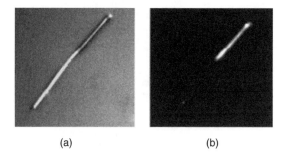

Figure 6.41 Functionalization of a two-segment Au−Ni nanowire. (Bauer *et al.* 2003)

(a)   (b)

Figure 6.42 Nickel−gold nanowire that has been functionalized with HemIX and nonylmercaptan. The diameter of the nanowire is about 350 nm, and the length of the nanowire is about 22 μm. (a) Reflection image of the nanowire; (b) fluorescent image of the nanowire. (Bauer *et al.* 2003)

Though the opposite reactions can also be found, the studies on planar surfaces clearly indicate that the soft−soft and hard−hard interactions are preferred. High selectivity can be achieved by optimizing the reaction conditions and by simultaneously binding two acids to the two-segment nanowires.

To demonstrate the selective functionalization on two-segment nickel−gold nanowires, Bauer *et al.* (2003) used HemIX as a fluorescent probe to quantify the coordination of carboxylic acids to the native oxide on nickel. HemIX has two carboxylic acid groups linked to the porphyrin ring by a flexible ethane spacer. When two-segment nickel−gold nanowires are reacted with HemIX, the nickel segment shows uniform fluorescence, while the gold segment shows weak and non-uniform fluorescence. The gold segment could be made nonfluorescent by adding a long-chain thiol to the reaction mixture, and the thiol likely displaces any weakly bound HemIX. Figure 6.42 shows optical images of a nickel−gold nanowire with a nickel−gold segment ratio of 2:3, functionalized with HemIX and nonylmercaptan. Figure 6.42(a) is a reflection image of the nanowire, and Figure 6.42(b) is a fluorescence image of the same nanowire.

6.5.2 *Biological Functionalization*

Many efforts have been made on the development of methods for biological functionalization of multi-segment nanowires with biomolecules (Prime and Whitesides 1991; Birenbaum *et al.* 2003; Bauer *et al.* 2004). Proteins are among the most frequently used biomolecules in biological functionalization. However, attaching proteins to individual segments of nanowires to achieve differential functionalization is particularly challenging because proteins tend to bind to most surfaces (Wildt *et al.* 2006).

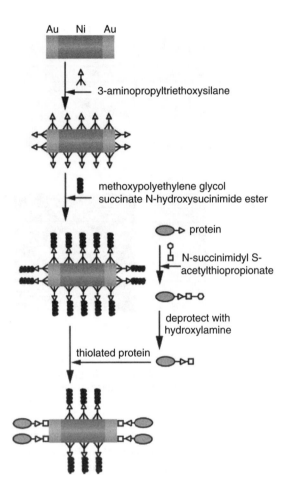

Figure 6.43 Selective protein functionalization of multi-segment nanowires. Thiolated KE2 antibody selectively bound to the Au segments on Au/Ni/Au nanowires. (Wildt *et al.* 2006)

Wildt *et al.* (2006) developed several methods for selectively functionalizing nanowires with proteins through the formation of strong covalent linkages. Figure 6.43 shows an approach for selective protein functionalization of multi-segment nanowires. In this approach, the coupling of primary amine groups to thiols is used as a method to attach a protein to Au segments, and the Ni segment functionalized with PEG is used as the protein-resistant segment.

To visualize the selective binding of the thiolated protein on the nanowire, fluorescent antimouse IgG is used to selectively bind to the KE2 antibody bound to the Au end-segments. Figure 6.44 shows light microscope and fluorescence images for a typical Au/Ni/Au nanowire functionalized by this method. The fluorescence image reveals the nanowire with two bright ends, and the fluorescence intensity is proportional to the concentration of protein present on the surface of the nanowire. It is clear from the image that more protein is present at the ends of the wire than in the middle, and the PEG-terminated Ni segments show good protein resistance. More approaches for protein functionalization can be found in the work of Wildt *et al.* (2006).

(a)   (b)

Figure 6.44   (a) Bright-field and (b) fluorescence images of an Au/Ni/Au nanowire with a KE2 antibody bound to the Au end-segment. (Wildt *et al.* 2006)

6.5.3   Assembly by Surface Chemistry

Nanoscale building blocks can be assembled based on their distinct surface chemistry. By extending the concept of selective functionalization on multi-segment nanowires, further selective chemistry can be performed through amide, thiourea, and thioether coupling reactions, and self-assembly taking advantage of these affinity binding events can be realized (Hurst *et al.* 2006). This can be taken as a bottom-up approach for building materials.

Chen *et al.* (2006) achieved the directed end-to-end linkage of Au/Ni/Au multi-segment nanowires using a biotin–avidin linkage, as shown in Figure 6.45. In these systems, the central Ni segments are passivated with palmitic acid to avoid nonspecific binding and minimize lateral nanowire assembly. The Au segments are functionalized with biotin-terminated 1-undecanethiol. The assembly of these rods can be achieved by two different methods. In one method, avidin-terminated nanowires are prepared by exposing a solution of biotin-terminated nanowires to avidin. End-to-end assembly, as dictated by the biotin–avidin linkages on the Au segments, can be observed when equal amounts of

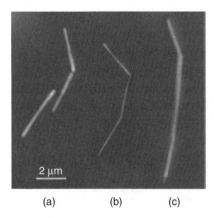

(a)   (b)   (c)

Figure 6.45   Optical microscope images of Au–Ni–Au nanowires. a) A single nanowire (left) and two linked nanowire chains (right); b), c) chains of three and four nanowires, respectively. (Chen *et al.* 2006)

avidin- and biotin-terminated nanowires are mixed. In the other method, end-to-end assembly can be achieved through the addition of free avidin to biotin-terminated nanowires.

6.6   Magnetic Nanowires in Suspension

In most of the biomedical applications, magnetic nanowires are suspended in water or organic solvents. Usually magnetic nanowires are synthesized using templated electrodeposition methods. To make nanowire suspensions, the template in which the nanowires are deposited should be dissolved in a suitable solvent. For example, dichloromethane is often used to dissolve polycarbonate. To obtain nanowires with a clean surface, usually sequential centrifugation and addition of clean solvent are needed to remove the residual polycarbonate. The obtained clean nanowires are then suspended in a suitable solvent.

The manipulation of magnetic nanowires in suspension is crucial for the realization of their biomedical applications under the control of external magnetic fields. In the following, we discuss the responses of magnetic nanowires in suspension, and the typical manipulations of magnetic nanowires in suspension, including the magnetic trapping of nanowires, self-assembled magnetic nanowire arrays and electromagnetic micromotor based on the rotation of magnetic nanowires.

6.6.1   Responses of Magnetic Nanowires in Suspension

The magnetic interactions between the magnetic nanowires in a suspension affect the stability of the suspension, and can be exploited in assembling magnetic nanowires into larger scale structures. For instance, in a suspension with magnetic nanowires with high remanence magnetizations, due to the attractive wire–wire interactions, the magnetic nanowires can form one-dimensional chains (Bauer *et al.* 2001). Due to their potential multi-functionality, multilayer ferromagnetic/nonmagnetic (FM/NM) nanowires are often used for biomedical applications. The magnetic properties of multilayer FM/NM nanowires could be tailored by adjusting the size and aspect ratio of each segment. For example, their remanence is dependent on the aspect ratio and thickness of the FM and NM layers, and this can be used for the control of aggregation and assembly of the nanowires in a suspension (Chen *et al.* 2003).

The responses of the magnetic nanowires in a suspension are crucial to their biomedical applications. For a given solvent, the responses of magnetic nanowires in suspension are mainly determined by their magnetic properties. For example, in a suspension with multilayer FM/NM nanowires, the responses of the magnetic nanowires to a small external magnetic field can be designed by adjusting the aspect ratio and thickness of the FM and NM layers of the nanowires. Figure 6.46 shows an example of the different responses of the multilayer Ni/Cu nanowires in two suspensions to a small external magnetic field (about 10 Oe). The multilayer nanowires in two suspensions have the same diameter and overall length, and have the same amount of nickel, but they have different geometrical structures. Their different responses to the external magnetic field are mainly due to their different layer structures. In Figure 6.46(a), the nanowires have a structure of $[Ni(1000\,nm)/Cu(1000\,nm)]_3$. As the nanowires have a diameter $d = 100\,nm$, FM segments of nanowires are rod-shaped. Under the external magnetic field, the nanowires align parallel to the magnetic field, because the easy magnetization axis of the nanowires is parallel to the nanowire axis. In Figure 6.46(b), the nanowires have a structure $[Ni(10\,nm)/Cu(10\,nm)]_{300}$. As their diameter is $d = 100\,nm$, the FM segments of

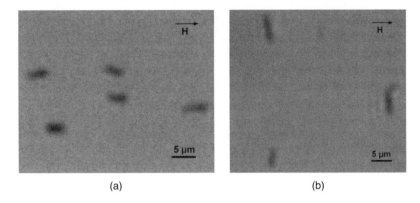

(a)   (b)

Figure 6.46 Responses of multilayer Ni/Cu nanowires, suspended in a 1:1 octadecane and hexde-
cane mixture, to external magnetic field. The nanowires in the two suspensions have the same
diameter (100 nm) and length (about 6 mm), and have the same average composition of nickel
(50 at. %). Nevertheless, they have different layer structures: (a) [Ni(1000 nm)/Cu(1000 nm)]$_3$, and
(b) [Ni(10 nm)/Cu(10 nm)]$_{300}$. (Chen *et al.* 2003)

the nanowires are disk-shaped. Because the easy magnetization axis of these multilayer
nanowires is perpendicular to the nanowire axis, under the application of an external
magnetic field, these nanowires align perpendicular to the magnetic field.

6.6.2 Magnetic Trapping of Nanowires

Figure 6.47 schematically illustrates the essential features of the magnetic trapping of
magnetic nanowires in a suspension (Reich *et al.* 2003). The magnetic nanowires in a
suspension can settle onto a substrate that contains elliptical micromagnets fabricated by

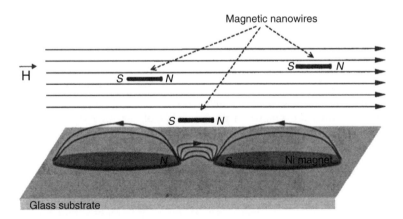

Figure 6.47 Process for magnetic trapping of nanowires suspended in water. A small external
magnetic field H aligns the nanowires parallel to the long axis of the micromagnets fabricated by
lithography. The nanowires are attracted to and trapped in the regions of strong local magnetic
gradients between the poles of the micromagnets. (Reich *et al.* 2003)

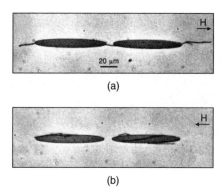

(a)

(b)

Figure 6.48   Effect of an external magnetic field (H = 10 Oe) on magnetic trapping. (a) The dipole moments of the nanowires are pre-oriented parallel to the moments of the elliptical micromagnets. In this case, the nanowires are trapped in the gaps; (b) the dipole moments of the nanowires are pre-oriented anti-parallel to the moments of the micromagnets. In this case, the nanowires are repelled from the gaps. (Reich *et al.* 2003)

lithography. The distribution of the nanowires on the substrate is affected by the local magnetic fields produced by the elliptical micromagnets on the substrate influence. The magnetic nanowires are usually attracted to the regions of strong local field gradients, such as the gap between two closely spaced micromagnets. If the gap between two micromagnets is smaller than the length of the nanowires, single nanowires can bridge the gap and are subsequently trapped.

As shown in Figure 6.48, the trapping efficiency can be improved by the application of a weak uniform external magnetic field (2–20 Oe) parallel to the long axis of the micromagnets. The applied magnetic field pre-orients the suspended nanowires. This aligning field reduces aggregation of nanowires in the suspension, and optimizes the configuration of the magnetic trapping. If there is no external magnetic field, the nanowires are attracted to the edges of the micromagnet, and the trapping events in this case are rare. When an external magnetic field is applied, the nanowires are repelled from the bodies of the micromagnets and concentrate at the ends of the micromagnets. As shown in Figure 6.48(a), in this case, magnetic nanowires are mostly trapped in the gaps, and chains of magnetic nanowires are formed.

The influence of the external pre-orienting magnetic field is confirmed by reversing its direction, so that the suspended magnetic nanowires are pre-aligned anti-parallel to the magnetic moments of the micromagnets. As shown in Figure 6.48(b), in this case, the magnetic nanowires are repelled from the ends of the micromagnets, and they are attracted to the bodies of the micromagnets.

6.6.3   Self-assembled Magnetic Nanowire Arrays

Templates can be used to provide a certain level of control over the driving forces for self-assembly processes, and magnetic interaction is extremely promising for templating in self-assembly processes mainly due to following reasons (Liu *et al.* 2007; Reich *et al.* 2003; Roberts *et al.* 2004; Yellen *et al.* 2005). First, based on the wide varieties of magnetic materials and the many feasible approaches for manipulating their magnetic properties such as saturate magnetization, remanent magnetization, anisotropy and coercivity, diverse self-assembly processes and nanosystems can be designed through magnetic interactions. Second, due to the dipole nature of magnetism, both the attractive

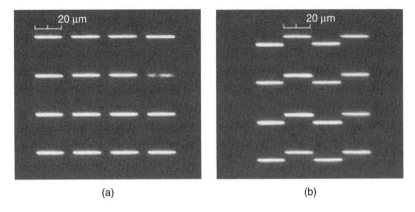

(a)   (b)

Figure 6.49 SEM images of the templates with different nanomagnet patterns. (a) Pattern (I); (b) pattern (II). (Liu *et al.* 2007)

and repulsive forces on the magnetic elements can be used to control the self-assembly processes. Third, due to their long-range characteristic, magnetic interactions typically do not interfere with the biological and chemical interactions, which is crucial for practical applications. Last, unlike some of the biologically or chemically active templates, magnetic templates can be used many times without concerns about poisoning, and they can be easily refunctionalized with a magnetization process.

In the self-assembly processes on magnetic templates driven by external magnetic forces, many parameters can be controlled. For example, the geometry and magnetic properties of the magnetic nanoelements and the nanomagnets on the magnetic templates, as well as the nanomagnet array on the magnetic templates, can be designed for special purposes. This is helpful for the diverse applications of the self-assembly processes, such as biological applications and nanocontacts. By using a combination of the external magnetic field and the local dipolar magnetic field, Liu *et al.* (2007) demonstrated controlled one-dimensional and two-dimensional self-assemblies of magnetic nanowire arrays on magnetic templates.

The nickel nanowires used have a diameter of 200 nm and a length of 10 μm, and their concentration in alcohol solutions is about 2.0×10^5/ml. Templates of cobalt nanomagnet arrays are fabricated by patterning Co full films with a thickness of 100 nm. As shown in Figure 6.49, the cobalt nanomagnets have lengths in the range of 20–25 μm, and their widths are in the range of 100–400 nm. Pattern (I) is designed for applied parallel external field and pattern (II) is designed for perpendicular external field.

The template is magnetically initialized by applying a 5 kOe magnetic field along the long axis of nanomagnets. By immersing the template with Co nanomagnet arrays in the alcohol solution of suspended magnetic nanowires under a parallel or perpendicular external magnetic field, self-assembled magnetic nanowire arrays can be achieved. Because the dipolar magnetic interaction between nanowires is dependent on their relative orientations, applying an external field can suppress the tendency of aggregation, pre-align the initially random nanowires and lead to the formation of extended head-to-tail nanowire chains (Liu *et al.* 2007). When an external magnetic field (10 Oe) is applied parallel to the magnetization vector direction of the nanomagnets on the template with pattern (I), the suspended magnetic nanowires will first be aligned by the external magnetic field, and then attracted and trapped in the gap of the two nanomagnets by the local dipolar magnetic field generated by the nanomagnets. Figure 6.50(a) shows a nanowire array self-assembled between the gaps of cobalt nanomagnet pairs. It is interesting to

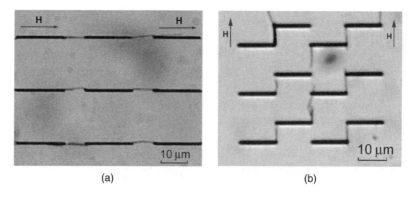

(a)   (b)

Figure 6.50   (a) Optic micrograph of self-assembly nanowire arrays when an external magnetic field (10 Oe) is applied parallel to the magnetization direction of the magnet template with pattern (I); (b) optic micrograph of self-assembly nanowire arrays when an external magnetic field (10 Oe) is applied perpendicular to the magnetization direction of the magnet with pattern (II). (Liu *et al.* 2007)

note that only one nickel nanowire was trapped in each gap of the cobalt nanomagnet pairs.

If the external magnetic field is perpendicular to the magnetization of the magnet template with pattern (II), another self-assembled nanowire array pattern can be achieved with an approximately 100 % trapping rate, as shown in Figure 6.50(b). Since the gaps between nanomagnet pairs show magnetic fringing fields that are anti-parallel to each other, only one of the two magnetic gaps that have a fringing field parallel to the external field can be filled with one magnetic nanowire when a 10 Oe external field perpendicular to the magnet is applied (Liu *et al.* 2007). The gaps for trapping the magnetic nanowires are designed to be 10 and 15 μm. The 10 μm gaps are all filled with one magnetic nanowire, while the 15 μm gaps are either filled with one nanowire or two nanowires due to the mismatch of the dimensions between the magnetic nanowires and the gaps. This indicates that the dimension matching between the nanowires and the gap is important for achieving a low magnetostatic energy and thus a high trapping rate for one nanowire per gap.

6.6.4   Electromagnetic Micromotor

Linear and rotational movements are required in almost all machines. Therefore, the control of magnetic nanowires for linear and rotational movements is crucial for the development of nanoscale machines. Usually the linear movements of magnetic nanowires are achieved by applying magnetic gradients, while the control of magnetic nanowires for continuous rotation is more complicated.

Barbic *et al.* (2001) demonstrated an approach for generating the rotation of a non-pivoted magnetic rotor in a fluid. The magnetic rotor is a single-domain magnetic wire with length less than 100 μm, and the rotation of the magnetic wire is driven by a stator outside the fluid. Figure 6.51 shows the operation principle of the micromotor. The stator of the micromotor is a set of integrated microcoils and microtips. Each microcoil has approximately 10 turns of soft magnetic wire with diameter 25 μm wound over a soft magnetic material with diameter 50 μm. A set of three microcoils and microtips is arranged into an equilateral triangle with the microtips placed 100 μm apart. Microcoils

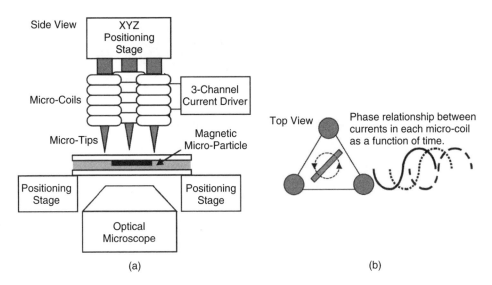

Figure 6.51 Block diagram of the electromagnetic micromotor. The stator microcoils and microtips are positioned outside the fluid above the single-domain magnetic wire inside the fluid. The currents through the three microcoils are 120° out of phase with respect to one another, resulting in rotational motion of the rotor. (Barbic *et al*. 2001)

of the stator are individually connected to three separate current amplifiers controlled by independent digital/analog (D/A) channels. The control channels are programmed so that the three microcoils in the stator are driven by three sinusoidal electrical currents, which are 120° out of phase with respect to each other. Therefore, the stator can apply sinusoidal attractive and repulsive forces to the magnetic rotor, and thus the magnetic rotor rotates under the driving force from the stator.

Motion of the rotator is observed with an inverted optical microscope and recorded with a CCD camera system. The nine sequent images shown in Figure 6.52 demonstrate one full rotation of the rotor of the micromotor in a thin glycerol film (Barbic *et al*. 2001).

This technique can be used for microfluidic application. Figure 6.53 shows two potential applications of micromotors in microfluidics: micropump and microvalve.

6.7 Biomedical Applications of Magnetic Nanowires

The special properties of magnetic nanowires make them attractive for biological applications. Due to their large magnetic moments and shape anisotropy, strong forces and torques can be applied by external magnetic fields. The diameter and length of magnetic nanowires can be independently controlled, so nanowires with suitable sizes for biological entities with different length-scales can be synthesized. Furthermore, usually magnetic nanowires are biocompatible. Magnetic nanowires do not disrupt the growth cycle of cells, and biologically active molecules can be functionalized on them so that the defined biomedical functions can be performed (Bauer *et al*. 2003). In the following subsections, we discuss the typical applications of magnetic nanowires, including confinement of magnetic nanoparticles, manipulation of biomolecules, suspended biosensing systems, drug and gene delivery and hybrid devices with magnetic nanowires.

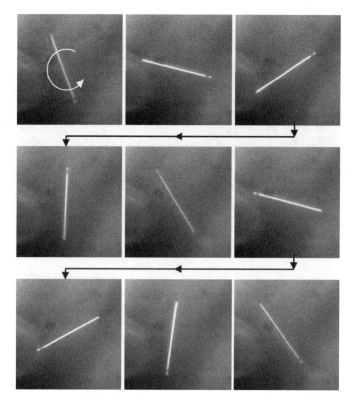

Figure 6.52 Nine sequent images demonstrate one full rotation of the rotor. The faint image of the stator coils can be seen out of focus in the background. (Barbic *et al.* 2001)

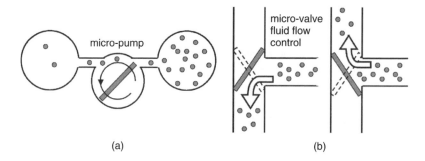

Figure 6.53 Potential applications of the micromotor in microfluidics: (a) micropump, (b) microvalve. (Barbic *et al.* 2001)

6.7.1 Confinement of Magnetic Nanoparticles

Magnetic nanoparticles have been used for separating biomolecules from mixtures, mechanically stimulating cells and enhancing MRI contrast. To effectively use magnetic nanoparticles for chemical or mechanical stimulation of biological processes at the subcellular level, magnetic particles should be positioned with submicron resolution, and thus the magnetic force attracting the magnetic particles must also have submicron region. The force exerted on a magnetic particle by a magnetic field is given by:

$$F = \Delta\chi V (\nabla \cdot \mathbf{B})\mathbf{B}\mu_0^{-1} \qquad (6.24)$$

where $\mathbf{B}$ is the flux density of the external magnetic field, $\Delta\chi$ is the difference of the susceptibilities of the magnetic particle and its surroundings, V is the volume of the magnetic particle, and μ_0 is the permeability of vacuum. For a particle with a given size, the gradient term $(\nabla \cdot \mathbf{B})$ is the only term that depends on distance. Therefore, confining magnetic particles in submicron region, requires structures that generate large magnetic gradients in submicron region.

Figure 6.54(a) shows the structure of a multi-segment nanowire. It has a diameter of about 80 nm, and it consists of ferromagnetic (CoNi) segments with length of about 350 nm separated by diamagnetic (Au) segments with length of 20–160 nm. When such a multi-segment nanowire is magnetized, the high magnetic field gradients at the boundaries between these segments attract and trap magnetic nanoparticles. As the aspect ratio of the ferromagnetic segments is greater than one, the easy magnetization axis of the multi-segment nanowire is parallel to the axis of the nanowire. To demonstrate the confinement of magnetic nanoparticles at submicron region, the multi-segment nanowires are

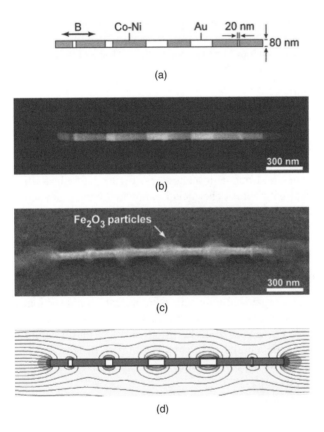

(a)

(b)

(c)

(d)

Figure 6.54 (a) Schematic of a metallic nanowire containing five diamagnetic sections (white), with thicknesses (from left to right) 40, 80, 160, 160 and 20 nm, respectively; (b) SEM image of a metallic nanowire; (c) SEM image of a metallic rod after treatment with 8 nm γ-Fe$_2$O$_3$ particles; (d) magnetic simulation of the field surrounding the nanowire. The contours represent field lines, and the shading represents flux density, B. (Urbach *et al.* 2003)

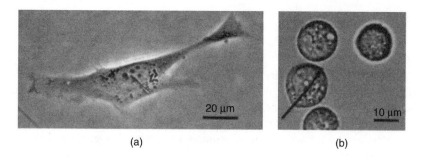

(a)   (b)

Figure 6.55 (a) A 15 μm Ni nanowire bound to a 3T3 mouse fibroblast cell after 24 h co-incubation; (b) suspended 3T3 cells, with one bound to a 20 μm Ni nanowire. The nanowires are coated with ECM proteins. (Tanase *et al.* 2005)

suspended in dichloromethane by ultrasonication and then deposited onto a Si wafer. During the deposition, a magnet is placed under the Si wafer with its dipole parallel to the Si surface and the magnetic nanowires are magnetized along their easy magnetization axes. Figure 6.54(b) shows an SEM image of such a nanowire. The magnetized nanowires, deposited on a surface, are treated with a dispersion of γ-Fe$_2$O$_3$ particles with a diameter of 8 nm. As shown in Figure 6.54(c), the γ-Fe$_2$O$_3$ particles are confined predominantly near the diamagnetic sections in a radically symmetric fashion. The sizes and shapes of the magnetic nanoparticle clusters closely match that of the gradients obtained by theoretical prediction, as shown in Figure 6.54(d) (Urbach *et al.* 2003).

6.7.2 Biomolecular Manipulation

Biomolecules can be manipulated using magnetic nanowires under external magnetic fields, and this is the basis of many biomedical applications of magnetic nanowires. Usually the biomolecule manipulation using magnetic nanowires is based on the bindings between biomolecules and magnetic nanowires. Figure 6.55 shows the binding between 3T3 cells and nickel nanowires. The nickel nanowires have a diameter of about 350 nm. Figure 6.55(a) shows a nickel nanowire bound to a 3T3 cell in culture, and Figure 6.55(b) a suspended 3T3 cell with a bound wire. In the following, we discuss spatial organization of cells and biomelecule separation.

6.7.2.1 Spatial Organization of Cells

The ability to spatially organize living cells is important for many biomedical applications, such as biosensing, the study of mechanotransduction and the exploration of the biochemistry of cell adhesion. Tanase *et al.* (2005) studied the spatial organization of mammalian cells using ferromagnetic nanowires in conjunction with patterned micromagnet arrays. In the following, we discuss the formation of cell chains and the magnetic trapping of cells.

(1) Formation of Cell Chains

Figure 6.56 illustrates the procedure for the self-assembly of cells into chains (Tanase *et al.* 2005). As shown in Figure 6.56(a) and (b), an external magnetic field is applied to align the nanowires parallel to each other. The cells descend very slowly through the culture medium, and the nanowires experience mutually attractive dipole-dipole forces due

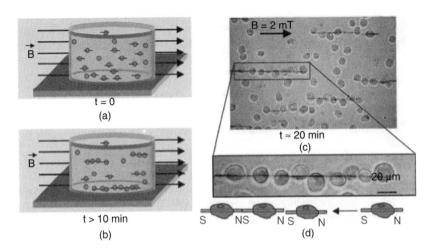

Figure 6.56 Process of magnetic chaining of cells. (a) Nanowires bound to suspended cells and aligned by an external magnetic field; (b) chain formation process due to magnetic dipole–dipole interactions between pre-aligned nanowires; (c) cell chains formed on the bottom of a culture dish with B = 2 mT; (d) close-up view of a single chain detailing wire–wire alignment. Interactions of the north and the south poles of adjacent nanowires are indicated schematically below. (Tanase *et al.* 2005)

to the interactions of their magnetic moments. The alignment of the magnetic moments of magnetic nanowires favors the formation of head-to-tail chains, where the north pole of one wire abuts the south pole of the next. As shown in Figure 6.62(c), these formations can encompass many cells, and extend over hundreds of micrometers. There are two mechanisms of chain formation. One approach is the aggregation in suspension, which leads to short chains. The other approach is the addition of descending cells or short chains to pre-existing chains on the chamber bottom. Because the interwire forces are not sufficiently strong to move the cells along the substrate, the chaining process ceases when all the cells settle on the substrate. Cells without wires are randomly distributed on the substrate.

(2) Magnetic Trapping of Cells

Once the cells with magnetic nanowires are brought close to a micromagnet array, for example, by sedimentation or by fluid flow, they are attracted to the ends of the micromagnets. As shown in Figure 6.57(a), 3T3 cells with magnetic nanowires are trapped at the ends of six ellipses. Tanase *et al.* (2005) studied the magnetic trapping with cell concentrations in the range $1 \times 10^4 - 1 \times 10^5$ cells ml^{-1}, and found that the optimum cell concentration is around 2.5×10^4 cells ml^{-1}. The trapping efficiency increases with the increase of cell concentration. However, if the cell concentration is higher than the optimum concentration, significant clumping and chaining may occur in suspension before the magnetic trapping.

Figure 6.57(b) shows the calculated sedimentation trajectories for a cell with a magnetic nanowire settling over the centerline of an isolated micromagnet. In the calculation, it is assumed that the spherical cell has a diameter of 16 μm, and the bound nanowire has a diameter of 350 nm, and a length of 20 μm. The concentration of the trajectories indicates the attractive action of the trap.

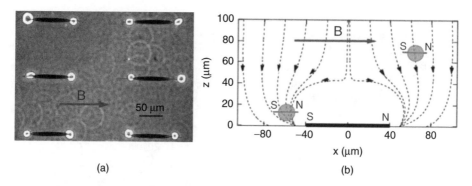

(a)   (b)

Figure 6.57   (a) Trapping of single cells by ellipsoidal micromagnets under an aligning magnetic field B = 2 mT; (b) calculated settling trajectories for a spherical cell with density 1.08 g cm^{-3} over the centerline of one of the ellipses in (a). (Tanase *et al.* 2005)

6.7.2.2 Biomolecular Separation

Magnetic nanowires can be used in performing high-yield separations of biomolecules. Both single-segment and multi-segment magnetic nanowires have been used for cell separations. Generally speaking, magnetic nanowires outperform magnetic spherical beads in cell separations.

(1)   Cell Separations using Single-segment Nanowires

Hultgren *et al.* (2004) studied the applications of Ni nanowires in cell separation. It was found that high-purity separations can be achieved for nanowires over a wide range of sizes, while the optimum separation yield is achieved when the average length of the nanowires matches the average diameter of the cells. This effect suggests the potential to magnetically separate cell populations based on their sizes.

Figure 6.58 shows transmitted light images of both treated and untreated trypsinized 3T3 cells attached to nickel nanowires. As shown in Figure 6.58(b), if the diameter of the cell is less than the length of the nanowire, the nanowire protrudes from the cell, whereas, as shown in Figure 6.58(c), if the diameter of the cell is larger than the length of the nanowire, the nanowire is enclosed by the cell. Usually, the cells with larger diameter are able to engulf longer nanowires. In Figures 6.58(a) and (d), the diameters of the cells are close to the lengths of the corresponding cells.

Two parameters, purity and yield can be used to compare the effectiveness of cell separations using beads and nanowires (Hultgren *et al.* 2004). Purity refers to the percentage of cells that have a magnetic particle attached in all the captured populations. For the beads and the 5 μm nanowires, the purity is about 40 %, while for nanowires longer than 15 μm, the purity is increased up to 80 %. The percent yield is the number of captured cells tagged with magnetic particles normalized by the initial number of cells tagged with magnetic particles. Figure 6.59 shows the percent yields of nanowires and beads for the separations of 15 and 23 μm diameter cell populations. For both populations, when the length of the nanowires is equal to the diameter of the cells, the percent yield reaches its maximum value, and the maximum yield is about four times larger than the percent yield when the beads are used.

(2)   Cell Separations using Multi-segment Nanowires

Histidine (His) has a high binding affinity for nickel, and this interaction can be used in Ni columns to separate His-tagged proteins from biological solutions (Stiborova *et al.* 2003).

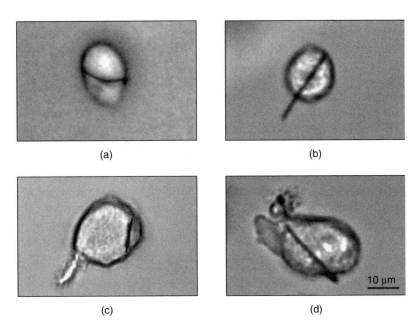

Figure 6.58 Optical images of trypsinized 3T3 cells. Top row: normal cells (average diameter 15 μm) with (a) 15 μm nanowire and (b) 22 μm nanowire. Bottom row: treated cells (average diameter 23 μm) with (c) 10 μm nanowire and (d) 22 μm nanowire. (Hultgren *et al.* 2004)

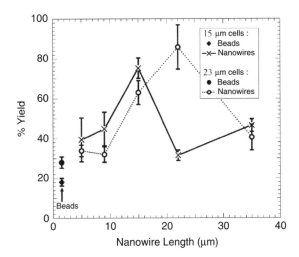

Figure 6.59 Percent yield as a function of nanowire length for separations of 3T3 cell populations with average diameters d = 15 μm and 23 μm. (Hultgren *et al.* 2004)

As shown in Figure 6.60, Lee *et al.* (2004) demonstrated that Au/Ni/Au multi-segment nanowires have more rapid binding kinetics in the separation of His-tagged proteins. When the Au/Ni/Au multi-segment nanowires are introduced to a solution containing both His-tagged and untagged proteins, the His-tagged proteins attach to the nickel segment of the nanowire, and can be removed from solution by applying an external magnetic field. In a similar way, the Au/Ni/Au multi-segment nanowires functionalized with poly-His

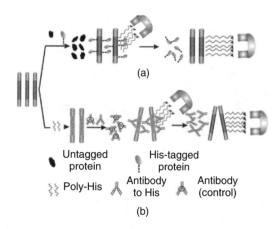

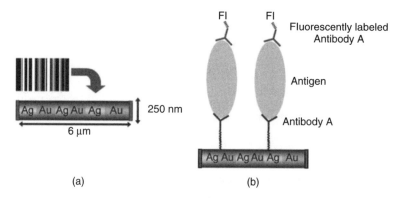

Figure 6.60 Schemes for separation of His-tagged proteins from untagged proteins (route a) and for separation of antibodies to poly-His from other antibodies (route b); (Lee *et al.* 2004)

can be used to efficiently separate mixtures of anti-His proteins from other antibodies. In these experiments, the Au segments of the nanowires are passivated with thiolated poly(ethylene glycol) (PEGSH) to minimize the nonspecific binding of proteins to Au surfaces and to minimize aggregation which may be caused by bare gold surface–surface interactions (Hurst *et al.* 2006).

6.7.3 Suspended Biosensing System (Tok et al. 2006)

As schematically shown Figure 6.61, multilayer nanowires can be used as a substrate in a biosensing platform for sandwich immunoassays. The multilayer nanowires consist of submicrometer layers of different metals, and are usually synthesized by electrodeposition within a porous alumina template. As a lot of variations can be realized in the synthesis of nanowires, a large number of unique yet easily identifiable encoded nanowires can be included in a multiplex array format. Tok *et al.* (2006) studied the applications of multilayer metallic nanowires in a suspended format for rapid and sensitive immunoassays.

Figure 6.61 a) Analogy between a conventional barcode and a metallic segment-encoded nanowire. Ni segments (50 nm) are deposited at both ends of the magnetic nanowire (not drawn to scale); b) schematic of the sandwich immunoassay performed on a nanowire. (Tok *et al.* 2006)

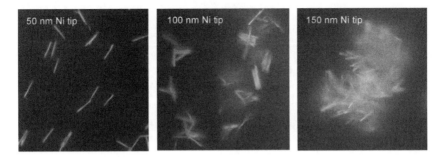

Figure 6.62 Images of the 'residual magnetic clumping' effect on various Ni segments deposited at both ends of pure Au nanowires (stripe pattern 30000003; 0 = Au, 3 = Ni). The images are taken after the nanowires are exposed to a magnetic field. Nanowires with Ni segments greater than 100 nm (total Ni amount is twice the segment length) led to a marked increase in the residual clumping upon exposure to magnetic fields. (Tok *et al.* 2006)

The basic working principle of a suspended biosensing system is illustrated in Figure 6.61(b). In a suspended biosensing system, the target analytes are captured and hybridized in solution. To ensure that the nanowires can be manipulated by external magnetic fields, an appropriate ferromagnetic metallic component, for example, nickel, is incorporated. Usually the nickel segments integrated at both ends of the nanowires have a length of 25–150 nm. The easy magnetization axis for these disk-shaped magnetic segments is perpendicular to the nanowire axis, and thus these nanowires align perpendicular to the externally applied magnetic field. As shown in Figure 6.62, the longer Ni segments result in stronger magnetophoresis, but also increase the aggregation of the magnetic nanowires after exposure to external magnetic fields, which hinders subsequent imaging.

Furthermore, if the length of the nickel segment is larger than its diameter, the nanowire exhibits an easy magnetization axis parallel to the nanowire axis, causing the nanowire to align parallel to the magnetic field. This alignment reduces hydrodynamic drag and should further increase particle mobility. Thus, to ensure the ease of magnetic manipulation by a magnetic field and the minimization of clumping due to the magnetization of the nanowires, the ideal length of the nickel segment is about 50 nm. The introduction of the nickel segments at the ends of the nanowires is helpful for ensuring both the structural robustness and the decoding process of the nanowires, while the presence of the nickel segments at the ends does not affect the overall biosensing capabilities of the nanowires (Tok *et al.* 2006).

6.7.4 Gene Delivery

The gene delivery using multi-segment magnetic nanowires exhibits obvious advantages. As the properties of conventional gene delivery systems could not be controlled on the nanoscale, they are limited by their relatively low transfection efficiency, which limits the ability of the system to incorporate foreign DNA inside a target cell (Hurst *et al.* 2006). However, in the fabrication of multi-segment nanowires, the materials of each segment and their properties can be precisely controlled in nanosized dimensions. Furthermore, multi-segment nanowires can be endowed with different functionalities in spatially defined regions, and so the precise control of antigen placement and the stimulation of multiple immune responses can be achieved.

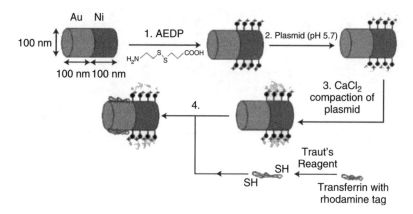

Figure 6.63 Functionalization of Au–Ni nanowires. 1. Nanowires are incubated with AEDP. Ni segment is bound with carboxylate end-group. 2. Plasmids are electrostatically bound to the protonated amine groups of the AEDP. 3. The surface-immobilized plasmids are compacted by $CaCl_2$. 4. The gold segment is selectively bound with hodamine-taged transferrin. (Salem *et al.* 2003)

Salem *et al.* (2003) investigated the application of electrochemically synthesized Au/Ni nanowires for therapeutic purposes. Figure 6.63 shows the procedure for selectively binding DNA plasmids and transferrin to Au/Ni nanowires. After nanowires are removed from the template, the nickel segment of the nanowires is functionalized with 3-[(2-aminoethyl) dithiol]-propionic acid (AEDP) through its carboxylic acid terminus. Plasmid DNA is then electrostatically bound to the protonated amine groups of the AEDP. The Au segment of the nanowires is then functionalized with transferrin, a cell-targeting protein, which has been chemically modified with a thiol.

Transfections using these multi-functionalized nanowires are performed on human embryonic kidney (HEK293) mammalian cell lines (Salem *et al.* 2003). It has been confirmed that multi-segment nanowires are more efficient in transfection than transferrin-modified single-component nanowires.

6.7.5 Hybrid Devices

Hybrid devices, based on molecular biology and micro/nano fabrication, can be used in the development of advanced biosensors and force bioactuators for medical and therapeutic applications. Among various hybrid systems, special attention is paid to motor proteins, such as Adenosine Tri-Phosphate synthase (ATPase), which can convert the chemical energy derived from the ATP hydrolysis into mechanical work (Noji *et al.* 1997). ATPase motor, ever-present in organisms from bacteria to man, is among the best characterized proteins in terms of its atomic structure and biochemistry. It consists of two rotary assemblies, F_0 (ab_2c_n) and F_1 ($\alpha_3\beta_3\gamma\delta_\varepsilon$) connected by a common elastic shaft, subunit γ. For an isolated F_1-ATPase, the central γ subunit rotates against the surrounding $\alpha_3\beta_3$ subunits during the hydrolysis of ATP in the three catalytic β subunits of the F_1-ATPase. It is difficult to realize the F_1-ATPase rotation experimentally. To expand the applications of F_1-ATPase motor to the physical sciences, it is necessary to develop a method to fabricate and functionalize the propellers for the F_1-motors.

Ren *et al.* (2006) integrated nanowires with F_1-ATPase motors. The multi-segment (Ni/Au/Ni) nanowires are synthesized by electrochemical deposition. The thiol group

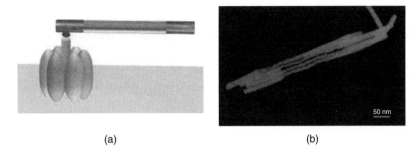

(a)   (b)

Figure 6.64 Multi-component nanowire driven by an F_1-ATPase motor. (a) Schematic drawing of the structure. The nickel segment on the nanowire is used as the region attached to the rotating shaft of the F_1-ATPase motor; (b) TEM picture of three-segment nanowires. (Ren *et al.* 2006)

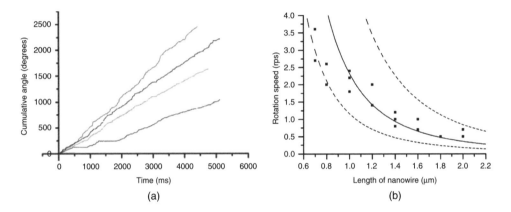

(a)   (b)

Figure 6.65 (a) Time course of F_1-ATPase rotation. Each line represents data from a rotating nanowire; (b) rotational speed in revolutions per second *vs* the nanowire lengths. The solid line shows the rotational speed under a constant torque of $20\,pN \cdot nm$. The upper dotted line and the lower dotted line show the rotational speed under a constant torque $40\,pN \cdot nm$, and $10\,pN \cdot nm$ respectively. (Ren *et al.* 2006)

modified ssDNA and the biotinylated peptide are selectively bound to the gold and nickel segments, respectively. The F_1-ATPase motor only attaches to the nickel segment of the nanowire by biotin-streptavidin linkage. Figures 6.64(a) and (b) show the schematic of the motor device and the TEM pictures of the multi-segment nanowires, respectively. The multi-component nanowires can be used as the propellers, and the rotations of the multi-component nanowires driven by F_1-ATPase motors can observed.

Figure 6.65(a) shows the cumulated angle versus the different length of the propellers, and Figure 6.65(b) shows the relationship between the rotational speed and the nanowire length. The rotation of the propeller driven by motors indicates that the multi-component nanowires accompanied by the different biomolecules can be designed to construct hybrid devices (Ren *et al.* 2006).

References

Bao, J.C., Tie, C.Y., Xu, Z., Suo, Z.Y., Zhou, Q.F. and Hong, J.M. (2002). A facile method for creating an array of metal-filled carbon nanotubes, *Advanced Materials*, **14**, 1483–6.

Barbic, M., Mock, J.J., Gray, A.P. and Schultz, S. (2001). Electromagnetic micromotor for microfluidics applications, *Applied Physics Letters*, **79**(9), 1399–401.

Bauer, L.A., Birenbaum, N.S. and Meyer, G.J. (2004). Biological applications of high aspect ratio nanoparticles, *Journal of Materials Chemistry*, **14**, 517–26.

Bauer, L.A., Reich, D.H. and Meyer, G.J. (2003). Selective functionalization of two-component magnetic nanowires, *Langmuir*, **19**, 7043–8.

Bauer, L., Tanase, M., Hultgren, A., Sun, L., Searson, P.C., Reich, D.H. and Meyer, G.J. (2001). Magnetic alignment of fluorescent nanowires, *Nano Letters*, **1**, 155–8.

Bera, D., Kuiry, S.C. and Seal, S. (2004). Synthesis of nanostructured materials using template-assisted electrodeposition, *JOM*, **56**(1), 49–53.

Birenbaum, N.S., Lai, T.B., Reich, D.H., Chen C.S. and Meyer, G.J. (2003). Selective noncovalent adsorption of protein to bifunctional metallic nanowire surfaces, *Langmuir*, **19**, 9580–2.

Blondel, A., Meier, J.P., Doudin, B. and Ansermet, J.P. (1994). Giant magnetoresistance of nanowires of multilayers, *Applied Physics Letters*, **65**, 3019–21.

Blondel, A., Doudin, B. and Ansermet, J.P. (1997). Comparative study of the magnetoresistance of electrodeposited Co/Cu multilayered nanowires made by single and dual bath techniques, *Journal of Magnetism and Magnetic Materials*, **165**, 34–7.

Chen, M., Guo, L., Ravi, R. and Searson, P.C. (2006). Kinetics of receptor directed assembly of multisegment nanowires, *Journal of Physical Chemistry B*, **110**, 211–17.

Chen, M., Sun, L., Bonevich, J.E., Reich, D.H., Chien, C.L. and Searson, P.C. (2003). Tuning the response of magnetic suspensions, *Applied Physics Letters*, **82**(19), 3310–12.

Chen, M., Searson, P.C. and Chien, C.L. (2003a). Micromagnetic behavior of electrodeposited Ni/Cu multilayer nanowires, *Journal of Applied Physics*, **93**, 8253–5.

Doudin, B., Redmond, G., Gilbert, S.E. and Ansermet, J.P. (1997). Magnetoresistance governed by fluctuations in ultrasmall Ni/NiO/Co junctions, *Physical Review Letters*, **79**, 933–6.

Ebels, U., Radulescu, A., Henry, Y., Piraux, L. and Ounadjela, K. (2000). Spin accumulation and domain wall magnetoresistance in 35nm Co Wires, *Physical Review Letters*, **84**, 983–6.

Encinas, A., Demand, M., Vila, L., Huynen, I. and Piraux, L. (2002). Tunable remanent state resonance frequency in arrays of magnetic nanowires, *Applied Physics Letters*, **81**, 2032–4.

Encinas, A., Demand, M., Piraux, L., Huynen, I. and Ebels, U. (2001a). Dipolar interactions in arrays of nickel nanowires studied by ferromagnetic resonance, *Physical Review B*, **63**, 104415.

Encinas, A., Demand, M., Piraux, L., Ebels, U. and Huynen, I. (2001b). Effect of dipolar interactions on the ferromagnetic resonance properties in arrays of magnetic nanowires, *Journal of Applied Physics*, **89**, 6704–6.

Henry, Y., Ounadjela, K., Piraux, L., Dubois, S., George, J.M. and Duvail, J.L. (2001) Magnetic anisotropy and domain patterns in electrodeposited cobalt nanowires, *European Physical Journal B*, **20**, 35–54.

Fert, A. and Piraux, L. (1999). Magnetic nanowires, *Journal of Magnetism and Magnetic Materials*, **200**, 338–58.

Frei, E.H., Shtrikman, S. and Treves, D. (1957). Critical size and nucleation field of ideal ferromagnetic particles, *Physical Review*, **106**, 446–55.

Goglio, G., Pignard, S., Radulescu, A., Piraux, L., Vanhoenacker, D., Vandervorst, A. and Huynen, I. (1999). Microwave properties of metallic nanowires, *Applied Physics Letters*, **75**, 1769–71.

Hernandez-Velez, M., Pirota, K.R., Paszti, F., Navas, D., Climent, A. and Vazquez, M. (2005). Magnetic nanowire arrays in anodic alumina membranes: Rutherford backscattering characterization, *Applied Physics A*, **80**, 1701–6.

Heydon, G.P., Hoon, S.R., Farley, A.N., Tomlinson, S.L., Valera, M.S., Attenborough, K. and Schwarzacher, W. (1997). Magnetic properties of electrodeposited nanowires, *Journal of Physics D*, **30**, 1083–93.

Hultgren, A., Tanase, M., Chen, C.S. and Reich, D.H. (2004). High-yield cell separations using magnetic nanowires, *IEEE Transactions on Magnetics*, **40**(4), 2988–900.

Hurst, M.J., Payne, E.K., Qin, L. and Mirkin, C.A. (2006). Multisegmented one-dimensional nanorods prepared by hard-template synthetic methods, *Angewandte Chemie, International Edition*, **45**, 2672–92.

Lee, K.B., Park, S. and Mirkin, C.A. (2004). Multicomponent magnetic nanorods for biomolecular separations, *Angewandte Chemie, International Edition*, **43**, 3048–50.

Liu, M., Lagdani, J., Imrane, H., Pettiford, C., Lou, J., Yoon, S., Harris, V.G., Vittoria, C. and Suna, N.X. (2007). Self-assembled magnetic nanowire arrays, *Applied Physics Letters*, **90**, 103105.

Love, J.C., Urbach, A.R., Prentiss, M.G. and Whitesides, G.M. (2003). Three-dimensional self-assembly of metallic rods with submicron diameters using magnetic interactions, *Journal of the American Chemical Society*, **125**, 12696–7.

Mallet, J., Yu-Zhang, K., Chien, C.L., Eagleton, T.S and Searson, P.C. (2004). Fabrication and magnetic properties of fcc Co_XPt_{1-X} nanowires, *Applied Physics Letters*, **84**(19), 3900–2.

Martin, B.R., St. Angelo, S.K. and Mallouk, T.E. (2002). Interactions between suspended nanowires and patterned surfaces, *Advanced Functional Materials*, **12**, 759–65.

Martin, B.R., Dermody, D.J., Reiss, B.D., Fang, M., Lyon, L.A., Natan, M.J. and Mallouk, T.E. (1999). Orthogonal self-assembly on colloidal gold-platinum nanorods, *Advanced Materials*, **11**(12), 1021–5.

Masuda, H. and Fukuda, K. (1995). Ordered metal nanohole arrays made by a 2-step replication of honeycomb structures of anodic alumina, *Science*, **268**, 1466–8.

Maurice, J.L., Imhoff, D., Etienne, P., Durand, O., Dubois, S., Piraux, L., George, J.M., Galtier, P. and Fert, A. (1998). Microstructure of magnetic metallic superlattices grown by electrodeposition in membrane nanopores, *Journal of Magnetism and Magnetic Materials*, **184**, 1–18.

Nicewarner-Pena, S.R., Freeman, R.G., Reiss, B.D., He, L., Pena, D.J. Walton, I.D., Cromer, R., Keating, C.D. and Natan, M.J. (2001). Submicrometer metallic barcodes, *Science*, **294**(5540), 137–41.

Noji, H., Yasuda, R., Yoshida, M. and Kinosita, K. Jr. (1997). Direct observation of the rotation of F-1-ATPase, *Nature*, **386**, 299–302.

O'Barr, R. and Schultz, S. (1997). Switching field studies of individual single domain Ni columns, *Journal of Applied Physics*, **81**, 5458–60.

Piraux, L., Encinas, A., Vila, L., Mátéfi-Tempfli, S., Mátéfi-Tempfli, M., Darques, M., Elhoussine, F. and Michotte, S. (2005). Magnetic and superconducting nanowires, *Journal of Nanoscience and Nanotechnology*, **5**, 372–89.

Piraux, L., Dubois, S., Ferain, E., Legras, R., Ounadjela, K., George, J.M., Maurice, J.L. and Fert, A. (1997). Anisotropic transport and magnetic properties of arrays of sub-micron wires, *Journal of Magnetism and Magnetic Materials*, **165** (1–3), 352–5.

Piraux, L., George, J.M., Despres, J.F., Leroy, C., Ferain, E., Legras, R., Ounadjela, K. and Fert, A. (1994). Giant magnetoresistance in magnetic multilayered nanowires, *Applied Physics Letters*, **65**, 2484–6.

Prime, K.L. and Whitesides, G.M. (1991). Self-assembled organic monolayers – model systems for studying adsorption of proteins at surfaces, *Science*, **252**, 1164–7.

Reich, D.H., Tanase, M., Hultgren, A., Bauer, L.A., Chen, C.S. and Meyer, G.J. (2003). Biological applications of multifunctional magnetic nanowires, *Journal of Applied Physics*, **93**(10), 7275–80.

Ren, Q., Zhao, Y.P., Yue, J.C. and Cui Y.B. (2006). Biological application of multi-component nanowires in hybrid devices powered by F_1-ATPase motors, *Biomedical Microdevices*, **8**, 201–8.

Roberts, L.A., Crawford, A.M., Zappe, S., Jain, M. and White, R.L. (2004). Patterned magnetic bar array for high-throughput DNA detection, *IEEE Transactions on Magnetics*, **40**, 3006–8.

Salem, A.K., Searson, P.C. and Leong, K.W. (2003). Multifunctional nanorods for gene delivery, *Nature Materials*, **2**, 668–71.

Sarkar, J., Khan, G.G. and Basumallick, A. (2007). Nanowires: properties, applications and synthesis via porous anodic aluminium oxide template, *Bulletin of Materials Science*, **30**(3), 271–90.

Stiborova, H., Kostal, J., Mulchandani, A. and Chen, W. (2003). One-step metal-affinity purification of histidine-tagged proteins by temperature-triggered precipitation, *Biotechnology and Bioengineering*, **82**, 605–11.

Sun, L., Hao, Y., Chien, C.L. and Searson, P.C. (2005). Tuning the properties of magnetic nanowires, *IBM Journal of Research and Development*, **49**(1), 79–102.

Tanase, M., Felton, E.J., Gray, D.S., Hultgren, A., Chen, C.S. and Reich, D.H. (2005). Assembly of multicellular constructs and microarrays of cells using magnetic nanowires, *Lab on a Chip*, **5**, 598–605.

Tanase, M., Silevitch, D.M., Hultgren, A., Bauer, L.A., Searson, P.C., Meter, G.J. and Reich, D.H. (2002). Magnetic trapping and self-assembly of multicomponent nanowires, *Journal of Applied Physics*, **91**, 8549–51.

Tanase, M., Bauer, L.A., Hultgren, A., Silevitch, D.M., Sun, L., Reich, D.H., Searson, P.C. and Meyer, G.J. (2001). Magnetic alignment of fluorescent nanowires, *Nano Letters*, **1**, 155–8.

Terrier, C., Abid, M., Arm, C., Serrano-Guisan, S., Gravier, L. and Ansermet, J.P. (2005). Fe_3O_4 nanowires synthesized by electroprecipitation in templates, *Journal of Applied Physics*, **98**, 086102.

Tok, J.B.H., Chuang, F.Y.S., Kao, M.C., Rose, K.A., Pannu, S.S., Sha, M.Y., Chakarova, G., Penn, S.G. and Dougherty, G.M. (2006). Metallic striped nanowires as multiplexed immunoassay platforms for pathogen detection, *Angewandte Chemie, International Edition*, **45**, 6900–4.

Tsymbal, E.Y. and Pettifor, D.G. (2001). Perspectives of giant magnetoresistance, in *Solid State Physics*, Vol. **56** (eds. H. Ehrenreich, F. Spaepen), Academic Press, pp. 113–237.

Urbach, A.R., Love, J.C., Prentiss, M.G. and Whitesides, G.M. (2003). Sub-100nm confinement of magnetic nanoparticles using localized magnetic field gradients, *Journal of the American Chemical Society*, **125**, 12704–5.

Vazquez, M., Pirota, K., Torrejon, J., Navas, D. and Hernandez-Velez, M. (2005). Magnetic behaviour of densely packed hexagonal arrays of Ni nanowires: Influence of geometric characteristics, *Journal of Magnetism and Magnetic Materials*, **294**, 174–81.

Vazquez, M., Nielsch, K., Vargas, P., Velazquez, J., Navas, D., Pirota, K., Hernandez-Velez, M., Vogel, E., Cartes, J., Wehrspohn, R. and Gosele, U. (2004). Modelling hysteresis of interacting nanowires arrays, *Physica B*, **343**, 395–402.

Vila, L., Piraux, L., George, J.M., Fert, A. and Faini, G. (2002). Multiprobe magnetoresistance measurements on isolated magnetic nanowires, *Applied Physics Letters*, **80**, 3805–7.

Wegrowe, J.E., Gilbert, S.E., Kelly, D., Doudin, B. and Ansermet, J.P. (1998). Anisotropic magnetoresistance as a probe of magnetization reversal in individual nano-sized nickel wires, *IEEE Transactions on Magnetics*, **34**, 903–5.

Wernsdorfer, W., Doudin, B., Mailly, D., Hasselbach, K., Benoit, A., Meier, J., Ansermet, J.P. and Barbara, B. (1996). Nucleation of magnetization reversal in individual nanosized nickel wires, *Physical Review Letters*, **77**, 1873–6.

Whitney, T.M., Jiang, J.S., Searson, P.C. and Chien, C.L. (1993). Fabrication and magnetic properties of arrays of metallic nanowires, *Science*, **261**, 1316–19.

Wildt, B., Mali, P. and Searson, P.C. (2006). Electrochemical template synthesis of multisegment nanowires: fabrication and protein functionalization, *Langmuir* **22**, 10528–34.

Xu, L., Zhang, W., Ding, Y., Peng, Y., Zhang, S., Yu, W. and Qian, Y. (2004). Formation, characterization, and magnetic properties of Fe_3O_4 nanowires encapsulated in carbon microtubes, *Journal of Physical Chemistry B*, **108**, 10859–62.

Yellen, B.B. Hovorka, O. and Friedman, G. (2005). Arranging matter by magnetic nanoparticle assemblers, Proceedings of the National Academy of Science of the United States of America, 102, 8860–4.

Zhang, D., Liu, Z., Han, S., Li, C., Lei, B., Stewart, M.P., Tour, J.M. and Zhou, C. (2004). Magnetite (Fe_3O_4) core–shell nanowires: synthesis and magnetoresistance, *Nano Letters*, **4**(11), 2151–5.

7

Magnetic Nanotubes and their Biomedical Applications

7.1 Introduction

It has been witnessed that microtechnology, which is based on thin film-structured materials, is naturally being converted to nanotechnology in which the size of the active electronic components is reduced to nanosize. This is mainly due to the fact that single molecules or atoms coupled with strong quantum-size effects limit further miniaturization. The unique physical, chemical and thermal properties of quasi one- and two-dimensional materials may lead to novel inventions, new products and fresh technology additions to existing knowledge.

Nanotubes have become a fast developing research area of nanotechnology for several decades. Theoretical simulation indicated that a narrow tube might be less energetically favored than a finite strip because of the strain energy coupled with bending. However, when the diameter is reduced to a critical value (e.g. at nanoscale), the strain in the nanotubes becomes smaller than the energy associated with the edges (dangling bonds) in the layered strips, and the self-closed cylindrical geometry, which is free of dangling bonds, becomes the most stable structure (Seifert *et al.* 2002). In general, the advantage of nanotubes with hollow structures is that they can be used as pipes, cavities or capsules in nanosize. In addition, nanotube-embedded nanostructured hybrid systems with extremely large surfaces have considerable advantages over nanoparticle-based systems in many applications such as catalysis and sensor technology. Carbon nanotubes have been demonstrated to have unique properties and potential applications. However, carbon is by no means always ideal as an all-purpose material for popular applications. In view of recent findings on carbon nanotubes, the potential of nanotubes made of other materials, such as polymers, metals, semiconductors or oxide, is also expected. The cylindrical geometry of carbon nanotubes is paralleled by the structure of numerous inorganic nanotubes (Remškar 2004). The first applications for non-carbon nanotubes were demonstrated by Martin *et al.* in such areas as the separation of racemic mixtures, sensors, substance separation, or in membranes for selective ion transport (Steinhart *et al.* 2004).

In the field of magnetic nanotechnology, the search for new geometries is always an important aspect. Various magnetic nanostructures such as nanodots, nanowires and antidotes have been developed. As an emerging field, synthesis of magnetic tubular nanostructures was pioneered in inorganic chemistry. Magnetic nanotubes have a low mass density, a high porosity and an extremely large surface to weight ratio due to their cylindrical geometry. The potential applications of magnetic nanotubes include data storage devices in nanocircuits, scanning tips for magnetic force microscopes, drug delivery, tunable fluidic channels for tiny magnetic particles, biosensors, biochemical separation, etc. Compared with nanowires or nanorods with only one surface, magnetic nanotubes with two separate surfaces could be used as multifunctional nanomaterials. By surface modification of the inner or outer nanotube walls respectively with oxides, polymers, biomolecules or metals, physical and chemical properties may be tailored within a single structure. Additionally, multilayer magnetic nanotubes, with concentric cylinder layers are expected to have unique magnetoelectronic properties.

Up to now, chemical synthesis and template-directed growth have been widely used to prepare magnetic nanotubes. Novel and effective synthetic techniques are expected to tune the magnetic properties of magnetic nanotubes in a controlled manner because magnetic quantities (anisotropy, coercivity, etc.) are important for many applications in permanent magnetism, spin electronics and magnetic recording (Sui *et al.* 2004a).

7.2   Magnetism of Nanotubes

The magnetic properties of magnetic nanotubes depend upon several factors such as the elemental composition, crystallinity, shape, diameter, wall thickness and orientation. For instance, for nanotubes with small and large diameters, the preferential arrangements of the magnetic moments for magnetic nanotubes are the parallel configuration and vortex state, respectively (Escrig *et al.* 2007). As the nanotube pore diameter decreases, its coercivity and the remanence increase accordingly. In the case of closely-spaced magnetic nanotube arrays, the preferential magnetization direction is perpendicular to the nanotube axis due to the strong interactions and alignment between the nanotubes (Bao *et al.* 2004).

Magnetic nanotubes exhibit unique magnetic properties. Compared with the bulk counterparts, magnetic nanotubes have higher coercivity coupled with enhanced anisotropic magnetic behavior owing to their shape anisotropy. Additionally, magnetic nanotubes have lower weight density than solid magnetic nanowires. This property gives them higher stability in solutions with less possibility of precipitation. Also, magnetic nanotubes and magnetic nanowires display difference in remanent state, thus showing different magnetic or magnetoresistive behaviors.

Investigation of the magnetic behaviors of magnetic nanotubes is a basis for their potential applications. Investigation of the nanotube magnetism starts with the analysis of the static distribution of magnetic moments in nanotubes. It is well known that, when subjected to an applied external magnetic field, a magnetic material will generate an additional magnetic field, called a demagnetization field. Considering a two-dimensional (2D) magnetic thin film, the demagnetization field is known to push down the magnetic moments in the plane of the film surface. In the case of one-dimensional (1D) magnetic nanowires and quasi-one-dimensional (quasi-1D) nanotubes, the demagnetization field is believed to align the magnetic moments along the axis of the nanotubes. And magnetic moment distribution in magnetic nanotubes is determined by the competition between magnetocrystalline anisotropy energy (E_k) and demagnetization energy (E_{demag}) (Li *et al.* 2007). In nanotubes, the magnetic moment distribution is isotropic with respect to the tube axis, and a paramagnetic quadrupole doublet can be induced. There are two possible reasons: 1) the surface effect of the nanotube leads to a spin disorder on the nanotube surface due to its large surface ratio; 2) the exchange coupling may not be strong enough to suppress superparamagnetism due to the nanograins with a smaller size.

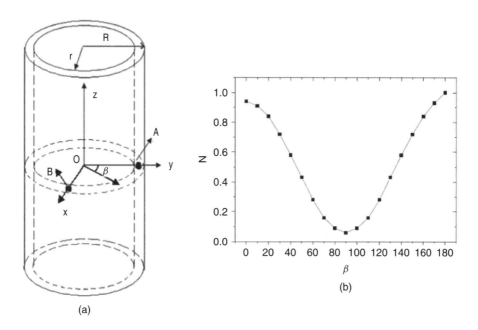

(a)

(b)

Figure 7.1 (a) Schematic structure of a magnetic nanotube; (b) the relationship between the demagnetization factor and angle β. (Wang *et al.* 2006)

To study the static distribution of magnetic moments, a theory of magnetic charge was used to calculate the demagnetization factor of the magnetite nanotubes (Fe_3O_4), (Wang *et al.* 2006). The demagnetization factor is defined as the ratio of the negative of the demagnetizing field to the magnetization of a magnetic sample. Figure 7.1a illustrates a schematic nanotube with a three-dimensional orthogonal coordinate set up in the center of the nanotube. Based on the TEM images of nanotubes, the values of the outer and inner radius are defined as 100 and 88 nm, respectively. In the calculation, the length of the nanotube is defined as 10 μm. An external magnetic field along the Y axis is assumed to be applied to magnetize the nanotube to saturation. Calculations based on the theory of magnetic charge gave the demagnetization factors in different areas of the nanotube wall in the X-Y plane. The result is plotted in Figure 7.1b. The X axis is β, which is the angle between the Y axis and the line connecting the original point O and the point on the wall. From Figure 7.1b, it can be seen that the demagnetization factor reaches the highest value of 0.94 when β is 0 (point A in the nanotube wall) while the lowest value of 0.06 is achieved at $\beta = 90°$ (point B in the nanotube wall). In the range of 0–90°, the factor decreases with the increase of the angle.

Considering a random point in the wall, the final state comes from the minimization of the total energy, which is a sum of three parts:

$$E_{total} = E_{exchange} + E_K + E_{demag} \tag{7.1}$$

where $E_{exchange}$, E_K, and E_{demag} are exchange, magnetocrystalline anisotropy, and demagnetization energies, respectively. In the calculation, the magnetocrystalline anisotropy and saturation magnetization of the magnetite nanotube are identical to those of bulk magnetite. The demagnetization energy E_{demag} tends to put the magnetic moments in the nanotube wall, and its highest value is $0.94 \times 2\pi Ms^2 = 1.38 \times 10^6 \, erg \, cm^{-3}$. The magnetocrystalline anisotropy energy E_K tends to align the magnetic moments along the easy

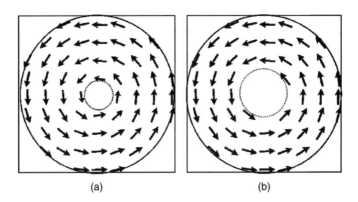

(a)   (b)

Figure 7.2 Snapshots of the magnetization on the cross-section of a magnetic nanotube. The size of the arrows indicates the magnitude of the component of the magnetization in the cross-sectional plane. (a) Dimension ratio $\xi = 0.1$, and (b) $\xi = 0.3$. (Escrig *et al.* 2007)

axis of grain, and its value is only $-1.35 \times 10^5 \, \text{erg cm}^{-3}$. Due to the much lower value of E_K, compared with E_{demag}, it can be ignored. Thus, equation (7.1) indicates that the magnetic moments of the Fe_3O_4 nanotube preferentially align parallelly to the nanotube wall with a minimum magnetostatic energy.

A continuous model was adopted for the internal magnetic structure of the nanoparticles, which provided a good basis to investigate their magnetic properties. Via this model, analytical results for total energy in different configurations of nanotubes can be obtained. For simplification purposes, a continuous distribution of magnetic moment, denoted by magnetization $\mathbf{M(r)}$, was used to displace the discrete one. Where δV is the volume of the analyzed elements centered at $\mathbf{r}$, the total magnetic moment within the element is $\mathbf{M(r)}\delta V$. Dipolar interactions within the magnetic nanotube induce shape anisotropy, which is the main anisotropy in this system. Several geometrical parameters for nanotubes considered are height (or length) H; external radius R; internal radius a; ratios $\xi = a/R$ and $\gamma = H/R$. Figure 7.2a and 7.2b depict the simulation results for ξ values of 0.1 and 0.3, respectively. Arrows in the plots represent the ϕ component of the magnetization, and a core is produced by the magnetization component along the z direction, shown with short arrows. M_0 and M_z are saturation magnetization and the component responsible for the core magnetization, respectively. Simulation showed a ratio M_z/M_0 of 0.11 for ξ of 0.1 and around 10^{-6} for ξ of 0.3, which implied that the effect of core can be negligible for thin-walled nanotubes. Considering the parallel configuration and vortex configuration, the interaction between nanotubes is ignored and the expression for the dipolar contribution to the energy in the parallel configuration can be described as:

$$E_{dip}^F = \pi \mu_0 M_0^2 R^3 \int_0^\infty \frac{1 - e^{-\gamma y}}{y^2} (J_1(y) - \xi J_1(\xi y))^2 \, dy \qquad (7.2)$$

where $J_1(z)$ are Bessel functions of first order and first kind. And the dipolar energy in the vortex configuration is equal to zero.

$$E_{ex}^V = 2\pi H A \ln \frac{1}{\xi} \qquad (7.3)$$

By equating the expressions for the energy in both configurations, the transition line as well as the magnetic phase diagram can be obtained (Escrig *et al.* 2007).

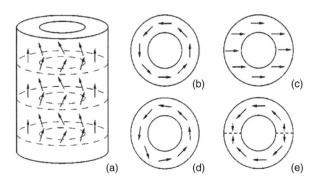

Figure 7.3 Schematic reversal modes in magnetic nanotubes. (a) and (b) curling, (c) coherent rotation, (d) perturbed curling, and (e) low-lying noncurling mode. (Sui *et al.* 2004b)

In magnetic nanotubes, the mechanism of magnetization reversal needs to be discussed. Several reversal mechanisms, induced by a homogeneous external magnetic field, are shown in Figure 7.3 (Sui *et al.* 2004b). Flux closure during magnetization reversal induces magnetization curling, which can be seen in Figure 7.3(a) and 7.3(b). Figure 7.3(c) depicts a coherent rotation which leads to surface poles. For magnetic nanomaterials, as the radius R is larger than a critical value, a transition takes place from coherent rotation to curling. In particular, this transition only occurs for nanotubes with very small diameters. For curling mode in tubes having wall thicknesses much smaller than the tube radius ($t \ll R$), the exchange energy H_n can be expressed as:

$$H_n = H_a + \frac{A}{\mu_0 M_s R^2} \tag{7.4}$$

where H_a is the anisotropy field, A is the exchange stiffness and M_s is the spontaneous magnetization. From this formula, the certain value of radii for curling occurring is $2\zeta_0$, and ζ_0, the exchange length of the system, equals to $(A/\mu_0 M_s^2)^{1/2}$. Magnetic nanotubes have a lower exchange energy (smaller by a factor of 5) compared with other magnetic nanomaterials. The reason is believed to be the absence of curling related vortices. In the case of polycrystalline nanotubes, the curling character of the reversal mode may be distorted by polycrystallinity. A mode without curling-type flux closure is schemed in Figure 7.3(d). In this mode, the curling mode switches to a localized mode at radius R_{rand} which is dependent upon the magnetocrystalline anisotropy. The reversal mode shown in Figure 7.3(d) is favorable for anisotropy, but unfavorable for both exchange and magnetostatics. The R_{rand} can be expressed by applying standard random anisotropy analysis:

$$R_{rand} \approx \delta_B^2 t^{1/2}/a^{3/2} \tag{7.5}$$

where a is the polycrystalline grain size and δ_B, the Bloch-wall thickness of the corresponding bulk material, is equal to $(A/|K_1|)^{1/2}$. With estimated values of a, t and δ_B, the value of R_{rand} can be obtained for different magnetic nanotubes, which shows whether the nanotube is curling-like or not. It should be pointed out that deduction of Equation (7.5) does not include magnetostatic self-interaction. When this is involved, the random-anisotropy analysis displays that internal poles, such as around the dashed lines in Figure 7.3(e), enhance R_{rand}.

7.3   Multifunctionality of Magnetic Nanotubes

Magnetic nanotubes have the capability of multifunctionality due to their inner and outer surfaces. For nanospheres/nanoparticles, there is only one surface available for modification, hence it is complex to generate multifunctional particles (Eisenstein 2005). By contrast, nanotubes with different functionalizations on both surfaces could serve as a type of multifunctional nanomaterials for several research problems.

7.3.1   Inner and Outer Surfaces of Nanotubes

Magnetic nanomaterials have been extensively explored for their extraordinary properties as well as their potential applications, especially in biomedical and biotechnological applications. Typical examples include enzyme immobilization or encapsulation, DNA transfection, drug delivery, biochemical separation, contrast enhancement in magnetic resonance imaging and biosensors. In most applications, magnetic nanoparticles have been used, mainly because of the ease of their synthesis and dimension control. However, the structural limitation of spherical nanoparticles causes problems when multifunctionality is needed for certain bioapplications. Magnetic nanotubes possessing a number of attributes are potential candidates for certain bioapplications in such situations.

Self-assembling lipid tubules have been used for this purpose. However, their tubule diameters as well as the surface morphology are difficult to control. The rapid development of nanotechnology has brought a number of chemical or physical synthesis approaches which are able to control readily the dimensions, properties and crystallinities of nanotubes. Additionally, it is possible to obtain any material in nanotube structure by general synthesis techniques. Hence, synthetic magnetic nanotubes with desirable characteristics are ready for further surface modifications for biocompatibility and subsequent bioapplications.

Regarding the multifunctionality of magnetic nanotubes, there are two aspects. First, nanotubes have inner voids, which can be filled and immobilized with biological species ranging in size from large proteins to small molecules, followed by a controlled release (Mitchell *et al.* 2002). Secondly, the distinct inner and outer surfaces of nanotubes render them a novel type of nanomaterial with multifunctional properties after being functionalized differently. The schematic of multifunctional magnetic nanotubes is illustrated in Figure 7.4. For instance, the inner surface of nanotubes can be chemically functionalized to obtain a hydrophilic surface while the outer surface can be modified to exhibit hydrophobic properties.

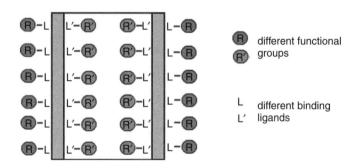

Figure 7.4   Functionalization of a magnetic nanotube for which L and L$'$ represent ligands that bind selectively to inner and outer surfaces, and R and R$'$ represent spatially separated functional groups

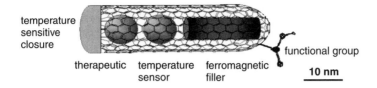

temperature
sensitive
closure

functional group

therapeutic   temperature   ferromagnetic
sensor   filler

therapeutic   temperature   ferromagnetic   10 nm
sensor   filler

Figure 7.5   Schematic of a magnetically-filled carbon nanotube as a nanocontainer. (Mönch *et al.* 2007)

7.3.2   Magnetic Encapsulated Nanotubes

While chemical or biological species can be encapsulated inside the inner void of magnetic nanotubes, their distinctive outer surfaces can be differentially functionalized with environment-friendly and/or probe molecules for specific targeting purposes. Therefore, the magnetic nanotube combining attractive loading capabilities and magnetic properties can be an ideal candidate for the multifunctional nanomaterial for biomedical applications, especially for magneto-controlled targeted drug delivery and magnetic resonance imaging (Son *et al.* 2005).

Another type of magnetic-encapsulated nanotube are non-magnetic nanotubes with encapsulated magnetic species. As shown in Figure 7.5, multifunctional nanocontainers can be realized by filling carbon nanotubes with various substances (Mönch *et al.* 2007). When magnetic nanoparticles/nanocrystals are encapsulated, nanotubes exhibit magnetic properties. At the same time, the outer surface of carbon nanotubes can be easily functionalized for specific targeting purposes in biomedical applications.

7.4   Synthesis and Characterization of Magnetic Nanotubes

Magnetic nanostructures are scientifically interesting and the research efforts on their properties and potential applications are technologically important. Despite the development of magnetic nanodots, nanowires and antidote structures, searching for magnetic nanomaterials with novel geometries has been an important aspect of magnetic nanotechnology (Sui *et al.* 2004b). Magnetic nanotubes with magnetic properties coupled with multifunctionality have stimulated much research interest. For common applications of magnetic nanomaterials such as magnetic recording, spin electronics, biomedical and biological applications, synthesis of magnetic nanotubes is always in high demand. To date, various chemical and physical methods have been used for magnetic nanotube synthesis, such as thermal decomposition of precursors, hydrothermal synthesis and galvanic displacement reactions. Without exception, any technique employed in magnetic nanotube synthesis aims to obtain magnetic nanotubes with controlled dimension, morphology, crystallinity and tailored magnetic properties.

Among a number of approaches, template-assisted synthesis is a versatile and inexpensive technique (Bao *et al.* 2001). It is Martin's group who first demonstrated template synthesis of various nanotubes using membranes with nanochannels (Hulteen and Martin 1997). By varying the physical and chemical properties of the used template and controlling the preparation conditions, the sizes, shapes and structural properties of nanotubes can be tailored. Importantly, this method is a general synthesis strategy which can be used to prepare metallic, polymeric, semiconducting and carbon nanotubes (Tourillon *et al.* 2000). In this section, the synthesis of various magnetic nanotubes will be reviewed.

7.4.1 Single Element Magnetic Nanotubes

Single element magnetic nanotubes are made of transition metals including Co, Fe and Ni. Tourillon *et al.* demonstrated using the template method to synthesize cobalt and iron nanotubes (2000). Commercially available track-etched polycarbonate membranes with an average 30 nm pore diameter and 6 mm in thickness were used as the template. For electrodeposition, a gold thin film was coated on the bottom surface of the templates by evaporation. This gold film, a saturated calomel reference electrode and a platinum grid served as the working electrode, reference electrode and counter electrode in a three-electrode electrochemical cell, respectively. The electrolyte for Co nanotubes synthesis was prepared by mixing 0.1 M $CoSO_4$, 0.1 M H_3BO_3 and high pure water (>18 MΩ · cm). For Fe nanotubes, $CoSO_4$ was replaced by $FeSO_4$. In a typical synthesis process, the prepared electrolyte should be degassed by bubbling argon flow for 15 minutes as the first step. Then, pulse cycles were applied for electrodeposition of Co and Fe nanotubes inside the template. In the case of Co, the deposition started with the application of an overpotential of -1.5 V for 0.1 s followed by an 'off' state for 2 s. This cycle lasted for 8 minutes. For Fe, -0.13 V/0.1 s cycle was continued for 6 minutes.

Figure 7.6a is a low resolution TEM image of iron nanotubes. An aperture at one end of a nanotube can be seen clearly from a high resolution TEM image, as shown in Figure 7.6b. The wall thickness is estimated to be 1–2 nm, which may consist of a few atomic iron planes. Electron diffraction patterns of Fe nanotubes indicate a mixture of Fe bcc and fcc phases. The as-prepared cobalt nanotubes also have similar wall thickness and dimension. The chemical complexation of the metal cations on the porous membrane walls through carbonate functions was expected to induce the formation of Co and Fe nanotubes. The nanotube formation can be hypothesized to occur in two steps. In the first step, Fe^{2+} and Co^{2+} metal ions are complexed with the $-CO_3^{2-}$ carbonate function of the polymeric membranes in a favored process due to the strong interaction between the chemical species. Thus, the inner surface of nanochannels is coated with metal ions. The following step is the electrochemical reduction of metal ions Fe^{2+} and Co^{2+} to Fe^0 and Co^0, respectively, resulting in the formation of nanotubes. It was found that longer deposition time produced nanotubes with thicker wall thickness and eventually solid nanowires.

Synthesis of highly ordered nickel nanotubes was achieved by electrodeposition inside the nanopores of alumina membranes with chemical surface modification (Bao *et al.* 2001). Before electrodeposition, anodic alumina oxide (AAO) membranes were immersed into a 1 % methyl-diethylenetriaminopropyl dimethoxysilane in anhydrous nonane under

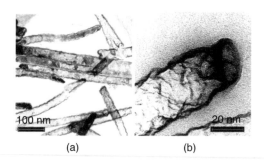

(a)   (b)

Figure 7.6 (a) TEM image of prepared iron nanotubes; (b) TEM image shows the tip of a Fe nanotube. The thickness of the nanotube wall is about 1–2 nm. (Tourillon *et al.* 2000)

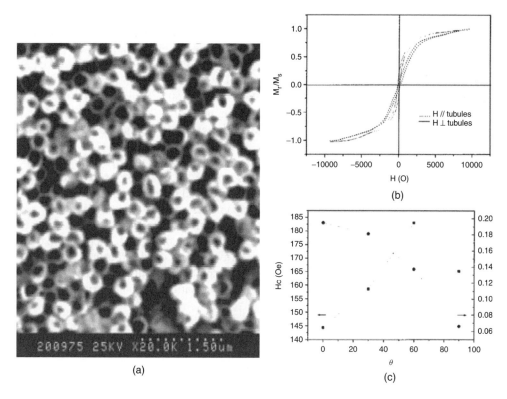

Figure 7.7 (a) SEM image of an as-prepared Ni nanotube array, (b) hysteresis loops of the Ni nanotubes embedded in the template, (c) plot of the relationship between coercivities, remanent/saturation magnetization ratio and θ. (Bao *et al.* 2001)

sonication for 1 minute. Treated with this pore-wall modifying agent produce silanized membranes. Again, a gold film was evaporated on one side of the membrane as a working electrode during the electrodeposition. A mixture of $NiSO_4 \cdot 6H_2O$, H_3BO_3 and purified water (resistivity $> 18\,M\Omega \cdot cm$) was used as the electrolyte. The working electrode, a saturated calomel electrode (SCE) and a nickel counter electrode, were immersed in the electrolyte forming a three-electrode electrochemical cell. During the electrodeposition, a constant current density of $0.3 \pm 0.6\,mA/cm^2$ was applied without heating and stirring. After deposition, the top layer of the alumina template was etched by sodium hydroxide solution. A top-view SEM micrograph of synthesized nickel nanotubes is shown in Figure 7.7a. Open ends of the Ni nanotubes can be seen. Experiments indicated the necessity of the inner surface modification with methyl-c-diethylenetriaminopropyl-dimethoxysilane for nickel nanotube formation. After modification, the inner surface of nanopores in membranes was covered with amino groups, which exhibit strong affinity for nickel ions. Consequently, preferential nickel deposition occurs with formation of a nanotube structure. Comparative study showed only nanowires were obtained without the modification step before electrodeposition.

Magnetic properties of Ni nanotubes with an average 160 nm outer diameter and 218 aspect ratio were measured. Figure 7.7b shows the hysteresis loops of Ni nanotube arrays. From the hysteresis loops, coercivities of nickel nanotubes are $H_{c//} \approx 165\,Oe$ and $H_{c\perp} \approx 144\,Oe$ ($1\,Am^{-1} \cong 4\pi \times 10^{-3}\,Oe$). Compared with bulk nickel materials with their coercivities of around 0.7 Oe, the nanotube morphology enhances the coercivities of nickel nanotubes. Both highly sheared hysteresis curves and very low remanence magnetization of Ni nanotubes, which is less than 20 % $M_{sat//}$, indicate a strong interaction among nanotubes. The SEM image reveals densely packed nanotube arrays thus resulting in such a strong intertubule interaction. Dependence of the coercivity H_c and remanent/saturation magnetization ratio M_r/M_s on the angle between the surface of the membrane and the applied magnetic field is plotted in Figure 7.7c. At $\theta = 60°$ the coercivity of the nanotube array reached a maximum value while maximum magnetization ratio of M_r/M_s was obtained at $\theta = 0°$, when the applied field is perpendicular to the axis of the nickel nanotubes.

Tao *et al.* reported an improved method to prepare a highly ordered array of magnetic Ni nanotubes by modified template-assisted electrodeposition (2006). By using this method, highly ordered arrays of Ni nanotubes can be prepared and their wall thickness and length can be controlled by adjusting experimental parameters such as the current density and electrodeposition time. The electrolytes used in the synthesis comprised of $NiSO_4 \cdot 6H_2O$, H_3BO_3, and Pluronic P123 ($EO_{20}PO_{70}EO_{20}$, EO: ethylene oxide, PO: propylene oxide), with proper concentrations in deionized water. A two-electrode electrochemical cell with a gold coated AAO template and nickel ring as the working and counter electrodes, respectively, was used for electrodeposition. The deposition current density was set to $0.13\,mAcm^{-2}$.

The outstanding feature of this method is the addition of a small amount of an amphiphilic triblock copolymer, P123. P123 is a surfactant and forms micelles with a concentration-dependent dimension in electrolyte solutions. For example, dynamic light scattering (DLS) measurements show about 20 nm micelles with 37 g/l P123 solution. Figure 7.8a illustrates a proposed formation mechanism of nickel nanotubes in this synthesis. The affinity between Ni^{2+} and oxygen in P123 induced gathering of Ni^{2+} on the surface of the P123 micelles. In the electrochemical cell, the Ni^{2+} attached P123 micelles entered into the nanochannels of the template and concentrated in the vicinity of the pore walls due to the strong adsorption ability of AAO. Therefore, the electrochemical deposition generated a preferential nickel deposition on the wall resulting in nanotube formation. A top-view SEM image of a nickel nanotube array after partial removal of the AAO template is shown in Figure 7.8b. A high filling ratio is obvious from the observation of an open-ended nanotube in each pore. The hysteresis loops of the synthesized nickel nanotube array reveal that its typical coercivities $H_{c//}$ and $H_{c\perp}$ are 106.1 and 93.72 Oe, respectively. The hysteresis loops confirm the uniaxial magnetic anisotropy with the easy axis perpendicular to the nanotubes. When the magnetic field was applied perpendicular to the nanotubes, measured values of the saturated magnetization (Ms) and remanent magnetization (Mr) were 0.0193 and 0.0090 emu, respectively. In the case of the magnetic field parallel to nanotubes, these two values changed to 0.018 and 0.0009 emu. Tao's method is a simple, convenient and effective synthesis technique for nickel nanotube synthesis.

In addition to template-assisted electrodeposition, template-assisted atomic layer deposition (ALD) has been explored to prepare magnetic nickel and cobalt nanotubes (Daub *et al.* 2007). In a typical ALD synthesis, two different vapor-phase reactants are used and deposit materials on the substrate with one monolayer or less in one cycle. Multicycle deposition is usually applied to obtain nanomaterials with desired layer thickness. Thus, the major advantages of ALD are its precise control of the growth rate and its ability to produce conformal coating on 3D structures. Alumina membranes, prepared by a two-step anodization process of aluminum, were used as the template for ALD deposition of Ni and Co. The nanotube synthesis consisted of three steps. First, nickelocene ($NiCp_2$) or cobaltocene ($CoCp_2$) vapor was used as the first precursors and formed a submonolayer on the sample surface. The precursor temperature was kept at 90 °C for both cases. The

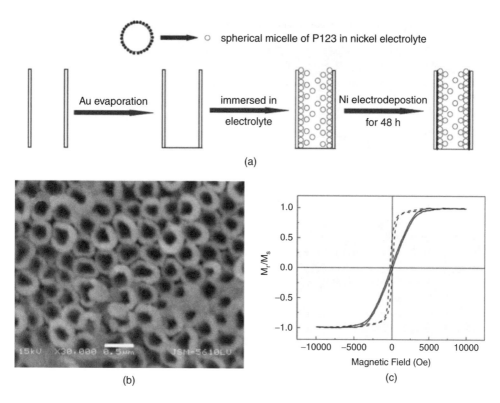

(a)

(b)   (c)

Figure 7.8 (a) Schematic of the preparation procedure for Ni nanotubes and the proposed mechanism for nanotube growth; (b) SEM image of Ni nanotube array after etching the alumina template membrane completely; (c) magnetic hysteresis loops of the Ni nanotube array without template removal. The solid line was obtained as the applied magnetic field was parallel to the Ni nanotube array ($H_{//}$), and the dashed line was obtained as the applied magnetic field was applied perpendicularly to the nanotube array ($H_{\perp}$). (Tao *et al.* 2006)

deposition temperatures for Ni and Co are 270–330 °C and 240–330 °C, respectively. In the second step, O_3 was introduced into the chamber, and its reaction with the adsorbed layer produced metal oxide. The last step involved thermal reduction in an oven at 400 °C under Ar + 5 % H_2 atmosphere for 5 hours. For all ALD cycles, the pulsing time, exposure time and purging time were set to 1 s, 30 s and 30 s, respectively. To improve their stability, TiO_2 was deposited on the nanotubes. Figure 7.9 shows a SEM image of nickel nanotubes obtained by 500 ALd cycles with O_3. The thickness of nickel nanotubes is estimated about 11–12 nm. As shown in Figure 7.9a, the TiO_2/Ni/TiO_2 top layer can be seen after complete removal of the alumina membrane. After ion milling, the interconnecting layer was etched away and the trilayer structure can be observed from the top-view perspective (see Figure 7.9b). A superconducting quantum interference device (SQUID) was used to characterize the dimension-dependent magnetic properties of synthesized Ni nanotubes. Hysteresis loops for nanotubes with 35, 55 and 85 nm diameters are shown in Figure 7.9c. All these samples should share the same Ni thickness (about 11–12 nm) since they were prepared with the same number of ALD cycles. The comparison suggested that the coercivity and the remanence increased as the pore diameter decreased. On the other hand, the saturation magnetization decreased with the pore diameter. To test the dependence of the magnetic properties on the Ni thickness, three samples were prepared with 150, 300 and 500 ALD cycles corresponding to 3–4, 6–7 and 11–12 nm-thick

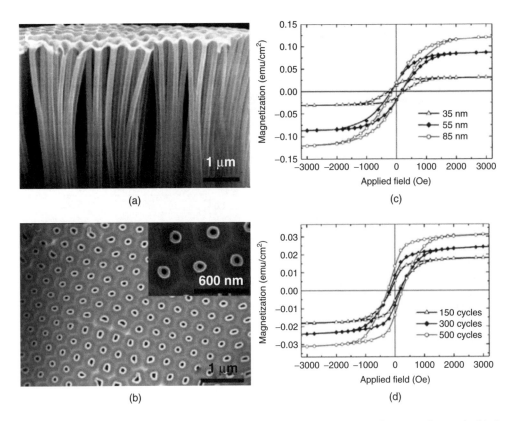

Figure 7.9 (a) SEM micrograph of obtained $TiO_2/Ni/TiO_2$ nanotubes; (b) nanotubes embedded in the membrane; (c) hysteresis loops for Ni nanotubes with different pore diameters (the magnetic field was applied in a parallel direction); (d) hysteresis loops for Ni nanotubes with different wall thickness: 3–4 nm, 6–7 nm, and 11–12 nm. (Daub *et al.* 2007)

Ni, respectively. Figure 7.9d indicates that both saturation magnetization and coercivity increases as the thickness rises.

7.4.2 Magnetic Oxide Nanotubes

As discussed in Chapter 3, there are a number of magnetic oxides with attractive properties. For iron oxide, hematite (α-Fe_2O_3), maghemite (γ-Fe_2O_3) and magnetite (Fe_3O_4) are three common phases under investigation.

α-Fe_2O_3 is a thermodynamic stable crystallographic phase of iron oxide. It is the most stable phase under ambient conditions. Its non-toxicity, low cost and relatively good stability render them a type of desirable candidate materials for various applications such as gas sensors, photoanodes for photo oxidation of water, and for photocatalytic oxidation, etc. (Shen *et al.* 2004). Hematite is also the raw material for the synthesis of maghemite. Magnetic hematite nanotubes have been prepared using several techniques including the template method and hydrothermal synthesis.

Shen *et al.* reported their synthesis of α-Fe_2O_3 nanotube arrays by the template-assisted chemical vapor deposition method (2004). They used commercially available AAO templates with a pore diameter of 100 nm. $Fe(acac)_3$ was used as the single-source molecular precursor in chemical vapor deposition, which took place in a two-zone tube furnace

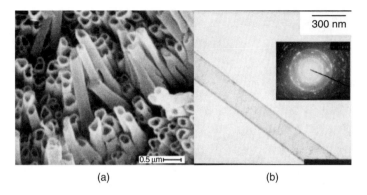

(a)   (b)

Figure 7.10   (a) SEM image of obtained hematite (α-Fe$_2$O$_3$) nanotubes; (b) TEM image of an individual α-Fe$_2$O$_3$ nanotube. The selected area electron diffraction (SAED) pattern is shown in the inset. (Shen *et al.* 2004)

with separate temperature controllers. In a typical synthesis, the organometallic precursor was placed in the low temperature zone at 150 °C. After evaporation, the organometallic vapor was carried by an oxygen gas to the high temperature zone at 300 °C, where AAO templates were placed vertically. After the reaction had lasted for one hour, thermal annealing was conducted at 500 °C for 4 hours. Throughout the whole process, the pressure of the system was kept at approximately 3 kPa. The morphology of as-prepared hematite nanotubes was checked by SEM. From Figure 7.10(a), it can be seen that open-ended hematite nanotubes are in parallel and well-ordered alignment. The high density of nanotubes should be identical to the pore density of the AAO template, which is about 1.0 × 10^9 cm^2. A TEM image, shown in Figure 7.10(b), reveals the smooth surface of the nanotube wall. The nanotube is quite straight with a uniform diameter along the axis. High-resolution TEM observation indicated that the thickness of the nanotube wall was about 20 nm and it consisted of very fine α-Fe$_2$O$_3$ nanoparticles with 10 nm in diameter. The inset of Figure 7.10(b) is a selected area electron diffraction pattern showing the polycrystallinity of hematite nanotubes.

A rational synthesis of single-crystalline hematite nanotubes was reported by Jia and his team (2005). They demonstrated that a conventional hydrothermal method could be a feasible, controllable and large-scale synthesis of single-crystalline hematite nanotubes. In hydrothermal reactions, a coordination-assisted dissolution process takes place resulting in nanotube formation. It was found that phosphate ions were essential for the formation of tubular structure. They can be selectively adsorbed on the surfaces of hematite particles and coordinated with ferric ions. In a typical solution phase-based procedure, a solution with 0.02 M FeCl$_3$ and 7.2 × 10^{-4} M NH$_4$H$_2$PO$_4$ was hydrothermally treated at 220 °C for two days. The morphology of synthesized hematite nanotubes is shown in Figure 7.11(a). The as-prepared product consists almost entirely of nanotubes with outer diameters, inner diameters and lengths of 90–110 nm, 40–80 nm, and 250–400 nm, respectively. The high resolution SEM image, as shown in the inset, reveals some nanotubes having one open end and one closed end. Starting with hematite nanotubes, maghemite nanotubes can be obtained by a reduction and re-oxidation process. Figure 7.11(b) is a SEM micrograph of obtained maghemite nanotubes. It can be seen that the size, morphology and structure of hematite nanotubes were maintained after the process. Regarding the nanotube growth mechanism, a time-dependent morphology evolution was envisaged, as illustrated in Figure 7.11(c). It was proposed that the sharp tips of spindle precursors are highly

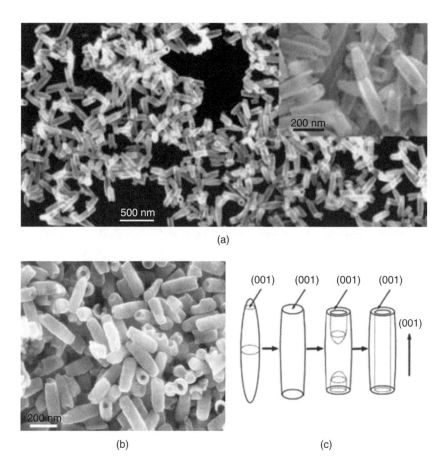

Figure 7.11 (a) SEM image of the prepared hematite nanotubes with a magnified view in the inset; (b) SEM image of the obtained maghemite nanotubes; (c) schematic process of the nanotube formation. (Jia *et al.* 2005)

reactive and the protons in acidic solution (pH = 1.8) could attack the spindle tips readily. Then dissolution of the spindle from the tip toward the interior along the long axis took place resulting in rod-like nanocrystals, semi-nanotubes and eventually hollow nanotubes. It was found that the dissolution process was not uniform resulting in partial dissolution during the hematite nanotube formation, and this non-uniform dissolution varied from spindle to spindle and even for a single spindle. Since the final hematite nanotubes have larger dimensions and smoother surfaces than those spindle precursors, recrystallization may occur with the dissolution process. A similar growth mechanism also applies for zinc oxide hexagonal rings, forming from the etching process from the center of the hexagonal disk precursors.

An efficient and convenient method using a soft template-assisted hydrothermal approach was developed for hematite nanotube synthesis (Liu *et al.* 2006). In this synthesis, rod-like surfactant micelles were used as a soft template. To prepare hematite nanotubes, about 0.2 g $FeCl_3$, an appropriate amount of carbamide, 10 ml of distilled water, 1.5 ml of the surfactant span80 and 10 ml of butanol were put into a Teflon-lined stainless steel autoclave, which was maintained at 150 °C for 12–15 hours without shaking or stirring followed by cooling to room temperature. After washing with water and

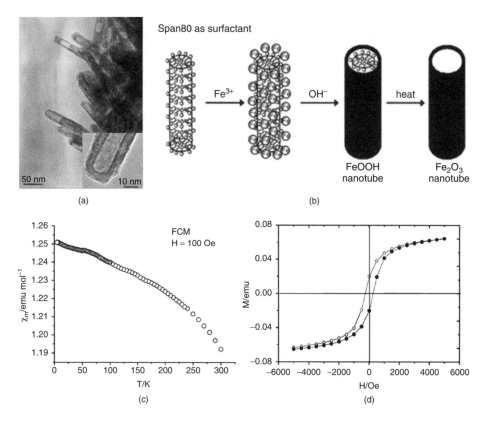

Figure 7.12 (a) TEM image of hematite (α–Fe_2O_3) nanotubes with a magnified view as the inset; (b) schematic of the proposed synthetic mechanism of nanotube formation; (c) results from the FCM measurement in 100 Oe for nanotubes; (d) hysteresis loop of the hematite nanotubes. (Liu *et al.* 2006)

ethanol, red-brown precipitate was collected. A transmission electron microscope was used to investigate the nanostructure of the as-prepared hematite nanotubes. Figure 7.12a is a typical TEM image of nanotubes showing smooth surfaces of nanotubes and their dimensions are 18–29 nm in diameter, 3–7 nm in wall thickness and 110–360 nm in length. The inset high resolution TEM image clearly reveals a multilayered nanostructure with approximately 2.76 Å of interlayer spacing. According to the JCPDS Card No. 33–0644, this interlayer spacing is attributed to the separation between (104) planes.

In this hydrothermal synthesis, the surfactant plays an essential role for nanotube formation. With concentrations higher than a critical value, surfactant molecules would self-assemble to form rod-shaped micelles or inverse micelles, which can serve as soft templates for nanotube formation with appropriate chemical reactions. In hematite synthesis, the surfactant span80 forms rod-shaped micelles between an organic phase, n-butanol, and an aqueous phase, water. These micelles are attached with hydrophilic functional groups such as C = O and –NH providing coordination sites for metal ions. As Fe^{3+} ions enter the system, they are apt to bind with the heterogeneous nucleation sites and form metal complexes and hydrophilic functional groups. At 90 °C, the carbamide agent decomposes and generates ammonia hydroxide (NH_4OH), which supply –OH^- ions. As –OH^- ions react with Fe^{3+} ions, FeOOH crystalline nucleus form on the coordination sites. If the FeOOH molecules are connected with FeOOH monomers by hydrogen bonds along the micelles, at $150\,^{\circ}\text{C}$, dehydration occurs for FeOOH molecules resulting in one-dimensional Fe_2O_3 nanowires or nanorods with a simple crystalline structure. When the hydrogen bonds are around a molecule of the surfactant or along the exterior surface of micelles, single crystalline tubular structures would form after solvent removal. The whole formation process is illustrated in Figure 7.12b. The field-cooled magnetization measurement was done in an applied field of 100 Oe showing a constant increase of χ_m and no maximum down to 5 K, as shown in Figure 7.12c. At higher temperatures, e.g. 300 K, an abrupt increase of the magnetization value can be attributed to the presence of magnetic spontaneous magnetization. The hysteresis loop of hematite nanotubes measured at 5 K is shown in Figure 7.12d. It reveals a coercive field of 280 Oe of soft magnetic nanotubes. The magnetic measurements also suggested the existence of a long-range magnetic ordering which suppressed the Morin transition. Similarly, this long-range magnetic ordering was also found in mesoporous α-Fe_2O_3 with disordered walls. This abnormality can be explained as the small crystalline particles in a few regions of the sample. Thus, it was concluded that, like the mesoporous α-Fe_2O_3, the as-prepared hematite nanotubes contained small crystalline particles. The curl of layers might cause significant defects in the nanotubes, and these defects result in the magnetic phase transition. Further investigation is required to clarify the dependence of magnetizations on sizes and shapes of hematite nanotubes.

Magnetite Fe_3O_4 is an important magnetic oxide material with strong ferromagnetic properties. Magnetite nanoparticles have been widely used in ferrofluids for hyperthermia, bioseparation, MRI, etc. Magnetite nanotubes are attractive for biomedical and biological applications because of the combination of tubular structure and ferromagnetic behaviors.

A template-assisted synthesis of ferromagnetic Fe_3O_4 nanotubes has been developed (Sui $et\ al.$ 2004b). Other than Fe_3O_4 nanotubes, FePt nanotubes were also synthesized by this simple approach. In the synthesis, commercially available AAO membranes, as the template, were first treated by thermal annealing at $600\,^{\circ}\text{C}$. After cooling, they were wetted with alcohol. Meanwhile, 65 wt % of $Fe(NO_3)_3 \cdot 9H_2O$ in ethanol solution was prepared and filtered through the membranes. In this step, the solution passed through the nanochannels inside the AAO membranes and some substances of the solution bound to the inner wall of the nanochannels. In the next step, the solution loaded AAO membranes were placed vertically in an oven and underwent a thermal decomposition of $Fe(NO_3)_3$ at $250\,^{\circ}\text{C}$, giving iron oxide solid materials. Another thermal reduction was conducted by introducing hydrogen flow into the oven at the same temperature for two and half hours. The final step involved an etching process to remove the AAO template by 0.3 M NaOH aqueous solution. A bundle of released Fe_3O_4 nanotubes is shown in Figure 7.13a, a TEM image. Both hollow structure and uniform nanotube diameter were confirmed. The XRD pattern of nanotubes reveals a cubic crystal structure of Fe_3O_4, as indicated in Figure 7.13b. Figure 7.13c shows the hysteresis loops of nanotubes measured by a SQUID system at room temperature. Measurements along the parallel and perpendicular directions, in relation to the magnetic field, resulted in a pronounced difference. Also the coercivity of as-prepared magnetite nanotubes was estimated to be 0.61 KOe.

The magnetite nanotubes prepared by Sui's method are polycrystalline. Single crystalline Fe_3O_4 nanotubes were successfully synthesized by Liu $et\ al.$, who used MgO nanowires as the template for magnetite nanotube formation (2004). Their synthesis strategy is coating Fe_3O_4 on the surface of MgO to form MgO/Fe_3O_4 core–shell nanowires, followed by etching the inner cores. The three-step process is depicted in Figure 7.14a. The merit of this synthesis is that it is able to control the length, diameter and wall thickness of the homogeneous magnetite nanotubes. In the first step, MgO nanowires were grown on Si/SiO_2 substrates, and these nanowires were found to be single crystalline.

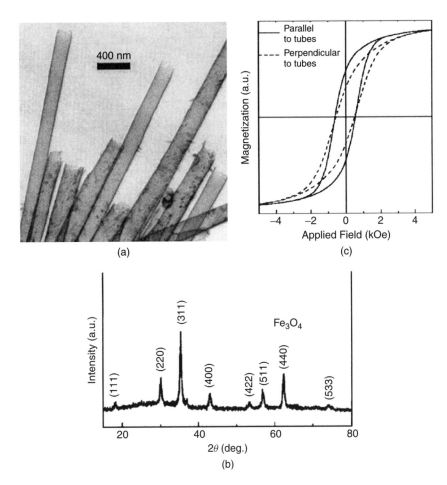

Figure 7.13 (a) TEM image of magnetite (Fe$_3$O$_4$) nanotubes; (b) X-ray diffraction pattern of as-prepared Fe$_3$O$_4$ nanotubes; (c) hysteresis loops of the Fe$_3$O$_4$ nanotubes under the external fields applied in parallel and perpendicular to the nanotubes. (Sui *et al.* 2004b)

Then a pulsed laser deposition (PLD) technique was applied to deposit a conformal layer of Fe$_3$O$_4$ on the nanowire surfaces and MgO/Fe$_3$O$_4$ core–shell nanowires were obtained. In the PLD process, a Nd:YAG laser (wavelength = 532 nm) was focused onto a Fe$_3$O$_4$ target at pulse mode to generate plume for Fe$_3$O$_4$ deposition on nanowires. The deposition conditions of PLD are temperature = 350 °C, Ar gas flow = 5 sccm, and pressure = 70 mTorr. The final step included a selective etching of the MgO inner cores of the MgO/Fe$_3$O$_4$ core–shell nanowires in (NH$_4$)$_2$SO$_4$ solution (10 wt %, pH ≈ 6.0) at an elevated temperature (80 °C). Usually, to get micrometer-long single crystalline Fe$_3$O$_4$ nanotubes, the etching should last for about one and a half hours, which is sufficient to remove the inner cores completely. After rinsing with deionized water, Fe$_3$O$_4$ nanotubes without residual contaminants were obtained. As shown in Figure 7.14b, a typical TEM image, the nanotubule structure is verified by the phase contrast between the tube wall and the inside hollow region. A Fe$_3$O$_4$ nanotube is very straight with a smooth and uniform sidewall along the whole length. The outer diameter of a Fe$_3$O$_4$ nanotube is about 30 nm. The TEM image also shows an open end of the nanotube from which the etchant entered the nanotube for the etching purpose. The high-resolution TEM image of the sidewall of

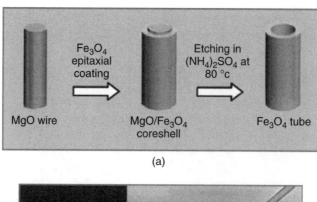

(a)

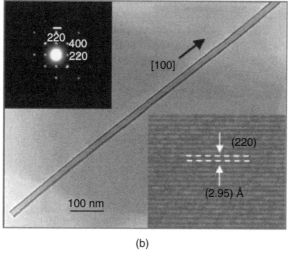

(b)

Figure 7.14 (a) Schematic of the Fe_3O_4 nanotubes fabrication process; (b) low-magnification TEM image of a single Fe_3O_4 nanotube, with the SEAD shown in the upper inset and a high-resolution image showing the multilayer structure in the lower inset. (Liu *et al.* 2004)

the nanotube is shown in the lower inset. The lattice distance between crystalline planes is estimated to be around 0.295 nm, which is exactly the spacing between two (220) planes for single crystalline magnetite. Additional crystallinity information was obtained from the selected area electron diffraction (SAED). The top inset of Figure 7.14b indicates the single-crystal nature of the magnetite nanotubes.

Another spinel-type compound of 3d transition-metal (iron, cobalt, nickel) oxides is Co_3O_4, which is an important ceramic oxide with unique electronic and chemical properties for electrochemical, magnetic and catalytic applications. Synthesis of stoichiometric Co_3O_4 porous nanotubes has been reported by using a simple modified microemulsion method (Wang *et al.* 2004). The prepared Co_3O_4 nanotubes have a large number of catalytic active sites because of their large surface-to-volume ratio. In a typical microemulsion synthesis, a solution was prepared by dissolving 0.2379 g $CoCl_2 \cdot 6H_2O$, 5.5756 g didecyl benzene sulfonate (DBS), and 0.1 ml of ethylene glycol in 30 ml xylene under stirring and ultrasound sonication. Next, 5 ml of hydrazine monohydrate ethanol solution (20 %) was dropped into the prepared solution and a strong agitation was maintained for one hour at room temperature till the color of the solution changed to turbid brick red. In the following step, the resulting solution underwent a refluxing process at about 139 °C

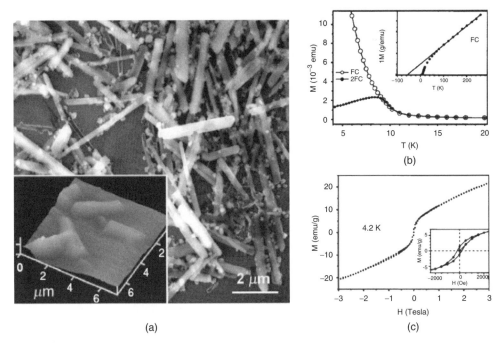

(a)   (b)

                                                (c)

Figure 7.15   (a) SEM image of Co_3O_4 nanotubes. The inset is a DFM image; (b) FC and ZFC DC magnetization curves. The magnetization property exhibiting well-behaved Curie–Weiss law at T > 70 K is shown in the inset; (c) hysteresis loop measured at 4.2 K with a magnified hysteresis loop shown in the inset. (Wang *et al.* 2004)

for one day, and its color changed to violet. At that time, royal purple precipitation was observed and collected by centrifugation followed by ethanol rinsing. After the solvent evaporation in vacuum at 80 °C, a loose royal purple powder was obtained and used for characterizations. A low magnification SEM image of the obtained sample is shown in Figure 7.15(a), which reveals a large quantity of rod-like nanomaterials with relatively uniform dimensions. A dynamic force microscope (DFM) was used to check the surface morphology of the sample. As shown in the inset, a three-dimensional tubular structure can be identified.

The magnetic properties of Co_3O_4 nanotubes were measured by using a Quantum Design SQUID magnetometer, and Figure 7.15(b) are plots of field-cooled (FC) (2T) and zero-field-cooled (ZFC) DC magnetization measurements. For the ZFC curve, a sharp maximal value is shown at about 8.4 K, which is expected as the freezing temperature of residual spin moments. A big difference was observed for FC and ZFC at a temperature lower than 8.4 K. In this range, the FC magnetization rapidly increases as the temperature decreases. The inset of Figure 7.15b reveals the magnetization of nanotubes exhibiting well-behaved Curie-Weiss law at a temperature over 70 K. The field dependence M–H hysteresis loop at 4.2 K is depicted in Figure 7.15(c) with a magnified M–H hysteresis loop shown in the inset. The loops indicate no anisotropic exchange interaction existing for Co_3O_4 nanoparticles. Higher magnetization moments were expected for the nanotubes compared with the nanoparticles. Measurements showed about 20 emu/g for Co_3O_4 nanotubes, 6 emu/g for 3–4 nm nanoparticles and 2 emu/g for 20 nm nanoparticles under the same applied field. This can be explained as the larger surface-to-volume ratio favoring the moment from the uncompensated surface Co ions.

7.4.3   Alloyed Magnetic Nanotubes

Alloyed magnetic nanotubes exhibit unique magnetic properties which can be tuned by adjusting the element composition. A common synthesis technique employed for alloyed magnetic nanotube synthesis is the template-assisted method. Compared with single element magnetic nanotubes, alloyed magnetic nanotubes are more difficult to prepare. This is mainly due to the difficulty in controlling the different diffusion and growing rates for different metal atoms. Consequently, deposition of different metals into the nanochannels of the template usually results in solid nanowires. Thus, synthesis of alloyed magnetic nanotubes needs to be conducted with special care.

Similar to the synthesis procedure for magnetite nanotube synthesis, Sui and his team reported their success in preparing hard magnetic $L1_0$ FePt nanotubes by the template method (2004a). Before the deposition, AAO membranes need to be treated in air at 600 °C for 10 minutes to remove water and transform to pure alumina. A mixture of $H_2PtCl_6.6H_2O$ and $FeCl_3.6H_2O$ was prepared in 1:1 atomic ratio of Fe:Pt to ensure the stoichiometric composition. Then the solution was loaded into the AAO membranes followed by reduction annealing with flowing hydrogen gas at 560 °C for 1.5 hours in an oven. Released nanotubes were obtained by a template etching process with 0.3 M NaOH aqueous solution for 30 minutes. The sample was checked by a SEM, which is shown in Figure 7.16(a). Close-packed nanotube arrays can be seen with deformed donut shapes for FePt nanotubes. During the reduction annealing, the loaded AAO membranes were kept inside an oven with horizontal nanochannels. Thus, it is possible that the liquid Fe and Pt chloride mixture near the channel ends may flow out of the tubes and accumulate leading to the somewhat asymmetrical cross-section of the FePt nanotubes. To prepare samples for TEM observation, released FePt nanotubes were dispersed in acetone and dropped on a TEM grid. Figure 7.16(b) shows a composite tube consisting of a FePt tube and alumina coating. The driving force for the formation of FePt nanotubes inside the AAO template was expected to be the chemical bonding between the interface of the FePt alloy and the inner walls of the nanochannels. Superconducting quantum interference device magnetometry reveals the hysteresis behavior of the FePt nanotubes. There is a difference in the hysteresis loops when measured in parallel to the tube axes at different temperatures, as can be seen from Figure 7.16(c). FePt nanotubes have a high coercivity, about 26.5 kOe at 5 K. When heated to room temperature (~300 K), the FePt nanotubes

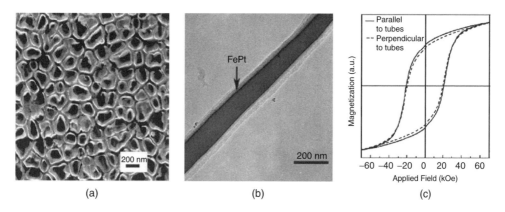

Figure 7.16   (a) SEM micrograph of FePt nanotubes after membrane removal in 0.3 M NaOH aqueous solution; (b) TEM image of an alloy nanotube of FePt surrounded by alumina; (c) hysteresis loop of FePt nanotubes measured with applied field in different directions at 300 K. (Sui *et al.* 2004a)

exhibited lower saturation magnetization, remanent magnetization and coercivity. After thermal annealing at 900 °C, the coercivities of FePt nanotubes are 38 kOe and 29.1 kOe at 5 K and 300 K, respectively.

Alloyed soft magnetic materials have been widely used in magnetic shielding, magnetic heads, voltage transformers, etc. Fe-based transition metal alloys FeNi and FeCo are two examples. Fe_xNi_{1-x} nanotubes can be used as carriers of pharmaceutical particles in drug delivery systems. Xue and others developed a modified template-assisted electrodeposition synthesis to prepare $Fe_{0.32}Ni_{0.68}$ alloy nanotubes (2005). They used 3 % vinyl-triethoxylated silane solution (in hexane) to modify the walls of the channels in commercially available polycarbonate membranes, which featured cylindrical pores of diameter 200 nm, thickness of 6–10 μm, and pore density of $\sim 10^8$ pores cm^{-2}. The surface modification of the nanochannel walls is essential for nanotube formation. A thin copper layer 30 nm in thickness was deposited on one side of the membranes by the sputtering method. Attached with a graphite electrode, they served as the cathode in electrodeposition. The electrodeposition was done in a two-electrode cell, with another graphite electrode as the anode. The electrolyte was an aqueous solution consisting of 0.02 M $FeSO_4 \cdot 7H_2O$, 0.04 M $NiSO_4 \cdot 6H_2O$, 0.05 M H_3BO_3, and 0.01 g/liter sodium laurylsulfonate (SDS). A constant current density of 1.5 mA/cm^2 was applied with an Ar gas stirring at room temperature. After deposition, the polycarbonate template was etched away using dichloromethane, and synthesized nanotubes were obtained by centrifugation, washing and drying. The morphology of released FeNi alloyed nanotubes is shown in Figure 7.17(a). SEM observation revealed the nanotubes had a diameter of 200 nm and a length of 7–8 μm. The average aspect ratio of the synthesized nanotubes was about 40. To confirm the hollow structure, TEM was performed and a typical TEM image is shown in Figure 7.17(b). Hollow tubular structures with stuffed tops can be seen. In template-assisted electroplating, the diameter of the nanotubes is determined by the pore size of the template and their length can be controlled by adjusting the reaction time and other factors. Since Ni is able to dissolve mutually in Fe, a solid solution alloy can be formed. It is necessary to figure out the ration of Ni and Fe in alloyed nanotubes. EDX analysis gave information about the elemental composition of the nanotubes. As shown in Figure 7.17(c), the weight and atomic percentage for Fe and Ni are 30.97 %, 32.05 % and 69.03 %, 67.95 %, respectively. The atomic ratio of Fe and Ni is close to 1:2, which is the ratio between precursors of $FeSO_4 \cdot 7H_2O$ and $NiSO_4 \cdot 6H_2O$ in the electrolyte. It was assumed that diffusion of Fe^{2+} to Ni^{2+} in the electrolyte is the determining step in the electrodeposition. Characterization results suggested a $Fe_{0.32}Ni_{0.68}$ composition for the alloy nanotubes.

The magnetic behavior of the $Fe_{0.32}Ni_{0.68}$ alloy nanotubes embedded inside the membrane was characterized at room temperature with an application of a magnetic field of 12 kOe. The result is shown in Figure 7.17(d). When the magnetic field was applied in parallel to nanotubes, the coercivity (Hc) of the nanotubes was about 21 Oe. In the case of the field perpendicular to nanotubes, the value was about 55 Oe. It is well known that single iron and single nickel only have about 1 Oe and 6 Oe coercivities, and bulk Fe_xNi_{1-x} alloy has a coercivity less than 1 Oe. Thus, $Fe_{0.32}Ni_{0.68}$ alloy nanotubes are superior to these materials. In a parallel applied field, the saturation of the magnetization of the alloy nanotubes takes place in a field of 5800 Oe, while a field of 6500 Oe is the point where magnetization saturation occurs in a perpendicularly applied field. This suggests an easier magnetization in a parallel field than that in a perpendicular field. As a soft magnetic material, the prepared $Fe_{0.32}Ni_{0.68}$ alloy nanotubes would be able to respond quickly to a relatively weak external magnetic field, which is ideal for carrier purposes. Therefore, the

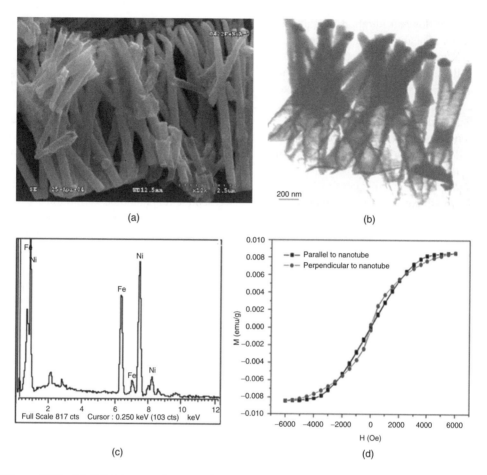

Figure 7.17 (a) SEM image of prepared $Fe_{0.32}Ni_{0.68}$ nanotubes; (b) TEM image of fractured $Fe_{0.32}Ni_{0.68}$ nanotubes with stuffed tops and a conical morphology; (c) EDX spectrum of $Fe_{0.32}Ni_{0.68}$ nanotubes showing the atomic ratio and mass ratio of 32.05:67.95 and 30.97:69.03, respectively; (d) hysteresis loop of $Fe_{0.32}Ni_{0.68}$ nanotubes embedded inside the membrane measured at room temperature. (Xue *et al.* 2005). Xue, S.; Cao, C.; Wang, D. and Zhu, H. (2005). "Synthesis and magnetic properties of Fe0.32Ni0.68 alloy nanotubes", Nanotechnology, 16, 1495–1499. Reproduced with permission of the Institute of Physics

as-prepared alloy nanotubes are expected to serve as carriers of pharmaceutical particles due to their inner void and the corresponding magnetic properties.

Magnetic FeCo alloys show high saturation magnetizations and high Curie temperatures. Thus it is also an important magnetic material for certain applications. FeCo alloyed nanotubes could be obtained by a wetting template method and hydrogen reduction (Li *et al.* 2007). The AAO template was first prepared by anodic oxidation of aluminum foil in phosphoric acid. A solution containing precursors was made by dissolving $Fe(NO_3)_3 \cdot 9H_2O$ and $Co(NO_3)_3 \cdot 6H_2O$ in water. The atomic ratio of Fe and Co was set to 2:1. After immersing the prepared templates in the prepared solution for a proper time, metal nitrates covered the walls of nanochannels. The loaded templates were transferred to an oven and heated to 400 °C for 3 hours. In this step, metal nitrates decomposed. The last step was done by hydrogen reduction at 350 °C for additional 3 hours to form FeCo alloy materials. The morphology and nanostructure of prepared FeCo nanotubes were studied by TEM. Figure 7.18(a) is a representative TEM image showing nanotubes with uniform diameter

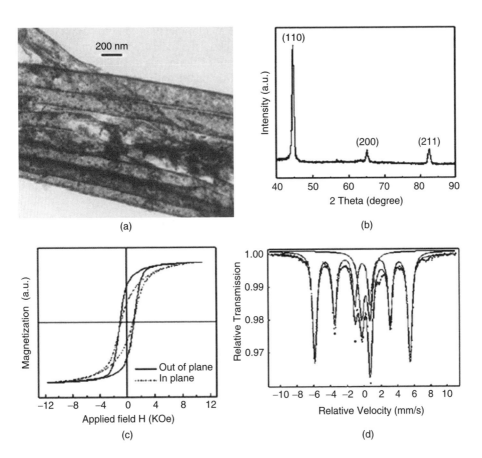

Figure 7.18 (a) TEM image of a bundle of FeCo nanotubes; (b) X-ray diffraction pattern of CoFe nanotubes embedded inside the AAO membrane; (c) hysteresis loops for FeCo nanotube arrays measured at room temperature; (d) transmission Mössbauer spectrum of the FeCo nanotube array measured at room temperature. (Li *et al.* 2007). Reproduced by permission of the American Institute of Physics

about 200 nm. Nanoparticles forming the nanotube walls can be seen. FeCo nanotubes after hydrogen reduction and thermal annealing displayed a body-centered cubic crystal structure corresponding to an FeCo alloy, as shown in Figure 7.18(b). Atomic absorption spectroscopy was conducted and the measurement result gave an atomic ratio of Fe:Co of about 2:1.

Room temperature hysteresis loops for FeCo nanotube arrays, measured at room temperature, are plotted in Figure 7.18(c). The solid and dashed lines correspond to the measured loops with applied magnetic field in parallel and perpendicular to the nanotube axes, respectively. As measured, both coercivity and squareness ratio M_r/M_s are much higher in the case of an applied parallel field than in the case of an applied perpendicular field. The former measurement gave about 1.01 kOe and 0.61 for coercivity and M_r/M_s value, respectively, while the values are 870 Oe and 0.32 in the latter situation. This implies that the nanotubes could be magnetized more easily when the external magnetic field was applied out of the plane of nanotubes than in-plane. The isotropic magnetic moment distribution can be further confirmed by a transmission Mössbauer spectrum of the FeCo nanotubes, which was measured at room temperature. In the measurement, the incident γ photons were in perpendicular to the membrane, thus were in parallel respective to the long axis of nanotubes. Figure 7.18(d) shows the measured spectrum as well as the fitting curve. The spectrum shows a good fitting of the experimental data by a magnetic splitting sextet and a paramagnetic doublet. Compared with the bulk FeCo materials, a smaller hyperfine field was obtained, which was about 35.5 T. In Mössbauer experiments, the ration M_r/M_s value can be estimated from the peak ratio obtained from the magnetic splitting sextet. There is a formula relating the peak intensity ratios and the angle between the incident photos and the magnetization direction:

$$\frac{I_{2,5}}{I_{1,6}} = \frac{4\sin^2\theta}{3(1+\cos^2\theta)} \tag{7.6}$$

where $I_{2,5}$ and $I_{1,6}$ are the relative intensity ratios of the 2,5 and 1,6 peaks in the magnetic splitting sextet, respectively. In the case of random arrangement of the magnetic moments, 2/3 should be obtained as the intensity ratio. The experiment measurement gave a ratio of 0.64, which is close to 2/3. Thus, it can be concluded that the magnetic moment distribution was approximately isotropic in tube axis direction in space.

7.4.4 Doped Magnetic Nanotubes

Doped magnetic nanotubes are non-magnetic nanotubes doped with transition metal species exhibiting magnetic properties. Introduction of magnetic atoms into carbon, silicon and polymer nanotubes offers the possibility of magnetism in these non-magnetic structures.

Singh and his team simulated the stability and magnetic effects of transition metal doped silicon nanotubes (Singh *et al.* 2004). It was expected that the magnetically doped semiconducting nanotubes could be interesting for nanospintronics applications due to their half-metallic and magnetic characteristics. Encapsulation of transition metals into silicon nanostructures gives rise to large embedding energies and hence improves their stability. 3d transition metals (Fe, Co, Ni) possess smallest dimension mismatch to minimize the strain in the Si–Si bonds, so they seem to be the best option for silicon nanotube doping. As far as the basic unit of doped silicon nanostructure is concerned, the encapsulated magnetic atoms usually exhibit quenched magnetic moments due to the hybridization between the d orbitals of transition metals and the sp orbitals of silicon atoms. Concurrently, the strong interaction between silicon clusters could be another reason. For Fe- and Mn-doped silicon nanotubes, the local magnetic moments are determined by the extents of doping. As depicted in Figure 7.19a), for a certain number of silicon atoms, a larger number of dopants lead to larger local magnetic moments. Particularly for Fe doping, antiferromagnetic behavior transfers to ferromagnetic behaviour during the increment of the number of dopants. Since the charge on silicon atoms is shared with surrounding Si atoms while interacting with the encapsulated metal atom, the local moments of Fe and Mn increase from the ends to the center part. For Fe, the local moment varies from 1.0 to 2.6 μB and the value of Mn is 0–3.6 μB. Both of them could find magnetic device applications because of their high magnetic moments. In contrast, Ni-doped nanotubes have completely quenched magnetic moments and Co-doped ones show relatively low moments. When a nanotube is formed from an $Si_{12}M$ cluster, the increments of the bond length of the Si–Si and TM–Si bonds result in a weakened covalent bonding between the transition metal atom and surrounding Si atoms. Thus, on the transition metal atoms, especially for those elements with large atomic moments, magnetic moments can be developed.

Simulation results indicated that a small or zero local moment exists at the edges of the finite nanotubes while on atoms away from the edges large moments occur. Another

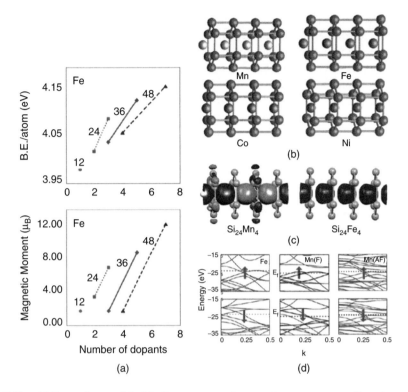

Figure 7.19 (a) Dependence of the binding energy and magnetic moment per cell on the number of dopant atoms in Fe-doped silicon nanotubes with finite length; (b) schematic structures of different $Si_{24}M_4$ infinite nanotubes; (c) schematic of spin density of $Si_{24}Mn_4$ antiferromagnetically-coupled nanotubes and $Si_{24}Fe_4$ ferromagnetically-coupled nanotubes. Blue and red isosurfaces show spin-up and down densities, respectively; (Please refer to the color image in the original article) (d) band structures of ferromagnetic infinite $Si_{24}Fe_4$ nanotubes, and ferromagnetic as well as antiferromagnetic infinite $Si_{24}Mn_4$ nanotubes. (Singh *et al.* 2004). Reproduced by permission of the Royal Society of Chemistry

general trend is that from Mn to Ni, the moments decrease. Transition metals with higher local moments exhibit larger atomic magnetic moments. When the number of transition metal atoms increases, the stability of the doped silicon nanotube is also improved. In the doped infinite nanotubes, $Si_{12}M$, the distance between transition metals, Mn, Fe, Co, and the hexagonal ring are 0.98, 0.79 and 0.27 Å, respectively. Figure 7.19(b) depicts the structures of $Si_{24}M_4$ infinite nanotubes with obvious shift in the positions of metal atoms. The localizations of moments around the doped metals are illustrated. The ferromagnetic phase of Fe-doped nanotubes is stable and the magnetic moment is about 2.4 μB per Fe atom. Thus, this type of material could be used as a nanoscale magnet. In Mn-doped nanotubes, the antiferromagnetic coupling between ferromagnetically coupled Mn atoms and their surrounding pairs cancels the moment. Figure 7.19(c) are schematic diagrams for this antiferromagnetic coupling for Mn-doped nanotubes and ferromagnetic coupling for Fe-doped nanotubes. Another important result obtained from the simulation is that the transformation between antiferromagnetic to ferromagnetic coupling could be achieved by a weak external magnetic field. The band structures of Fe- and Mn-doped nanotubes are plotted in Figure 7.19(d). It can be seen from all plots that, for both the spin-up and spin-down components, metallic behaviors are displayed because of the band crossing at the Fermi level. A gap exists above the Fermi energy in the band structure of Mn-doped ferromagnetic nanotubes. It can be envisaged that half-metallic nanotubes could be achieved with the introduction of a small shift in the Fermi energy. In the case of Co- and Ni-doped silicon nanotubes, both configurations result in nanomagnetic solutions.

Another example of theoretic studies is Rahman's work on Fe-filled single-walled carbon nanotubes (SWNTs) (2004). Both electronic and magnetic properties of these magnetically doped carbon nanotubes were investigated by using *ab initio* spin-polarized density functional theory. According to the study, the filling of magnetic metal atoms into SWNTs improves their stability. And it also seems the properties of filled SWNTs are determined by their diameters. It was concluded that (4, 4), (5, 5), (6, 6), (6, 0) SWNTs with filled Fe nanowires exhibited metallic behavior and magnetic moments, while Fe-filled SWNTs (3, 3), (5, 0) did not show spin polarization due to their smaller diameters. In their calculations, the generalized gradient approximation (GGA) for the exchange-correlation energy was used with plane waves and pseudopotentials. As a metal-lic (3, 3) SWNT is considered, there are 12 carbon atoms and one Fe atom in a unit with 2.48 Å in length along the tube axis. In order to study the magnetism and conductivity due to the C–Fe bonds, it is assumed that there is a vacuum region about 10 Å in diameter so that the interaction among SWNTs can be ignored. The initial magnetic moment of Fe is set as $2.4\,\mu_B$. Optimal structure of a Fe-encapsulated SWNT unit was obtained at the minimum total energy with the residual forces smaller than $0.05\,eV\,Å^{-1}$. Besides the 12C/1Fe unit, a super-cell unit with 4.96 Å in length along the nanotube axis, consisting of 24 carbon and 2 Fe atoms, was also simulated. In this case, the linear arrangement of Fe atoms can be treated as a nanowire with both ferromagnetic and antiferromagnetic states. Figure 7.20(a) is a schematic of (3, 3) SWNT with a filled Fe nanowire after structure optimization. The Fe atom is apart from the nearest C atom by about 1.9 Å, and about 0.6–0.8 electron $Å^{-3}$ of the charge density between a Fe atom and surrounding C atoms was estimated from Figure 7.20(b). Because the Fe atom may form σ bonds with four (or six) C atoms nearby, the orbitals of Fe should be divided into bonding and anti-bonding orbitals. Meanwhile, C atoms also have sp^3 hybrid-like orbitals accordingly for Fe–C bonding. As a result, the Fe atom does not possess any magnetic moment, and neither the Fe atom nor C atoms exhibit spin polarization. In addition, the Fe-filled (3, 3) SWNT shows semiconducting behavior.

In the following study, similar simulation was applied on Co-filled (3, 3) SWNTs to investigate their electrical and magnetic properties (Rahman *et al.* 2005). Simulation results indicated that the majority-spin and minority-spin electrons showed metallic and semiconducting behaviors, respectively, which is a so-called half metallic ferromagnetic property. It was de Groot *et al.* who first predicted the half metallic state of the fer-romagnet, which is featured by a combination of the electrons in both metallic spin and opposite insulating spin (de Groot *et al.* 1983). Unlike ferromagnetic metals, this half metallic magnet exhibits large giant magnetoresistance and tunneling magnetore-sistance effects, which can be used in spintronics. By minimizing the total energy, an optimized structure of Co nanowire-filled (3, 3) SWNT can be constructed, as illustrated in Figure 7.20(c). The super-cell is comprised of 24 C atoms and 2 Co atoms. It was found that the filled Co nanowire remained straight, and was not located in the center of the nanotube. Figure 7.20(d) is a schematic diagram showing the electron charge density on the cross-section of the Co-filled carbon nanotube. After studying the cases with ferro-magnetic, paramagnetic and antiferromagnetic states, they confirmed that the most stable state is the ferromagnetic one. A conclusion was made on the half metallic ferromagnetic behavior of Co-filled (3, 3) SWNT based on the 100 % spin polarization at the Fermi level.

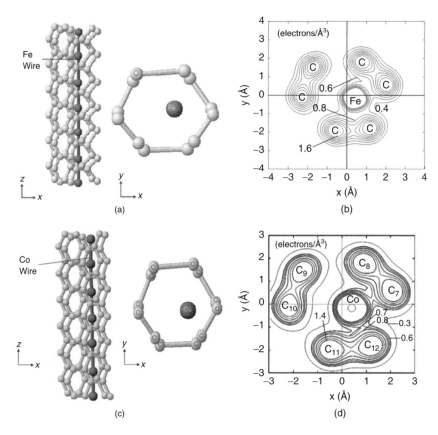

Figure 7.20 (a) Schematic of the side view and the cross-sectional view of a Fe-filled (3, 3) single-walled carbon nanotube; (b) the corresponding contour lines showing the distribution of electron charge density in its cross-section (Rahman *et al.* 2004); (c) schematic of the side view and the cross-sectional view of a Co-filled (3, 3) single-walled carbon nanotube; (d) the corresponding contour lines showing the electron charge density distribution in its cross-section. (Rahman *et al.* 2005)

Theoretical simulations based on *ab initio* total energy calculations have suggested the possibility of using ferromagnetic nanotubes in spintronic devices. Magnetically doped III–V semiconductors have also displayed attractive ferromagnetic behaviors. Early research has produced Mn-doped GaAs with around 150 K as the Curie temperature. For practical usage, it is desirable to have ferromagnetic semiconductors at room temperature. Prompted by this motivation, extensive research efforts have been made on Mn- or Co-doped ZnO and Co- or V-doped TiO_2. Co-doped anatase powders can be prepared by a sol-gel method. In the conventional method, titanium *n*-butoxide and cobalt acetate in the appropriate ratio were dissolved in 2-methoxyethanol. The solution was then refluxed at 130 °C for 2 hours. After drying at 120 °C, the resulting blue gel underwent thermal annealing at 500 °C for 1 hour to obtain crystalline powders. According to the XRD analysis and inductively coupled plasma (ICP) resonance spectroscopy, the prepared anatase powder has a composition of $Ti_{0.93}Co_{0.07}O_2$. Synthesis of Co-doped anatase nanotubes was reported by Kasuga *et al.* (1998). A simple hydrothermal method was developed to synthesize room temperature ferromagnetic Co-doped titanate nanotubes (Wu *et al.* 2005). In their synthesis, 100 mg prepared Co-doped TiO_2 powders and 20 ml 10 M NaOH were added into a Teflon-lined autoclave. The reagent-loaded autoclave was kept in an oven at

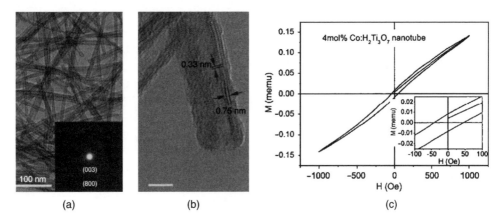

Figure 7.21 (a) Low-resolution bright field TEM image of Co-doped $H_2Ti_3O_7$ nanotubes with a selected area electron diffraction pattern shown in the inset; (b) high-resolution TEM image of an individual nanotube showing the interlayer lattice; (c) measured hysteresis loop of Co-doped $H_2Ti_3O_7$ nanotubes by a SQUID at room temperature. The inset shows the magnified curve near the origin. (Wu *et al.* 2005)

130 °C for hydrothermal reaction. After 36 hours, the resulting solid products were collected followed by washing with deionized water. The nanotube morphology was studied by TEM. Figure 7.21(a) is a bright field TEM image of the collected material showing a hollow tubular structure. The outer and inner diameters of nanotubes are around 10 and 5 nm, respectively, and the average length is about 300 nm. A high resolution TEM image of an individual nanotube is shown in Figure 7.21(b). A 0.78 nm of interlayer spacing was estimated for the nanotube. From this image, it is possible to resolve another set of fringe with a spacing of 0.33 nm in the direction of 103 ° respective to the tube axis. These two lattice distances are believed to correspond to the (200) and (003) planes of a monoclinic trititanate compound $H_2Ti_3O_7$ (C2/m, a = 1.603, b = 0.375, c = 0.919 nm, $\beta = 101.47$ °). In the crystalline structure of $H_2Ti_3O_7$, H^+ ions separate the stepped Ti-O octahedron layers. The tube axis of nanotubes was found to be coincident with the [001] direction of monoclinic $H_2Ti_3O_7$. The crystallinity information can be also obtained from selected area electron diffraction. The inset in Figure 7.21(a) shows the diffraction rings with periodicities of 0.33 and 0.19 nm, which are attributed to the lattice distance of (003) and (800) planes of a monoclinic layered $H_2Ti_3O_7$.

About 22 mg of 4 % Co-doped $H_2Ti_3O_7$ nanotubes were utilized to measure their magnetic response with respect to varied magnetic field strength at room temperature. Figure 7.21(c) plots the measured hysteresis loop suggesting the nanotubes are ferromagnetic even. A magnified plot of the loop, shown in the inset, gives estimated coercivities of 40 and 46 Oe, respectively. The magnetic characterization indicated a combination of a paramagnetic component with a ferromagnetic one. Apparently, this property should not be counted as the contribution from the Co oxide because of the antiferromagnetic behavior of both CoO and Co_3O_4 at room temperature. It is noteworthy to point out that the prepared Co-doped $H_2Ti_3O_7$ nanotubes have diluted Co doping. Incorporating high concentration Co ions, e.g. Co:Ti = 1:3.5, may result in antiferromagnetic behavior because of the Co−Co interaction favored by the large amount of Co ions. It is expected that the magnetic ion density plays an essential role for the ferromagnetism of Co-doped $H_2Ti_3O_7$ nanotubes.

Huang and his team also reported their synthesis of magnetic Co-doped TiO_2 nanotubes by using a template-assisted sol-gel method (Huang *et al.* 2006). Crystalline characterization revealed a pure anatase structure. It was identified that no metallic Co was incorporated inside the anatase TiO_2 structure, instead of Co^{2+}, which substituted the places of Ti^{4+}. Magnetic characterization indicated a ferromagnetic behavior of Co-doped anatase nanotubes at room temperature. This ferromagnetism was expected to be related to the oxygen vacancies induced by the substitution of Ti^{4+} sites with Co^{2+} ions.

Another type of doped magnetic nanotubes is magnetic carbon nanotubes incorporated with magnetic nanoparticles. Carbon nanotubes exhibit extraordinary electrical, chemical and thermal properties. Doping carbon nanotubes with transition metal encapsulation is one of the common methods used to introduce magnetic properties of carbon nanotubes. Combining of magnetic properties of the magnetic nanoparticles and the unique characteristics of carbon nanotubes, magnetic doped carbon nanotubes display unique electronic, magnetic and nonlinear optical properties, which render them an ideal material candidate in potential applications such as magnetic data storage, xerography and magnetic resonance imaging (Liu *et al.* 2005).

Ferromagnetic metals (Fe, Ni, Co)-incorporated carbon nanotubes offer a fascinating novel nanomaterial. Paramagnetic needles could be obtained by simply encapsulating paramagnetic particles into carbon nanotubes, and the movement of these needles can be manipulated by an applied magnetic field. The *in situ* synthesis of iron-filled carbon nanotubes has been demonstrated by several research groups (Leonhardt *et al.* 2003; Liu *et al.* 2005; Mönch *et al.* 2005; Su and Hsu 2005; Leonhardt *et al.* 2005; Geng and Cong 2006). It was found from previous studies that the covering graphite layers of nanotubes played a dual role: they protected filled metal nanoparticles from oxidation and enhanced the magnetic coercivities of the magnetic nanoparticles.

A common method to prepare magnetic doped carbon nanotubes is the thermal chemical vapor deposition method. In this method, metallocenes of Fe, Co or Ni and oxidized silicon wafers coated with Fe, Co or permalloy ($Ni_{80}Fe_{20}$) can be used as the starting substances and the substrates, respectively. Leonhardt *et al.* used a two-stage furnace to synthesize Fe-, Co-, or Ni-filled multi-walled carbon nanotubes (2003). The former two were obtained by using bis(cyclopentadienyl) iron and bis(cyclopentadienyl) cobalt, respectively, as metallocenes. And Ni-filled carbon nanotubes were prepared by pyrolysis of methane or benzene on crystalline Ni particles. The advantage of using metallocenes as the precursors is that they provide both carbon atoms for nanotube growth and ferromagnetic nanoparticles to be encapsulated. The chemical vapor deposition can be controlled by adjusting three reaction conditions: the temperature of the first furnace for metallocene sublimation, the total gas flow rate and the temperature of the second furnace for deposition of metal-filled carbon nanotubes. In a typical synthesis of Co- and Fe-filled carbon nanotubes, a quartz boat with a metallocene was loaded into the first furnace with temperature kept at 120 and 180 °C, respectively. And the sublimed metallocene was carried by a controlled gas flow (e.g. Ar) into the second hot furnace where the temperature was set to 900–1150 °C. Decomposition of metallocene occurred and filled carbon nanotubes formed. To synthesize Ni-filled carbon nanotubes, a hydrocarbon was introduced and decomposed to deposit a carbon atom on Ni-coated silicon wafers.

Figure 7.22(a) is a SEM image of aligned Fe-filled carbon nanotubes. A typical TEM image of a single nanotube is shown in Figure 7.22(b). A uniform metal core with diameters ranging from 15 to 30 nm can be observed. The carbon nanotube wall consists of a number of cylindrical grapheme layers with a lattice distance of 0.34 nm. The thickness of the nanotube wall is estimated to be 3–5 nm. Magnetization curves of Fe-filled carbon nanotubes were obtained by AGM measurement parallel and perpendicular to the substrate plane. Because of the geometrical characteristics of the nanotubes, uniaxial anisotropy was observed. Magnetization along the direction parallel to the substrate is

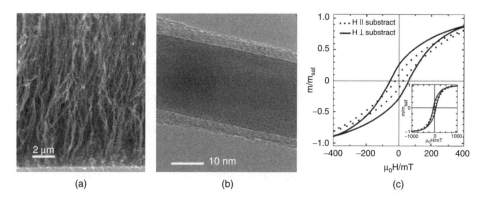

Figure 7.22 (a) SEM image of vertically aligned Fe-filled multi-walled carbon nanotubes; (b) HRTEM image of an individual Fe-filled nanotube; (c) hysteresis loops of the aligned Fe-filled tubes grown on an oxidized Si substrate in different directions. (Leonhardt *et al.* 2003)

easier than that along the direction which is perpendicular to the substrate. The measured hysteresis loops along both directions are shown in Figure 7.22(c). The coercivity for the field direction parallel to the nanotube axis is approximately 56 mT, while it is 25 mT for the other direction. Both values are significantly higher than that of bulk iron amounts, which is around 0.09 mT. Thus, it is apparent that the magnetic nanoparticle-filled carbon nanotubes exhibit an enhanced magnetization.

Paramagnetic particle-encapsulated carbon nanotubes in AAO templates were prepared by Korneva *et al.* (2005). Carbon nanotubes were synthesized by chemical vapor deposition inside the nanopores of AAO templates. The prepared carbon nanotubes have open ends and they are hydrophilic. These properties make it possible to encapsulate water- or organic-based magnetic ferrofluids. Commercially available ferrofluid used for nanotube filling purposes included EEG 508 (water-based) and EMG 911 (organic-based) from Ferrotec Corporation. Both of them consist of magnetite nanoparticles with average diameter of 10 nm. A schematic diagram of the whole procedure to prepare ferrofluid-filled carbon nanotubes is depicted in Figure 7.23(a). For comparison purposes, two methods were applied, which are filling ferrofluid in carbon nanotubes embedded inside the template and filling ferrofluid in released carbon nanotubes.

Figure 7.23(b) is a representative TEM image showing magnetic grain encapsulation inside the carbon nanotubes. In both synthesis methods, a large amount of magnetite nanoparticles was found to be able to enter the inner cavity of nanotubes. An alternating gradient magnetometer (AGM, Princeton Measurements Inc.) was used to characterize the magnetic properties of magnetic doped carbon nanotubes. To prepare the sample for magnetic characterization, the as-prepared membranes were washed to get rid of the magnetic particles on the surface. Then under a microscope, the area of a cut membrane was measured. Figure 7.23(c) shows a hysteresis loop of a load membrane with an area of 5.9177×10^{-6} m^2. The average magnetic moments of the magnetic-encapsulated carbon nanotubes at an applied field of 0.007 T can be calculated via the formula:

$$m = M \frac{d^2 \pi}{4\, pA} \qquad (7.7)$$

where M is the total magnetization value of the sample (9.138351×10^{-8} Am^2), A is the sample area, d is the nanotube diameter (300 nm) and p is the fraction of the membrane area occupied by the nanotubes (0.25). Formula (7.7) gives about $4.37 \times$

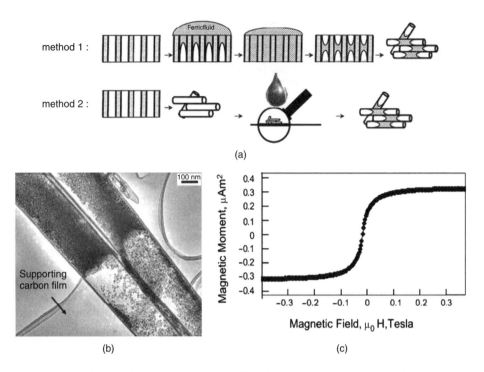

(a)

(b)   (c)

Figure 7.23 (a) Schematic of two methods to fill carbon nanotubes with magnetic nanoparticles; (b) TEM image of magnetic nanoparticle-filled carbon nanotubes; (c) hysteresis loop of magnetic nanoparticle-loaded nanotubes inside the membrane without a magnet. (Korneva *et al.* 2005)

10^{-15} Am^2 for the average magnetic moment of the magnetic-encapsulated nanotubes. Considering each magnetic particle has 4.48×10^{-20} Am^2 of magnetic moment, the number of magnetic nanoparticles encapsulated inside the nanotubes can be estimated: c.a. 7×10^4. Further, when taking account each nanoparticle's volume, about 5.24×10^{-25} m^3, and the average volume of the nanotubes $\sim 8.84 \times 10^{-19}$ m^3, about 11 % of magnetic nanoparticles filling the nanotube volume can be estimated. After being encapsulated with ferrofluid, carbon nanotubes exhibited magnetic behavior. When applying a rotating magnetic field on released magnetic carbon nanotubes in suspensions, they rotated with a steady state accordingly.

Bao *et al.* applied a similar method to prepare nickel-filled carbon nanotubes (2002). Unlike the previous methods, they used carbon nanotubes embedded inside the template as the 'second-order template' for the electroplating that followed. Thus, magnetic Ni nanowire-filled carbon nanotubes were obtained. Several steps were involved in the synthesis procedure. First, a small amount of cobalt was electrodeposited inside the nanopores of AAO templates, serving as the catalyst for carbon nanotube growth. Secondly, chemical vapor deposition was used to deposit carbon nanotubes inside the template. After a pretreated template was loaded inside a horizontal quartz tube in a furnace, the temperature was raised to 600 °C with a flowing gas mixture (2 % H$_2$ and 98 % argon). After one hour dwelling time, the furnace temperature was further raised to 700 °C, at which 50 sccm of the source gas containing 2 % C$_2$H$_2$, 2 % H$_2$ and 96 % Ar was introduced. The deposition lasted for 20 minutes followed by an annealing process with 2 % H$_2$ and 98 % argon for an additional 4 hours. The third step was the electrodeposition of Ni nanowires

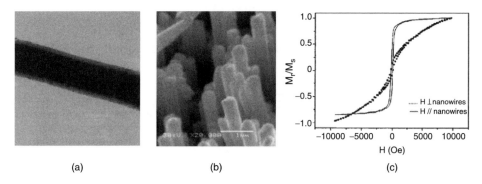

Figure 7.24 (a) TEM image of a carbon nanotube filled with Ni; (b) SEM image of the Ni-filled carbon nanotube array after the template removal; (c) magnetization curves of the Ni/C composite nanostructure array with the template. (The dashed line was obtained with the applied field perpendicular to nanotubes while the solid line was for the applied field parallel to the nanotubes.) (Bao *et al.* 2002)

inside the as-prepared carbon nanotubes. The electroplating solution is an aqueous solution of $NiSO_4.6H_2O$, 50 g/l H_3BO_3 in proper concentration. Carbon nanotubes served as the second-order template for Ni nanowire formation. From a TEM image, Figure 7.24(a), one can see that a carbon nanotube is completely filled with nickel. A nickel nanowire was formed inside the carbon nanotube. Also, no nickel deposition is shown on the outer surface of the nanotube. Due to the template characteristics, the outer diameter of carbon nanotubes was determined by the pore size of the AAO template. And the thickness of the carbon nanotube walls can be adjusted by the reaction time. For instance, carbon nanotubes have about 22 nm thick walls with 20 minutes' thermal deposition while 10 minutes' deposition resulted in 6 nm thick nanotube walls. Consequently, the outer diameter of encapsulated nickel nanowires, which is determined by the inner diameter of carbon nanotubes, can be controlled by the chemical vapor deposition as well. In addition, the length of both the Ni nanowires and carbon nanotubes is about 50 μm, which is exactly the same as the thickness of the alumina membranes. Figure 7.24(b) is a SEM image of Ni-encapsulated carbon nanotubes after removal of the alumina template. It can be seen from the SEM image that highly oriented nanostructures can be obtained, which are superior to randomly oriented nanomaterials for many applications. Also, it seems that every carbon nanotube was filled with a nickel nanowire suggesting a high yield. This modified template method offers a simple and effective approach to synthesize aligned magnetically doped carbon nanotube arrays with other common synthesis techniques such as the wet chemical method and arc-discharge technique.

Nickel nanowire filled carbon nanotubes with aspect ratio of 270 and 185 nm internal diameter were tested for magnetic characterization. For this measurement, the template was not removed. Figure 7.24(c) shows the hysteresis loops of this composite nanostructure. The measured coercivities with the magnetic field parallel and perpendicular to the membrane are 184 Oe ($H_{c//}$) and 134 Oe ($H_{c\perp}$), respectively. Figure 7.24(c) also confirms the uniaxial magnetic anisotropy with the easy axis parallel to the nanotubes. Compared with the bulk nickel coercivity about 0.7 Oe, the nickel encapsulated carbon nanotubes exhibit much higher coercivities. Furthermore, the feature of highly sheared loops suggests a strong interaction among closely-packed nanowire-filled carbon nanotubes. As shown in the SEM image, nanotubes are very close to each other resulting in strong interaction. Further evidence is the very low remanent magnetization compared with the saturated magnetization, as indicated in the hysteresis loop.

A novel bionanotechnological method for fabricating magnetite-filled peptide nanotubes with the aid of magnetic bacteria was reported (Banerjee *et al.* 2005). Figure 7.25a illustrates the fabrication process of the magnetite-filled peptide nanotubes. Briefly, the fabrication process consists of extracting magnetic nanocrystals from the cells followed by incorporating them into peptide nanotubes. There are many types of bacteria that are able to produce intercellular magnetite nanocrystals. And the magnetic nanocrystals' physical characteristics, including crystal size and morphology, are dependent upon the bacterial species. Therefore, it is possible to obtain bigenerated magnetic nanocrystals with specific properties by carefully selecting the bacteria. There are several advantages of using magnetic nanocrystals produced by bacteria. First, they exhibit an exceptionally high affinity

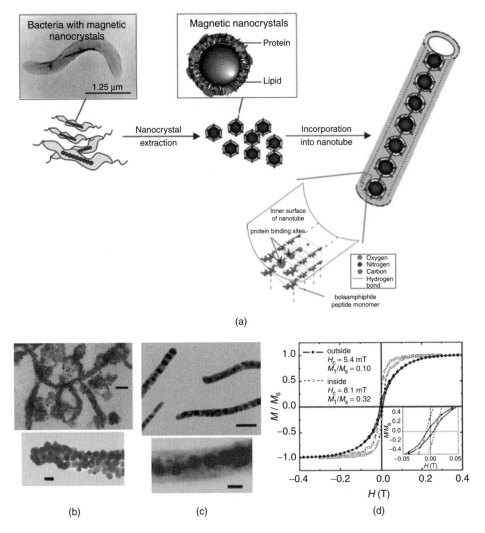

Figure 7.25 (a) Schematic of the synthesis procedure for the magnetic nanotubes; (b) TEM images showing the coating of the bacterial magnetic nanocrystals on the entire peptide nanotube surface; (c) TEM images showing the bacterial magnetic nanocrystals incorporated inside the peptide nanotube. Scale bars are 300 nm; (d) comparison of hysteresis curves between two types of magnetic nanotubes. (Banerjee *et al.* 2005)

to the peptide nanotubes. The magnetic nanocrystals generated by bacteria are usually coated with proteins or lipids. Thus a regular biofunctionalization step which is truly necessary for magnetic nanocrystals produced by any chemical method can be omitted. Since there are unused amide groups in peptide nanotubes which provide the anchor site for other proteins via hydrogen bonding, protein coated magnetic nanocrystals could be able to bind with the nanotube readily. Another major advantage of using bacterial magnetic nanocrystals is the peptide nanotubes' capability of incorporating nanocrystals selectively at an optimized concentration. Therefore, it is possible to control the magnetic properties of nanotubes by varying the nanocrystal concentration. When magnetic nanocrystals are incorporated into a peptide nanotube, they are apt to form a linear chain. In this case, magnetic nanocrystals can be aligned to exhibit uniaxial magnetic anisotropy.

In Banerjee's fabrication, *magnetospirillum magneticum* strain AMB-1 was selected as a model for the investigation on the incorporation of bacterial magnetic nanocrystals on the peptide nanotubes. After bacteria growth anaerobically in fermentor, the cells were disrupted and the generated magnetic nanocrystals were collected by using a magnet followed by washing ten times with 2-[4-(2-hydroxyethyl)-1-piperazinyl]ethane sulfonic acid buffer (pH 7.2). The obtained magnetic nanocrystals had an average diameter of 70 nm. In the next step, incubation was conducted at 4 °C by mixing the bacterial magnetic nanocrystals in buffer solution (pH = 7.0) and the peptide-nanotube solution. The concentration of the bacterial magnetic nanocrystals was about 330 μg/ml. After a 2-day incubation, magnetic nanocrystals were immobilized on the surface of the nanotubes, according to the TEM observation. Two TEM images at different magnifications shown in Figure 7.25b reveal a uniform coating of the magnetic nanocrystals on the nanotube surface. An interesting phenomenon was observed when a nanocrystal solution at much lower concentration (33 μg/ml) was incubated with the peptide-nanotube solution. As shown in Figure 7.25(c), magnetic nanocrystals entered inside the peptide nanotubes and aligned into a linear chain.

It is reasonable to expect different magnetic behaviors for nanocrystal coated peptide nanotubes and nanocrystal incorporated peptide nanotubes. A vibrating sample magnetometer was used to investigate the magnetic properties of both samples. Measurements were done on aqueous solutions containing magnetic nanotubes with an applied magnetic field up to 0.5 T at room temperature. Magnetic parameters such as coercivity, H_c and the remanent saturation magnetization, M_r/M_s, can be obtained from the measured hysteresis loops. In the case of magnetic nanocrystal immobilization on the nanotube surface, its hysteresis loop is shown as the filled circles in Figure 7.25(d). A coercivity of 5.4 mT and a M_r/M_s of 0.10 can be extracted and the magnetic nanotubes appear somewhat ferromagnetic. For magnetic nanocrystal incorporated peptide nanotubes, the hysteresis loop, shown as the open circles in Figure 7.25(d), give larger values of H_c and M_r/M_s, which are 8.1 mT and 0.32, respectively. The magnetic measurements indicated that the linear alignment of magnetic nanocrystals played a significant role in the overall magnetic behavior while the local disorder in the monolayer assembly exhibited quite different magnetic properties. The magnetic nanocrystal chain inside the peptide nanotubes could be coupled and behave as a nanowire. Consequently, higher values of H_c and M_r/M_s were resulted.

7.4.5 Hybrid Magnetic Nanotubes

Hybrid nanomaterials are heterogeneous nanomaterials such as inorganic–organic one-dimensional nanowires/nanotubes, core–shell nanoparticles, organic–inorganic nanocomposite, etc. Compared with homogeneous nanomaterials, hybrid nanomaterials exhibit enhanced structural, thermal, optical, electrical and mechanical properties. Hybrid magnetic nanotubes have several advantages over other types of magnetic nanotubes as discussed above including higher structure design flexibility, higher multifunctionality, wider material selection, etc.

Ferromagnetic polymer composite (Co/polystyrene, Co/poly-l-lactide) nanotubes have been prepared by Nielsch *et al.* (2005). They used alumina templates with 160 nm-sized pores and 1.5 μm thickness. As usual, the nanopores in the template were arranged in a hexagonal array with 500 nm interpore distance. The synthesis started with preparations of 2–5 wt % polystyrene (PS) and poly-l-lactide (PLLA) in dichloromethane (CH_2Cl_2). A metallo-organic precursor, $Co_2(CO)_8$, was dissolved in the prepared solutions with a variable weight ratio of metallo-organic precursor over the polymer (ranging between 0.4 to 10). To avoid the unnecessary Co oxidation, an inert gas (nitrogen or argon) protection was applied during the solution preparation and the following infiltration process. The next step was wetting the membrane pores with the solution by infiltration. Then, the dichloromethane evaporated and a phase-separation process occurred resulting in the formation of $Co_2(CO)_8$/polystyrene or $Co_2(CO)_8$/poly-l-lactide nanotubes inside the inner surface of the nanochannels. Subsequently, vacuum annealing was conducted to treat the sample at 180 °C for 1–3 days. During this thermal treatment, the metallo-organic precursor decomposed and metallic Co precipitated to form a Co tube with a very small thickness in the interface between the template walls and the polymer nanotubes. To remove the polymer and cobalt residing on the top surface of the template, an ion milling process was applied. Figure 7.26(a) is a typical SEM image of PLLA/Co composite nanotubes in an alumina template after thermal annealing. Each nanopore was filled with a composite

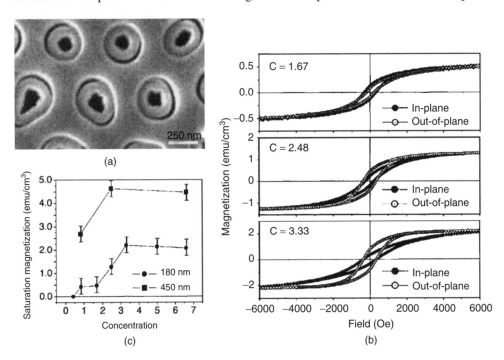

Figure 7.26 (a) SEM image of the cobalt/PLLA composite nanotube array with the outer diameter of 300 nm; (b) in-plane (perpendicular to tube axis) and out-of-plane (parallel to axis) hysteresis loops for cobalt nanotube arrays prepared with different precursor concentration C (interpore distance: 500 nm, diameter: 180 nm); (c) plot of the relationship between the saturation magnetization per unit volume of the nanotube arrays within the alumina membrane and the $Co_2(CO)_8$ concentration C in the polystyrene matrix. (Nielsch *et al.* 2005)

nanotube. The polymer nanotubes' wall thickness ranges from 40 to 70 nm, whereas it is difficult to tell the wall thickness of the Co nanotubes. The SEM image also reveals a tiny gap between the polymer nanotubes and the template pores. Probably, this is induced by the thermal-contraction mismatch between the polymer and alumina during the cooling process from $180\,°C$.

Hybrid Co/polymer nanotubes prepared with a different precursor ($Co_2(CO)_8$) concentration C were used for magnetic characterization at room temperature on a DMS vibrating sample magnetometer. Figure 7.26(b) shows a comparison of measured hysteresis loops for hybrid nanotubes. For nanotubes prepared at a low precursor concentration C (e.g. 1.67), the in-plane and out-of-plane hysteresis loop do not show obvious difference. Both share a similar shape. As the concentration increases, the prepared hybrid nanotubes exhibit significant difference for the in-plane and out-of-plane loops, suggesting a significant anisotropy. Figure 7.26(c) plots the dependence of the saturation magnetization M (per unit volume of the array of tubes) on the precursor concentration C. A similar trend can be seen from this plot. As C increases, the saturation magnetization of hybrid nanotubes also increases and reaches a saturation value, which varies with the pore size of the membrane. The reason for these trends for both saturation magnetization and hysteresis anisotropy is believed to be the amount of Co deposited on nanotubes at different concentrations. At low precursor concentrations, discontinuous Co nanoparticles formed on the nanotube walls leading to isotropic hysteresis loops. On the other hand, continuous Co nanotubes formed at high precursor concentrations, thus the shape-induced easy axis induced anisotropic hysteresis loops. From Figure 7.26(c), the saturation magnetization for hybrid nanotubes of 180 nm diameter is estimated to 2.2 emu/cm^3. Considering the saturation magnetization of pure Co as 1420 emu/cm^3, the estimated value may correspond to a 0.3 nm thick Co nanotube wall. Atomic absorption spectroscopy (AAS) was conducted to measure the effective saturation magnetization for hybrid nanotubes prepared with different C. The measured values are 620–640, ~400, and <200 emu/cm^3 for high concentration (3.3–6.6), intermediate range (0.8–1.6) and low C <0.5, respectively.

A hybrid double walled nanotube (HDWNT) array comprised of conducting polypyrrole encapsulated ferromagnetic nickel nanotubes was developed (Park *et al.* 2007). The prepared HDWNTs showed enhanced anisotropic magnetization. A typical synthesis included three steps: (1) electropolymerization of pyrrole to form polypyrrole (PPy) nanotubes inside the template; (2) electrodeposition of Ni to form Ni nanotubes between the outer walls of existing polypyrrole nanotubes and inner walls of the template; (3) template removal to release the synthesized HDWNTs. AAO membranes with nanopores about 100–200 nm in diameter were used as the template throughout the synthesis procedure. For polypyrrole nanotube synthesis, the electrolyte was made by mixing a monomer (purified pyrrole), a solvent (acetonitrile) and a dopant (Tetrabutylammonium hexafluorophosphate). The molar ratio of the monomer over dopant was set as 5. A well developed template-assisted electropolymerization was conducted to deposit polypyrrole nanotubes inside the nanochannels of the AAO membrane. For Ni nanotube deposition, an aqueous solution containing $NiSO_4 \cdot 6H_2O$, $NiCl_2 \cdot 6H_2O$, and H_3BO_3 buffered at pH 3.5 prepared in proper concentrations was used as the electrolyte. And electrodeposition of Ni was performed at a potential of $-1.0\,V$ (vs Ag/AgCl reference electrode) to deposit Ni nanotubes inside the PPy nanotube-filled AAO template. Sequentially, either hydrofluoric acid or sodium hydroxide solution was utilized to remove the alumina template. Figure 7.27(a) is a schematic of the procedure for PPy-Ni HDWNTs synthesis. It is rather interesting to notice that, instead of depositing inside the PPy nanotubes, Ni atoms deposit outside of the PPy nanotubes and form Ni nanotubes between PPY nanotubes and pore walls. It is expected that Ni compounds in the electrolyte diffuse into the space between the PPy

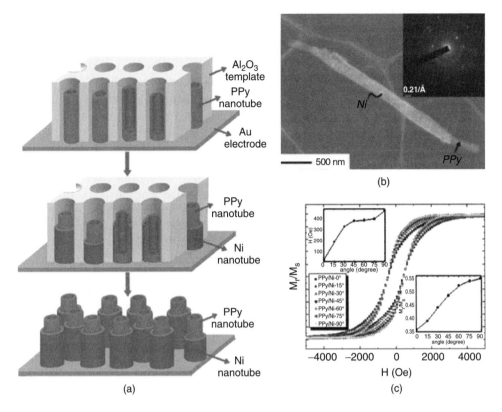

Figure 7.27 (a) Schematic illustration of the synthetic procedure of the PPy–Ni hybrid double walled nanotube array; (b) TEM image of an individual PPy–Ni hybrid nanotube; (inset: the electron diffraction pattern) (c) hysteresis loops of the PPy–Ni hybrid nanotubes with different angles respective to the direction of an applied magnetic field at room temperature. Top inset: coercivity of the nanotubes as a function of the different angles of the applied magnetic field. Bottom inset: remanent-saturation magnetization as a function of the angles of the applied magnetic field. (Park *et al.* 2007)

nanotubes and the alumina nanopores and electrochemically deposit on the area of the gold electrode at the bottom. Because of the higher conductivity of gold than the existing PPy, Ni nanotubes form on the gold electrode and nanotube growth starts.

TEM image of an individual HDWNT is shown in Figure 7.27(b). The hybrid morphology of PPy-encapsulated Ni nanotube can be differentiated. The outer diameter of the HDWNT is about 150–200 nm, corresponding to the pore size of the AAO membrane. An electron diffraction pattern of the HDWNT, as shown in the inset of Figure 7.27(b), suggests the polycrystalline phase of the outer Ni nanotube. The direction of the incident electrons was kept perpendicular to the hybrid nanotube orientation. For magnetic property measurements, magnetic hysteresis loops were recorded with an applied magnetic field applied at different angles in relation to HDWNTs. The angle changed from 0°, where H was perpendicular to the tubes' axis, to 90°, where H was parallel to the tubes' axis. So the angular dependence of magnetization of HDWNTs can be determined. Figure 7.27(c) shows a comparison of measured hysteresis loops along a number of directions. The remanent saturation magnetization, M_r/M_s, is the ratio of the remanent magnetization to the saturation magnetization. Apparently, the magnetic hysteresis, coercivity and remanent magnetization confirm the ferromagnetic behavior of the HDWNTs along all directions. The difference among hysteresis loops along different directions indicates an ordered magnetic anisotropy with the parallel direction as the easy axis for magnetization. The top inset of Figure 7.27(c) plots the trend of coercivities in different directions. As can be seen, the coercivity value increases from 415 to 457 Oe as the direction of H changes from $0°$ to $90°$. The bottom inset of Figure 7.27(c) shows that the remanent saturation magnetization increases as the angle increases. The value changes from 0.353 to 0.563 as H varies from $0°$ to $90°$. Experimental results confirmed that the maximum values of H_c and M_r/M_s were obtained when the applied magnetic field was applied in parallel to HDWNTs. Both values are relatively higher than those of Ni nanotubes ($H_c = 185$ Oe, $M_r/M_s = 0.19$) reported previously.

Multi-segmented nanowires have been developed by the template-assisted synthesis approach. One example is metal-CdSe-metal nanowires with switchable photoconductivity (Peña *et al.* 2002). Magnetic multi-segmented nanotubes offer a unique capacity for structure tailoring which is of importance for applications in the field of catalysis, microfluidics, magnetic sensors and bioseparation, etc. Multi-segmented magnetic nanotubes with an Au–Ni stacking configuration were synthesized by the modified template-assisted electrochemical method (Lee *et al.* 2005). The synthesis adopted an established sensitization-preactivation process to deposit metal nanoparticles homogeneously over the entire surface of the inner walls of the templates. This step proved to be essential for the formation of metal nanotubes in the electrodeposition. Typically, AAO membranes were immersed into a $SnCl_2 + HCl$ solution. After a thorough wash step and then drying, the resulting membranes were dipped into an $AgNO_3$ aqueous solution. In this step, the Ag^+ was reduced by Sn^{2+} to Ag and the generated Ag nanoparticles attached to the inner walls of nanopores. Multiple cycles of the sensitization-preactivation process can be conducted to adjust the size and the number of Ag nanoparticles. TEM observation indicated a discontinuous distribution of Ag nanoparticles which did not form a conducting thin film inside the nanopores. For electrodeposition, a thin gold layer was coated on one side of the nanoparticle-modified membrane by the sputtering method and served as the working electrode. The electrolytes for gold and nickel plating are commercially available gold plating solution and a mixture solution containing $NiCl_2 \cdot 6H_2O$, $Ni(H_2NSO_3)_2 \cdot 4H_2O$, H_3BO_3, and sodium acetate buffer (pH $= 3.4$), respectively. The current density was kept constant as 2.4 mA $\cdot$ cm^{-2} during the electrodepositions.

Like a common template-assisted electroplating process, the formation of metal nanotubes must start from the gold electrode at the bottom of the nanopores. The Ag nanoparticles on the pore walls ensure a preferential deposition of metal atoms on the nanochannel surface. It is expected that the nanotube growth will be determined by both the deposition rate and the metal ion diffusion rate. Meanwhile, the preferential deposition makes it possible to prepare multi-segmented metal nanotubes by changing the electrolyte accordingly. For example, Au-Ni-Au-Ni-Au multi-segmented nanotubes were synthesized by alternating gold and nickel electroplating solutions for electroplating. A typical SEM image of the cross-section of the multi-segmented nanotubes inside the AAO template is shown in Figure 7.28(a). The bimetal starching configuration of multi-segmented nanotubes can be seen clearly. Ni segments about 800 nm long are sandwiched between adjacent Au segments with different image contrast, which comes from the different intensities of backscattered electrons from Ni and Au. A beauty of the template-assisted electroplating method is that the length of each segment can be readily tuned by exact

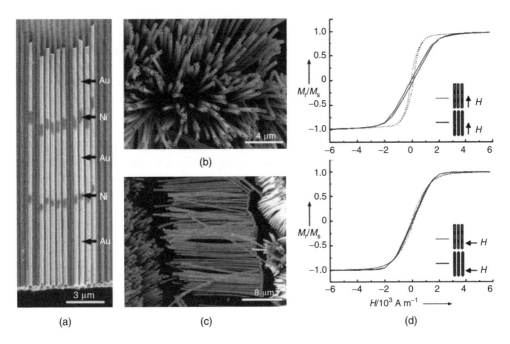

Figure 7.28 (a) SEM image of cross-section view of the as-prepared Au-Ni-Au-Ni-Au nanotube inside the AAO template; (b) and (c) SEM images of multi-segmented magnetic nanotubes after the template removal; (d) hysteresis loops for Au-Ni-Au-Ni-Au nanotube array (dashed line) and ferromagnetic Ni nanotube array (solid line), which were measured by applying a magnetic field in parallel (top) and perpendicularly to (bottom) the nanotube axis. (Lee 2005)

control of the total charges in the reaction. Figure 7.28(b) and 7.28(c) are SEM micrographs of released nanotubes after etching the AAO template by a 1.0 M NaOH solution.

Figure 7.28(d) shows hysteresis loops of multi-segmented nanotubes with Au-Ni-Au-Ni-Au configuration and pure Ni nanotubes measured by a superconducting quantum interference device (SQUID) magnetometer at 298 K. Both nanotubes have similar diameter, length, spacing and wall thickness. For comparison purposes, a magnetic field was applied on both nanotubes along the nanotube axis and the direction perpendicular to the nanotube axis. In the former case, close values of coercivities $H_{c//}$ and remanence values for both nanotubes were estimated to be 85–90 Am^{-1} and 4, 17 %, respectively. And reversible magnetic behavior was observed for both nanotube samples because of the high saturation field $H_{s//}$. Further, multi-segmented nanotube arrays can be completely magnetized at lower saturation field, $H_{s//} \approx 750$ Am^{-1}, than that required by Ni nanotubes which is around 2300 Am^{-1}. Based on this result, it was assumed that, compared with Ni nanotubes, multi-segmented nanotubes possessed lower dipolar interactions. This is reasonable because the intertube distance in Ni nanotubes is less than the average distance between the neighboring Ni segments in the others. On the other hand, in the case of perpendicular field applications, both nanotube samples exhibited highly similar hysteresis loops with reversible magnetic behaviors. The saturation magnetizations for both samples are around 2000 Am^{-1}. As far as the magnetic properties of two nanotube samples are concerned, the multi-segmented nanotube array shows magnetic anisotropy while the Ni nanotubes display magnetic isotropy.

7.5   Biomedical Applications of Magnetic Nanotubes

As discussed in previous chapters, magnetic nanoparticles and magnetic nanowires exhibit unique properties which have attracted attentions for both scientific research and practical applications. As extensive research efforts on magnetic nanoparticles and nanowires have generated plenty of results, an emerging area of the relatively new magnetic geometry, magnetic nanotubes, has been explored. Magnetic nanotubes inherit the paramagnetic nature of magnetic nanoparticles and they even display much higher saturation magnetization than their bulk counterparts. Generally speaking, magnetic nanotubes could be used in almost all applications for magnetic nanoparticles and nanowires. Additionally, the higher surface area, lower density and multifunctionality of magnetic nanotubes render them a promising candidate for many other applications in special situations. Magnetic nanotubes may have possible applications in ultrahigh-density magnetic storage devices, nanoelectromechanical systems (NEMS), sensors, catalysts, etc. (Li *et al.* 2007). Recently, the use of magnetic nanotubes in the biomedical and biological fields has been envisaged (Sui *et al.* 2004b). Major bioapplications of magnetic nanotubes include bioseparation, cell manipulation, targeted drug or gene delivery, neuron regulation, tips for magnetic force microscopes, etc.

7.5.1   Bioseparation

Mitchell and his team have proved the concept of bioseparation for multifunctional silica nanotubes (2002). In their experiments, silica nanotubes were synthesized by a template method and application of different functional groups on the inner and outer walls of nanotubes was achieved. In their simple protocol, the inner nanotube surface reacted with a first hydrophilic silane while the nanotubes were still embedded inside the template. Since the outer walls of nanotubes were still bound tightly with the nanopore walls, this first reaction did not happen to the masked outer walls of nanotubes. After removal of the template, a second silane was applied to react with the outer walls of nanotubes to attach different silane groups there. Via this simple technique, multifunctional nanotubes with hydrophilic behavior on their inner surfaces and hydrophobic behavior on their outer surfaces could be obtained. Demonstration was made on this type of multifunctional silica nanotubes to prove their capability of extracting lipophilic molecules from aqueous solution.

A similar multifunctionality concept can also be applied for magnetic nanotubes. In addition, magnetic-field-assisted bioseparation can be achieved by using magnetic nanotubes. The feasibility of this magnetic bioseparation was demonstrated by Son *et al.* (2005). The magnetic nanotubes were synthesized according to the following procedure. First, silicon oxide nanotubes were obtained by a well developed template-assisted 'surface sol-gel' method. The templates used were porous alumina membrane with 60 and 200 nm in pore sizes. Then the silica nanotube loaded membranes were dipped in a solution containing 1 M $FeCl_3$ and 2 M $FeCl_2$. In this step, iron ions entered the inner cavity of the silica nanotubes. After drying in an argon flow, the membranes were immersed in 1 M NH_4OH for 5 minutes followed by a thorough washing with deionized water. After this process, a thin layer of magnetite nanoparticles was coated on the inner surface of silica nanotubes hence magnetic nanotubes were obtained. While nanotubes were still embedded inside the template, octadecyltriethoxysilane (C18-silane) was used to treat the magnetic nanotubes for inner surface functionalization. The resulting magnetic nanotubes possessed both hydrophobic chemistry on the inner surface, due to the functionalization, and hydrophilic chemistry on the outer surface due to the silica layer. To obtain

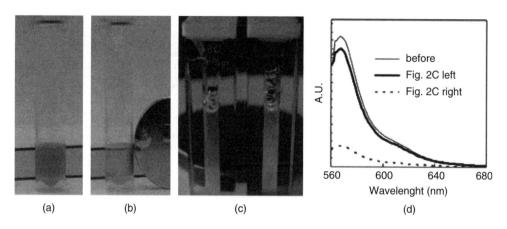

(a) (b) (c) (d)

Figure 7.29 Digital pictures of a vial containing 38 μM DiIC18 and 10^9 C-18-modified magnetic nanotubes before (a) and after (b) the magnetic separation for 2 minutes; (c) a digital picture of the solution containing green FITC-labeled anti-bovine IgG and red Cy3-labeled anti-human IgG after the magnetic separation. (Please refer to the color image in the original article). Incubations were done for both solutions with BSA-magnetic nanotubes (left) and IgG-incorporated magnetic nanotubes (right) before the magnetic separation; (d) fluorescence spectra operating at 552 nm wavelength corresponding to (c) for measuring the amount of remaining red Cy3-labeled anti-human IgG in solution. (Son *et al.* 2005)

free-standing magnetic nanotubes, mechanical polishing was applied on both sides of the template and removal of the membrane was done by etching with a 0.1 N NaOH solution.

In a typical experiment to test the chemical extraction and separation, C18 functionalized magnetic nanotubes were treated with a solution of 1,1'-dioctadecyl-3,3,3',3'-tetramethylindocarbocyanine perchlorate (DiIC$_{18}$, 38 μM) in a mixture of water and methanol (9:1 v/v). Due to the strong hydrophobic interaction, these dye molecules could enter the magnetic nanotubes. And these treated nanotubes can be uniformly dispersed in aqueous solutions readily owing to the hydrophilic property of the outer silica surface. As shown in Figure 7.29(a), the uniform suspension containing magnetic nanotubes displays red color, transparent and clear. (Please refer to the color image in the original article.) When a Nd-Fe-B magnet (ca. 0.3 T) was placed close to the bottle, the red color moved towards the magnet (see Figure 7.29(b)). This phenomenon clearly indicated the capability of magnetic separation of magnetic nanotubes by applying an external magnetic field. Further UV-vis spectroscopy and fluorescence microscopy investigations confirmed the removal of >95 % of the DiIC$_{18}$ from the solution. In a control experiment, such a phenomenon was not observed with magnetic nanotubes without C-18 functionalization indicating the importance of the inner wall functionalization. Using this technique, it is possible to extract, separate, release and further analyze trace amounts of hydrophobic biohazards (e.g. polychlorinated biphenyls and polycyclic aromatic hydrocarbons) in water.

A sequential experiment was carried out to test the capability of magnetic nanotubes for magnetic bioseparation using antigen–antibody coupling. In this experiment, human IgG was incorporated inside magnetic nanotubes via a similar process discussed as above. This IgG loaded magnetic nanotubes were added into a mixture solution of phosphate-buffered saline containing fluorescence-labeled anti-bovine IgG in green color (0.71 μM, pH = 7.4) and Cy3-labeled anti-human IgG in red color (0.67 μM). For comparison, a nonspecific biointeraction was demonstrated as well by using BSA-derivatized magnetic nanotubes. Meanwhile, poly(ethylene glycol) silane was used to functionalize the outer surfaces of the magnetic nanotubes so that the color change of the suspension could be completely due to the substance inside the magnetic nanotubes. For IgG incorporated magnetic nanotubes, the color of the solution changed from the original pink to greenish-blue after the magnetic separation. Figure 7.29(c) is a digital image showing this color change. It was expected that specific binding between Cy3-labeled anti-human IgG and human IgG-incorporated magnetic nanotubes would enable magnetic bioseparation. Conversely, the control experiment did not show the similar color change. Fluorescence spectra of the solution before separation, the control solution and red Cy3-labeled anti-human IgG after magnetic separation are plotted in Figure 7.29(d). The percentages of Cy3-labeled anti-human IgG separated by human IgG- and BSA-magnetic nanotubes are 84 % and 9 %, respectively. This comparison indicated the effect of specific antibody–antigen interaction on magnetic separation.

The feasibility of using magnetic nanoparticle-incorporated carbon nanotubes for DNA separation has been demonstrated. Korneva's experiment suggested the possibility of freezing nanotubes and changing the intertube spacing by applying a magnetic field (2005). By this technique, DNA coils can be unraveled and separated effectively. In their experiments, magnetic nanotubes were suspended in liquids. Then a magnetic field of $\mu_0 H = 0.01$ T or 0.03 T was applied to investigate the response of magnetic nanotubes. Figure 7.30 contains digital images showing the typical behavior of magnetic nanotubes responding to the applied field. As shown in Figure 7.30(a) and 7.30(b), when the magnetic field was applied in parallel to the substrate, the magnetic nanotubes were oriented along the substrate planes. When a magnetic field of 0.03 T was applied perpendicular to the substrate, magnetic nanotubes were frozen and aligned perpendicularly to the wafer, as can be seen from Figure 7.30(c). These magnetic nanotubes could be used to replace the nanoposts in fluidic chips for DNA separation. Thus, this work may give useful information for research efforts toward the nanoengineering of complex multifunctional nanosystems.

7.5.2 Cell Manipulation

Biomanipulations including cell, gene and DNA manipulations are of great importance in bioengineering, biomedical therapy and other related areas. Magnetic beads have been widely used in cell and cell membrane manipulations. It was reported that magnetic nanowires offered higher purity and separation efficacy than magnetic beads in cell

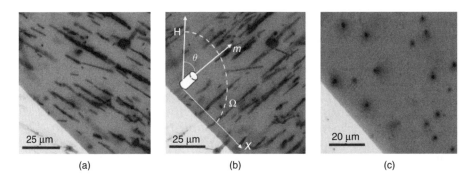

(a)   (b)   (c)

Figure 7.30 Images showing the manipulation of the magnetic nanotubes by a magnetic field. (a) and (b) magnetic nanotubes aligned along in the substrate plane ($\mu_0 H = 0.01$ T); (c) magnetic nanotubes frozen in a magnetic field which was applied perpendicularly to the substrate ($\mu_0 H = 0.03$ T). The inset shows a schematic illustration of the orientations of the applied magnetic field and the magnetization with respect to the axis of nanotubes. All vectors are concerned as they lie in the same plane. (Korneva et al. 2005)

manipulation due to their larger surface area (Hultgren *et al.* 2003). This work also suggested that magnetic nanotubes be a better material candidate for enhanced cell manipulation because nanotubes have even higher surface area than nanowires. Therefore, it is believed that magnetic nanotubes can be used as the magnetic handle to manipulate cells efficiently by applying a magnetic field. To prove this concept, work was conducted by Gao and his team (2006), where the manipulation of sheep red blood cells using magnetic carbon nanotubes was demonstrated.

They first developed a facile and effective approach to preparing carbon nanotube-based magnetic nanotubes. The convex surface of carbon nanotubes is chemically active, hence various species, such as macromolecules, metals, oxides and quantum dots, can be attached or bound to the convex of carbon nanotubes via covalent bonding, physical adsorption and other interactions. Magnetic nanotubes can be prepared by attaching magnetic nanoparticles on the surface of carbon nanotubes. To achieve this, an electrostatic self-assembly approach was adopted. After functionalization of carbon nanotubes with Poly(2-diethylaminoethyl methacrylate) (PDEAEMA) and quaternization with methyl iodide (CH_3I), cationic polyelectrolyte-grafted carbon nanotubes (PAmI) resulted. The electrostatic self-assembling between magnetite nanoparticles and PAmI enabled the loading of magnetite nanoparticles on the nanotube surface. The obtained magnetic nanotubes exhibit paramagnetic properties and the magnetic density can be adjusted by varying the load concentration of magnetic nanoparticles.

Manipulation of sheep red blood cells in buffer solutions using the obtained magnetic nanotubes was demonstrated. A relatively strong magnetic force is required to manipulate the sheep red blood cells because of their large dimension (approximately 4 µm in diameter). At a certain applied magnetic field, higher loading concentration of magnetic nanoparticles leads to higher magnetic force. In the first experiment, the magnetic behavior of magnetic nanotubes was tested in the buffer solution. With the help of an optical microscope equipped with a high-speed camera, the movement of magnetic nanotubes can be recorded. It should be pointed out that only the magnetic nanotube bundles can be resolved by the microscope due to its magnification limit. The successive micrographs clearly revealed the magnetic response of nanotubes to the magnetic field. By rotating the magnetic field, the magnetic nanotube bundles can be rotated easily while the rotation speed was determined by the frequency of the applied field. Figure 7.31(a) contains a set of snapshots showing this rotation of magnetic nanotube bundles. When attached with sheep blood red cells, the magnetic behavior of magnetic nanotubes reserved. In addition to rotation, the magnetized blood cells can be aligned or relocated in two-dimensional directions from one area to another by changing the direction of the magnetic field. Snapshots in Figure 7.31(b) clearly show the rotation manipulation for one blood cell. Under a magnetic field with 0.5 Hz frequency and 12.7 kA/m strength, the magnetized cell rotated clockwise at a speed of 3 s per cycle. Interestingly, two blood cells can be bridged by magnetic nanotubes and are able to rotate at the same speed under an identical magnetic field, as shown in Figure 7.31(c).

Such cell manipulation with magnetic nanotubes strongly suggested an opportunity of taking advantage of their magnetic behaviors to manage and operate individual biological entities.

7.5.3 Drug and Gene Delivery

For magnetic nanomaterials, targeted drug delivery is one of the most important bioapplications. In modern biopharmaceutical drug delivery, there has been a strong demand for biologically stable drug-loaded carriers capable of entering living bodies and facilitating

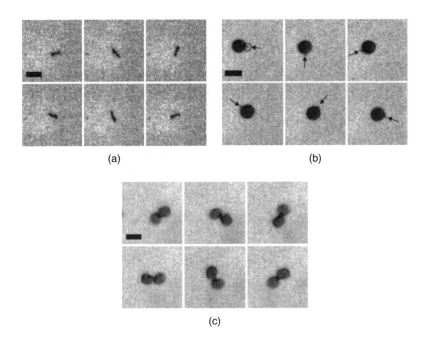

(a)   (b)

(c)

Figure 7.31   (a) Snapshot pictures of a bundle of magnetic nanotubes' rotation under an applied
rotational magnetic field. The applied magnetic field was 0.5 Hz of frequency and 12.7 kA/m of
intensity. The rotation of the magnetic nanotube bundles was in the clockwise direction as shown
in the snapshots corresponding to 0th, 0.2th, 0.4th, 0.6th, 0.8th and 1.0th cycle of the rotational
magnetic field (scale bar: 5 μm); (b) snapshot pictures of a sheep red blood cell's rotational motion.
The cell was attached with magnetic nanotubes, indicated by an arrow. Again, the blood cell rotated
in the clockwise direction and the snapshots were taken at 0th, 0.2th, 0.4th, 0.6th, 0.8th and 1.0th
cycle of the rotational magnetic field. (scale bars: 5 μm); (c) snapshot pictures showing the rotation
of two sheep red blood cells bridged with magnetic nanotubes. (Gao *et al.* 2006)

the drug development itself even at an early stage of toxicity screening. An ideal drug
carrying system should be able to preserve the protein secondary and tertiary structures,
prevent aggregation and eliminate any possible chemical or enzymatic degradation. In
Chapter 4, magnetic nanoparticle-based drug delivery was discussed. Compared with magnetic nanoparticles, magnetic nanotubes have both inner surfaces and outer surfaces for
drug incorporation and drug immobilization, respectively. Most importantly, the magnetic
nanotubes have the capability to enhance biointeractions between the outer surface of nanotubes and the target biospecies. For this reason, using magnetic nanotubes could enhance
the efficiency of drug delivery. It is believed that magnetic nanotubes with drug-friendly
interiors and target-specific exteriors are an ideal drug carrier in versatile bioapplications.

A proof-of-concept experiment was conducted to demonstrate the enhanced drug delivery performance (Son *et al.* 2005). The inner surface of magnetic nanotubes with 60 nm
diameter was functionalized by FITC while their outer surface was modified and attached
with rabbit IgG antibodies. These multifunctional magnetic nanotubes were added to an
antirabbit IgG-modified glass slide and incubation was maintained for 10 minutes with
an applied magnetic field from the bottom of the glass slide. A control experiment was
also done with identical conditions except for the absence of a magnetic field during
the incubation. To check the effect of the magnetic field for biointeraction, a fluorescence microscope was used to image the number of bound nanotubes after washing the

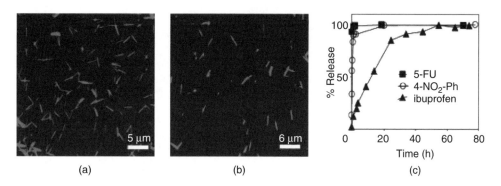

(a)   (b)   (c)

Figure 7.32 Fluorescence microscope images showing the binding between magnetic nanotubes and the surface of anti-rabbit IgG-modified glass after antigen-antibody interaction with (a) and without (b) magnetic field under the glass substrate. The magnetic nanotubes had 60 nm diameter and 3 μm length, and their inner surface was FITC-modified and outer surface was modified by rabbit IgG. The long bright green spots are individual magnetic nanotubes. (Please refer to the color image in the original article.) Bundles of a few nanotubes display brighter spots in the images; (c) comparison of the *in vitro* release of ibuprofen, 4-nitrophenol (4-NO$_2$Ph) and 5-fluorouracil (5-FU) from magnetic nanotubes with 60 nm diameter and 250 nm length). (Son *et al.* 2005)

unbound ones. Figure 7.32(a) and 7.32(b) are fluorescence microscopy images derived from averaging from five different areas on the glass slide. The comparison between two images indicated that the antibody–antigen interaction was enhanced about 4.2-fold with an applied magnetic field. Thus the efficiency of the biointeraction can be controlled spatially by means of an external field.

Son and his team also tested the performance of magnetic nanotubes in drug loading and drug releasing. Unlike the antibody–antigen bonding, the hydrogen bonding exists between the drugs and the inner surface of the magnetic nanotubes. Inner wall functionalization of magnetic nanotubes was conducted with amino-silane (aminopropyl triethoxysilane, APTS). The model drug molecules, selected in the experiments, were 5-fluorouracil (5-FU), 4-nitrophenol and ibuprofen. For drug loading, amine functionalized magnetic nanotubes were immersed in the drug solutions in either hexane (ibuprofen) or ethanol (5-FU, 4-nitrophenol). The strong ionic and/or hydrogen bonding interactions between the amine groups of the inner wall of magnetic nanotubes and the acid groups of the drug molecules enabled an effective drug loading. The efficacy of the drug loading was checked using a UV-vis spectrophotometer. Based on the measured results, the rough numbers of ibuprofen molecules, 4-nitrophenol and 5-fluorouracil per magnetic nanotube were $\sim 10^7$, $\sim 10^6$ and 10^7, respectively. In all cases, the magnetic nanotubes exhibited high drug loading. For example, the value of effective ibuprofen loading is about twice the monolayer coverage of the inner surface area of the magnetic nanotubes. The drug release for drug loaded magnetic nanotubes were also tested in PBS solution (pH = 7.4). The test results are plotted in Figure 7.32(c). It was noticed that more than 90 % of loaded 5-fluorouracil and 4-nitrophenol were released within one hour. However, the release of ibuprofen was much slower than the other two drugs. In a release test for ibuprofen, less than 10 % was released within one hour and about 80 % release occurred within one day. This drug release test suggested the strongest interaction between the amine group of the magnetic nanotubes' inner surface and ibuprofen among three drugs resulting in the slowest drug release.

Carbon nanotube-related nanomaterials have been considered as a potential delivery system because of their capability of entering various cells. Both their inner cavities and

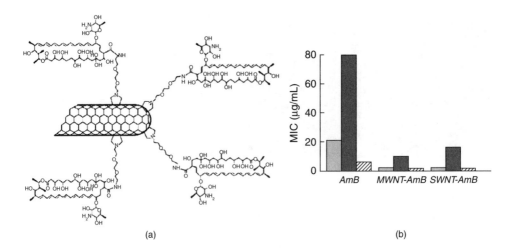

(a)   (b)

Figure 7.33 (a) Schematic illustration of an amphotericin B-modified carbon nanotube; (b) comparison of the antifungal activities of amphotericin B in solution and when it was attached to single- and multi-walled carbon nanotubes. Results are shown for *Candida parapsilosis* (gray bar), *Candida albicans* (black bar) and *Cryptococcus neoformans* (dashed bar). (Malmsten 2006)

outer surfaces can be used to load drugs and their outer surfaces can be functionalized for various covalent bindings owing to the full chemistry of carbon nanotubes. Figure 7.33(a) schemes a covalent-linked amphotericin B, for fungal infection treatment, to a functionalized carbon nanotube (Malmsten 2006). The antifungal activities of amphotericin B in solution and attached to carbon nanotubes are shown in Figure 7.33(b). It suggests a significantly enhanced effect of amphotericin B when loaded in carbon nanotubes. Using the same approach, magnetically decorated carbon nanotubes could be used for drug delivery as well. More interestingly, their magnetic properties offer them additional magnetic function in potential drug delivery systems when the situation arises.

Magnetic carbon nanotubes filled with iron nanoparticles have been demonstrated to deliver lipid-nanostructure complexes into the cytoplasm (Mönch *et al.* 2005). Fe is important in cell culture, complex formation and lipid-mediated delivery, etc. In the experiments, EJ28, a human bladder cancer cell line, was cultured in the presence of Lipofectin, a cationic lipid formation, with the addition of a suspension containing magnetic carbon nanotubes. After two hours, the cultured cells were harvested followed by centrifugation, washing and fixation with 4 % buffered formalin. Neither adhesion of nanotubes to the cell membrane nor uptake of nanotubes by the tumor cells was observed after the incubation. Conversely, Figure 7.34(a) reveals an effective lipid/nanotube complex delivery into the cytoplasm instead of the nucleus. Figure 7.34(b) is an EDX spectrum indicating the existence of pure Fe inside the cells. It was believed that a strong complexation occurred between nanotubes and lipid, and the resulting complexes were stable introcytoplasmatically.

Therapeutic agents, such as carboplatin, can be incorporated into the Fe-filled carbon nanotubes for drug delivery purposes. However, a big concern which needs to be considered is the biocompatibility of the drug carrier when a therapeutic system is introduced inside human bodies. Mönch *et al.* injected Fe-filled carbon nanotubes into mice intraperitoneally or intravenously to test their toxicity effects (2007). Both electron microscopy and light microscopy studies revealed that magnetic carbon nanotubes aggregated to form clusters after injection. The resulting magnetic clusters may diffuse through capillaries to reach various tissues or organs so that they are observed to be present in lung, heart,

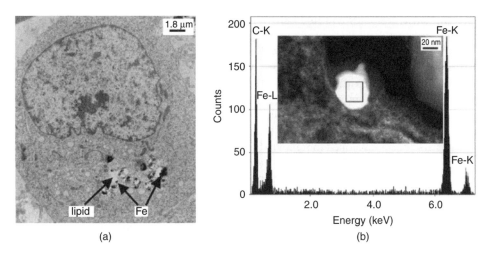

(a)   (b)

Figure 7.34   (a) TEM image showing the presence of the internalized nanotubes (black) complexed by lipid (white) in EJ28 cells; (b) EDX analysis confirming Fe-nanoparticle (white) in an EJ28 cell (particle diameter: $\sim$50 nm). (Mönch *et al.* 2005)

liver, colon, etc. However, these clusters did not display any adverse effect on the health and survival of the mice under the investigation. This biocompatibility study suggests that the Fe-filled carbon nanotubes can be used as the nanocontainer for drug delivery for therapeutic treatments.

7.5.4   Neuronal Applications

Application of magnetic nanotubes in neuron science and neuron technology have stemmed from the carbon nanotubes' substratum behavior for neuron growth. Carbon nanotubes have been demonstrated as a promising substrate for neuron growth, as they boosted neurite growth and neuronal electrical signaling (Mattson *et al.* 2000; Hu *et al.* 2004). The surface chemical properties, carbon composition coupled with their hollow structure, and the unique electrical properties of carbon nanotubes tend to make them ideally suited for use in neuroscience applications. Prompted by this promising study, researchers have explored using magnetic nanotubes in such applications. Besides their hollow structure, nanoscale dimension and the possibility of surface chemical modification, magnetic nanotubes have the advantage of their inherent magnetic moment as well as their interaction with an external magnetic field.

The nervous system is one of the most important systems for an animal, as the survival of an animal depends greatly on its ability to respond appropriately to the stimuli from its environment. The nervous system enables animals to sense and respond to environmental cues by generating and conducting nerve impulses. The neuron, uniquely specialized to generate and conduct nerve signals, is the structural and functional unit of nervous tissue. Many investigations have attempted to unravel the working mechanism of neurons and the modulations of neuronal activities. Injuries and disorders of the nervous system may result in severe disabilities and diseases, such as spinal cord injuries and Parkinson's disease, and the treatments of such disabilities and diseases are formidable challenges in biology and medicine. Though great progress has been made in understanding the molecular mechanisms which interfere with neuronal regeneration, effective strategies to overcome those obstacles and find cures for the disabilities arising from the injuries and neurodegenerative disorders are yet to be found.

Chen *et al.* tested the neuron survival and neurite growth on a magnetic nanotube mat (2006). Uniform magnetic nanotubes were synthesized by a two-step process inside an anodic aluminum oxide membrane. The process involved pressure filtering of iron nitrate/ethanol solution in proper concentration through an AAO membrane with 200 nm pore size followed by thermal annealing in air at 250 °C. The synthesized magnetic nanotubes were hematite nanotubes, according to the XRD analysis. A post treatment technique has also been developed to remove the template membrane by 3M sodium hydroxide aqueous solution, followed by the deposition of magnetic nanotubes on a track etch membrane (0.6 μm pore size) to form a 'magnetic nanotube mat'. These as-prepared magnetic nanotube mats are ready for testing as the substratum for neurite growth. Dorsal root ganglion (DRG) neurons were established from one-day-old pups from Sprague-Dawley, and timed pregnant rats were purchased from the Harlan Co. (Indianapolis, IN). All media and supplements for culture were purchased from GIBCO (MD). The newborn rats were sacrificed by placing on ice to anesthetize and the cervical, thoracic and lumbar DRGs were dissected out under sterile conditions. The dissected ganglia were enzymatically dissociated by first treating with collagenase in Leibowitz 15 (L-15) medium for 30 minutes at 37 °C followed by treating with trypsin (0.25 %) in calcium and magnesium-free Hank's balanced salt solution for 20 minutes at 37 °C. Ganglia were then washed three times in L-15 + 1 mg/ml of bovine serum albumin (BSA) followed by a wash in DMEM containing 10 % of heat inactivated fetal bovine serum. The ganglia were then triturated with a pipette with fire polished tip, centrifuged at 500 rpm for two minutes at 4 °C to separate cells. The cell suspension was then passed through a nylon filter (Nitex HD3-15, Tetko Inc.) to remove undissociated cells and preplated on uncoated 100 mm sterile dishes in DMEM containing 20 ng/ml of β-NGF. The dishes were incubated at 37 °C in 5 % CO_2 for 30 minutes. The unattached neurons were collected and preplated again and incubated for 30 minutes. This step was repeated one more time. Then the unattached neurons in the preplated dishes were collected and centrifuged again and the resulting neuronal suspension was plated (50 μl/dish) on different nanotube mats coated with poly-D-lysine (50 μg/ml) kept inside the sterile 35 mm culture dishes after determining the plating density (1.2 × 10^5 cells/ml) using a hemocytometer. The neurons were incubated for 24 to 72 hours in serum free Neurobasal medium containing B-27 supplement, β-NGF (15 ng/ml) and a mixture of 5-fluro-5'-deoxyuridine and uridine (10 μM, to prevent proliferation of any supporting cell). The nanotube mats with plated neurons were fixed with 4 % paraformaldehyde and processed for SEM observations. A thin gold film was sputtered on the samples for better SEM imaging.

No obvious neurite growth was observed on neurons plated on the blank polycarbonate membrane. However, projection of neuronal membrane resembling growth cone can be seen as shown in Figure 7.35(a). Observation of neurons cultured for two days revealed differences between those plated on magnetic nanotubes and those plated on blank polycarbonate membrane. Not only growth cone was observed but also growing neurites could be distinguished clearly. The growth of neurites could be attributed to the influence of magnetic substrate during neuron culture. However, according to their SEM observations, not all the neurons exhibited similar neurite growth. It was interesting to find two closely plated neurons trying to establish a morphological connection as shown in Figure 7.35(b).

The tubular structure of nanotubes provides many advantages for their applications in biomedicine. The inner void of a nanotube can be used for capturing, concentrating and releasing various species ranging in size from large proteins to small molecules. Besides, the diameter and length of a nanotube are independently tunable. Furthermore,

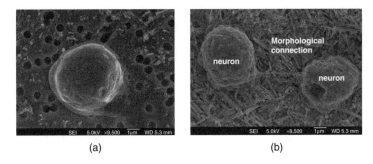

(a)   (b)

Figure 7.35   (a) SEM micrograph of a neuron on the blank polycarbonate membrane after two days in culture; (b) a morphological connection between two neurons. (Chen *et al.* 2006)

due to the one-dimensional structures of nanotubes, when external forces and torques are applied to them, they could penetrate the membranes of neurons and get inside the cell bodies of the neurons. This property is important for delivering drugs directly into the cell body of a neuron, and activating or regulating the activities of a neuron by targeting intracellular second messengers. The effects of magnetic fields on neuron activities have been extensively investigated. Intracellular recordings indicated that both static and alternating magnetic fields affect the bioelectric activities of neurons. As magnetic nanotubes can get very close to or inside of a neuron, the magnetic fields produced by the remanent magnetizations of the magnetic nanotubes will affect the activities of the neuron. If an external magnetic field is applied, the magnetic nanotubes have magnetizations much stronger than the remanent magnetizations of the magnetic nanotubes. Usually the magnetic fields produced by magnetic nanotubes have large gradients. Due to their ferromagnetic and superparamagnetic properties, and their shape anisotropies, magnetic nanotubes usually have large magnetic moments. Therefore strong magnetic forces and torques can be applied to the magnetic nanotubes by external magnetic fields, and this is crucial for nanotubes to get close to or inside of a neuron and then activate or modulate the function of the neuron. The biocompatibility of magnetic nanotubes is a must for their use in neuroscience applications. The basic requirement for biocompatibility is that the magnetic nanotubes are non-toxic to neurons, and the magnetic nanotubes are stable when exposed to biological systems. It has been found that, with suitable surface functionalization, magnetic nanowires are compatible with living cells. They do not disrupt normal cell functions including cell adhesion and gene expression. Therefore, survival of neurons will not be affected when suitably functionalized magnetic nanotubes get close to or inside of the neuron (Chen *et al.* 2007). In addition to the magnetic fields, the movements of magnetic nanotubes may also affect the neuron activities. When an external magnetic field is applied, the magnetic nanotubes might make movements in micron scale. A uniform magnetic field gives rise to a torque, so the magnetic nanotubes may rotate while a magnetic field gradient exerts a force at a distance, resulting in the translational actions of the magnetic nanotubes. As the magnetic nanotubes may directly contact the membranes of the neurons on the magnetic nanotube mat, the movements of the magnetic nanotubes adjacent to the neurons may cause the changes of the bioelectric signals of the neurons. Because of the forces and torques applied on magnetic nanotubes due to the external magnetic field, the magnetic nanotubes may penetrate the membrane of a neuron, getting inside the cell body of the neuron. With suitable surface functionalization on magnetic nanotubes, small molecules can be delivered into the cell body of a neuron using the magnetic nanotubes targeting these molecules to bind with specific intracellular enzymes or second messengers. After magnetic nanotubes get inside the cell body of a neuron, the magnetic field produced by the magnetizations of the magnetic nanotubes and the movements of the magnetic nanotubes due to the external magnetic field are anticipated to have much stronger effects on the bioelectric activities of the neuron, than the effects of the magnetic nanotubes outside the neuron. Such investigations can lead to finding means of modulating the neuron signal.

Further investigation was conducted to explore the effect of magnetic nanotubes in PC12 cell differentiation and neurite growth (Xie *et al.* 2008). Instead of a magnetic nanotube mat, NGF-incorporated magnetic nanotubes were present in the PC12 cell culture. There are two goals for this research: 1) to test the accessibility of NGF from magnetic nanotubes in PC12 cell differentiation into neuron; 2) to study the neurite growth possibly regulated by magnetic nanotubes. For PC12 cell culture, 100 μL of 50 mM MES buffer pH 6.3 was added to magnetic nanotubes. Then 200 ng/ml of NGF was added for incorporation. Any unbounded NGF was removed in the supernatant from the nanotubes by centrifugation. A second wash was done by resuspending the NGF-immobilized magnetic nanotubes in 100 μL of PBS Saline pH 7.4 again by pulse sonication and then removing by centrifugation. As the final step, the nanotubes were resuspended in 100 μL of Neurobasal A by pulse sonication before being laid on sterile glass coverslips which were coated with poly D-lysine (60 μg/ml) and cells were plated on them. PC12 cells were plated on nanotube-laid glass coverslips kept in sterile culture dishes. Culture medium Neurobasal A (GIBCO) containing 1 % penicillin-streptomycin, 2 % B27 supplement, 2 mM of L-Glutamine and glutamax mixture was added to all dishes. After 2-day cell culture, the dishes were fixed with 3 % paraformaldehyde to be examined by microscopes.

Both differentiation of PC12 cells and the neuritic process growth in the presence of functional magnetic nanotubes with incorporated NGF were observed. Figure 7.36a reveals a random distribution of nanotubes on the glass substrate and an appreciable neurite outgrowth. Major components of a growing neuron including soma, neurites and growth cone can be discerned, suggesting that the functional magnetic nanotubes are not toxic for neuron survival and neither do they impair neurite outgrowth. It also confirms that the bioactive nanotube-bound NGF was available to the PC12 cells to induce neuronal differentiation. The enlarged SEM image (inset of Figure 7.36a) shows the growth cone area, located on the tip of the axon, where slender extensions, the filopodia, were formed

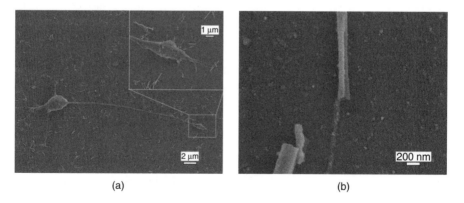

(a)   (b)

Figure 7.36 (a) SEM micrograph of a typical PC12 cell after culture showing the cell body and neurites (the inset is a magnified image of the growth cone area); (b) high-resolution SEM micrograph of a NGF-incorporated magnetic nanotube and its interaction with a filopodium. (Xie *et al.* 2008)

towards the nearby magnetic nanotubes. Figure 7.36b shows a nanotube split longitudi-
nally contacted by the tip of a growing neurite, a filopodium and, it seems, the neuritic
process growing into the nanotube.

According to the microscope observation, a possible neurite growth mechanism was proposed. The growth cone was apt to send filopodia towards the NGF-immobilized magnetic
nanotubes. These filopodia were rather stable, which implies an existence of attractive
guidance cues coming from the NGF-incorporated hematite nanotubes for future growth
cone navigation. These stable filopodia could be selected as the path for the consequent
neurite outgrowth. The remanent magnetization of a magnetic nanotube generates a localized magnetic field, which may affect the process of filopodia extension. Intracellular
recordings indicated that both static and alternating magnetic fields affect the bioelectric
activities of neurons. Due to the structural anisotropy of nanotubes, the remanent magnetization of hematite nanotubes could provide strong magnetic fields and large magnetic field
gradients in the regions quite close to the nanotubes. As hematite nanotubes can get very
close to a neuron, the magnetic fields and the magnetic field gradients produced by the
remanent magnetizations of the magnetic nanotubes will affect the activities of the neuronal elements. Furthermore, due to their ferromagnetic or superparamagnetic properties,
and their shape anisotropies, magnetic nanotubes usually have large magnetic moments.
Therefore strong magnetic forces and torques can be applied on the magnetic nanotubes
by an external magnetic field to activate or modulate the function of neuron. NGF-coated
nanoscale magnetic beads have been reported to promote neurite outgrowth (Naka *et al*.
2004). It is expected that NGF immobilization inside nanotubes will improve the stability
of neurotrophins and the release process could also be better controlled. Since magnetic
nanotubes could be manipulated on certain locations with the help of an external magnetic
field, they provide a potential of confining neurite outgrowth within a selected area or
directing neurite extension to the target place by carefully designing an applied external
magnetic field.

7.5.5 Magnetic Force Microscope

As a common piece of equipment in the research of magnetic nanomaterials and nanodevices, a magnetic force microscope (MFM) is popularly used to observe the magnetic
domains in magnetic films and evaluate the magnetic storage media. Besides its convenience, a MFM is characterized by a high resolution. From an early stage, with pretty low
resolution, the lateral resolution of the MFM has been improved continuously. Typically, a
MFM probe consists of a cantilever with a mounted magnetic tip. The MFM resolution is
determined by the size and shape of the tips and their distance from the sample surface. A
MFM with normal probes (larger than 70 nm in diameter) is able to resolve sub-100 nm
magnetic nanostructures. The fast technological development in the field of magnetic
nanomaterials has demanded an even higher resolution in magnetic domain observation
by MFMs. To achieve this goal, MFM magnetic probes with decreased tip diameter are
required, since the tip diameter is crucial for the tip-sample interaction and consequently
the lateral resolution (Kuramochi *et al*. 2005).

Magnetically coated carbon nanotubes have been used as the magnetic probe in MFMs
with enhanced resolution (Deng *et al*. 2004). For magnetic probe preparation, a wafer-scale
chemical vapor deposition was conducted to deposit carbon nanotubes on commercially
available silicon tips. After an electrical cutting process, the shortened carbon nanotubes
as well as the pyramidal tip and the cantilever were coated with a few angstroms of
Fe by an electron beam metal evaporator. Figure 7.37a is a SEM micrograph showing a
magnetically coated nanotube tip located at the apex of the pyramid on a MFM cantilever.

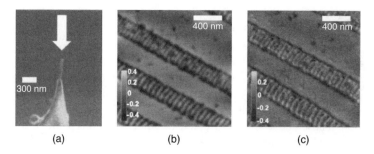

Figure 7.37 (a) Scanning electron microscope micrograph of the apex of the pyramid with a magnetically coated nanotube tip on a MFM cantilever (the arrow: the direction of the metal evaporation); (b) phase image of the tracks in the DC-erased area, taken with a commercial tip; (c) phase image of the tracks taken with the magnetic nanotube tip. (Deng *et al.* 2004)

Custom written tracks on a perpendicular recording mediium were the target for imaging by MFM. To compare the resolutions, a DC-erased was created before writing the tracks. Figure 7.37b and 7.37c are MFM phase images of tracks in the DC-erased area taken with a commercial tip and magnetic nanotube tip, respectively. Obviously, Figure 7.37c shows finer details with a higher spatial resolution. And irregular features smaller than 20 nm can be differentiated with the magnetic nanotube tip indicating about 20 nm lateral resolution for this novel magnetic probe in MFM.

A similar demonstration was done by Kuramochi *et al.* (2005). According to their experiments, CoFe-coated carbon nanotube probes in an MFM showed improved imaging performance. The MFM cantilever with a microfabricated silicon tip was modified by attaching an individual multi-walled carbon nanotube with around 11 nm diameter and 300–400 nm length. Afterwards, CoFe was deposited on the carbon nanotube surface by the radio-frequency sputtering technique. The SEM image of a CoFe coated carbon nanotube probe is shown in Figure 7.38(a). After coating, the overall diameter was about 40 nm indicating about 15 nm in thickness of the CoFe layer. The SEM observation also revealed a smooth and uniform magnetic coating. EDX was performed to detect the elemental composition of the coating layer and the results confirmed the presence of CoFe. For MFM imaging, an environmental controlled scanning probe microscope was used and operated in a magnetic force mode at room temperature. To avoid interferences and to achieve a high sensitivity, the whole imaging process was conducted in vacuum. The sample for MFM imaging was an ultra-high-density perpendicular magnetic storage medium with different densities ranging from 600 to 1100 kFCI. With the magnetic nanotube-attached MFM probe, the optimal working parameters were found as ∼10 nm in amplitude and 3000–10 000 in Q factor. Figure 7.38(b) is a MFM image obtained by using a commercially available magnetic probe. Significant image blurring was observed suggesting a low resolution for this probe. On the other hand, the MFM image obtained by using a magnetic nanotube probe shows much clearer details of the domain structure thus resulting in much higher image resolution, as shown in Figure 7.38(c). The cross-sectional profile along the line indicated in the image is also shown. With regard to the attainable resolution by magnetic nanotube probes, the linear relationship between the normalized intensity and the recording density is illustrated in Figure 7.38(d). The average noise level line of the recorded area was set at an intensity of 0.04, while an intensity of 0.01 was set for the dc-erased area. The noise level in the former came from the whole system under the imaging situation. And in the latter case, the noise level was due to both the magnetic probe and the detection system. From the plot in Figure 7.38(d), it is possible to estimate the lateral resolution limit by extrapolation. The intersects of the fitting line with these

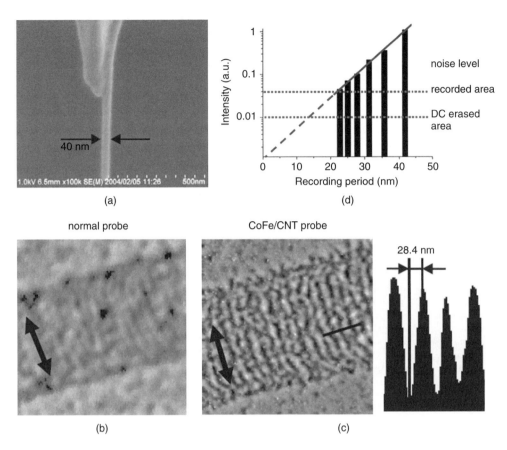

normal probe   CoFe/CNT probe

(b)   (c)

Figure 7.38 (a) A high resolution SEM image of a CoFe coated carbon nanotube attached with the probe; (b) MFM image obtained by a conventional probe ($40\,N\,m^{-1}$); (c) MFM image obtained by a CoFe-coated nanotube probe and a cross-sectional profile corresponding to the line indicated in the image; (d) plot of the normalized intensity of the power spectrum as the function of various recording densities (600–1100 kFCI). (Kuramochi *et al.* 2005)

two noise levels at about 22 and 14 nm correspond to the recording density at 1200 and 1700 kFCI for the perpendicular magnetic storage media, respectively. With conventional MFM probes, 1200 kFCI is the observation limit in the experiments. The attainable resolution is expected to be improved to 1700 kFCI as long as the noise level can be reduced. Consequently, using the ferromagnetic nanotube probes, higher MFM imaging resolution can be achieved which is capable of observing higher recording density in a perpendicular magnetic storage medium. This type of novel magnetic nanotube MFM probe may offer an effective means of determining the exact states of the recorded area, which is of great help in the development of new recording technology.

References

Banerjee, I.A., Yu, L., Shima, M., Yoshino, T., Takeyama, H., Matsunaga, T. and Matsui, H. (2005). Magnetic nanotube fabrication by using bacterial magnetic nanocrystals, *Advanced Materials*, **17**, 1128–31.

Bao, J., Tie, C., Xu, Z., Zhou, Q., Shen, D. and Ma, Q. (2001). Template synthesis of an array of nickel nanotubules and its magnetic behavior, *Advanced Materials*, **13**, 1631–3.

Bao, J., Zhou, Q., Hong, J. and Xu, Z. (2002). Synthesis and magnetic behavior of an array of nickel-filled carbon nanotubes, *Applied Physics Letters*, **81**, 4592–4.

Bao, J., Xu, Z., Hong, J., Ma, X. and Lu, Z. (2004). Fabrication of cobalt nanostructures with different shapes in alumina template, *Scripta Materialia*, **50**, 19–23.

Chen, L.F., Xie, J., Srivatsan, M. and Varadan, V.K. (2006). Magnetic nanotubes and their potential use in neuroscience applications, *Proceedings of the SPIE*, **6172**, 61720J.

Chen, L.F., Xie, J., Yancey, J., Srivatsan, M. and Varadan, V.K. (2007). Experimental investigation of magnetic nanotubes in PC-12 neuron cells culturing, *Proceedings of the SPIE*, **6528**, 65280L.

Daub, M., Knez, M., Goesele, U. and Nielsch, K. (2007). Ferromagnetic nanotubes by atomic layer deposition in anodic alumina membranes, *Journal of Applied Physics*, **101**, 09J111–1-3.

Deng, Z., Yenilmez, E., Leu, J., Hoffman, J.E., Straver, E.W.J., Dai, H. and Moler, K.A. (2004). Metal-coated carbon nanotube tips for magnetic force microscopy, *Applied Physics Letters*, **85**, 6263–5.

Eisenstein, M. (2005). An attractive alternative, *Nature Methods*, **2**, 484.

Escrig, J., Landeros, P., Altbir, D., Vogel, E.E. and Vargas, P. (2007). Phase diagrams of magnetic nanotubes, *Journal of Magnetism and Magnetic Materials*, **308**, 233–7.

Gao, C., Li, W., Morimoto, H., Nagaoka, Y. and Maekawa, T. (2006). Magnetic carbon nanotubes: synthesis by electrostatic self-assembly approach and application in biomanipulations, *Journal of Physical Chemistry B*, **110**, 7213–20.

Geng, F. and Cong, H. (2006). Fe-filled carbon nanotube array with high coercivity, *Physica B*, **382**, 300–4.

de Groot, R.A., Muller, F.M., van Engen, P.G. and Buschow, K.H.J. (1983). New class of materials: half-metallic ferromagnets, *Physical Review Letters*, **50**, 2024–7.

Hu, H., Ni, Y., Montana, V., Haddon, R.C. and Parpura, V. (2004). Chemically functionalized carbon nanotubes as substrates for neuronal growth, *Nano Letters*, **4**, 507–11.

Huang, C., Liu, X., Liu, Y. and Wang, Y. (2006). Room temperature ferromagnetism of Co-doped TiO2 nanotube arrays prepared by sol-gel template synthesis, *Chemical Physical Letters*, **432**, 468–72.

Hulteen, J.C. and Martin, C.R. (1997). A general template-based method for the preparation of nanomaterials, *Journal of Materials Chemistry*, **7**, 1075–87.

Hultgren, A., Tanase, M., Chen, C.S., Meyer, G.J. and Reich, D.H. (2003). Cell manipulation using magnetic nanowires, *Journal of Applied Physics*, **93**, 7554–6.

Jia, C.-J., Sun, L.-D., Yan, Z.-G., You, L.-P., Luo, F., Han, X.-D., Pang, Y.-C., Zhang, Z. and Yan, C.-H. (2005). Single-crystalline iron oxide nanotubes, *Angewandte Chemie International Edition*, **44**, 4328–33.

Kasuga, T., Hiramatsu, M., Hoson, A., Sekino, T. and Niihara, K. (1998). Formation of titanium oxide nanotube, *Langmuir*, **14**, 3160–3.

Korneva, G., Ye, H., Gogotsi, Y., Halverson, D., Friedman, G., Bradley, J.-C. and Kornev, K.G. (2005). Carbon nanotubes loaded with magnetic particles, *Nano Letters*, **5**, 879–84.

Kuramochi, H., Uzumaki, T., Yasutake, M., Tanaka, A., Akinaga, H. and Yokoyama, H. (2005). A magnetic force microscope using CoFe-coated carbon nanotube probes, *Nanotechnology*, **16**, 24–7.

Lee, W., Scholz, R., Nielsch, K. and Gösele, U. (2005). A template-based electrochemical method for the synthesis of multisegmented metallic nanotubes, *Angewandte Chemie International Edition*, **44**, 6050–4.

Leonhardt, A., Ritschel, M., Kozhuharova, R., Graff, A., Mühl, T., Huhle, R.; Mönch, I., Elefant, D. and Schneider, C.M. (2003). Synthesis and properties of filled carbon nanotubes, *Diamond and Related Materials*, **12**, 790–3.

Leonhardt, A., Ritschel, M., Elefant, D., Mattern, N., Biedermann, K., Hampel, S., Müller, Ch., Gemming, T. and Büchner, B. (2005). Enhanced magnetism in Fe-filled carbon nanotubes produced by pyrolysis of ferrocene, *Journal of Applied Physics*, **98**, 074315–1-5.

Li, F.S., Zhou, D., Wang, T., Wang, Y., Song, L.J. and Xu, C.T. (2007). Fabrication and magnetic properties of FeCo alloy nanotube array, *Journal of Applied Physics*, **101**, 014309–1-3.

Liu, B., Wei, L., Ding, Q. and Yao, J. (2005). Synthesis and magnetic study of Fe-doped carbon nanotubes (CNTs), *Journal of Crystal Growth*, **277**, 293–7.

Liu, L., Kou, H.-Z., Mo, W., Liu, H. and Wang, Y. (2006). Surfactant-assisted synthesis of α–Fe$_2$O$_3$ nanotubes and nanorods with shape-dependent magnetic properties, *Journal of Physical Chemistry B*, **110**, 15218–23.

Liu, Z., Zhang, D., Han, S., Li, C., Lei, B., Lu, W., Fang, J. and Zhou, C. (2004). Single crystalline magnetite nanotubes, *Journal of the American Chemical Society*, **127**, 6–7.

Malmsten, M. (2006). Soft drug delivery systems, *Soft Matter*, **2**, 760–9.

Mattson, M.P., Haddon, R.C. and Rao, A.M. (2000). Molecular functionalization of carbon nanotubes and use as substrates for neuronal growth, *Journal of Molecular Neuroscience*, **14**, 175–82.

Mitchell, D.T., Lee, S.B., Trofin, L., Li, N., Bevanen, T.K., Söderlund, H. and Martin, C.R. (2002). Smart nanotubes for bioseparations and biocatalysis, *Journal of the American Chemical Society*, **124**, 11864–5.

Mönch, I., Meye, A., Leonhardt, A., Krämer, K., Kozhuharova, R., Gemming, T., Wirth, M.P. and Büchner, B. (2005). Ferromagnetic filled carbon nanotubes and nanoparticles: synthesis and lipid-mediated delivery into human tumor cells, *Journal of Magnetism and Magnetic Materials*, **290–1**, 276–8.

Mönch, I., Leonhardt, A., Meye, A., Hampel, S., Kozhuharova-Koseva, R., Elefant, D., Wirth, M.P. and Büchner, B. (2007). Synthesis and characteristics of Fe-filled multi-walled carbon nanotubes for biomedical application, *Journal of Physics: Conference Series*, **61**, 820–4.

Naka, Y., Kitazawa, A., Akaishi, Y. and Shimizu N. (2004). Neurite outgrowths of neurons using neurotrophin-coated nanoscale magnetic beads, *Journal of Bioscience and Bioengineering*, **98**, 348–52.

Nielsch, K., Castaño, F.J., Ross, C.A. and Krishnan, R. (2005). Magnetic properties of template-synthesized cobalt/polymer composite nanotubes, *Journal of Applied Physics*, **98**, 034318–1-6.

Nielsch, K., Castaño, F.J., Matthias, S., Lee, W. and Ross, C.A. (2005). Synthesis of cobalt/polymer multilayer nanotubes, *Advanced Engineering Materials*, **7**, 217–21.

Park, D.H., Lee, Y.B., Cho, M.Y., Kim, B.H., Lee, S.H., Hong, Y.K., Joo, J., Cheong, H.C. and Lee, S.R. (2007). Fabrication and magnetic characterizations of hybrid double walled nanotube of ferromagnetic nickel encapsulated conducting polypyrrole, *Applied Physics Letters*, **90**, 093122–1-3.

Peña, D.J., Mbindyo, J.K.N., Carado, A.J., Mallouk, T.E., Keating, C.D., Razavi, B. and Mayer, T.S. (2002). Template growth of photoconductive metal-CdSe-metal nanowires, *Journal of Physical Chemistry B*, **106**, 7458–62.

Rahman, M.M., Kisaku, M., Kishi, T., Matsunaka, D., Dino, W.A., Nakanishi, H. and Kasai, H. (2004). Ab initio study of magnetic and electronic properties of Fe-filled single-walled carbon nanotubes, *Journal of Physics: Condensed Matter*, **16**, S5755–S5758.

Rahman, M.M., Kisaku, M., Kishi, T., Roman, T.A., Dino, W.A.; Nakanishi, H. and Kasai, H. (2005). Electric and magnetic properties of Co-filled carbon nanotubes, *Journal of the Physical Society of Japan*, **74**, 742–5.

Remškar M. (2004). Inorganic nanotubes, *Advanced Materials*, **16**, 1497–504.

Seifert, G., Köhler, T. and Tenne, R. (2002). Stability of Metal Chalcogenide Nanotubes, *Journal of Physical Chemistry B*, **106**, 2497–501.

Shen, X.-P., Liu, H.-J., Pan, L., Chen, K.-M., Hong, J.-M. and Xu, Z. (2004). An efficient template pathway to synthesis of ordered metal oxide nanotube arrays using metal acetylacetonates as single-source molecular precursors, *Chemistry Letters*, **33**, 1128–9.

Singh, A.K., Kumar, V. and Kawazoe, Y. (2004). Metal encapsulated nanotubes of silicon and germanium, *Journal of Materials Chemistry*, **14**, 555–63.

Son, S.J., Reichel, J., He, B., Schuchman, M. and Lee, S.B. (2005). Magnetic nanotubes for magnetic-field-assisted bioseparation, biointeraction, and drug delivery, *Journal of the American Chemical Society*, **127**, 7316–17.

Steinhart, M., Wehrspohn, R.B., Gösele, U. and Wendorff, J.H. (2004). Nanotubes by template wetting: a modular assembly system, *Angewandte Chemie International Edition*, **43**, 1334–44.

Su, Y.-C. and Hsu, W.-K. (2005). Fe-encapsulated carbon nanotubes: nanoelectromagnetics, *Applied Physics Letters*, **87**, 233112–1-.

Sui, Y.C., Skomski, R., Sorge, K.D. and Sellmyer, D.J. (2004a). Magnetic nanotubes produced by hydrogen reduction, *Journal of Applied Physics*, **95**, 7151–3.

Sui, Y.C., Skomski, R., Sorge, K.D. and Sellmyer, D.J. (2004b). Nanotube magnetism, *Applied Physics Letters*, **84**, 1525–7.

Tao, F., Guan, M., Jiang, Y., Zhu, J., Xu, Z. and Xue, Z. (2006). An easy way to construct an ordered array of nickel nanotubes: the triblock-copolymer-assisted hard-template method, *Advanced Materials*, **18**, 2161–4.

Tourillon, G., Pontonnier, L., Levy, J.P. and Langlais, V. (2000). Electrochemically synthesized Co and Fe nanowires and nanotubes, *Electrochemical and Solid-State Letters*, **3**, 20–3.

Xie, J., Chen, L., Varadan, V.K., Yancey, J. and Srivatsan, M. (2008). The effects of functional magnetic nanotubes with incorporated nerve growth factor in neuronal differentiation of PC12 cells, *Nanotechnology*, **19**, 105101–1-7.

Xue, S., Cao, C., Wang, D. and Zhu, H. (2005). Synthesis and magnetic properties of $Fe_{0.32}Ni_{0.68}$ alloy nanotubes, *Nanotechnology*, **16**, 1495–9.

Wang, R.M., Liu, C.M., Zhang, H.Z., Chen, C.P., Guo, L., Xu, H.B. and Yang, S.H. (2004). Porous nanotubes of Co_3O_4: synthesis, characterization, and magnetic properties, *Applied Physics Letters*, **85**, 2080–2.

Wang, T., Wang, Y., Li, F., Xu, C. and Zhou, D. (2006). Morphology and magnetic behavior of an Fe_3O_4 nanotube array, *Journal of Physics: Condensed Matters*, **18**, 10545–51.

Wu, D., Chen, Y., Liu, J., Zhao, X., Li, A. and Ming, N. (2005). Co-doped titanate nanotubes, *Applied Physics Letters*, **87**, 112501–1-3.

8

Magnetic Biosensors

8.1 Introduction

8.1.1 Biosensors and Magnetic Biosensors

Due to their extensive applications, biosensors are under intense investigation and development (Rife *et al.* 2003; Baselt *et al.* 1997; Chung *et al.* 2004; Could 2004; Jaffrezic-Renault *et al.* 2007; Whitesides and Wong 2006). Biosensors usually utilize biological reactions for detecting target analytes. As shown in Figure 8.1, a biosensor is an analytical device mainly consisting of a biological recognition element (or bioreceptor) and a signal transducer (Turner 2000; Wang 2000). When the analyte interacts with the bioreceptor, the resulting complex produces a change which can be converted into a measurable effect by the transducer. Common types of bioreceptor/analyte complexes are based on antibody–antigen interactions, nucleic acid interactions, enzymatic interactions, cellular interactions or the interactions using biomimetic materials. The most prevalent signal transduction methods include optical measurements, electrochemical and mass-sensitive measurements (Vo-Dinh *et al.* 2001). For practical applications, biosensors should have high sensitivity, small size, low power consumption, stability of operation parameters, quick response, resistance to aggressive medium and low price (Kurlyandskaya and Levit 2005; Kriz *et al.* 1998; Lagae *et al.* 2005; Prinz 1998).

Usually a magnetic biosensor detects the stray magnetic fields of magnetic nanoparticles or magnetic beads tagged to targets. Biosensing that utilizes magnetic micro- or nanoparticles as labels of specific biomolecular interactions is a particularly fruitful area of research (Landry *et al.* 2004). The idea of using a magnetic field sensor in combination with magnetic particles working as magnetic labels for detecting molecular recognition events was first reported by Baselt *et al.* (1998). Magnetic labels have obvious advantages in biosensing (Rife *et al.* 2003; Landry *et al.* 2004; Lany *et al.* 2005; Chung *et al.* 2005; Arakaki *et al.* 2004; Babb *et al.* 1995). First, the size of magnetic nanoparticles can range from a few nanometers to several micrometers and thus is compatible with biological entities ranging from proteins (a few nm) to cells and bacteria (several µm). By coating the magnetic nanoparticles with specific ligands, the nanoparticles can selectively bind to a target material of interest. Second, the magnetic properties of nanoparticles are very stable, and they are not affected by reagent chemistry or subject to photo-bleaching. Third, magnetic fields are not screened by aqueous reagents or biomaterials. Since most biological systems do not exhibit ferromagnetism and typically have only small magnetic

Nanomedicine: Design and Applications of Magnetic Nanomaterials, Nanosensors and Nanosystems
V. Varadan, L.F. Chen, J. Xie
© 2008 John Wiley & Sons, Ltd

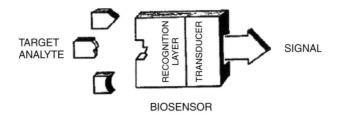

Figure 8.1   Basic structure of a biosensor. (Wang 2000)

susceptibilities revealing dia- or paramagnetism, the magnetic moment from a magnetic nanoparticle can be detected with little noise from the biological environment. Fourth, magnetic nanoparticles can be remotely manipulated using field gradients even when they are embedded in a biological environment. Finally, there are many highly sensitive magnetic field detection devices that have been developed and that could be used for biosensing applications, such as giant magnetoresistive (GMR) sensors, anisotropic magnetoresistive (AMR) sensors, inductive sensors, miniature Hall crosses and superconducting quantum interference devices (SQUIDs).

However, it should be noted that the development of magnetic biosensors is at an early stage and many challenges remain. In this chapter, we concentrate on discussing different types of magnetic transducers, and sometimes we do not distinguish magnetic biosensors and magnetic transducers. The discussion about the target recognition by the recognition layer could be found in the sections on surface functionalizations in the chapters discussing nanospheres, nanowires and nanotubes.

8.1.2 Magnetic Biosensing Schemes

Most magnetic particle labels in biomedical applications are in micron or submicron size. In order to achieve single molecule detection, the dimension of magnetic particle labels should be comparable to that of biomolecules (Li *et al.* 2006). In the case of detecting DNA fragments, it is ideal to have the particle labels at 20 nm or smaller in diameter. Such small nanoparticles would not block bimolecular interactions. Moreover, one nanoparticle label may be conjugated with one or at most a few DNA fragments, which is important for establishing a quantitative relationship with sufficient accuracy between the number of captured particle labels and the actual biorecognition events. However, it is difficult to do so with microbeads because of their large size mismatch with biomolecules. The monodispersity of crystalline magnetic nanoparticles in both size and magnetic moment also benefits the signal quantification, in contrast to the large variations of the microbeads in magnetic moment. Given all these desirable properties, magnetic nanoparticles with a diameter of 20 nm or smaller become desirable biomolecular labels for ultrasensitive, highly quantitative magnetic biodetection technology. On the other hand, such tiny magnetic nanoparticles are a great challenge to the detectors, because their magnetic moments are very low due to their limited physical volume, relatively large surface area and significant thermal disturbance to magnetic moments, i.e., superparamagnetism (Li *et al.* 2006).

There are two general sensing schemes utilizing magnetic nanoparticles: one is the substrate-free approach which is based on the change in the Brownian relaxation time of the particles suspended in liquids upon their binding to the target molecules, and the other approach detects the stray magnetic fields of the magnetic particles upon their binding to the biomolecule-functionalized surface of a solid-state sensor (Mihajlovic *et al.* 2005).

8.1.2.1 Solid-state Based Sensors

In a solid-state based sensor, the analyte to be detected is usually attached to a substrate. Two schemes are often used: the direct detection scheme and secondary detection scheme. Based on the sensitivity of the sensor, the size of a sensor can be tailored according to the size of the particles required for a particular bioassay, enabling effective detection of a very low number of particles. Provided that each particle can label only one target molecule, detection of even a single nanometer-size particle is possible, which can lead to the ultimate detection limit of a single biomolecule.

(1) Direct Detection Scheme

As shown in Figure 8.2, in a direct detection scheme, an array of probe molecules is immobilized over a magnetic field sensor, and the biomolecules (target analytes) to be detected are magnetically labeled and pass over the probe array. The target analytes are bound to the array, and the unbound biomolecules are then washed away (Graham *et al.* 2004). A common strategy is to coat both the magnetic nanoparticle and a magnetic field sensor with ligands such that both have an affinity for the desired target. In this manner the target material acts as a link between the nanoparticles and the magnetic field sensor, and the presence of the target is confirmed by detecting the stray fields from the nanoparticles (Chung *et al.* 2005).

(2) Secondary Detection Scheme

This scheme is based on a secondary detection step, performed after the interrogation of the probe array with the target molecules (Graham *et al.* 2004). As shown in Figure 8.3,

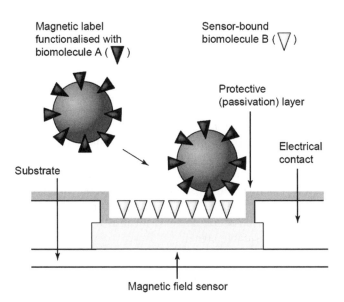

Figure 8.2 Schematic illustration of the direct detection scheme. The magnetic label, functionalized with biomolecule A, interacts with biomolecule B which is bound to the magnetic field sensor. The stray magnetic field produced by the magnetic label causes a signal in the magnetic field sensor. After washing away the unbound biomolecules, the signal from the magnetic field sensor indicates the molecular recognition between biomolecules A and B. (Graham *et al.* 2004)

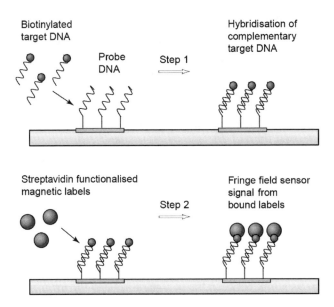

Figure 8.3   Schematic illustration of the secondary detection scheme. In step one, the biotinylated target DNA is hybridized with the probe DNA immobilized over a magnetic sensor. In step two, the hybridized DNA is detected by binding the magnetically labeled streptavidin to the biotinylated hybridized target DNA. The fringe magnetic field of the magnetic labels is detected by the magnetic sensor. (Graham *et al.* 2004)

this scheme consists of two steps. In the first step, a small biochemical label, such as biotin, is tagged to the target molecules, and the tagged target molecules are bound to complementary probe molecules. In the second step, magnetic labels functionalized with streptavidin, which is complementary to biotin, are introduced, and the magnetic fields produced by the magnetic labels are detected.

8.1.2.2   Substrate-free Sensors

In a substrate-free sensor, the target analytes labeled by magnetic particles are not immobilized to a substrate; instead, they are suspended in a liquid, forming a ferrofluid. The detection of the magnetic particles, or the target analytes, is deduced from the properties of the ferrofluid, such as relaxation time and magnetic susceptibility. This scheme offers an excellent method for distinguishing between multiple potential target molecules in the solution based on their size difference, but is limited in terms of the size of the magnetic particles that can be used for detection and also requires relatively high concentration of magnetic particles for a measurable signal.

8.1.3   Magnetic Properties of a Magnetic Bead

The microbeads used in biomedical applications have a diameter of several micrometers, and usually consist of thousands of nanometer-sized maghemite or magnetite particles dispersed in a polymer matrix. These microbeads are specially functionalized so that they can selectively attach to the desired targets, such as bacteria, cells and nucleic acids (Besse *et al.* 2002). The essence of magnetic biosensing is the detection of the presence of beads selectively attached on a surface.

The use of magnetic beads as labels in biosensing applications requires detailed knowledge of the magnetic properties of the beads. The magnetic properties often used in magnetic biosensors can be generally classified into three types: magnetic fields, magnetic relaxations and magnetic susceptibility.

8.1.3.1 Stray Magnetic Fields

The magnetization $\mathbf{M}$ of a magnetic bead consisting of superparamagnetic nanoparticles dispersed in a nonmagnetic matrix can be expressed by (Besse *et al.* 2002):

$$\mathbf{M(H)} = \mathbf{M}_{sat} \cdot \left[\coth \left(\frac{\mu_0 m_p H}{kT} \right) - \left(\frac{kT}{\mu_0 m_p H} \right) \right] \cdot \frac{\mathbf{H}}{H} \qquad (8.1)$$

where $\mathbf{H}$ is the total magnetic field in the bead, $\mathbf{M}_{sat}$ is the magnetization of the bead at saturation, k is the Boltzmann constant, T is the temperature, μ_0 is the permeability of a vacuum and m_p is the average magnetic moment of each nanoparticle. The beads are magnetized when a magnetic field is applied, and they lose their magnetization immediately after the magnetic field is removed. In a homogeneous sphere, the magnetic field is expressed as $\mathbf{H} = \mathbf{H}_e - \mathbf{M}/3$, where $\mathbf{H}_e$ is the field applied externally. At the position $\mathbf{r}$ (measured from the center of the bead), the magnetic induction is (Besse *et al.* 2002):

$$\mathbf{B(r)} = \mu_0 \mathbf{H}_e + \frac{\mu_0}{4\pi} \cdot \frac{3(\mathbf{r} \cdot \mathbf{M}V)\mathbf{r} - (\mathbf{r} \cdot \mathbf{r})\mathbf{M}V}{r^5} \qquad (8.2)$$

where the second term represents the magnetic induction generated by a bead of volume V.

Usually solid-state based sensors detect the stray fields of magnetic markers. The magnetic markers, being superparamagnetic, have to be magnetized in order to produce a sensor signal. We assume that an external magnetic field is applied along the z direction. As shown in Figure 8.4, for a bead separated from a sensor by an overlayer with thickness t, the magnetic field B_x at a distance d along the x-axis relative to the center of the bead is (Rife *et al.* 2003):

$$B_x = \mu_0 M \frac{a^3(a+t)d}{[(a+t)^2 + d^2]^{5/2}} \qquad (8.3)$$

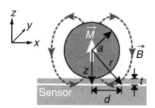

Figure 8.4 Cross-section of a microbead with radius a on top of a sensor separated by an overlayer with thickness t. The bead is under an external magnetic field in the direction of $\mathbf{M}$. The magnetic field induced by the bead is indicated by dashed lines and arrows. (Rife *et al.* 2003). Reprinted from: Rife, J. C. Miller, M. M. Sheehan, P. E., Tamanaha, C. R. Tondra, M. Whitman, L. J. (2003). "Design and performance of GMR sensors for the detection of magnetic microbeads in biosensors", *Sensors and Actuators A*, **107**, 209–218., with permission from Elsevier

where a and **M** are the radius and magnetization of the bead respectively. At distance $d = (a + t)/2$, this field reaches its maximum value:

$$B_x^0 = 3.6 \left(\frac{\mu_0}{4\pi}\right) \frac{M}{[1 + (t/a)]^3} \tag{8.4}$$

The normal component field B_z is given by (Landry *et al.* 2004):

$$B_z(z, \text{d}) = \frac{M(2z^2 - d^2)}{(z^2 + d^2)^{5/2}} \tag{8.5}$$

where z is measured from the plane of the Hall carriers, and d is a transverse dimension, $d^2 = x^2 + y^2$, relative to the center of the cross.

8.1.3.2 Brownian and Neel Relaxations

After the magnetizing field is removed, the magnetic moment of a particle suspended in a fluid may relax in two different mechanisms (Connolly and St Pierre 2001). One is the Brownian motion which involves the bulk rotation of the particle within the fluid. This mechanism applies mainly to magnetically blocked particles, whose magnetic moments are fixed relative to the crystal axes of the particles. The relaxation time for a Brownian rotation is given by:

$$\tau_B = \frac{4\pi \eta r^3}{kT} \tag{8.6}$$

where r is the hydrodynamic radius of the particle, η is the dynamic viscosity of the fluid, k is Boltzmann constant and T is the absolute temperature. The other mechanism is Neel relaxation, which applies mainly to unblocked particles. For an unblocked particle in Neel relaxation, its magnetic moment vector rotates, while the particle remains stationary.

Generally speaking, there exists a critical volume above which the Brownian relaxation mechanism becomes dominant. For an iron oxide sphere at room temperature, its typical critical diameter is about 25 nm (Connolly and St Pierre 2001). Detailed discussion about Brownian and Neel relaxations can be found in Chapter 2.

8.1.3.3 Susceptibility of a Ferrofluid

The Debye's theory, which was originally developed for analyzing the dielectric dispersion of dipolar fluids, can be used to model the response of a dilute suspension of spherical magnetic particles to an alternating magnetic field (Connolly and St Pierre 2001). Due to the finite rate of change of magnetization with time, the magnetization of the fluid lags behind the application of a magnetic field. For a weak magnetic field, the magnetization of an object is linearly dependent on the magnetic field. In this case, the response of a magnetized object to an alternating magnetic field can be described by the complex magnetic susceptibility, $\chi = \chi' + i\chi''$. The real part χ' is the in-phase component of the magnetic susceptibility, related to the storage of the magnetic energy; while the imaginary part χ'' is the quadrature component, related to dissipation of magnetic energy.

According to Debye's theory, the spectrum of the complex magnetic susceptibility of the fluid is given by:

$$\chi'(\omega) = \frac{\chi_0}{1 + (\omega\tau)^2} \tag{8.7}$$

$$\chi''(\omega) = \frac{\chi_0\omega\tau}{1 + (\omega\tau)^2} \tag{8.8}$$

where χ_0 is the magnetic susceptibility of the fluid at $\omega = 0$. The real part $\chi'(\omega)$ decreases monotonically with the increase of frequency, whereas the imaginary part $\chi''(\omega)$ reaches its maximum at $\omega\tau = 1$.

8.1.4   Typical Types of Magnetic Biosensors

The magnetic properties discussed in Section 8.1.3 can be used in the development of magnetic biosensors. Correspondingly, the magnetic biosensors generally fall into three categories: biosensors based on stray magnetic fields, biosensors based on relaxation and biosensors based on susceptibility.

8.1.4.1   Detection of Stray Magnetic Field

The working principle of most magnetic sensors based on stray magnetic fields is the galvanomagnetic effects due to the Lorentz force on moving electrons in a metal, semi-conductor or an insulator in one way or another (Mahdi *et al.* 2003). Three types of galvanomagnetic effects are widely used in magnetic sensors: the Hall effect, magnetoresistance and flux quantization in superconducting loop. The Hall effect produces an electric field perpendicular to the magnetic induction vector and the original current direction, and the magnetoresistive effect is the change of electrical resistivity due to exposure to the magnetic field. SQUID is based on the quantum mechanical galvanomagnetic effects occurring in Josephson junctions between superconducting materials (Mahdi *et al.* 2003).

Galvanomagnetic effects occur when a material carrying an electric current is exposed to a magnetic field. They are manifestations of the charge-carrier transport phenomena occurring in the matter when the carriers are subject to the action of the Lorentz force $\mathbf{F}$:

$$\mathbf{F} = e\mathbf{E} + e[\mathbf{v} \times \mathbf{B}] \tag{8.9}$$

where e is the carrier charge (for electrons $e = -q$, and for holes $e = q$ and $q = 1.6 \times 10^{-19}$ C), $\mathbf{E}$ is the electric field, $\mathbf{v}$ the carrier velocity and $\mathbf{B}$ the magnetic induction. The first term on the right-hand side of Equation (8.8) represents a Coulomb force and the second term is the Lorentz force law. For non-degenerate semiconductors exposed to transverse electrical and magnetic fields ($\mathbf{E} \cdot \mathbf{B} = 0$), the current transport equation for one type of carrier is:

$$\mathbf{J} = \mathbf{J}_0 + \mu_H[\mathbf{J}_0 \times \mathbf{B}] \tag{8.10}$$

where $\mathbf{J}$ is the total current density and the term $\mathbf{J}_0$ is the current density due to only the electric field and carrier-concentration gradient ∇n:

$$\mathbf{J}_0 = \sigma\mathbf{E} - eD\nabla n \tag{8.11}$$

It should be noted that $\mathbf{J}_0 \neq \mathbf{J}$ ($\mathbf{B} = 0$), since a magnetic field generally influences the electric potential and carrier-concentration distributions. The transport coefficients μ_H (the Hall mobility which has the sign of the corresponding charge-carrier), σ (the conductivity) and D (the diffusion coefficient) are determined by the carrier scattering processes and generally depend on electric and magnetic fields. Both the Hall and the magnetoresistive effects can be derived from the solutions of Equation (8.10) subject to the appropriate boundary conditions (Mahdi *et al.* 2003).

(1) Hall Effects

To understand the Hall effect in semiconductors, consider a special case of carrier transport in a very long strip with narrow cross-section of a strongly extrinsic and homogeneous ($\nabla n = 0$) semiconductor material, subjected to a magnetic field described by a magnetic flux density $\mathbf{B} = (0, B_y, 0)$. As shown in Figure 8.5, the strip axis is along the x-axis, and the strip plane is the xz-plane. If the strip is exposed to an external electric field $\mathbf{E}_{ex} = (E_x, 0, 0)$, a current $\mathbf{I}$ flows through it with current density $\mathbf{J} = (J_x, 0, 0)$. Since $J_z = 0$, an internal transverse electric field $\mathbf{E}_H$ must build up to counteract the magnetic part of Lorentz force, the second term in Equation (8.9). The field $\mathbf{E}_H$ is known as the Hall field and can be determined from Equation (8.10) by substituting $\mathbf{E} = \mathbf{E}_{ex} + \mathbf{E}_H$, under the condition that the transverse current density vanishes, i.e. $\mathbf{E}_H = (0, 0, E_z)$ and $E_z = -\mu_H B E_x$ (Mahdi *et al.* 2003).

A macroscopic consequence of the Hall field is the appearance of a measurable transverse voltage, known as the Hall voltage, V_H, which is proportional to the magnetic field:

$$V_H = R_H J_x B w \tag{8.12}$$

where w denotes the strip width, R_H denotes the Hall coefficient given by:

$$R_H = \mu_H / \sigma = r / (e \cdot n) \tag{8.13}$$

where r is the Hall scattering factor of carriers and n is the carrier density, and the negative sign is omitted.

The presence of the Hall field also results in the inclination of the total electric field in the sample, with respect to the external field, by the Hall angle θ_H (Mahdi *et al.* 2003):

$$\tan \theta_H = E_z / E_x = -\mu_H B \tag{8.14}$$

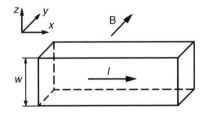

Figure 8.5   A material carrying an electric current is exposed to an external magnetic field

(2) Magnetoresistance Effect

The magnetoresistive effect can be understood based on the assumption of zero Hall electric field ($E_z = 0$), and this condition can be approximately realized by considering a short strip with wide cross-section (Mahdi *et al.* 2003). As shown in Figure 8.5, when an electric field $E_{ex} = (E_x, 0, 0)$ is applied, a lateral component J_z, will be produced, resulting in a rotation of the current lines:

$$\frac{J_z}{J_x} = \mu_H B = \tan\theta_L \tag{8.15}$$

where θ_L is the carrier deflection angle. The longer carrier drift path caused by this deflection leads to the transversal geometric magnetoresistance effect:

$$\frac{\rho_B - \rho_0}{\rho_0} = (\mu_H B)^2 \tag{8.16}$$

where ρ_0 is the electrical resistivity at $B = 0$, and ρ_B is the resistivity enhanced by the magnetic field. Equation (8.16) shows that the relative change in resistivity increases with the square of the mobility-induction product, and the sensors based on this effect require high mobility.

For ferromagnetic transition metals and alloys, there exists anisotropic magnetoresistance which is a stronger physical effect than the geometrical magnetoresistance effect discussed above. In these materials, the magnetization vector determines the direction along which the current normally flows. The application of an external magnetic field rotates the magnetization vector in the sample, and thus the current path, by an angle θ. The specific resistivity of the sample $\rho(\theta)$, is given by:

$$\rho(\theta) = \rho_\perp + (\rho_\| - \rho_\perp)\cos^2\theta = \rho_\perp + \Delta\rho\cos^2\theta \tag{8.17}$$

where $\rho_\|$ is the resistivity of the sample when $\theta = 0$, $\rho_\perp$ is the resistivity of the sample when $\theta = 90°$, θ being the angle between the internal magnetization and the direction of the current. The quotient $\Delta\rho/\rho_\perp$ is called the anisotropic magnetoresistance effect, and usually the resistivity ratio $\Delta\rho/\rho_0$ is used. The anisotropic magnetoresistance effect can be easily realized using thin film technology and, hence lends itself to sensor applications (Mahdi *et al.* 2003).

(3) Flux Quantization in Superconducting Loop

The principle of measuring a magnetic field using superconducting materials is based on the Meissner effect and flux quantization which result in the constancy of the magnetic flux through a superconducting closed loop (Mahdi *et al.* 2003). If such a superconducting loop is placed in an external field H_{ex}, a shielding current, known as the supercurrent I_s, will circulate around the inner surface of the ring such that the total magnetic flux, Φ_i, inside the ring is quantized:

$$\Phi_i = LI_s + \Phi_{ex} = m\Phi_0 \tag{8.18}$$

where $\Phi_0 = 2.07 \times 10^{-15}$ Wb is the flux quantum, m is an integer, L is self inductance of the loop around which the supercurrent flows, and Φ_{ex} is the flux of the external field. Equation (8.18) indicates that the superconducting ring responds to any change in external flux by setting up an equal but opposite flux. Provided that the supercurrent value does not exceed the critical current density, I_c, for as long as the superconducting specimen remains superconducting, Φ_i will remain constant, and quantized, at the same value. This behavior, coupled with the Josephson effects, provides the operational basis of the SQUID.

Equations (8.10)–(8.18) represent some physical principles for the development of magnetic sensors detecting the stray magnetic fields of magnetic nanoparticles (Mahdi et al. 2003). Sensors in this category mainly include the Hall effect sensors, the GMR sensors and the SQUID magnetometers.

8.1.4.2   Detection of Magnetic Relaxations

The magnetic relaxations of ferrofluids can be used in sensing the existence of magnetic particles in the ferrofluids. As discussed earlier, the magnetic relaxations of ferrofluids generally fall into two categories: Neel relaxation and Brownian relaxation. As will be discussed later, the relaxations of ferrofluids can be detected by fluxgate and SQUID.

8.1.4.3   Detection of Ferrofluid Susceptibility

The existence of magnetic particles in ferrofluids can also be determined by measuring magnetic susceptibility. In magnetic sensing, usually the frequency at which $\chi''(\omega)$ reaches its maximum is measured experimentally, and the hydrodynamic radius of blocked particles can be derived based on the relationship $\omega\tau = 1$ and Equation (2.23) (Connolly and St Pierre 2001). Generally speaking, when biological macromolecules are bound to a magnetic particle, the hydrodynamic radius of the particle is increased, and the frequency of the maximum of $\chi''(\omega)$ is thus decreased.

8.1.5   Sensor Sensitivity and Dynamic Range

The sensitivity and dynamic range of a biosensor can be expressed magnetically or biologically (Graham et al. 2004). The magnetic sensitivity of a biosensor is defined as the minimum magnetic moment this sensor can detect. For a particular magnetic composition, the magnetic sensitivity can also be defined as the smallest magnetic label the sensor can detect. While the biological sensitivity of a biosensor is defined as the smallest number of target biomolecular interactions the biosensor can detect, or the lowest biomolecule concentration required for producing a signal that can be detected by the sensor. Usually, the dynamic range of a sensor is the range of the number of the labels this sensor can detect, or the concentration range of the target biomolecules this sensor can quantitatively distinguish.

Besides the inherent magnetic sensitivity and dynamic range of a sensor, the signal obtained from a sensor also depends on many physical parameters, such as the magnetic moment of the label, the size ratio of label vs sensor, the distance between the label and the sensing layer, the electric current used for sensing and whether signal amplification techniques are used (Graham et al. 2004). The moment of the magnetic label is related to the magnetic composition and content of the label; it is also related to the strength and direction of the external magnetization field. Generally speaking, the magnetic moment of a label increases with the increase of the magnetization field until the field becomes saturating. After saturation is technically achieved, the moment no longer changes with the increase of the magnetic field.

8.2   Magnetoresistance-based Sensors

Magnetoresistance (MR) is the change of the electrical resistance of a material due to an external magnetic field. Though the MR effect was first discovered in the 1850s, its widespread technological applications were not realized until the late twentieth century, due to the advances in solid-state technology. Magnetoresistance-based sensors provide a highly sensitive sensing technology with wide dynamic range. In principle, a single molecular interaction can be detected using this technology (Graham *et al.* 2004). Typical MR sensors include large GMR sensors, anisotropic magnetoresistance ring (AMR) sensors, spin valves and magnetic tunnel junction sensors.

8.2.1   Giant Magnetoresistance Sensor

Giant magnetoresistance (GMR)-based magnetic biodetection technology has received increasing research and development efforts, because the GMR biosensors are promising for sensitive, large-scale, inexpensive and portable biomolecular identification. Moreover, given the nature of the solid-state thin film sensors and the IC compatible fabrication, the GMR biosensors can be integrated into a very high density, and hence such a GMR sensor array will be well suited for multi-analyte biodetection (Li *et al.* 2006).

A basic GMR structure mainly consists of a pair of magnetic thin films separated by a non-magnetic conducting layer. When the magnetizations of the magnetic layers are aligned by an external magnetic field, the electrical resistance of the structure decreases due to the reduction of spin-dependent electron scattering within the structure. A GMR sensor can be of microscopic size, and can sensitively detect the presence of a magnetic particle with micron or smaller size in close proximity (Rife *et al.* 2003).

Rife *et al.* (2003) developed a biosensor system, the Bead ARray Counter (BARC). In a BARC system, magnetic microbeads are captured and detected on a chip containing GMR sensors. As illustrated in Figure 8.6, a BARC sensor is made from a multilayer GMR film with a large saturation field and GMR effect. The GMR film consists of four ferromagnetic layers separated by three non-ferromagnetic layers. Each ferromagnetic layer has three sub-layers, consisting of a layer of NiFeCo, sandwiched between two thin films of CoFe. The thicknesses of the films are optimized to ensure anti-parallel exchange coupling across the CuAgAu layers and meanwhile maintaining the high sensitivity and linearity. Due to the shape anisotropy, the magnetization of each GMR trace naturally lies in the plane of the film, therefore only the planar components of the induced microbead field cause appreciable magnetoresistance (Rife *et al.* 2003).

To detect magnetic microbeads using a BARC system, an AC magnetic field, H_z^0, is applied normal to the chip (the z direction). As schematically shown in Figure 8.6(a), when a single bead resting above the GMR sensor is magnetized by an external magnetic field, it will generate a local dipole field, B, whose planar components are strong enough to cause a magnetoresistance effect. The overall GMR signal, $\Delta R/R$, is mainly determined by the sensor geometry and the cumulative local magnetoresistance changes associated with individual microbeads. Usually a Wheatstone bridge is used in the measurement of the GRM signal, $\Delta R/R$. As shown in Figure 8.7, the changes of resistance due to individual beads are independent, and additive until a saturation level is achieved.

Wood *et al.* (2005) proposed a technique for characterizing the response of GMR sensors, using a scanned magnetic probe that generates a highly localized magnetic field with a spatial distribution and field strength similar to that expected in biomagnetic applications. The sensors were fabricated from a GMR multilayer structure. As shown in Figure 8.8, the bulk material had a GMR response of 13 % at the saturation field of 300 Oe, and the

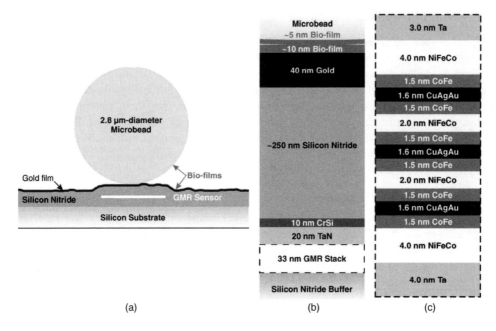

Figure 8.6 Illustration of a BARC sensor. (a) The arrangement of a bead and a sensor; (b) the multilayer structure of the sensor; (c) detailed structure of the GMR stack. (Rife *et al.* 2003). Reprinted from: Rife, J. C. Miller, M. M. Sheehan, P. E., Tamanaha, C. R. Tondra, M. Whitman, L. J. (2003). "Design and performance of GMR sensors for the detection of magnetic microbeads in biosensors", *Sensors and Actuators A*, **107**, 209–218., with permission from Elsevier

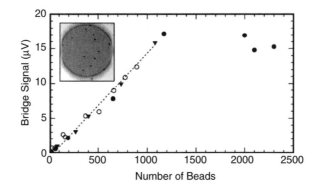

Figure 8.7 The relationship between the Wheatstone bridge signal and the number of beads deposited onto a BARC sensor. The open circles, filled circles and filled triangles indicate the measurement data of three different chips. The dashed line is a linear fit to the triangles corresponding to 15 nV per bead. The inset is an optical micrograph sensor with 14 beads. (Rife *et al.* 2003). Reprinted from: Rife, J. C. Miller, M. M. Sheehan, P. E., Tamanaha, C. R. Tondra, M. Whitman, L. J. (2003). "Design and performance of GMR sensors for the detection of magnetic microbeads in biosensors", *Sensors and Actuators A*, **107**, 209–218., with permission from Elsevier

sensors are single straight wires etched from the bulk GMR material. The four sensors were all 2 μm in length, with widths of 100, 150, 250 and 400 nm, and resistances in the anti-parallel magnetic states of 14.4 kΩ, 1.62 kΩ, 501 Ω and 417 Ω, respectively. All the sensors showed a strong GMR response.

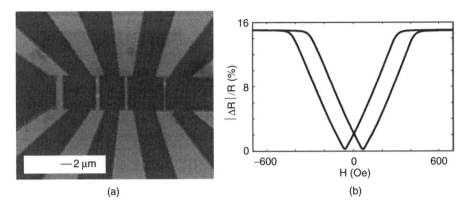

(a)   (b)

Figure 8.8 (a) Electron micrograph of four sensors. The active region comprises the narrow wires in the center; (b) response of the bulk material at 300 K. The maximum $|\Delta R|/R$ was 13 %, which corresponds to an overall decrease in resistance at higher fields, and the saturation field was 300 Oe. (Wood *et al.* 2005)

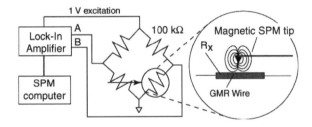

Figure 8.9 Bridge circuit integrated with SPM hardware to allow measurement of GMR response as a function of tip position. (Wood *et al.* 2005)

To simulate the presence of a magnetically tagged biomolecular analyte in the vicinity of the GMR sensor, a magnetized scanning probe microscope (SPM) tip was scanned over the surface of the patterned sensor while monitoring the sensor's electrical response. In the far field region, magnetic SPM tips do not behave as magnetic dipoles because of extended magnetic material. On the size scale of the sensors, however, the peak responses are only recorded for near field region, where the dipole fit is nearly exact. The fields from a dipole of this size are comparable to those from biomagnetic beads commonly used in biomagnetic applications (Wood *et al.* 2005). As shown in Figure 8.9, resistance measurements were made using a balanced Wheatstone bridge, and the bridge error signal was read out using a lock-in amplifier.

The line scans shown in Figure 8.10 were taken with the tip scanning perpendicular to the sensor long axis, near the device center line. The peak fractional resistance changes for the 400, 250 and 150 nm devices were 0.25, 0.31 and 0.15 %, respectively. The spatial resolution is seen to be a small fraction of the linewidth, showing the highly local nature of the magnetic response. The scans display the strong response of the sensors to the tip field as the tip moved over the device. In addition, a hysteretic response is visible as a shift of the curve maximum as a function of tip position between the scan trace and retrace. The significantly smaller response observed for the 100 nm wide line may be due to significant etch-induced thinning of the GMR structure (Wood *et al.* 2005).

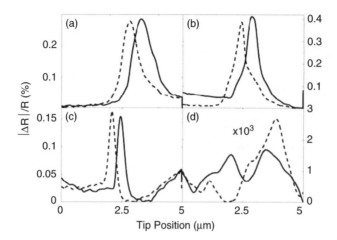

Figure 8.10 Resistance change for magnetic SPM tip trace (solid line) and retrace (dashed line). Data are shown for (a) 400 nm, (b) 250 nm, (c) 150 nm, and (d) 100 nm devices; note scale change for the last. The peaks are close to the geometric center of the etched lines. Line scans were taken near the midpoint of the sensor long axis. (Wood *et al.* 2005)

The sensitivity of a GMR sensor increases with decreasing sensor volume, while the number of biomolecules bound to the sensor surface, and thus the available signal, increases with the increase of the surface area. Thus a large array of small, tightly spaced sensors optimizes these two quantities for maximum sensitivity, as shown in Figure 8.11. The high sensitivity of GMR sensors to small magnetic fields has been used for detection of biomolecules tagged with magnetic labels (Wood *et al.* 2005). GMR sensors are favored over competing optical detection schemes due to their higher sensitivity, lower background, compact size and easy integrability with existing semiconductor electronics. This technology has implications in many areas of biological and medical research, including disease detection, treatment and prevention.

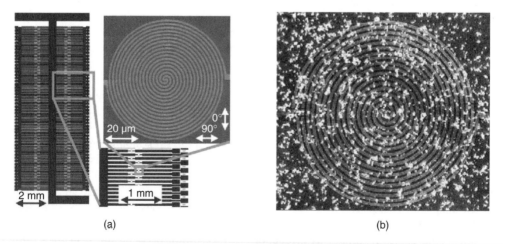

Figure 8.11 (a) Micrograph zoom series of the magnetoresistive sensor layout; (b) single element covered by DNA sample. (Schotter *et al.* 2004)

8.2.2 Anisotropic Magnetoresistance Sensor

As discussed earlier, a uniformly magnetized sphere (bead) can be described as a pure dipole with a dipole moment m at the center of the sphere. As shown in Figure 8.12, for applied magnetic fields H_z, the xy-planar component of the induced field from the bead is axially symmetric. In cylindrical coordinates, this purely radial component is given by (Miller *et al.* 2002):

$$B_\rho = \frac{3mz\rho}{r^5} \tag{8.19}$$

The distance of the bead center from the sensor surface is $z = a + t$, where a is the bead radius and t the separation of the bottom of the bead and the sensor. In a biosensor application, t represents the thickness of passivation and chemical functionalization layers. B_ρ is strongly dependent upon ρ and z and reaches a maximum value at $\rho = z/2$ and a ring sensor can be optimized by sizing the ring such that the peak in the radial field is coincident with the ring annulus.

By this approach, Miller *et al.* (2002) developed a single layer $Ni_{80}Fe_{20}$ (NiFe) ring sensor based on current-in-plane (CIP) anisotropic magnetoresistance (AMR). The ring has an outer diameter of $5\,\mu m$ and inner diameters of $3.2\,\mu m$. The AMR of the ring is modulated by the radial fringing field. As shown in Figure 8.13(a), in the absence of a radial field, the magnetization is circumferential. If an electrical current passes through the ring, the current is mostly parallel or anti-parallel to the magnetization. Therefore, even if the magnetizations in the upper and lower halves of the ring have opposite chirality, the ring is in essentially the same high resistance state as the single chirality state. As shown in Figure 8.13(b), applying the radial field rotates the magnetization toward a radial direction, normal to the current, and the resistance is minimized.

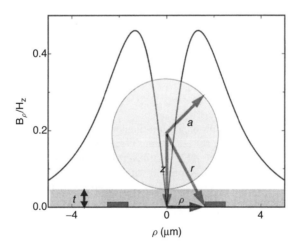

Figure 8.12 Definition of coordinates. This figure shows the radial field B_ρ normalized to the applied field H_z as a function of the radial position ρ. The ring sensor cross-section is schematically shown below the spherical magnetic bead. (Miller *et al.* 2002). Reprinted with permission from: Miller, M. M. Prinz, G. A. Cheng, S. F. and Bounnak, S. "Detection of a micron-sized magnetic sphere using a ring-shaped anisotropic magnetoresistance-based sensor: A model for a magnetoresistance-based biosensor", *Applied Physics Letters*, **81** (12), 2211–2213. (2002) American Institute of Physics

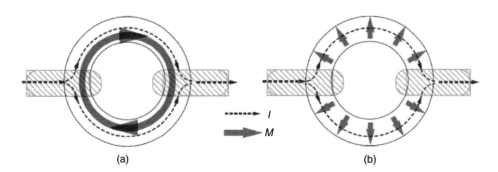

Figure 8.13 Schematic diagram of the AMR ring sensor. The hatched areas represent the contact fingers; (a) maximum resistance state in which the current I is mostly parallel or anti-parallel to the circumferential magnetization M; (b) minimum resistance state in which I is mostly normal to M. (Miller *et al.* 2002). Reprinted with permission from: Miller, M. M. Prinz, G. A. Cheng, S. F. and Bounnak, S. "Detection of a micron-sized magnetic sphere using a ring-shaped anisotropic magnetoresistance-based sensor: A model for a magnetoresistance-based biosensor", *Applied Physics Letters*, **81** (12), 2211–2213. (2002) American Institute of Physics

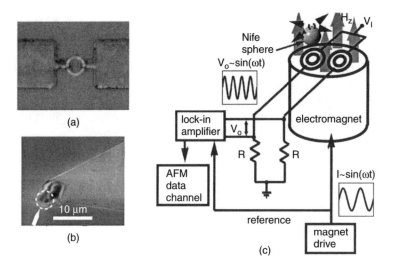

Figure 8.14 (a) Optical micrograph of a 5 μm-diameter AMR ring sensor. Electrical contact is made by the equatorial Cu fingers; (b) a scanning electron micrograph of the 4.3 mm-diam NiFe sphere (dashed circle) attached to the AFM cantilever; (c) schematic representation of the experimental apparatus. (Miller *et al.* 2002). Reprinted with permission from Miller, M. M. Prinz, G. A. Cheng, S. F. and Bounnak, S. "Detection of a micron-sized magnetic sphere using a ring-shaped anisotropic magnetoresistance-based sensor: A model for a magnetoresistance-based biosensor", *Applied Physics Letters*, **81** (12), 2211–2213. (2002) American Institute of Physics

Figure 8.14(a) shows the optical micrograph of an AMR ring sensor. To test the sensor, a 4.3 μm-diameter NiFe sphere was attached to an atomic force microscope (AFM) cantilever using a micromapulator and UV-curing epoxy, as shown in Figure 8.14(b). To detect the radial fringing field from the magnetic particle, a differential technique, as shown in Figure 8.14(c) is employed. Two AMR ring structures, separated by 50 μm, constitute the upper half of a Wheatstone bridge. The bridge is completed with off-chip

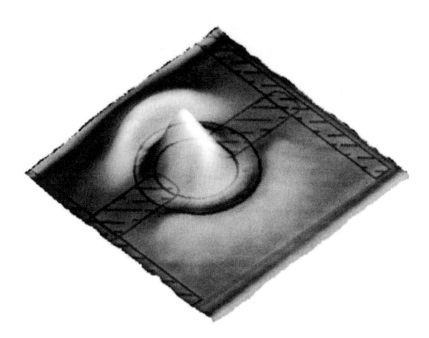

Figure 8.15 Surface plot of the AMR signal of the ring sensor for $H_{z,0} = 35\,\mathrm{Oe}$. The ring sensor outline is shown schematically for clarity. (Miller *et al*. 2002). Reprinted with permission from: Miller, M. M. Prinz, G. A. Cheng, S. F. and Bounnak, S. "Detection of a micron-sized magnetic sphere using a ring-shaped anisotropic magnetoresistance-based sensor: A model for a magnetoresistance-based biosensor", *Applied Physics Letters*, **81** (12), 2211–2213. (2002) American Institute of Physics

resistors $R \approx 65\,\Omega$, which is the approximate resistance of a NiFe ring. The sensors are placed on the pole face of an AC-driven homopolar electromagnet where $H_z = H_{z,0} \sin(\omega t)$. The bridge is measured by applying a DC bias V_i. The bead is scanned using an AFM and the topographic and bridge output signals are simultaneously recorded. When the bead is centered directly over the ring, the radial fringing field B_ρ rotates the magnetization of the ring from circumferential towards a radial outward (or inward) direction. The resistance then decreases, generating an AC bridge imbalance V_0 that can be detected. As the magnetoresistance has a roughly quadratic magnetic field dependence about $H = 0$, it is desirable to detect at 2ω because expansion of the magnetoresistance yields the dominant AC term (Miller *et al*. 2002):

$$V_0 \approx H_{z,0}^2 \frac{\partial^2 R}{\partial H^2} \cos(2\omega t) \tag{8.20}$$

Figure 8.15 shows a surface plot V_0 for a scanned bead with $H_{z,0} = 35\,\mathrm{Oe}$ and $V_i = 1\,\mathrm{V}$. The peak value occurs when the particle is above the center of the ring. As expected from Equation (8.19), V_0 fits a quadratic dependence on H_z very well, demonstrating that the sensor response is magnetoresistive in origin and not due to an extraneous effect such as capacitive coupling. The asymmetry of the signal from the upper to lower half of the ring is mainly due to a small misalignment of the Cu contacts from the equatorial position (Miller *et al*. 2002).

8.2.3 Spin Valve Sensor

Figure 8.16 shows the working principle of a spin valve sensor (Freitas *et al.* 2000; Li *et al.* 2004). As shown in Figure 8.16(a), a spin valve film is patterned into a stripe with dimensions L (magnetic length), W (track width, distance between pads) and h (height). The pinned layer easy axis is in the transverse orientation. The easy axis of the free layer is in the longitudinal orientation. When excited by a transverse signal field, the free layer magnetization rotates out of the longitudinal direction. For small signal fields, the magnetization of the free layer rotates away from the longitudinal orientation, and sensor resistance deviates linearly with field. Figure 8.16(b) shows the experimental transfer curves of a $6 \times 2\,\mu m^2$ sensor with a sense current of 5 mA.

Figure 8.17 schematically shows a spin valve sensor designed for detecting magnetic nanoparticles that are conceptually immobilized onto the sensor surface via DNA hybridization. The spin valve sensor is configured in an orthogonal magnetization state. Figure 8.17(a) shows a patterned rectangular spin valve sensor with its pinned magnetization M_p fixed in the transverse (y direction) and the free magnetization M_f rotating freely in the sensor plane. In this configuration, the sensor resistance can be expressed as (Li *et al.* 2006):

$$R = R_0 + \frac{1}{2}\Delta R_{max} \sin \theta_f \qquad (8.21)$$

where $\Delta R_{max} = R_{AP} - R_P$ is the sensor resistance change between the anti-parallel (R_{AP}) and parallel (R_P) magnetization configurations, $R_0 = R_P + \Delta R_{max}/2$ is the sensor resistance at the orthogonal configuration, and θ_f is the orientation angle of free layer magnetization with respect to the longitudinal or x direction. For an external field applied in the transverse direction within a certain range, $\sin\theta_f$ is linear or approximately linear with the field, leading to a linear dependence of sensor resistance on the field too.

Figure 8.18 shows (a) the optical micrograph of a fabricated spin valve sensor array of 60 sensors on a 7 mm $\times$ 8 mm chip, (b) the scanning electron microscope (SEM) image of one submicron sensor in the array, and (c) SEM image of a detection sensor with about 23 deposited Fe_3O_4 nanoparticles (Li *et al.* 2006).

The resistance of the sensor is usually measured by a differential technique (Li *et al.* 2003). The magnetic excitation field H_t can be DC or AC. Compared to a DC excitation, an AC excitation will generate an AC MR signal from a spin valve sensor, which therefore

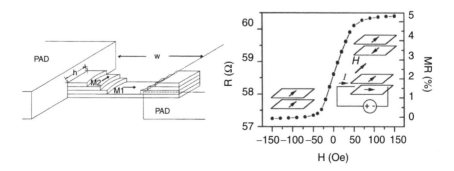

Figure 8.16 (a) Schematic drawing of a spin valve sensor element. M_1 is the free ferromagnetic layer, and M_2 is the pinned ferromagnetic layer; (b) resistance *vs* H transfer curve for a $6 \times 2\,\mu m^2$ spin valve element obtained with a 5 mA sense current. (Freitas *et al.* 2000)

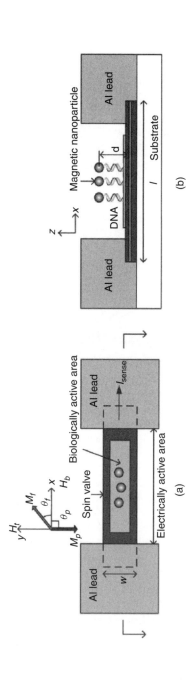

Figure 8.17 Schematic illustrations of (a) the top view of a spin valve sensor and (b) its cross-section. A few magnetic nanoparticle labels are conceptually shown bound to the sensor through hybridized probe and target DNAs in the biologically active area. M_f and M_p are the free and pinned magnetization of the spin valve sensor, respectively. H_t and H_b are the applied magnetic excitation and bias fields, respectively. (Li *et al.* 2006). Reprinted from: Li, G. X. Sun, S. H. Wilson, R. J. White, R. L. Pourmand, N. and Wang, S. X. (2006). "Spin valve sensors for ultrasensitive detection of superparamagnetic nanoparticles for biological applications", *Sensors and Actuators A*, **126**, 98–106., with permission from Elsevier

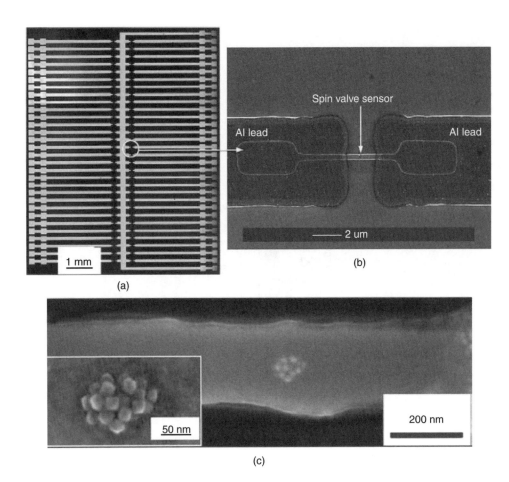

(a)

(b)

(c)

Figure 8.18 (a) An optical micrograph of a fabricated spin valve sensor array of 60 sensors on a ~7 mm × 8 mm chip and (b) the scanning electron microscope (SEM) image of one submicron sensor in the array; (c) SEM image of a detection sensor with ~23 deposited Fe_3O_4 nanoparticles that are also shown in the inset at a higher magnification. (Li *et al.* 2006). Reprinted from: Li, G. X. Sun, S. H. Wilson, R. J. White, R. L. Pourmand, N. and Wang, S. X. (2006). "Spin valve sensors for ultrasensitive detection of superparamagnetic nanoparticles for biological applications", *Sensors and Actuators A*, **126**, 98–106., with permission from Elsevier

can be measured with a lock-in amplifier to achieve a high signal-to-noise ratio (SNR). Figure 8.19 shows the peak-to-peak resistance difference ΔR_{p-p} versus the number of deposited Fe_3O_4 nanoparticles. The solid line is a linear fit to the data. The horizontal dashed line indicates the fluctuation range of reference sensor signals, and may be considered the noise level of this method. It is found that the sensor signal ΔR_{p-p} increases with the number of nanoparticles in a reasonably linear manner, and the fitting line intersects the reference signal level at 14 nanoparticles, which may be the minimum detectable number of nanoparticles in this detection scheme with a signal-to-noise ratio of about 1. The data scatter may be caused by factors such as the varying locations and distributions of nanoparticles on the sensors.

To improve the measurement sensitivity, a method has been developed to control the movement and placement of magnetically labeled biomolecules using two current lines with widths tapered adjacent to the sensor unit. Detailed discussion can be found in (Graham *et al.* 2005; Ferreira *et al.* 2003).

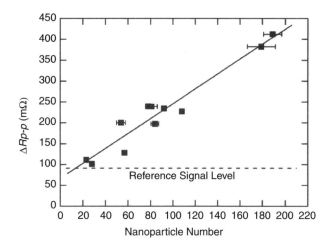

Figure 8.19   The peak-to-peak resistance difference *vs* the number of the deposited Fe_3O_4 nanoparticles. The solid line is a linear fit and the dashed horizontal line designates the average signal fluctuation range of the reference sensor. (Li *et al.* 2006). Reprinted from: Li, G. X. Sun, S. H. Wilson, R. J. White, R. L. Pourmand, N. and Wang, S. X. (2006). "Spin valve sensors for ultrasensitive detection of superparamagnetic nanoparticles for biological applications", *Sensors and Actuators A*, **126**, 98–106., with permission from Elsevier

8.2.4   Magnetic Tunnel Junction Sensor

Magnetic tunnel junctions (MTJs) can be used in the development of magnetic field sensors (Gallagher *et al.* 1997). Compared with GMR sensors, MTJs offer higher magnetoresistance ratios and therefore higher sensitivity at low fields. MTJs are better suited for the accurate detection of the small (< 1 Oe) fields, which are typically encountered in biological applications. In the following, we discuss the use of highly sensitive MTJ sensors for the detection of individual micron-sized magnetic labels (Shen *et al.* 2005).

Figure 8.20 shows a typical transfer curve of one MTJ sensor with a MR ratio of 15.3 % and a device resistance of 142 Ω (Shen *et al.* 2005). The MTJ sensors have the following layer structure: Pt (300 Å)/Py (30 Å)/FeMn (130 Å)/Py (60 Å)/Al_2O_3 (7 Å)/Py (120 Å)/Pt (200 Å). Py stands for permalloy. The sensors were patterned using standard optical lithography followed by Ar ion-beam etching.

Figure 8.21(a) is an optical image of the finished sensor die after integration with the microfluidic channel. Supermagnetic beads in solution were funneled toward the active area of the MTJ device through a microfluidic channel with a height of 50 μm and a width of 600 μm. The superparamagnetic beads used are polymer beads with an even dispersion of iron oxide (γ-Fe_2O_3) nanoparticles. Because the beads have a higher density (~ 1.3 g/cm³) than deionized water, the beads settle at the bottom of the channel at sufficiently slow flow rates. In this case, the beads will roll along the bottom surface of the channel, which is coincident with the top of the MTJ sensor die (Shen *et al.* 2005). Figure 8.21(b) shows an optical image of a single 2.8 μm diameter bead sitting on the sensor area.

The sensor is operated in an AC bridge configuration and the signal is extracted using a lock-in technique (Shen *et al.* 2005). Figure 8.22 shows the real-time voltage output of the MTJ sensor for an interval during which several beads pass over the MTJ one at a time. Each sharp signal drop (points A, C, D, H and I) corresponds to an event where a

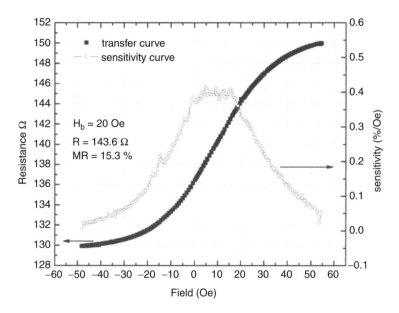

Figure 8.20 A typical MTJ transfer curve showing the device resistance of a MTJ sensor in an external applied field. The most sensitive region occurs over the range 0–15 Oe. (Shen *et al*. 2005)

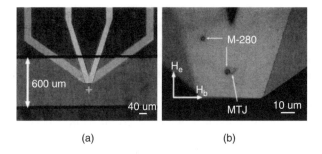

(a)   (b)

Figure 8.21 Optical images of (a) a single $2 \times 6\,\mu$m MTJ sensor sealed inside a 600 mm wide microchannel, and (b) an identical sensor with two single M-280 beads in close proximity. The orientation of the two external fields H_e and H_b is also shown in (b). (Shen *et al*. 2005)

single bead passes by the sensor quickly. Plateaus are observed (points B and G) when the beads stick to the sensor area for a longer interval. Points E and F correspond to a situation where two beads are attached to the junction simultaneously. Starting from point E, a single bead sits on the sensor for about 20 seconds, during which a voltage of ~80 mV is observed. At point F, another bead approaches the sensor and sticks to it along with the first one for about 3 s. This results in an additional contribution to the sensor voltage of ~60 mV (point E). As shown in Figure 8.22, a consistent average signal of 80 mV was observed for single beads over many measurements. It should be noted that for some beads (points C, H and I) the signal level is less than 80 μV, and this can be due to the variation in the bead-to-sensor separation for different beads. The signal is the highest when beads cross over the central area of the sensor, and the amplitude of this signal variation is denoted by the shadowed region in the figure.

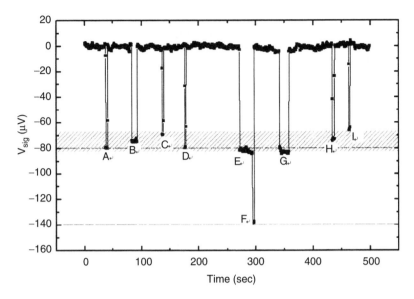

Figure 8.22   Real-time voltage data demonstrating single-bead detection. When a single bead passes by the sensor, a sharp signal drop is observed (points A, C, D, H and I). When a bead becomes stuck on the sensor area for an extended length of time, a plateau signal is obtained (points B and G). Two-step signals (points E and F) correspond to a situation wherein two beads are attached to the junction at the same time. The shadowed band indicates the typical signal range measured for a single bead. (Shen *et al.* 2005)

8.3   Hall Effect Sensors

Various types of Hall effect sensors have been developed for the detection of magnetic nanoparticles. The ordinary Hall effect is due to the Lorentz force acting on charge carriers in metals, semi-metals and semiconductors. Magnetic materials show additional Hall phenomena generated by spin–orbit interactions: extraordinary and planar Hall effects. Some heterostructures show quantum well Hall effect. In the following subsections, we discuss the silicon Hall effect sensor, planar Hall effect sensor, extraordinary Hall effect sensor and quantum well Hall effect sensor.

8.3.1   Silicon Hall Effect Sensor

A single magnetic microbead can be detected and characterized using a silicon Hall sensor fabricated with standard metal-oxide-semiconductor (CMOS) technology (Besse *et al.* 2002). As shown in Figure 8.23(a), using micromanipulators, a single microbead of diameter 2.8 μm was manually placed in the center of a cross-shaped silicon Hall sensor fabricated in a 0.8 μm CMOS technology. The distance between the sensor and the center of the bead is r = 7 μm. The active area of the sensor is $2.4 \times 2.4\,\mu m^2$. It has a sensitivity of 175 V/AT, a nominal bias current $I_b = 0.3$ mA and a resistance R = 8.5 kV. Above 100 Hz at room temperature, its magnetic field resolution is limited by the thermal noise to about $0.2\,\mu T\,H_z^{-1/2}$ (Besse *et al.* 2002).

As shown in Figure 8.23(b), to determine the magnetic characteristics of a single bead, the Hall sensor and the bead are placed in a static field H_0 in the z direction. A small AC field H_2 is applied parallel to H_0 at frequency f_0 to measure the small signal behavior of the magnetization. The Hall voltage V_H at frequency f_0 is measured using a low-noise

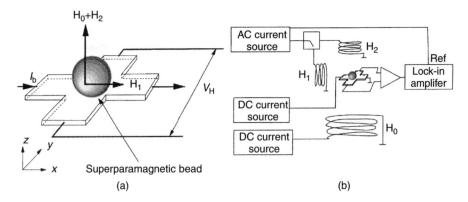

Figure 8.23 Schematic view of the measurement setup. A single superparamagnetic bead is centered on a Hall sensor, biased with the current I_b. A DC magnetic field H_0 is applied perpendicular to the sensor. An AC field is added either in the z direction (H_2) or in the x direction (H_1). The Hall voltage V_H, containing a component proportional to the magnetic induction produced by the bead, is measured with a lock-in amplifier. Reprinted with permission from: Besse, P. A. Boero, G. Demierre, M. Pott, V. and Popovic, R. (2002). "Detection of a single magnetic microbead using a miniaturized silicon Hall sensor", *Applied Physics Letters*, (2002) **80** (22), 4199–4201. (2002) American Institute of Physics

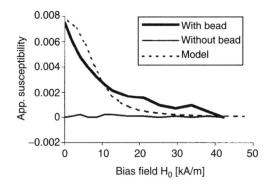

Figure 8.24 Measurement of the apparent susceptibility as defined by Equation (8.22). ($I_b = 0.3\,\text{mA}$, $H_2 = 0.18\,\text{kA/m}$, $\tau = 2\,\text{s}$, $\Delta f = 0.06\,\text{Hz}$ and $f_0 = 520\,\text{Hz}$). Reproduced by permission of American Institute of Physics

preamplifier and lock-in detector. The voltage V_H measures the magnetic induction at the location of the sensor as a function of the externally applied magnetic field. Since the magnetization of the bead saturates, the Hall voltage at high magnetic field is used to normalize the measured values. Here we define the apparent susceptibility χ_{app} by (Besse *et al.* 2002):

$$\chi_{app}(H_0) + 1 \equiv \frac{V_H(H_0)}{V_H(H_0 \to \infty)} \tag{8.22}$$

The value of χ_{app} depends on the geometry of the bead and on the distance between the bead and the sensor. It tends to vanish at high fields. Figure 8.24 shows the measured and modeled apparent susceptibility of a single bead.

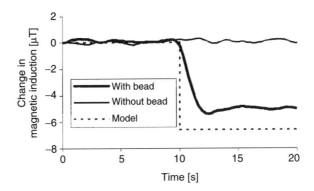

Figure 8.25 Detection of the bead, based on the apparent susceptibility measurement of Figure 8.23. An AC field H_2 is applied in the z direction. After 10 s, a bias magnetic field H_0 = 13 kA/m is switched on. The Hall voltage is recorded at frequency f_0 (I_b = 0.3 mA, H_2 = 0.9 kA/m, τ = 1 s, Δf = 0.12 Hz and f_0 = 520 Hz). (Besse *et al.* 2002). Reproduced by permission of American Institute of Physics

Two methods can be used to detect the presence or the absence of the bead: the first-harmonic detection method and second-harmonic detection method (Besse *et al.* 2002). The first-harmonic detection method is similar to the susceptibility measurement method in Figure 8.24. In the presence of the AC field H_2, the bias field H_0 is switched from 0 to 13 kA/m stepwise. The magnetic induction on the sensor, and therefore the Hall voltage V_H at frequency f_0, also vary, with a delay corresponding to the integration time of the lock-in amplifier, as shown in Figure 8.25. This variation is the signature of the presence of a bead. Compared to the susceptibility measurement, a higher AC field H_2 is used in order to increase the change in magnetic induction. It should be noted that the signal at frequency f_0 has a strong offset due to the parasitic inductive signals and to the direct Hall detection of the excitation field H_2. The detection of small variations superposed to this large offset requires a highly stable AC signal (Besse *et al.* 2002).

In order to suppress the large offset in the first method, the second-harmonic detection method detects the particle at the second harmonic of the excitation frequency. A bias field H_0 is applied perpendicular to the sensor. An AC field H_1 at frequency f_0 is applied in the x direction, as shown in Figure 8.23. The magnetization vector of the bead follows the total field and, hence, moves in the xz plane. The component M_z varies at frequency $2f_0$. The Hall voltage is detected with the lock-in amplifier at the second harmonic. Switching on the bias field H_0 leads to a step response, as shown on Figure 8.26. No signal is measured at $2f_0$ when the bead is absent, and so the offset has disappeared (Besse *et al.* 2002).

For both detection methods, the signal depends linearly on the sensor bias current I_b, and it is independent of the excitation frequency f_0 and disappears when the bead is removed (Besse *et al.* 2002). The choice of the detection method depends strongly on the applications. With the first-harmonic detection method, integrated planar coils could be used as AC field generators around each sensor. On the other hand, the second-harmonic detection method suppresses the offset and therefore releases the tolerances on the electronics, but external coils are needed to produce high AC fields.

8.3.2 Planar Hall Effect Sensor

As shown in Figure 8.27, a Hall sensor with a plate-shaped geometry is not only sensitive to a magnetic flux density component perpendicular to the plate B_z, but also to a

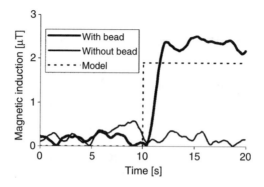

Figure 8.26 Detection of the bead, based on the second-harmonic measurement. An AC field H_1 is applied in the x direction. After 10 s, a bias magnetic field $H_0 = 3.7$ kA/m is switched on. The Hall voltage is recorded at frequency $2f_0$ ($I_b = 0.3$ mA, $H_1 = 62$ kA/m, $\tau = 1$ s, $\Delta f = 0.12$ Hz and $f_0 = 520$ Hz). (Besse *et al.* 2002). Reproduced by permission of American Institute of Physics

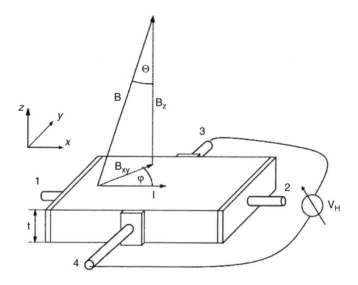

Figure 8.27 The planar Hall effect in a plate-shaped Hall sensor. (Schott *et al.* 2000)

component in the plane z of the plate B_{xy} (Schott *et al.* 2000; Montaigne *et al.* 2000). This phenomenon, called the planar Hall effect, is important for the measurements where the flux density vector is not perpendicular to the sensor plane, and is especially important for accurate measurements, when more than one magnetic field component is present.

In Figure 8.27, the voltage U_{34} between the electrodes 3 and 4 for a rectangular Hall device can be written as (Morvic and Betko 2005):

$$U_{34} = U_{asym} + U_H + U_{H,plan} \qquad (8.23)$$

where the Hall voltage is given by:

$$U_H = R_{H,s} I B_z \qquad (8.24)$$

and the planar Hall voltage is given by:

$$U_{H,plan} = R_{H,plan,s} I B_x B_y \tag{8.25}$$

U_{asym} is the voltage between electrodes 3 and 4 due to their asymmetry; $R_{H,s}$ and $R_{H,plan,s}$ are the transversal (magnetic field perpendicular to the sensor plane) and planar (magnetic field parallel with the sensor plane) sheet Hall coefficients, respectively; I, the current flowing from electrode 1 to electrode 2; $B_x = B \sin \theta \cos \phi$, $B_y = B \sin \theta \sin \phi$ and $B_z = B \cos \theta$ are the x, y and z components of the magnetic field, respectively; θ is the angle between the magnetic field vector **B** and the z-axis and ϕ is the angle between the planar component of the magnetic field B_{xy} and the x-axis.

The planar Hall effect is based on the AMR of ferromagnetic materials (Ejsing *et al.* 2004). The transverse voltage on a planar Hall cross depends on the orientation of the magnetization of the material with respect to the longitudinal current running through the material. Figure 8.28(a) shows a schematic drawing of a planar Hall sensor. Initially, the magnetization lies along the easy axis, which is also the direction of the applied current, I_x. When a magnetic field, H_y, is applied in the y direction, the magnetization rotates by an angle ϕ in the sensor plane, resulting in a change of the electrical output signal by the amount V_y. For small angles, the value of V_y is given by:

$$V_y = \frac{(\rho_\| - \rho_\perp) I_x}{t} \cdot \frac{H_y}{H_{an}} \tag{8.26}$$

where t is the metal layer thickness, H_{an} is the effective anisotropy field of the metal layer, $\rho_\|$ and $\rho_\perp$ are the resistivities when the magnetization is parallel and perpendicular to the current, respectively and $\Delta \rho = (\rho_\| - \rho_\perp)$ is responsible for AMR and the planar Hall effect. Equation (8.26) indicates that the response of the sensor is linear. For bead detection, H_y in Equation (8.26) represents the applied external field plus the sum of

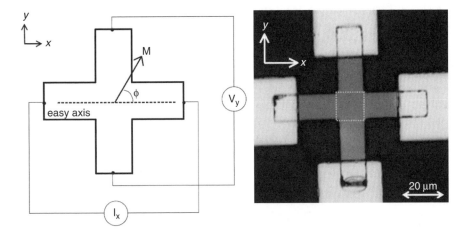

Figure 8.28 (a) Planar Hall geometry. A current is applied in the x direction and a voltage is measured in the y direction. The magnetization vector, **M**, lies in the x–y plane at an angle, ϕ, to the current direction. The easy axis is along the current direction; (b) top view micrograph of the planar Hall sensor. The cross is made of magnetic layers, and the central area marked by a dotted frame is the 10 μm × 10 μm sensitive area of the sensor. (Ejsing *et al.* 2005)

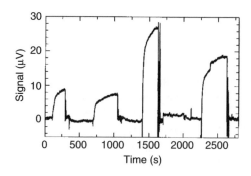

Figure 8.29 Bead detection measurements. A constant voltage ($V_0 = -464\,\mu V$) is subtracted from the signal to give a baseline at $0\,\mu V$

the y components of the field created by the homogeneously magnetized spherical beads weighed by the bead-to-sensor area fraction (Ejsing *et al.* 2004).

Figure 8.28(b) is a micrograph of a planar Hall sensor. The structure of the sensor layer is: Ta (30 Å)/NiFe (50 Å)/MnIr (200 Å)/NiFe (200 Å)/Ta (30 Å). The active sensing layer is the NiFe (200 Å) layer. To demonstrate bead detection, a droplet containing superparamagnetic beads (2 µm or 250 nm diameter) was placed on top of the sensor, and the beads were magnetized with an in-plane applied field generated by electromagnetic coils. Figure 8.29 presents the bead detection results (Ejsing *et al.* 2004). Current through the sensor is 5 mA and the applied field is -15 Oe. At time t = 100 s, the 2 µm beads are added onto the chip and the signal rises as the beads settle on the sensor. At time t = 300 s, the beads are washed away and the signal returns to the baseline. At times t = 700 s and t = 1050 s, respectively, the experiment is repeated yielding the same result. At time t = 1400 s, the 250 nm beads are added to the chip and at time t = 1650 s the beads are washed away. The experiment is repeated with the 250 nm beads at times t = 2250 s and t = 2650 s, respectively. The 250 nm beads give higher saturation signals than the 2 µm beads due to their higher susceptibility and higher number on top of the sensor area. Saturation of the signal occurs when the beads are piled up on top of the sensor and the addition of another bead is no longer sensed because it is too far away from the sensor to be detected.

8.3.3 Extraordinary Hall Effect Sensor

Figure 8.30 schematically represents a cobalt-carbon Hall device with submicrometer structures (Boero *et al.* 2005). The SiO$_2$ layer (about 100 nm) is obtained by dry oxidation. The electrodes are realized by lift-off using an Au (300 nm)/Ti (20 nm) layer. The Co–C deposit is obtained by decomposing dicobalt octacarbonyl [Co$_2$(CO)$_8$] with the electron beam of a scanning electron microscope. The deposited material consists essentially of cobalt nanoparticles embedded in carbonaceous matrix. The devices show a strong extraordinary Hall effect, whereas the ordinary and planar Hall effects are relatively small.

Figure 8.31 shows the SEM and AFM images of a Hall device. Each arm of the cross-shaped device has a length of about 7 µm, a full width at half maximum (FWHM) of 500 nm, and a thickness of about 50 nm. Though the deposition process produces a halo-deposit around the main cross-shaped deposit, the active area of the complete device is about (500 nm)^2 (Boero *et al.* 2005).

Figure 8.32 shows the Hall voltage V_H as a function of the externally applied magnetic induction **B** (Boero *et al.* 2005). The magnetic induction is applied 'out-of-plane' as shown

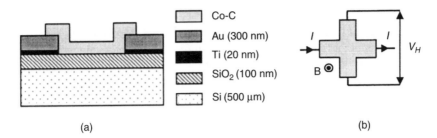

Figure 8.30 (a) Schematic cross-section of the devices; (b) the magnetic induction is applied: out-of-plane. (Boero *et al.* 2005)

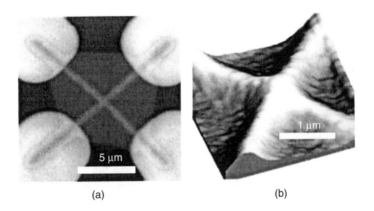

Figure 8.31 Images of a Hall device. (a)SEM image; (b)AFM image. (Boero *et al.* 2005)

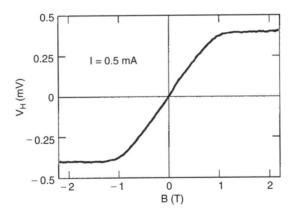

Figure 8.32 Hall voltage as a function of the out-of-plane magnetic induction. (Boero *et al.* 2005)

in Figure 8.30(b). The current sensitivity S_I is about 0.9 V/AT in the linear region. Since the device thickness is about 50 nm, this corresponds to a Hall coefficient R_H of about 5 $\times 10^{-8}$ Ωm/T. The absence of a significant hysteresis in the V_H–B curve indicates that the remanent magnetization is small, and this allows one to use this device for 'weakly invasive' quantitative magnetic imaging at low magnetic fields (Boero *et al.* 2005).

8.3.4 Quantum Well Hall Effect Sensor

The Hall coefficient and noise characteristics of a material depend on the carrier mobility μ and the inverse of carrier density, $1/n$. It is therefore important to choose a material, such as a quantum well heterostructure, that can provide the necessary characteristics (Mihajlovic *et al.* 2007). Landry *et al.* (2004) fabricated Hall sensors from a single quantum-well heterostructure: GaAs (substrate)/GaAs–AlSb buffer layer/300 $Al_{0.6}Ga_{0.4}Sb/15InAs/$ 12.5 $Al_{0.6}Ga_{0.4}Sb/5$ $In_{0.6}Al_{0.4}As/3$ InAs/80 sputtered SiN_x (the thickness units are in nanometers). The two-dimensional electron gas (2DEG), at the 15 nm InAs layer, is not intentionally doped but rather the conduction electrons originated, in part, in the InAlAs layer. The Hall sensors are defined as a cross shape with an arm width w of roughly 1 µm by photolithography and dry etch with an argon ion mill (Landry *et al.* 2004).

Figure 8.33 shows the experimental set-up. An individual magnetic bead is attached to an atomic force microscopy (AFM) cantilever and scanned over the sensor. This permits precise control of both the lateral and vertical position of the bead with respect to the Hall sensor, and also enables simultaneous topographic and Hall voltage imaging. The magnetic bead is a $Fe_{70}Ni_{30}$ nanocomposite with a radius a of approximately 2 µm. Because such a bead has virtually no remanence, detection requires application of an external magnetic field, **H**, so that the bead develops a magnetic moment **m** proportional to **H** (Landry *et al.* 2004).

The Hall sensor is placed at the center of the face of a cylindrical magnet that consists of a NdFeB core that is configured with soft-magnetic pole pieces in order to yield a number of discrete magnetic field values, up to 4000 Oe, normal to the face of the magnet. An AC current source provides a bias of 2–5 µA at 337 Hz and the Hall sensor output voltage V_H is measured with a lock-in amplifier. The signal from the lock-in amplifier is fed into an auxiliary AFM data channel so that the Hall voltage is recorded in registry with the topographical signal. These data are shown in Figures 8.33(b) and (c), respectively. The background magnetic field may be subtracted by moving the bead a sufficient distance

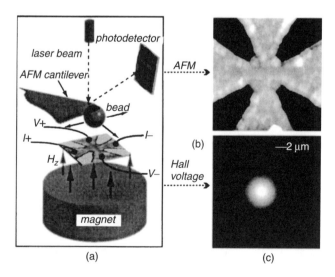

Figure 8.33 (a) Measurement set-up for Hall sensors in AFM; (b) topographical image of Hall sensor taken with magnetic bead; (c) intensity plot of Hall voltage V_H, recorded simultaneously with (b). (Landry *et al.* 2004)

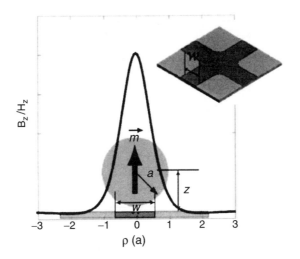

Figure 8.34 Normalized field profile B_z/H_z from a magnetized bead with a magnetic moment m = m_z over a Hall sensor with width w. Here ρ is the radial distance from the center point under the bead and is given in units of the bead radius a. z is the distance from the center of the bead to the quantum well. (Landry *et al.* 2004)

away from the sensor so that the Hall sensor only detects the external field H = H_z. (Landry *et al.* 2004)

The Hall sensor measures the field of the bead averaged over an effective area of the Hall sensor. A numerical study of diffusive transport in a 2DEG Hall cross showed that this area can be twice the geometric area. The field from a soft magnetic sphere is expected to be identical to the field from a point dipole having the same moment, located at the center of the sphere. As shown in Figure 8.34, when a bead is at position (x, y, z), the perpendicular component B_z has a gradient of values across the area of the sensor and the Hall output voltage is a response to the average value of this field ($V_H = R_H \times \langle B_z \rangle$). The average value of B_z is given by (Landry *et al.* 2004):

$$\langle B_z \rangle = \frac{16 H a^3 (w^2/2 + z^2)^{1/2}}{w^4 + 6w^2 z^2 + 8z^4} \tag{8.27}$$

The field of a bead is determined by rastering the bead across the Hall cross to get V_H, and taking $\langle B_z \rangle = V_H/R_H$ (Landry *et al.* 2004). The distance z from the center of the bead to the quantum well, z = a + t, includes the bead radius a and a free fitting parameter t. The latter includes the thickness of the top barrier of the quantum well and the passivation layer, plus additional spacing that may arise from a bead asperity or residue on the device and/or bead surface.

The fields from a 1.6 and a 2.1 µm radius bead were measured using Hall sensors with arm widths ranging from 500 nm to 2.5 mm (Landry *et al.* 2004). Since the magnetization of the bead is linearly proportional to the external field (inset in Figure 8.35), $\langle B_z \rangle/H_z$ is constant for a given value of a/w. Figure 8.35 shows that the detected bead field decreases as the arm width of the Hall sensor increases. This is expected because the Hall voltage is proportional to the average magnetic field over the sensor area. It follows that efficient bead detection requires values of a/w larger than 1, and that detection of nanometer scale magnetic labels will require sensors that are fabricated with comparable minimum feature size.

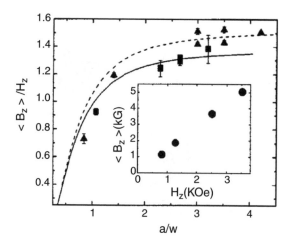

Figure 8.35 Magnetic field of $Fe_{70}Ni_{30}$ beads measured with Hall sensors. Data points are the measured bead field normalized to the external field for the 2.1 mm (triangle) and 1.6 mm (square) bead. The x axis is the bead radius divided by the nominal Hall sensor dimension. Solid and dashed lines are fits to Equation (8.26) for 1.6 and 2.1 mm beads, respectively, with 0.2 mm of passivation. The inset shows $\langle B_z \rangle$ vs H_z for the 2.1 mm bead measured with a 0.6 μm Hall cross. (Landry *et al.* 2004)

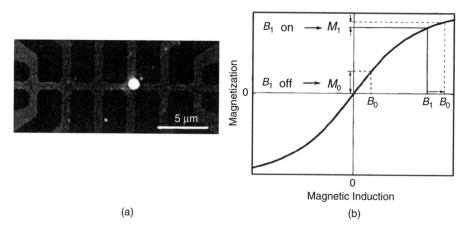

Figure 8.36 (a) SEM image of the central region of an InAs quantum well micro-Hall sensor with six Hall crosses and a superparamagnetic bead positioned on one of them. The Hall cross size was estimated to be 0.9 μm × 1.0 μm; (b) typical Langevin response of a superparamagnetic bead to an applied magnetic field and a sketch of the physical principle underlying the detection method. B_0, M_0 and M_1 represent rms values of the corresponding AC quantities. (Mihajlovic *et al.* 2005)

Here we discuss the single particle detection using an AC phase-sensitive detection method (Mihajlovic *et al.* 2005). The sensor was fabricated by photolithography and wet chemical etching from molecular beam epitaxy grown heterostructure consisting of GaAs substrate/GaAs buffer (100 nm)/AlAs (10 nm)/AlSb (30 nm)/$Al_{0.7}Ga_{0.3}$Sb (1000 nm)/AlSb (8 nm)/InAs QW (12.5 nm)/AlSb (13 nm)/GaSb (0.6 nm)/$In_{0.5}Al_{0.5}$As (5 nm). As shown in Figure 8.36(a), the central region of the sensor was an array of six Hall crosses, and a single superparamagnetic bead was placed on a Hall cross. The bead consisted of Fe_3O_4 nanoparticles embedded in a spherical latex matrix and its diameter was 1.2 μm.

In order to detect the presence of the bead immobilized on the Hall cross, the sensor chip was placed in a perpendicular AC excitation magnetic field B_0 with frequency f_0. The sensor was biased by a DC current I_0, and the Hall voltage was measured at the frequency f_0 with a lock-in amplifier. Since the bead is superparamagnetic, its magnetization follows Langevin behavior, as shown in Figure 8.36(b). The AC signal essentially measures the slope of the Langevin curve, so it depends on the DC magnetic state of the bead. When the bead is exposed to a DC magnetic field B_1 its magnetic state shifts towards lower susceptibility and it lowers the induced AC magnetization in the bead. This reduces the average AC stray field from the bead sensed by the cross, which manifests itself as a drop in the AC Hall voltage signal. The linearity of the Hall sensors ensures that B_1 does not induce any change in the AC Hall signal on an empty Hall cross without a bead on top. Therefore this drop is a definitive signal indicating the presence of a bead on the Hall cross. The magnitude of the drop in the AC Hall voltage can be expressed as:

$$\Delta V_H = R_H I_0 C \Delta M \tag{8.28}$$

where C is the coupling coefficient that quantifies the strength of the interaction between the magnetization of the bead and a sensing 2DEG in the Hall cross area, and $\Delta M = M_0 - M_1$ is the difference in the induced AC magnetizations of the bead before and after B_1 has been applied respectively, as shown in Figure 8.36(b). For a particular sensing application the values of B_0 and B_1 can be chosen to achieve optimal signal to noise ratio (S/N) based on the magnetic properties of the bead and the sensor's sensitivity (Mihajlovic *et al.* 2005).

Figure 8.37(a) shows an SEM image of two adjacent Hall crosses, an empty one and one with a bead. A constant DC current, $I_0 = 40$ μA, ran through the sensor in a horizontal direction and the Hall voltage was measured perpendicular to the current flow for both crosses. The AC excitation field was $B_0 = 2.13$ mT at $f_0 = 622$ Hz, and the DC field

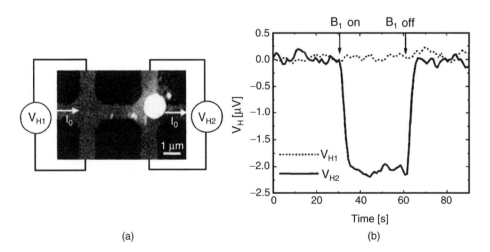

(a)   (b)

Figure 8.37 (a) SEM image of two adjacent Hall crosses adapted to show the actual detection measurement configuration; (b) AC Hall voltage as a function of time for the two crosses shown in (a). The drop in the signal from one cross upon applying the static field B_1 is due to the presence of the bead. (Mihajlovic *et al.* 2005)

$B_1 = 6.1 \, mT$ was applied and then removed at about 30 and 60 s, respectively, after the measurement had started. As shown in Figure 8.37(b), a clear drop in the Hall voltage was observed for the cross with the bead and was completely absent for the empty cross. The magnitude of the voltage decrease was $\Delta V_H = 2.0 \, \mu V$, which corresponds to a change in the sensed stray field of $C\Delta M = 80 \, \mu T$. Under the given experimental conditions the noise level was measured to be $V_{HN} = 105 \, nV$, corresponding to a minimum detectable change in the stray field $C\Delta M_{min} \sim 4.3 \, \mu T$ and S/N = 25.6 dB (Mihajlovic *et al.* 2005).

A weakness of this detection method is the large offset created by the direct sensor Hall response to the AC excitation field which is typically orders of magnitude larger than the small signal from the magnetic bead. This large background can be eliminated using a Hall gradiometry method (Mihajlovic *et al.* 2005).

8.4   Other Sensors Detecting Stray Magnetic Fields

Besides the magnetoresistance-based sensors and Hall-effect sensors, other sensors based on measuring the stray magnetic fields of magnetic beads mainly include the giant magnetoimpedance sensor, frequency-mixing method and SQUID method.

8.4.1   Giant Magnetoimpedance Sensor

The giant magnetoimpedance (GMI) can be considered as a high frequency analogy of giant magnetoresistance, and the GMI effect has received much attention as a candidate to develop new generation micro-magnetic sensors (Mahdi *et al.* 2003). The giant magnetoimpedance (GMI) effect includes a large and sensitive change in an AC voltage measured across a soft magnetic specimen subjected to a high frequency current under the effect of a DC, or lower frequency, magnetic field. To get a large change in impedance, two conditions should be satisfied (Kurlyandskaya and Levit 2005): (i) the frequency of the excitation current must be such that it ensures a strong skin effect, and (ii) the magnetic structure has to provide an AC transverse permeability sensitive to the external field.

Kurlyandskaya and Levit (2005) developed a prototype GMI sensor using an amorphous ribbon ($Co_{67}Fe_4Mo_{1.5}Si_{16.5}B_{11}$). As shown in Figure 8.38, a non-magnetic plastic bath was situated in the central part of the non-magnetic support of an imprinted circuit board. The ribbon sensitive element was installed in the bath through holes made in the centers of butt-ends and then plumbed by non-evaporating high quality vacuum resin to avoid mechanical stresses, which affect the magnetic element sensitivity. The ribbon sensitive element was flat and two surfaces of it were active and participated in the sensing process.

The magnetoimpedance was measured by a standard four point technique as shown in Figure 8.39 (Kurlyandskaya and Levit 2005). The magnetic conductive sensitive element was always connected in series with a reference resistor. Current amplitude was constant through each measurement regardless of impedance changes. Driving current intensities I_{rms} were kept small to prevent Joule heating. The voltage induced in the sample U, was measured by an oscilloscope and the impedance value was calculated according to Ohm's law. An external magnetic field in the interval of -150 to $+150$ Oe was applied by a pair of Helmholtz coils in the plane of the amorphous ribbon. In a majority of studies the external magnetic field was parallel to the wire axis and the flowing current. Some calibration measurements were taken under different angles, α, between the external magnetic field and the ribbon axis. To control the hysteresis of the magnetoimpedance with respect to the applied field, the measurements can be taken both in increasing and decreasing fields, always starting from the state of magnetic saturation. The response of

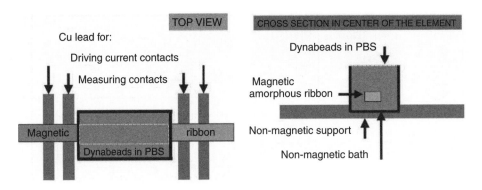

Figure 8.38 GMI-biosensor prototype with amorphous $Co_{67}Fe_4Mo_{1.5}Si_{16.5}B_{11}$ ribbon as a magnetic sensitive element. The driving AC current applied via the driving current contacts and induced voltage is collected from the measuring contacts. Two surfaces of the amorphous ribbon are active. (Kurlyandskaya and Levit 2005)

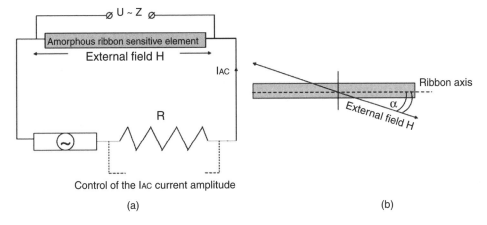

Figure 8.39 Basic diagram of the GMI circuit. External field can be applied (a) parallel to the axis of the amorphous ribbon or (b) at an angle α, to it. (Kurlyandskaya and Levit 2005)

the magnetoimpedance-based element depends on a number of parameters: frequency and intensity of the driving current, angle between the external field H and the ribbon long axis. GMI ratio, $\Delta Z/Z$, is defined as follows:

$$\frac{\Delta Z}{Z(H, f)} = \frac{100 \cdot [Z(H) - Z(H = 150\,Oe)]}{Z(H = 150\,Oe)}\% \tag{8.29}$$

Figure 8.40 shows a general testing procedure of the GMI-biosensor. A suspension containing magnetic labels coated with antibodies specific for the antigens of interest is mixed with a test solution, and all the free magnetic labels are removed after the formation of the complexes of antigen–antibody–magnetic label. The first response of the biosensor is measured in a solution containing no magnetic labels. Afterwards the response of the biosensor is measured in the solution containing antigen–antibody–magnetic label complexes. The difference in the induced voltage represents the quantity of the antigens of interest in the test sample (Kurlyandskaya and Levit 2005).

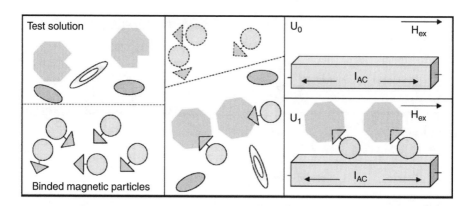

Figure 8.40 A general procedure for detection of biocompatible magnetic particles using a magnetoimpedance-based sensor. Magnetic particles coated with specific antibodies are mixed with the test solution, and all the particles that do not participate in the molecular recognition events are removed. The voltage U_0 induced in the sensor in a solution containing no magnetic labels is compared with the voltage U_1 induced when the solution contains antigen–antibody–magnetic label complexes. (Kurlyandskaya and Levit 2005)

As the sensitivity of GMI is very high, it is not necessary to attach a magnetic label at a short distance from passivation and/or chemical reactive layers. It makes an analysis of the sensor response more complex, and requires the use of elaborate mathematical models. It should be noted that the response of the magnetoimpedance-based element depends on a number of parameters: frequency and intensity of the driving current, angle between the external field H and the ribbon long axis. Detailed discussion on the effects of these parameters can be found in (Kurlyandskaya and Levit 2005).

8.4.2 Frequency Mixing Method

Meyer *et al.* (2007) developed a detection technique for magnetic nanoparticles, based on frequency mixing at the nonlinear magnetization curve of superparamagnets. Upon magnetic excitation at two distinct frequencies f_1 and f_2 incident on the sample, the response signal generated at a frequency representing a linear combination $mf_1 + nf_2$ is detected. The appearance of these components is highly specific to the nonlinearity of the magnetization curve of the particles. The low-frequency field component f_2 is used to periodically drive the magnetic particles into the nonlinearity regime. Therefore, its amplitude has to be sufficiently high, in the range of several milli-tesla. Each time the low-frequency field is close to its absolute maximum, the magnetic particles are magnetically saturated, whereas at the time of zero crossings, they are in their linear regime. Thus, the particles are switched between linear and nonlinear behavior with a frequency $2 \times f_2$, equal to twice the driver frequency (Meyer *et al.* 2007).

The high-frequency field component f_1 serves as a probe of the nonlinearity of the magnetization curve. A frequency around 50 kHz constitutes a good compromise, yielding a high field sensitivity of the detection coil and an acceptable probing field amplitude of a few hundreds of micro-tesla (Meyer *et al.* 2007). The response of the superparamagnetic particles to this probing field will differ depending on the respective state of the driving field. In case of saturation, the response will be lower than in the case of linearity. Thus, it is obvious that the simultaneous presence of both the driving field at frequency f_2 and the probing field at frequency f_1 leads to the appearance of a sum frequency at frequency $f_1 + 2 \times f_2$. In addition to this mixing component, other linear combinations also appear.

Figure 8.41 shows the experimental set-up. An 80 MHz quartz oscillator serves as a common reference. All the frequencies in the measurement system are derived from this clock to ensure that they are phase-locked, thus eliminating problems from frequency drift. Coaxial coils provide magnetic excitation fields at two distinct frequencies $f_1 = 49.38$ kHz and $f_2 = 61$ Hz incident on the sample. By means of a differential pickup coil, the response signal of the sample inside the coil at a frequency $f_1 + 2f_2$ is detected. This mixing component was chosen since it is maximal for a vanishing static offset field. It is important to balance the differential pickup coil well against the middle excitation coil in order to minimize direct induction at $f_1 = 49.38$ kHz in the detection coil. The signal is retrieved by first demodulating the probing frequency, f_1, and subsequently demodulating at twice the driving frequency, $2f_2$. The output signal is low-pass filtered in order to suppress the carrier frequencies (Meyer *et al.* 2007).

As an example this method was used to detect the c-reactive proteins (CRPs). CRP is a very significant human blood marker for inflammatory processes and is routinely determined for many clinical purposes. For the purpose of fast and easy CRP-detection combined with a high specificity, the two anti-CRP antibodies clone C2 and clone C6 are used. One of them (C2) was used as capture antibody, adsorptively bound to the solid phase of the measurement column. The second antibody (C6) served as a detection label antibody, meaning that it was biotinylated and attached to the streptavidin coated magnetic beads. As shown in Figure 8.42, this antibody–magnetic bead complex interacts with the captured antigen on the measurement column and can be quantified by the magnetic sensor (Meyer *et al.* 2007).

Figure 8.43 shows a standard calibration curve for the CRP detection based on magnetic beads. Measurements were performed in pH7.3 PBS-buffer (0.15 M). The measurable dynamic range lasts is from 25.0 ng CRP/ml (28.5 pM) to 2.5 µg (2.9 nM) CRP/ml. Higher signals are obtainable, but leave the linear dynamic range. The detection limit of 10 times the technical background noise is located between 10.0 and 25.0 ng CRP/ml (Meyer *et al.* 2007).

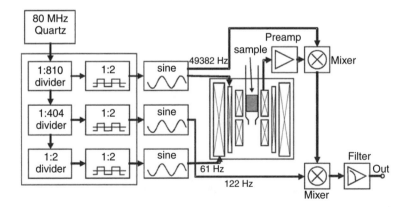

Figure 8.41 Set-up for the detection of magnetic nanoparticles using the frequency mixing technique. The frequency of an 80 MHz quartz oscillator is divided by three counters. The magnetic field B_1 of frequency $f_1 = 49382$ Hz is generated by a 1/1620 division of the oscillator frequency, the magnetic field B_2 of frequency $f_2 = 61$ Hz by an additional 1/808 division. After demodulation with f_1, the frequency component $2f_2$ is demodulated in a second stage and used as the output signal. (Meyer *et al.* 2007)

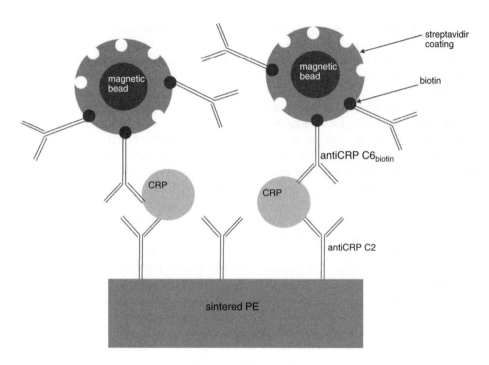

Figure 8.42   Magnetic sandwich bioassay for the detection of c-reactive protein (CRP). (Meyer *et al.* 2007)

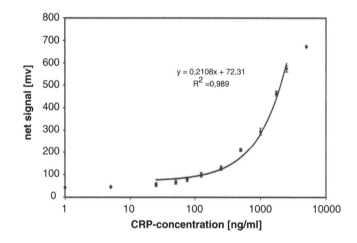

Figure 8.43   Calibration curve of magnetic bead-based CRP biosensor assay in pH 7.3 PBS-buffer (0.15 M). Dynamic detection range is linear, but appears exponential due to the logarithmic concentration axis. Net signal denotes the signal without the blank value signal. (Meyer *et al.* 2007)

8.4.3   Superconducting Quantum Interference Detectors

Superconducting quantum interference detectors are among the most sensitive sensors. In the following, we discuss the working principles of SQUID magnetometers, followed by the detection of magnetic particles bound to a substrate. SQUID can also be used as a sensor based on magnetic relaxations, as discussed in subsection 8.5.1.

8.4.3.1 Working Principles of SQUID Magnetometers

SQUID magnetometers have the ability to convert minute changes in current or magnetic field to a measurable voltage, and this ability is based on the principles of superconductivity, the Meissner effect, flux quantization and the Josephson effect (Clarke 1996). A Josephson junction is a weak link between two superconductors that is capable of carrying supercurrents below a critical value I_c. The weak link can be either a thin layer of insulator, an area where the superconductor itself narrows to a very small cross-section or a superconducting bridge between two superconducting sections. When a superconducting ring interrupted by a weak link is exposed to an external magnetic field, a shielding supercurrent flows around the inner surface of the ring via the weak link. In this case the supercurrent will be an oscillating function of the magnetic field intensity, such that it first rises to a peak as the field increases, then falls to zero then increases again and so on. In a SQUID, these periodic variations are exploited to measure the current in the superconducting ring and, hence, the applied magnetic field (Mahdi *et al.* 2003).

As depicted in Figure 8.44, there are two types of SQUIDs: RF and DC. An RF SQUID is actually a superconducting ring interrupted by one Josephson junction; while in a DC SQUID, the superconducting ring interrupted by two Josephson junctions. The difference between the two is in the nature of the biasing current being an RF or a DC. In either type, the special properties of the Josephson junction cause the impedance of the SQUID to be a periodic function of the magnetic flux threading the ring. This makes a SQUID function as a flux-to-voltage converter with the highest ever known magnetic sensitivity. Depending on the superconductors, we have low temperature (LTS) SQUIDs operating at liquid helium temperature (4.2 K), and high temperature (HTS) SQUIDs operating at liquid nitrogen temperatures (77 K and above). LTS devices are mostly DC SQUIDs, while HTS devices can be either DC or RF SQUIDs, fabricated from the ceramic oxide known as YBCO (Mahdi *et al.* 2003).

As the SQUID ring is too small to detect weak fields, a superconducting transformer with a pick-up loop forming a gradiometer is usually added to the device. The gradiometer discriminates strongly against distant noise sources, which have small gradient, in favor of locally generated signals and thus enhance the sensitivity of the SQUID magnetometer (Mahdi *et al.* 2003). A complete SQUID magnetometer comprises extra coupling and

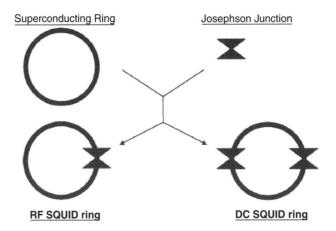

Figure 8.44 Basic components of DC and RF SQUID rings. (Mahdi *et al.* 2003)

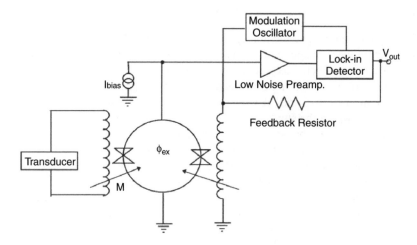

Figure 8.45   A simplified circuit of a flux-locked DC SQUID magnetometer. (Mahdi *et al.* 2003)

read-out units to the basic SQUID ring, as shown in Figure 8.45. The actual sensitivity of a SQUID magnetometer is dependent on the sensitivity of the added units and on the intrinsic and external noise. Different noise cancellation methods using gradiometers of various orientations and orders, with additional filtering techniques, have been developed and successfully used to improve the sensitivity of SQUID magnetometers, particularly in cases where weak magnetic signals immersed in noise are to be detected in unshielded environments (Bick *et al.* 2001).

8.4.3.2   Measurement of Magnetic Particles' Bond to a Substrate

Enpuku *et al.* (2005) have developed a magnetic immunoassay using a SQUID. The magnetic field from the marker that couples to the antigen is detected with the SQUID. Compared to the conventional optical system, the SQUID system is expected to have high sensitivity and the capability to detect in the liquid state. Figure 8.46 schematically shows the working principle of magnetic immuno-assay utilizing the magnetic marker and the SQUID. The antigen is detected by using its antibody that selectively couples to the antigen. The antibody is labeled with the marker made of magnetic nanoparticle with diameter d. The binding reaction between the antigen and its antibody is detected by measuring the magnetic field from the marker.

The magnetic signal from the marker depends on the magnetic property of the nanoparticle. The magnetic property of the nanoparticle is strongly affected by the thermal noise, i.e., relaxation of magnetization occurs due to thermal noise. The detection method of the nanoparticle must be chosen depending on the size of the particle. Using the anisotropy energy density K and the volume $V = (4\pi/3)(d/2)^3$ of the particle, the relaxation time τ can be expressed as:

$$\tau = \tau_0 \exp(KV/k_B T) \tag{8.30}$$

where $\tau_0 = 10^{-9}$ s is the characteristic time. For typical value of $K = 13\,\text{kJ/m}^3$, the relaxation time becomes $\tau = 2.6 \times 10^{-7}$ s for $d = 15\,\text{nm}$ and T = 300 K. In this case, remanence cannot be kept, and the particle shows superparamagnetic property. When the size is increased to $d = 25\,\text{nm}$, the relaxation time becomes $\tau = 144$ s. In this case,

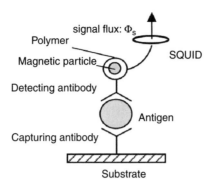

Figure 8.46 Magnetic immuno-assay using a SQUID. The antibody is labeled with the magnetic marker made of Fe_3O_4 nanoparticle. The binding reaction between an antigen and its antibody is detected by measuring the magnetic field from the marker. (Enpuku *et al.* 2005). Reproduced by permission of IEICE

the particle can keep remanence after magnetization. Here Fe_3O_4 nanoparticles with d = 25 nm are used (Enpuku *et al.* 2005).

Figure 8.47(a) shows a SQUID system for the magnetic immuno-assay. The sample is a square with side length 2a, and the number of the magnetic particles in the sample is N, and the volume of each particle is V. The sample is magnetized in the z direction outside the SQUID system, and each particle has the magnetic moment M per unit volume. The magnetic moment m_z per unit area is given by $m_z = NVM/(2a)^2$. As shown in Figure 8.47(b), the pickup coil for the SQUID is made of thin film with width w_p, and the outermost side length is 2b. The distance between the SQUID and the sample is z, and the z- component of the magnetic field B_z is detected.

Figure 8.48(a) schematically shows the sample used in the experiment configuration, where two antigens called IL8 and human IgE were used. To align the direction of the moment and to generate remanence of the sample, a field of 0.1 T was applied perpendicular to the substrate outside the SQUID system. Then, the sample was inserted into the SQUID system, and the remanence field of the sample measured, where the time interval

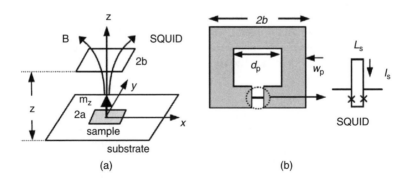

(a)   (b)

Figure 8.47 (a) Measurement of the magnetic field from the assembly of magnetic nanoparticles. The size of the sample is 2a, and the distance between the sample and the SQUID is z. The sample has the magnetic moment m_z per unit area; (b) directly coupled SQUID. Square pickup coil with side length 2b is made of thin film with width w_p. (Enpuku *et al.* 2005). Reproduced by permission of IEICE

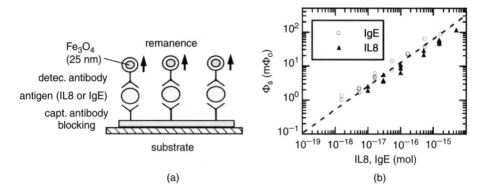

Figure 8.48 (a) Schematic figure of the sample. Two antigens called IL8 and human IgE are used for the experiment; (b) relationship between the signal flux Φ_s and the molecular concentration of IL8 and IgE. (Enpuku *et al.* 2005). Reproduced by permission of IEICE

between magnetization and measurement was about one minute. Figure 8.48(b) shows the relationship between the signal flux Φ_s and the molecular concentration of the antigen. A good relationship between Φ_s and the number of antigens was obtained for both samples. A few atto-mol of IL8 and IgE can be detected, which shows the high sensitivity of this method (Enpuku *et al.* 2005).

8.5 Sensors Detecting Magnetic Relaxations

Magnetorelaxometry (MRX) is a powerful analytical tool for the specific detection of biological molecules, such as proteins, bacteria or viruses. The basic idea of MRX is that moments of magnetic nanoparticles are aligned by a magnetic field and then, after switching off the field, the decay of the net magnetic signal as a function of time is analyzed (Ludwig *et al.* 2005; Perez *et al.* 2002). Mobile magnetic nanoparticles relax via the Brownian mechanism on a time scale of microseconds, whereas nanoparticles that are immobilized, e.g., by binding to a large biomolecule, relax via the Neel mechanism. Therefore, bound and unbound magnetic nanoparticles (MNPs) can be distinguished by their different relaxation times and time dependencies. Measurement times typically amount to a few seconds depending on the number of averages.

Usually MRX is performed using superconducting quantum interference devices (SQUIDs) known to be the most sensitive solid-state magnetic field sensors, while fluxgate is a simple method which does not require complicated instruments.

8.5.1 SQUID MRX

Grossman *et al.* (2004) developed a technique for the detection of magnetically labeled *Listeria monocytogenes* and for the measurement of the binding rate between antibody-linked magnetic particles and bacteria. Using this technique to quantify specific bacteria, the bacteria do not need to immobilized, and the unbound magnetic particles do not need to be washed away. In the measurement, a pulsed magnetic field is applied to align the magnetic moments, and when the pulsed magnetic field is turned off, a SQUID is used to detect the magnetic relaxation signal. Brownian rotation of unbound particles is too quick to be detected. On the other hand, the particles bound to *L. monocytogenes* relax in about one second by rotation of the internal dipole moment. Such a Neel relaxation process can be detected by SQUID, and the binding rate between the particles and bacteria can be obtained by time-resolved measurements.

As shown in Figure 8.49, 50 nm-diameter γ-Fe_2O_3 particles were coupled to polyclonal antibodies raised against the bacterial pathogen *Listeria monocytogenes* and added to a suspension of that organism. After allowing time for the particles to bind to the targets, the sample was placed in a SQUID and a pulsed magnetic field was applied to align the magnetic dipole moments. Each time the field is turned off, the SQUID detects the magnetic relaxation signal. Unbound particles relax in about 50 µs by Brownian rotation, and this time is too short for the SQUID system to measure. However, the particles bound to the relatively large bacteria undergo Neel relaxation, in which their internal dipole moments relax to the lowest energy state. The resulting magnetic decay, which occurs in about 1 s, is detected by the SQUID. Because the measured magnetic relaxation is due only to the bound particles, changes in the magnetic relaxation amplitude over time indicate the rate at which particles bind to bacteria (Grossman *et al.* 2004).

This method does not require immobilization of the targets or washing away of the unbound particles. The bound and unbound particles can be differentiated by the different mechanisms by which they relax after the removal of a magnetic field. Brownian relaxation is a physical rotation of the particles, with a relaxation time:

$$\tau_B = 3\eta V_H / k_B T \tag{8.31}$$

where η is the viscosity of the medium, V_H is the hydrodynamic volume, k_B is Boltzmann's constant and T is the temperature. Taking $T = 293\,K$ and $\eta = 10^{-3}\,kg \cdot m^{-1} \cdot s^{-1}$, τ_B is about 50 µs for sphere particles with a hydrodynamic diameter of 50 nm (Grossman *et al.* 2004).

Neel relaxation originates from the anisotropy of the crystalline lattice. Many magnetic materials have an easy axis of magnetization. When the crystal is magnetized along the easy magnetization axis, the energy is minimized. If an external field rotates the magnetization away from the easy axis, the magnetization eventually returns to its preferred

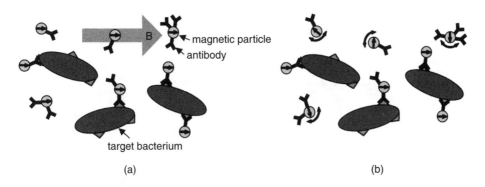

(a)   (b)

Figure 8.49 Procedure of magnetorelaxometry. A suspension of superparamagnetic particles, coupled to antibodies, is added to the liquid sample. (a) A magnetic field is applied to align the magnetic moments of the particles; (b) as the magnetic field is turned off at time t $\approx \tau_B$, the orientations of the unbound particles are randomized due to the Brownian rotation, whereas particles bound to bacteria are still aligned. The magnetic moments of the bound particles reorient slowly by means of Neel relaxation. (Grossman *et al.* 2004). Published with permission of the National Academy of Sciences, USA

direction on removal of the field. The Neel relaxation time for a single domain particle is:

$$\tau_N = \tau_0 \exp(K V_M / k_B T) \tag{8.32}$$

where τ_0 is about 10^{-9} s, K is the magnetic anisotropy constant and V_M is the magnetic core volume. The parameters τ_0 and K depend on the shape of the particle, and values of τ_0 vary by up to four orders of magnitude. By assuming that the magnetic core of each particle consisted of a cluster of 10 nm γ-Fe$_2$O$_3$ nanoparticles (K $\approx 2.5 \times 10^4$ J·m^{-3}), Equation (8.32) predicts $\tau_N \approx 25$ ns for an individual 10 nm nanoparticle at T = 293 K. As the magnetic interactions between the nanoparticles within each core slow down the overall relaxation rate, the Neel relaxation time of these particles fall within the 1 ms to 1 s measurement window of the SQUID system (Grossman *et al.* 2004).

Figure 8.50 shows the measurement configuration and the basic structure of the SQUID. The voltage across the current-biased SQUID oscillates quasi-sinusoidally as a function of the magnetic flux Φ threading the loop with a period of the magnetic flux quantum, $\Phi_0 = h/2e \approx 2 \times 10^{-15}$ T·m^{-2}. To linearize the flux-to-voltage conversion, the SQUID is operated in a flux-locked loop that maintains the flux through it at a constant value; the output voltage of this feedback circuit is proportional to Φ.

Figure 8.51 shows typical time traces for an *L. monocytogenes* sample and associated controls. These data were fitted to a sum of logarithmic and exponential functions. The logarithmic decay is characteristic of Neel relaxation for particles with a wide distribution of sizes, and therefore of relaxation times. The exponential decay comes from particle aggregates, formed after the filtration step, which are large enough to relax via Brownian rotation on a measurable timescale without being bound to targets. The fitting function is:

$$\Phi(t) = \Phi_{offset} + \Phi_s \ln(1 + \tau_{mag}/t) + \Phi_{exp} \exp(-t/\tau_{exp}) \tag{8.33}$$

where Φ_{offset} is an offset caused by the fact that the SQUID measures relative, rather than absolute, magnetic flux; Φ_s, the logarithmic decay amplitude, is proportional to the number of bound particles; $\tau_{mag} = 1$ s is the magnetization time; Φ_{exp}, the exponential

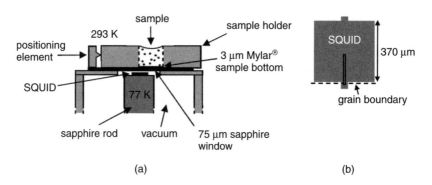

(a)   (b)

Figure 8.50 (a) Top portion of the SQUID microscope. The SQUID, inside a vacuum enclosure, is mounted on a sapphire rod thermally connected to a liquid nitrogen reservoir (not shown). A 75 μm-thick sapphire window separates the vacuum chamber from the atmosphere. The sample is contained in a Lucite holder, with a 3 μm-thick Mylar base, aligned against a positioning element; (b) configuration of the YBCO SQUID. The slit is 4 μm wide. (Grossman *et al.* 2004). Published with permission of the National Academy of Sciences, USA

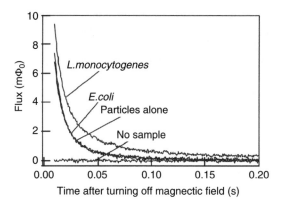

Figure 8.51 Example of magnetic decay signals. For the traces shown, the concentration of bacteria was 10^8 per ml, and the concentration of particles was 0.13 relative to the stock suspension. A 0.4 mT field was pulsed on for 1 s and off for 1 s, and data were recorded each time the field was turned off; 100 averages were taken. (Grossman *et al.* 2004). Published with permission of the National Academy of Sciences, USA

decay amplitude, depends on the number of unbound particle aggregates; and τ_{exp} is the exponential decay time constant. Fitting Equation (8.32) to the measurement results gives $\tau_{exp} \approx 15$ ms, corresponding to a hydrodynamic diameter of ≈ 340 nm for a sphere (Grossman *et al.* 2004).

To investigate whether the observed increase in response was due to the specific interaction between *L. monocytogenes* and magnetic particles, the *L. monocytogenes* was exchanged for *E. coli*. As shown in Figure 8.51, because the particle–antibody complexes show little cross-reactivity to *E. coli*, the 'E. coli' and 'particles alone' curves overlay each other.

8.5.2 Fluxgate MRX

Figure 8.52 schematically shows MRX set-ups using a single fluxgate magnetometer (a) and a differential fluxgate configuration (b). In both cases, the fluxgate magnetometers

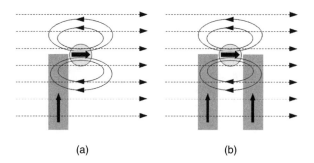

(a)   (b)

Figure 8.52 Schematic drawing of MRX set-ups using (a) a single fluxgate sensor and (b) a differential arrangement using two fluxgate sensors. Arrows in the fluxgate magnetometers indicate sensitive axes. Magnetizing field (dotted lines) is oriented perpendicularly to the sensitive axes of the fluxgate sensors. Arrows in the circularly shaped sample indicate direction of its magnetic moment. (Ludwig *et al.* 2005)

and the magnetic nanoparticle sample are located in the middle of a Helmholtz coil, which provides a homogeneous magnetic field over relatively large samples and over the magnetic field sensors. As shown in Figure 8.52(a), using a single fluxgate magnetometer with its sensitive axis oriented perpendicularly to the magnetizing field, the signal coupled by a magnetic dipole to the fluxgate is maximum when the dipole is located at one of the ends of the rod-like core. A disadvantage of the single fluxgate arrangement is that only nanoparticles that are relatively close to the fluxgate sensor contribute to the signal due to the $1/r^3$ decay of the magnetic field of a magnetic dipole. In the differential configuration shown in Figure 8.52(b), the sample is placed symmetrically between two parallel fluxgates with their sensitive axes being oriented perpendicularly to the magnetizing field and thus to the magnetic dipole axis. Thus, with no magnetic sample, ideally no magnetic field from the Helmholtz coil is seen by the magnetometers. Taking the difference of the two fluxgate magnetometer signals as output, spatially homogeneous environmental fields are cancelled whereas the magnetic signals from the sample add constructively. Thus, this set-up can perform MRX measurements in a magnetically disturbed environment. In addition, using the differential set-up, the total signal is expected to be twice that of a single fluxgate, resulting in an increase of the signal-to-noise ratio (SNR) by a factor of $\sqrt{2}$ compared to the single-sensor set-up (Ludwig et al. 2005).

Figure 8.53 shows the 20 times averaged relaxation curves measured on the highest concentration freeze-dried magnetite sample, without any magnetic shielding. The sample was magnetized in a field of 2 mT for 2 s and data were recorded for 1.5 s (Ludwig et al. 2005). For comparison, the difference signal is depicted along with the signals measured with the two individual fluxgates. The magnetized sample causes the signals in both fluxgates to have opposite signs, whereas the amplitude of the difference signal is twice that of each individual sensor. Furthermore, it can be seen that the differential fluxgate configuration effectively suppresses 50 and 150 Hz signals detected by the individual sensors.

One advantage of using fluxgates compared to SQUID magnetometers is that a fluxgate measures absolute flux densities. In spite of the lower resolution compared to SQUID MRX, the differential fluxgate set-up has the advantage of being a compact and robust measurement system that does not require any cryogenic cooling and magnetic shielding. This method has the potential for an on-line bio-analytical tool without the need to wash away unbound markers (Ludwig et al. 2005).

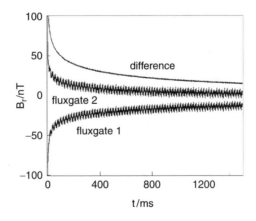

Figure 8.53 Magnetic relaxation signal measured with the two individual fluxgate magnetometers and the difference signal. Measurements were performed in a magnetically unshielded environment on a freeze-dried magnetite sample (Fe content $\approx 14\,\mu$mol). (Ludwig et al. 2005)

8.6 Sensors Detecting Ferrofluid Susceptibility

8.6.1 Theoretical Background

The binding of biomolecules to colloidal magnetic particles in suspension can be derived from the frequency dependence of the magnetic susceptibility of the magnetic colloids. This method is based on the fact that binding with biomolecules increases the hydrodynamic radii of the microbeads, and thus the frequency dependence of the magnetic susceptibility of the fluid will be shifted (Connolly and St Pierre 2001; de Oliveira *et al.* 2005).

The complex magnetic susceptibility of a fluid containing monodisperse magnetic particles is given by Equations (8.7) and (8.8). However, the magnetic particles in an actually magnetic fluid usually have a radius distribution. The susceptibilities then become (Connolly and St Pierre 2001):

$$\chi'(\omega) = \int_0^\infty \frac{\chi_0 p(r)}{1 + (c\omega r^3)^2} dr \tag{8.34}$$

$$\chi''(\omega) = \int_0^\infty \frac{\chi_0 c\omega r^3 p(r)}{1 + (c\omega r^3)^2} dr \tag{8.35}$$

where the parameter c is given by $c = 4\pi\eta/kT$, and p(r)dr is the probability of a particle having a radius between r and (r + dr). Usually the hydrodynamic radii in suspension follow a normal distribution:

$$p(r) = \frac{1}{\sigma\sqrt{2\pi}} \exp\left[\frac{-(r - r_m)^2}{2\sigma^2}\right] \tag{8.36}$$

where σ is the standard deviation (SD) and r_m is the mean hydrodynamic radius of the distribution.

Figure 8.54(a) shows the calculated frequency-dependent susceptibility of a magnetic suspension. The hydrodynamic radii of the magnetic particles in the suspension have a normal distribution with a mean of 75 nm and a standard deviation of 7.5 nm (Connolly and St Pierre 2001). As shown in Figure 8.54(b), if these particles are coated with a layer of biomolecules (1 nm), the spectrum of the susceptibility will be shifted by about 10 Hz. Figure 8.54 also indicates that a 1:1 mixture of the two types of particles with mean radii 75 nm (SD 7.5 nm) and 76 nm (SD 7.5 nm) respectively, has a susceptibility spectrum which is different from that of either pure type. The susceptibility spectrum of a mixture, consisting of two types of particles with very different distributions of hydrodynamic radii, will exhibit two discrete peaks corresponding to the rotational relaxations of the two types of particles.

Biomolecules could be detected by using AC magnetic susceptibility measurements to monitor the binding-induced modification of Brownian relaxation of magnetic nanoparticles suspended in liquids. This substrate-free detection scheme has several advantages (Connolly and St Pierre 2001; Chung *et al.* 2005): (i) it generates a useful signal in both the presence and absence of the target, thus providing an inherent check for integrity; (ii) it permits discrimination between several potential targets, since beside the binding affinity additional information about the target size can be obtained; and (iii) it can generate quantitative information about the target concentration. We discuss below three methods for the measurement of complex magnetic susceptibility: the slit toroid method, coil method and PPMS method.

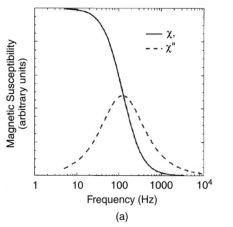 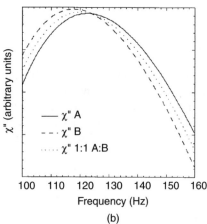

(a)   (b)

Figure 8.54   (a) Real part $\chi'(\omega)$ and imaginary part $\chi''(\omega)$ of the complex susceptibility for an aqueous suspension with blocked magnetic particles at 20 °C. The particles have a normal distribution of hydrodynamic radii, with mean radius 75 nm and standard deviation 7.5 nm; (b) $\chi''(\omega)$ for aqueous suspensions of blocked particles with hydrodynamic radii of (A) $r_m = 75$ nm and (B) $r_m = 76$ nm. The standard deviation for both (A) and (B) is 7.5 nm. The curve marked (1:1 A:B) represents the $\chi''(\omega)$ value for a 1:1 mixture of type (A) and type (B) particles. All the calculations are made according to Equations (8.34)–(8.36). (Connolly and St Pierre 2001)

8.6.2   Slit Toroid Method

Fannin *et al.* (1986) developed a slit toroid method to measure frequency dependence of $\chi'(\omega)$ and $\chi''(\omega)$, as shown in Figure (2.17). In this method, a mumetal toroid with a narrow slit wound with 20 turns of wire is used. The sample under test can be held in the slit by the surface tension of the fluid, and an alternating current runs through the wire, producing an alternating magnetic field in the toroid and the slit. An impedance analyzer is used to measure the impedance of the toroid.

The impedances of the coil-toroid system at three different states are measured as functions of frequency. In the first state, the slit is empty; in the second state, the slit is filled with magnetic fluid; and in the third state, the slit is filled with magnetic fluid plus biomolecules. From the impedances of the above three states, the information about the bindings between biomecules and magnetic particles can be obtained (Connolly and St Pierre 2001).

8.6.3   Coil Methods

8.6.3.1   Inductance Method

Kriz *et al.* (1996) developed a transducer concept in biosensors, utilizing measurements of magnetic permeability. As the magnetic permeability of a material inside a coil influences the inductance of the coil, it is possible to detect changes in magnetic permeability using inductance measurements. The inductance L for a coil with a magnetic material inside is described by:

$$L = (\mu_r \mu_0 A / l) N^2 \qquad (8.37)$$

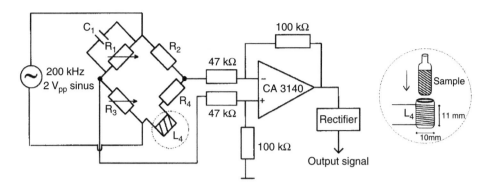

Figure 8.55 The measurement system. Measuring coil L_4 (transducer) is part of a balanced Maxwell bridge. The other components in the Maxwell bridge are $R_1 = 2860\,\Omega$, $C_1 = 1\,nF$, $R_2 = 180\,\Omega$, $R_3 = 27\,\Omega$, $R_4 = 0.44\,\Omega$ and $L_4 = 4.9\,\mu H$. A sinusoidal wave of 200 kHz and 2 Vp-p is fed into the bridge. The voltage difference measured over the bridge is further processed by a differential operational amplifier circuit, rectified and finally recorded. (Kriz *et al.* 1996)

where μ_r is the relative magnetic permeability of the material in the coil, μ_0 is the permeability of a vacuum, A is the cross-section area, l is the coil length and N is the turn number on the coil.

To measure the inductance, and thus indirectly the relative magnetic permeability, the coil can be placed in a Maxwell bridge (Kriz *et al.* 1996). As shown in Figure 8.55, the transducer measures the changes in the magnetic permeability of materials and comprises a coil which is a part of a balanced Maxwell bridge, with two variable resistances. These are needed because both phase and amplitude have to be balanced. A sinusoidal wave is fed into the bridge. The bridge is balanced when the following equations are fulfilled: $L_4 = R_2R_3C_1$ and $R_4 = R_2R_3/R_1$. The voltage difference measured over the Maxwell bridge is further processed by a differential operational amplifier circuit and rectified. The introduction of ferromagnetic materials inside the coil causes an increase in the voltage difference over the Maxwell bridge.

Figure 8.56 shows the response obtained from the Maxwell bridge as a function of the initial concentration of the ferromagnetic model analyte (dextran ferrofluid) in a sample solution. As expected, saturation is achieved for higher concentrations. To investigate whether the observed increase in response was due to the specific interaction between Con A Sepharose and dextran ferrofluid, the Con A Sepharose was exchanged for Sepharose. In this case, no response could be observed, as shown by the lower curve in Figure 8.56.

8.6.3.2 Resonant Frequency Method

Coated paramagnetic particles (PMPs) can be readily obtained with a variety of different coatings. A PMP consists of a paramagnetic core coated with a suitable polymer layer attached to which is the antibody–antigen layer depending on the particular application of the PMP. Here, we discuss an instrument capable of directly measuring the number of PMPs on a strip (Richardson *et al.* 2001). As shown in Figure 8.57, in the instrument, a plastic strip with immobilized PMPs on it is inserted into a coil of wire. The technique for detecting the PMPs relies on the paramagnetic nature of the particles and the effect they have on the self-inductance, L, of the coil.

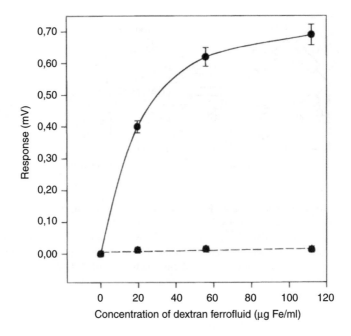

Figure 8.56 Response obtained from a Maxwell bridge as a function of various initial concentrations of dextran ferrofluid incubated with Con A Sepharose (upper curve) and with Sepharose (lower curve). (Kriz *et al.* 1996)

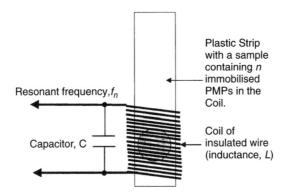

Figure 8.57 The number of paramagnetic particles, PMPs, immobilized on a plastic strip can be determined by placing the strip in a coil parallel with a capacitor and measuring the resonant frequency. (Richardson *et al.* 2001). Reprinted from: Richardson, J. Hill, A. Luxton, R. and Hawkins, P. (2001). "A novel measuring system for the determination of paramagnetic particle labels for use in magneto-immunoassays", *Biosensors and Bioelectronics*, **16**, 1127–1132, with permission from Elsevier

The self-inductance L of a uniform helical coil with a large number of turns wound on a former can be expressed as (Richardson *et al.* 2001):

$$L = L_0 + kn \qquad (8.38)$$

with L_0 corresponding to the inductance of the coil containing a plastic strip with just the buffer solution residue and no PMPs on it, and the constant k is given by:

$$k = c\mu_p\mu_0 m^2 dA \tag{8.39}$$

where μ_0 is the permeability of a vacuum, m is the turn number per unit length, d is the length, A is the cross-sectional area of the coil, c is a constant relating the permeability of individual PMPs and μ_p is the relative permeability of PMPs. Equation (8.38) predicts that the inductance of the coil increases linearly with the number of PMPs on the plastic strip.

Putting the coil in parallel with a capacitor, C, forms a resonant inductor–capacitor (LC) circuit. If the internal resistance of the coil is negligible, the resonant frequency of the circuit is given approximately by (Richardson *et al.* 2001):

$$f_n = f_0\left[1 - \frac{1}{2}(k/L_0)n\right] \tag{8.40}$$

where $f_0 = [2\pi(L_0C)^{1/2}]^{-1}$ corresponds to the resonant frequency of the circuit containing a plastic strip with just the residue buffer solution and no particles. Equation (8.40) predicts that the frequency of oscillation of the LC-resonant circuit decreases linearly with increasing numbers of paramagnetic particles present on the strip.

Figure 8.58 shows an oscillator circuit based on a phase-locked loop (PLL) circuit, which locks the output frequency of a voltage-controlled sine wave oscillator (VCO) onto the resonant frequency of the LC circuit (Richardson *et al.* 2001). The circuit makes use of the fact that at the resonant frequency, f_n, given by Equation (8.40), the current flowing

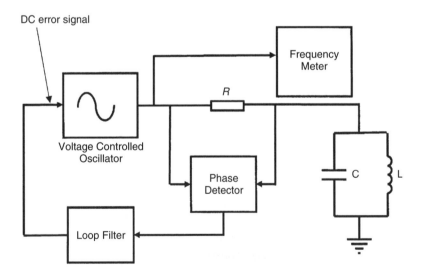

Figure 8.58 Phase-locked loop circuit used to determine the resonant frequency of the coil. (Richardson *et al.* 2001). Reprinted from: Richardson, J. Hill, A. Luxton, R. and Hawkins, P. (2001). "A novel measuring system for the determination of paramagnetic particle labels for use in magneto-immunoassays", *Biosensors and Bioelectronics*, **16**, 1127–1132, with permission from Elsevier

into the LC circuit and the applied voltage are in phase. At this frequency, the impedance of the LC circuit is entirely resistive and also very large. At resonance then, the voltages across the resistor, R, in series with L and C are in phase. The phases of the voltages across R are compared by a phase detector, which produces a DC-error signal that is fed back to the VCO. The error signal changes the output frequency of the VCO until there is no difference in phase in the voltages across R. The output frequency of the VCO is now locked onto the resonance frequency of the LC circuit. When a sample is placed in the coil, the resonant frequency, f_n, changes according to Equation (8.40) and the output frequency of the VCO follows it. The output frequency of the VCO is measured on a frequency meter.

The design of sample coil makes a great difference to the sensitivity of the instrument (Richardson *et al.* 2001). Figure 8.59 shows three coils with different designs. The coil shown in Figure 8.59(a) had a rectangular cross-section. A former for the coil was constructed from 1 mm-thick plastic pieces. The coil that does not require a former was based on a ferrite ring, as shown in Figure 8.59(b). Due to the high magnetic permeability of ferrite, a coil with a ferrite core requires a relatively small number of turns to have a high inductance. A 2 mm wide slot was cut into the ferrite ring so that a test strip could be inserted into the gap. As the gap between the two ends of the ferrite is small, little of the electromagnetic field produced by the coil is lost and most passes through the sample

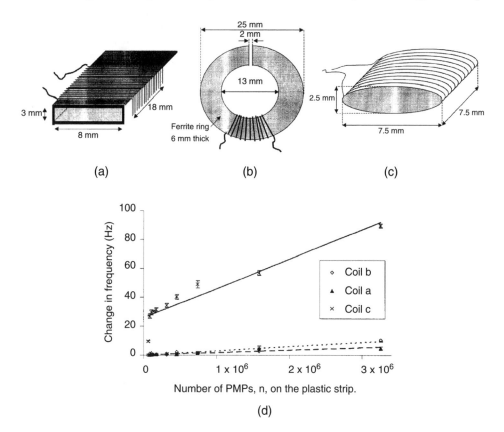

Figure 8.59 Three difference designs of sample coils (a)–(c); (d) the change in frequency of the PLL circuit for the three different coil designs. (Richardson *et al.* 2001). Reprinted from: Richardson, J. Hill, A. Luxton, R. and Hawkins, P. (2001). "A novel measuring system for the determination of paramagnetic particle labels for use in magneto-immunoassays", *Biosensors and Bioelectronics*, **16**, 1127–1132, with permission from Elsevier

strip. The presence of the PMPs in the slot increases the inductance of the coil as before. Equation (8.40) also predicts that the sensitivity of the measuring circuit would increase if the coil has a higher resonant frequency. A coil was designed that could be operated at a higher frequency, as shown in Figure 8.59(c). This coil has an oval cross-section and does not have a former. The dead space in the coil was kept to a minimum by making the width of the coil just large enough to allow easy access for the test strips into the coil. The responses of the PLL circuit with the above three types of sample coils are shown in Figure 8.59(d), and it is clear that coil c has the highest sensitivities.

8.6.3.3 Gradiometric Method

Here we discuss an inductive detection principle based on the nonlinearity of the magnetization of superparamagnetic particles. Figure 8.60 shows the block diagram of a system developed by Lany *et al.* (2005). Both the excitation and the detection coils are copper traces on a 0.8 mm-thick printed circuit board (PCB). The excitation coil is tuned at the oscillator frequency and matched to 50 Ω. A capacitance in parallel with the detection coil is adjusted to form a resonant circuit at the excitation frequency.

In order to reduce the signal in the absence of beads at the detection coil ends, two adjacent coils in series wound in opposite directions are used, as shown in Figure 8.61(a).

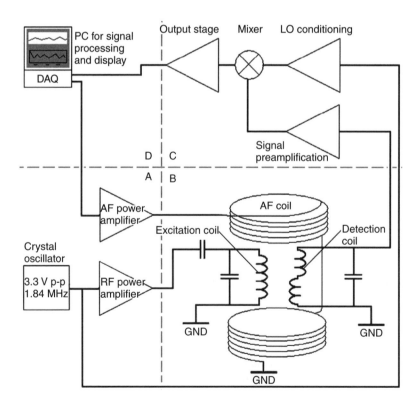

Figure 8.60 Block diagram of the system. (A) Electronics generating the RF excitation and audio frequency (AF) modulation currents; (B) AF, excitation and detection coils, and their mechanical assembly; (C) detection electronics performing amplification and demodulation; (D) the software running on a PC that processes the signal of the beads generates the AF signal, and displays the results. (Lany *et al.* 2005)

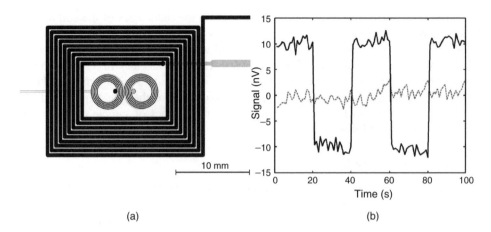

(a)   (b)

Figure 8.61   (a) Representation of the PCB coils. The layer in gray is on the opposite side of the PCB; (b) signal at the detection coils ends with the sample alternatively placed over each coil for 20 s. The gray dotted line is the result of the same procedure with an empty sample holder. The black line shows the signal obtained with the sample containing 3500 ± 500 microbeads. (Lany *et al.* 2005)

This gradiometric configuration ideally nulls the signal resulting from the direct inductive coupling between the excitation and detection coils. Figure 8.61(b) shows the measurement results of a sample containing approximately 3500 microbeads. The peak-to-peak signal amplitude is doubled by the measurement procedure. In the measurement, the integration time is 1 s, the modulation frequency is 37 Hz, the modulation amplitude is 22 mT and the excitation amplitude is about 0.5 mT.

8.6.4 PPMS Method

The AC magnetic susceptibilities of a solution can be measured using a Physical Property Measurement System (PPMS, Quantum Design, San Diego, CA). This system measures the sample's magnetic responses (amplitude and phase) by two detection coils while applying a small AC magnetic field to the sample by an AC drive and compensation coils.

Chung *et al.* (2005) studied biotinylated S-protein using this susceptibility method. As shown in Figure 8.62, upon adding biotinylated S-protein (6.3 μM) to the avidin coated magnetite nanoparticles, the peak frequency decreases from 210 to 120 Hz. Since S-protein does not exhibit any magnetic properties, this frequency shift has to be induced by the interaction of biotinylated S-protein with the avidin coated magnetic nanoparticles. Such an interaction will consequently lead to the increase of the nanoparticles' hydrodynamic radius, and the peak frequency of the AC magnetic susceptibility is inversely proportional to the particle volume.

To further test this approach, Chung *et al.* (2005) pretreated the biotinylated S-protein with T7 bacteriophage particles so that the biotinylated S-protein will be anchored to the surface of the T7 bacteriophage due to the specific interaction between the S-peptide and T7 bacteriophage. Subsequently such biotinylated T7 bacteriophage particles were added to the magnetic nanoparticle solution. As shown in Figure 8.61, the addition of biotinylated T7 bacteriophage suppresses the peak in the imaginary part of the AC susceptibility, indicating that the magnetic nanoparticles are immobilized upon binding to the T7 bacteriophage. The large phage particle cross-links the avidin coated magnetite

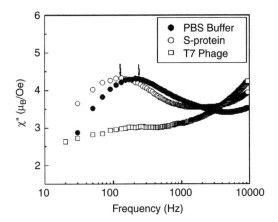

Figure 8.62 Imaginary part of the AC magnetic susceptibility of an avidin coated magnetic particle before (solid circles), and after (open circles) binding to S-protein, and after (open squares) binding to biotinylated T7 bacteriophage

nanoparticles, causing their aggregation, which immobilizes the particles. Therefore the rotational motion of the particles in the aggregate is blocked resulting in the disappearance of the Brownian relaxation frequency peak (Chung *et al.* 2005).

The above results demonstrate the feasibility of using the frequency peak of the AC susceptibility to monitor the attachment of the targeted molecules to magnetic nanoparticles, and thus provide a biosensor scheme based on the Brownian relaxation of magnetic nanoparticles in a liquid (Chung *et al.* 2005).

References

Arakaki, A., Hideshima, S., Nakagawa, T., Niwa, D., Tanaka, T. and Matsunaga, T. (2004). Detection of biomolecular interaction between biotin and streptavidin on a self-assembled monolayer using magnetic nanoparticles, *Biotechnology & Bioengineering*, **88**, 543–6.

Babb, C.W., Coon, D.R. and Rechnitz, G.A. (1995). Biological neurosensors. 3. noninvasive sensors using magnetic stimulation and biomagnetic detection, *Analytical Chemistry*, **67**, 763–9.

Baselt, D.R., Lee, G.U., Natesan, M., Metzger, S.W., Sheehan, P.E. and Colton, R. (1998). A biosensor based on magnetoresistance technology, *Biosensors and Bioelectronics*, **13**, 731–9.

Baselt, D.R., Lee, G.U., Hansen, K.M., Chrisey, L.A. and Colton, R.J. (1997). A high-sensitivity micromachined biosensor, *Proceedings of the IEEE*, **85**, 672–80.

Besse, P.A., Boero, G., Demierre, M., Pott, V. and Popovic, R. (2002). Detection of a single magnetic microbead using a miniaturized silicon Hall sensor, *Applied Physics Letters*, **80** (22), 4199–201.

Bick, M., Sternickel, K., Panaitov, G., Effern, A., Zhang, Y. and Krause, H.J. (2001). SQUID gradiometry for magnetocardiography using different noise cancellation techniques, *IEEE Transactions on Applied Superconductivity*, **11** (1), 673–767.

Boero, G., Utke, I., Bret, T., Quack, N., Todorova, M., Mouaziz, S., Kejik, P., Brugger, J., Popovic, R.S. and Hoffmann, P. (2005). Submicrometer Hall devices fabricated by focused electron-beam-induced deposition, *Applied Physics Letters*, **86**, 042503.

Chung, S.H., Hoffmann, A., Guslienko, K., Bader S.D., Liu, C., Kay, B., Makowski, L. and Chen, L. (2005). Biological sensing with magnetic nanoparticles using Brownian relaxation (invited), *Journal of Applied Physics*, **97**, 10R101.

Chung, S.H. Hoffmann, A., Bader, S.D., Liu, C., Kay, B., Makowski, L. and Chen, L. (2004). Biological sensors based on Brownian relaxation of magnetic nanoparticles, *Applied Physics Letters*, **85** (14), 2971–3.

Clarke, J. (1996). SQUIDs fundamentals, in *SQUID Sensors: Fundamentals, Fabrication and Applications*, (ed. H. Weinstock), Kluwer Academic Publishers, Dordrecht.

Connolly, J. and St Pierre T.G. (2001). Proposed biosensors based on time-dependent properties of magnetic fluids, *Journal of Magnetism and Magnetic Materials*, **225**, 156–60.

Could, P. (2004). Nanoparticles probe biosystems, *Materials Today*, **7** (2), 36–43.

de Oliveira, J.F., Wajnberg, E., Esquivel, D.M.S. and Alves, O.C. (2005). Magnetic resonance as a technique to magnetic biosensors characterization in Neocapritermes opacus termites, *Journal of Magnetism and Magnetic Materials*, **294**, e171–e174.

Ejsing, L., Hansen, M.F., Menon, A.K., Ferreira, H.A., Graham, D.L. and Freitas, P.P. (2005). Magnetic microbead detection using the planar Hall effect, *Journal of Magnetism and Magnetic Materials*, **293**, 677–84.

Ejsing, L., Hansen, M.F., Menon, A.K., Ferreira, H.A., Graham, D.L. and Freitas, P.P. (2004). Planar Hall effect sensor for magnetic micro- and nanobead detection, *Applied Physics Letters*, **84** (23), 4729–31.

Enpuku, K., Inoue, K., Yoshinaga, K., Tsukamoto, A., Saitoh, K., Tsukada, K., Kandori, A., Sugiura, Y., Hamaoka, S., Morita, H., Kuma, H. and Hamasaki, N. (2005). Magnetic marker and high T_c superconducting quantum interference device for biological immunoassays, *IEICE transactions on Electronics*, **E88C** (2), 158–67.

Fannin, P.C. , Scaife, B.K. and Charles, S.W. (1986). New technique for measuring the complex susceptibility of ferrofluid, *Journal of Physics E: Scientific Instruments*, **19**, 238–9.

Ferreira, H.A., Graham, D.L., Freitas, P.P. and Cabral, J.M.S. (2003). Biodetection using magnetically labeled biomolecules and arrays of spin valve sensors (invited), *Journal of Applied Physics*, **93** (10), 7281–6.

Freitas, P.P., Silva, F., Oliveira, N.J., Melo, L.V., Costa, L. and Almeida, N. (2000). Spin valve sensors, *Sensors and Actuators*, **81**, 2–8.

Gallagher, W.J., Parkin, S.S.P., Lu, Y., Bian, X.P., Marley, A., Roche, K.P., Altman, R.A., Rishton, S.A., Jahnes, C., Shaw, T.M. and Xiao, G. (1997). Microstructured magnetic tunnel junctions (invited), *Journal of Applied Physics*, **81** (8), 3741–6.

Graham, D.L., Ferreira, H.A., Feliciano, N., Freitas, P.P., Clarke, L.A. and Amaral, M.D. (2005). Magnetic field-assisted DNA hybridisation and simultaneous detection using micron-sized spin-valve sensors and magnetic nanoparticles, *Sensors and Actuators B*, **107**, 936–44.

Graham, D.L., Ferreira, H.A. and Freitas, P.P. (2004). Magnetoresistive-based biosensors and biochips, *Trends in Biotechnology*, **22**, 455–62.

Grossman, H.L., Myers, W.R., Vreeland, V.J., Bruehl, R., Alper, M.D., Bertozzi, C.R. and Clarke, J. (2004). Detection of bacteria in suspension by using a superconducting quantum interference device, *Proceedings of the National Sciences of the United States of America*, **101**, 129–34.

Jaffrezic-Renault, N., Martelet, C., Chevolot, Y. and Cloarec, J.P. (2007). Biosensors and bio-bar code assays based on biofunctionalized magnetic microbeads, *Sensors*, **7**, 589–614.

Kriz, K., Gehrke, J. and Kriz, D. (1998). Advancements toward magneto immunoassays biosensors, **13**, 817–23.

Kriz, C.B., Radevik, K. and Kriz, D. (1996). Magnetic permeability measurements in bioanalysis and biosensors, *Analytical Chemistry*, **68**, 1966–70.

Kurlyandskaya, G. and Levit, V. (2005). Magnetic Dynabeads® detection by sensitive element based on giant magnetoimpedance, *Biosensors and Bioelectronics*, **20**, 1611–16.

Lagae, L., Wirix-Speetjens, R., Liu, C.X., Laureyn, W., Borghs, G., Harvey, S., Galvin, P., Ferreira, H.A., Graham, D.L., Freitas, P.P., Clarke, L.A. and Amaral, M.D. (2005). Magnetic biosensors for genetic screening of cystic fibrosis, *IEE Proceedings – Circuits, Devices and Systems*, **152**, 393–400.

Landry, G., Miller, M.M., Bennett, B.R., Johnson, M. and Smolyaninova, V. (2004). Characterization of single magnetic particles with InAs quantum-well Hall devices, *Applied Physics Letters*, **85** (20), 4693–5.

Lany, M., Boero, G. and Popovic, R.S. (2005). Superparamagnetic microbead inductive detector, *Review of Scientific Instruments*, **76**, 084301.

Li, G.X., Sun, S.H., Wilson, R.J., White, R.L., Pourmand, N. and Wang, S.X. (2006). Spin valve sensors for ultrasensitive detection of superparamagnetic nanoparticles for biological applications, *Sensors and Actuators A*, **126**, 98–106.

Li, G.X., Wang, S.X. and Sun, S.H. (2004). Model and experiment of detecting multiple magnetic nanoparticles as biomolecular labels by spin valve sensors, *IEEE Transactions on Magnetics*, **40** (4), 3000–2.

Li, G.X., Joshi, V., White, R.L., Wang, S.X., Kemp, J.T., Webb, C., Davis, R.W. and Sun, S.H. (2003). Detection of single micron-sized magnetic bead and magnetic nanoparticles using spin valve sensors for biological applications, *Journal of Applied Physics*, **93** (10), 7557–9.

Ludwig, F., Mauselein, S., Heim, E. and Schilling, M. (2005). Magnetorelaxometry of magnetic nanoparticles in magnetically unshielded environment utilizing a differential fluxgate arrangement, *Review of Scientific Instruments*, **76**, 106102.

Mahdi, A.E., Panina L. and Mapps, D. (2003) Some new horizons in magnetic sensing: high-Tc SQUIDs, GMR and GMI materials, *Sensors and Actuators A*, **105**, 271–85.

Meyer, M.H.F., Hartmann, M., Krause, H.J., Blankenstein, G., Mueller-Chorus, B., Oster, J., Miethe, P. and Keusgen, M. (2007). CRP determination based on a novel magnetic biosensor, *Biosensors and Bioelectronics*, **22** (6), 973–9.

Mihajlovic, G., Xiong, P., von Molnar, S., Ohtani, K., Ohno, H., Field, M. and Sullivan, G.J. (2007). InAs quantum well Hall devices for room-temperature detection of single magnetic biomolecular labels, *Journal of Applied Physics*, **102**, 034506.

Mihajlovic, G., Xiong, P., von Molnar, S., Ohtani, K., Ohno, H., Field, M. and Sullivan, G.J. (2005). Detection of single magnetic bead for biological applications using an InAs quantum-well micro-Hall sensor, *Applied Physics Letters*, **87**, 112502.

Miller, M.M., Prinz, G.A., Cheng, S.F. and Bounnak, S. (2002). Detection of a micron-sized magnetic sphere using a ring-shaped anisotropic magnetoresistance-based sensor: A model for a magnetoresistance-based biosensor, *Applied Physics Letters*, **81** (12), 2211–13.

Montaigne, F., Schuhl, A., van Dau, F.N. and Encinas, A. (2000). Development of magnetoresistive sensors based on planar Hall effect for applications to microcompass, *Sensors and Actuators*, **81**, 324–7.

Morvic, M. and Betko, J. (2005). Planar Hall effect in Hall sensors made from InP/InGaAs heterostructure, *Sensors and Actuators A*, **120**, 130–3.

Perez, J.M., Josephson, L. O'Loughlin, T., Högemann, D. and Weissleder, R. (2002). Magnetic relaxation switches capable of sensing molecular interactions, *Nature Biotechnology*, **20**, 816–20.

Prinz, G.A. (1998). Device physics-magnetoelectronics, *Science*, **282**, 1660–3.

Richardson, J., Hill, A., Luxton, R. and Hawkins, P. (2001). A novel measuring system for the determination of paramagnetic particle labels for use in magneto-immunoassays, *Biosensors and Bioelectronics*, **16**, 1127–32.

Rife, J.C., Miller, M.M., Sheehan, P.E., Tamanaha, C.R., Tondra, M. and Whitman, L.J. (2003). Design and performance of GMR sensors for the detection of magnetic microbeads in biosensors, *Sensors and Actuators A*, **107**, 209–18.

Schott, C., Besse, P.A. and Popovic, R.S. (2000). Planar Hall effect in the vertical Hall sensor, *Sensors and Actuators A*, **85**, 111–15.

Schotter, J., Kamp, P.B., Becker A., Pühler, A., Reiss, G. and Brückl, H. (2004). Comparison of a prototype magnetoresistive biosensor to standard fluorescent DNA detection, *Biosensors and Bioelectronics*, **19**, 1149–56.

Shen, W.F., Liu, X.Y., Mazumdar, D. and Xiao, G. (2005). *In situ* detection of single micron-sized magnetic beads using magnetic tunnel junction sensors, *Applied Physics Letters*, **86**, 253901.

Turner, A.P. (2000). Biosensors – sense and sensitivity, *Science*, **290**, 1315–17.

Vo-Dinh, T. Cullum, B.M., Stokes, D.L. (2001). Nanosensors and biochips: frontiers in biomolecular diagnostics, *Sensors and Actuators B*, **74**, 2–11.

Wang, J. (2000). From DNA biosensors to gene chips, *Nucleic Acids Research*, **28** (16), 3011–16.

Whitesides, G.M. and Wong, A.P. (2006). The Intersection of Biology and Materials Science, *MRS Bulletin*, **31** (1), 19–27.

Wood, D.K., Ni, K.K., Schmidt, D.R. and Cleland, A.N. (2005). Submicron giant magnetoresistive sensors for biological applications, *Sensors and Actuators A*, **120**, 1–6.

9

Magnetic Biochips: Basic Principles

9.1 Introduction

Identifying materials and pathogens at the molecular level is important for many applications including detection, sensing and medical diagnosis. A range of powerful technologies exists for observing and detecting chemicals and biomaterials at the molecular level. Many scale-down analytical processes combined with advances in microfluidic devices and signal detection technologies motivated the development of biochip devices. These chips are formed by combining synthetic chemistry and biochemistry with microelectronic fabrication technologies by defining arrays of selected biomolecules immobilized on the surface. The biochips integrated with electrical, magnetic, optical and physical signal detection with microfluidic delivery systems provide very fast results using very low sample volumes at a lower cost. The development of these chip-based technologies is changing the nature of the experiments done at the clinical laboratories. This chapter presents the theory, design and development of biochip technology based on magnetic signal detection, particularly using nanomagnetic tools.

Pocket-sized analytical platforms integrated with microfluidics and signal detection devices are of recent interest due to their technical abilities in coupling biological and chemical systems with microelectronic circuits. These devices which generally fall under the category of lab-on-a-chip (LoC) devices, have many advantages such as high performance, very low quantity sample consumption, fast and automated analysis and batch fabrication. The driving force behind the miniaturization is the requirement to increase the processing power while reducing the cost per device and testing complexities. The LoC devices speed up the reaction time, while allowing massively parallel design. The LoC devices are also used to study the cell functions and its responses to an external stimulus. The common clinical tests that require hours or days and skilled personnel to process biological samples in regular laboratories could be replaced with the LoC devices. The LoC devices dramatically reduce the reagent and energy consumption and hence the waste output. Today, as shown in Figure 9.1, by integrating these devices with microelectronic signal detection circuits, it could be possible to examine the characteristics of a single cell either *in vivo* or *in vitro*. A typical LoC device consists of microfluidic channels to dispense the sample as well as for reagents, a biochemical reactor in which the reactions take place, electrical, optical or magnetic signal detection circuits to read/monitor

Nanomedicine: Design and Applications of Magnetic Nanomaterials, Nanosensors and Nanosystems
V. Varadan, L.F. Chen, J. Xie
© 2008 John Wiley & Sons, Ltd

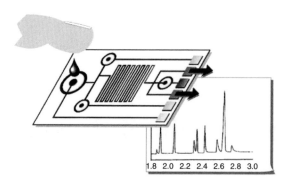

Figure 9.1 Schematic diagram of a lab-on-a-chip device for measurements of cations. Source: Tudos, A.J., Besselink, G.A.J and Schasfoort, R.B.M. (2001). Trends in miniaturized total analysis systems for point-of-care testing in clinical chemistry, *Lab on a Chip*, **1**(2) 83–95

the electrochemical signals and a processing and display unit to interpret the electrochemical signal. The use of the living cell as a sensor element is highly desirable and a very sensitive electrochemical sensor of biologically active substances could be achieved using biological cells integrated with a biochip. These devices provide higher sensitivity, improved speed and efficiency, less sample consumption, portability and cost-effective manufacturing.

The microminiature biochips, as shown in Figure 9.2(a) are small, compact and suitable to use in non-laboratory settings. Using the IC fabrication tools the sensor array integrated biochip could be further integrated with the microfluidic flow channels, temperature controllers and signal detection systems as shown in Figure 9.2(b). The LoC is a collection of miniaturized test sites arranged such that the tests can be performed simultaneously. The higher throughput is achieved by the higher density of the arrays and hence the reduction of required volume of the sample. The fabrication of LoC devices can benefit from the microelectronics industry fabrication that is geared to high-volume production at a lower cost. One of the key advantages of biochips is the opportunity to integrate all complex multistep analytical processes into a single device. The individual tasks such as sample addition, processing, analysis and read-out electronics are in a single chip. Liquid samples or reagents can be transported through the microchannels from the reservoirs to the reactors by a pump based on electrokinetic, magnetic or hydrodynamic principles.

Detection is one of the key features in analytical LoC platforms. A variety of detection strategies are developed in LoC devices based on different principles. Even though microscopy is a promising tool in the laboratory environment, it could not be implemented easily in LoC devices due to its disadvantages in size as well as higher cost. Electrochemical detection based on microfabricated electrodes promises higher levels of portability for the LoC devices. The electrical interface in such a system is achieved by using an array of planar metallic microelectrodes. The incorporation of an electrochemical detection mechanism into the LoC devices offers a well characterized and wide variety of analyte/electrode combinations as well as the ability to customize the materials and geometries. The development of customized intelligent sensors with wireless network capability will suit remote analysis and unsupervised monitoring of chemical/biological agents.

A common approach to detecting a biological molecule is to attach the target molecule with a label moiety that can produce a highly observable signal (Vo-Dinh *et al.* 2001). This is accomplished by using a target molecule and a receptor such as an antibody which is tagged with a label. In general, an antigen is a molecule that stimulates an immune

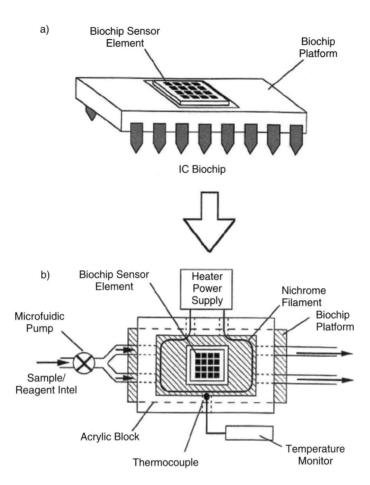

a) Biochip Sensor Element

Biochip Platform

IC Biochip

b) Biochip Sensor Element

Heater Power Supply

Nichrome Filament

Biochip Platform

Microfuidic Pump

Sample/ Reagent Intel

Acrylic Block

Thermocouple

Temperature Monitor

Figure 9.2 Schematic representation of a biochip: (a) array of sensing elements integrated on to the chip makes it compact and suitable for non-laboratory applications; (b) shows the integrated biochip with microfluidic channels, temperature controllers and sensor arrays. (Von-Dinh *et al.* 2001). Reprinted from: Graham D.L., Ferreira H. A. and Freitas P.P. (2004). Magnetoreisitive based biosensors and biochips, *Trends in Biotechnology*, **22**, 9, 455–462, with permission from Elsevier

response. Antibodies are proteins that are found in the blood or body fluids of vertebrates to identify foreign bodies such as bacteria and viruses. The structure of antibodies is in general very similar; however a small region at the tip of the protein is unique, allowing millions of possible antibodies with different molecular structures. The binding of antibody to antigen is very specific so that the antibody binds only to a specific antigen. Examples of such labeling molecules are enzymes, fluorescent molecules or charged molecules. An attached label is detected based on transaction mechanisms such as optical, electrical, electrochemical, thermal and piezoelectric methods.

If a fluorescently labeled particle passes through a detecting element, it produces scattered light with pulses of 3–20 μs depending on flow velocity and particle sizes. The fluorescence-based microarray system has a drawback in quantitative data analysis. The detection platforms generally suffer from high background fluorescence due to microarray substrates. Also, due to the photosensitive nature of the fluorescent labels, it bleaches when exposed to light.

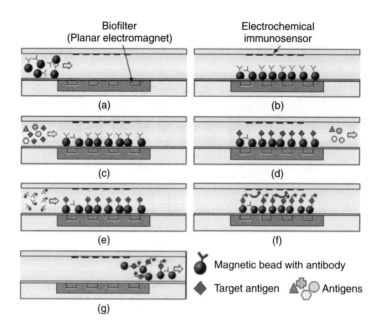

Figure 9.3 Illustration of working principle of a magnetic biochip; (a) injection of magnetic beads; (b) separation and holding of the beads; (c) insertion of sample; (d) immobilization of the target antigen; (e) insertion of label antibody; (f) detection; and (g) washing out the magnetic beads to make it ready for next test. (Choi *et al.* 2002)

Figure 9.3 presents the schematic representation of various magnetic biosampling and detection procedures. Antibody coated magnetic beads are introduced into the test chamber and are separated using an electromagnet. The antigen is then introduced into the chamber while holding the antibody coated beads in position. The antibody–antigen interaction permits only those target antigens to immobilize. Other antigens are washed out due to the flow. Enzyme-labeled secondary antibodies are introduced into the chamber. These antibodies are incubated along with the immobilized antigen. The chamber is then rinsed to remove all unbounded secondary antibodies. Substrate solution is injected into the channel, which will react with the enzyme, and the electrochemical detection can be performed. Finally, the magnetic beads are cleaned and cleared to the waste chamber and the chip is ready for the next set of testing.

Magnetism is a collective effect due to the movement of electrical charges in a material. Generally there are two cases of magnetic field generation: (1) due to the orbital and spin motion of the electrons; (2) the electric current flow through a conductor. Most of the materials exhibit magnetic properties, dictated by the atomic structure, temperature and pressure. Many microorganisms exhibit magnetism due to intracellular chains of nanoparticles called magnetosomes, which are referred to as magnetotatic bacteria.

Magnetic separation of biomolecules or bioparticles involves coupling of magnetic particles to a biorecognition agent such as an antibody or oligonucleotide. A magnetic separator unit retains the particles with targets bound to them. Magnetic biochips have evolved as a new class of devices capable of sensing a lower concentration of samples using the magnetic interaction. Micro- and nanosized magnetic nanoparticles are used in magnetic biochips to capture, concentrate and manipulate target analytes. The detection is mainly achieved by analyzing the magnetic field signatures generated by an external magnetic field. Magnetic detection is found to be useful because the signal can be easily taken out due to the nature of very low background magnetic noise, which is one of the advantages for magnetic biosensors. This chapter presents the principle behind the development of magnetic biochips and magnetic labs-on-chips (MLoCs) and an overview of the developments and applications of magnetic lab-on-a-chip biological systems. The MLoC devices are compatible with the already matured semiconductor processing technology and can directly give the voltage output for automated analysis.

Magnetic nanoparticles with a diameter of 20 nm or smaller become desirable for ultrasensitive biomolecular detection. However, the detection of magnetic moments of such tiny magnetic particles is very challenging due to the smaller physical volume, the relatively large surface area and the background thermal disturbances. Recently many magnetic nanosensors were developed based on magnetoresistive properties, magnetic tunnel junctions and spin valve sensors. These are presented in the following sections.

(1) Overview of Magnetism and Magnetic Nanoparticles

The magnetic force is observed as mutual attraction or repulsion of certain materials. The response of any material to an externally applied magnetic field is generally defined as *magnetic induction B*, which can be calculated from the permeability of the material, μ as

$$B = \mu H \tag{9.1}$$

where H is the magnetic field intensity. Since the permeability of free space is a constant, the permeability of any other medium will generally vary with the applied field. The relative permeability μ_r is defined as the ratio between μ and μ_0. The magnetic moment **m** of a magnetic dipole is a vector that is related to the torque τ acting on the dipole.

$$\tau = m \times B \tag{9.2}$$

The magnetic properties of a material are due to the magnetic induction generally defined as the magnetization, which is due to the individual atomic moments m which exists in the sample. The magnetization M can be written as

$$M = \frac{1}{V} \sum_V m \tag{9.3}$$

$$B = \mu_0(H + M) \tag{9.4}$$

The magnetic susceptibility of the material can be defined as

$$\chi = \frac{M}{H} \tag{9.5}$$

In general, all materials exhibit magnetic properties to some extent. The variation of magnetic property depends on its atomic structure, temperature and pressure. Based on the characteristic susceptibilities of the materials, they can be classified into ferrimagnetic, antiferromagnetic, ferromagnetic, diamagnetic, paramagnetic and superparamagnetic materials. The ordered states in the material structures contribute to ferromagnetism, antiferromagnetism and ferrimagnetism, while the diamagnetism, paramagnetism and superparamagnetism exist only due to the transient states as a result of applied external magnetic field.

Superparamagnetism occurs when the size of the materials approaches to the subdomain sized dimensions – typically of the order of nanometers. At these dimensions even though the thermal energy is not sufficient to overcome the coupling forces between neighboring atoms, the magnetization of the entire crystal can be changed due to the thermal energies at nanometer dimensions. Due to the fluctuations in the direction of magnetizations, the average field becomes zero and hence the material behaves like a paramagnetic substrate. However, the external magnetic field tends to align the magnetic moment of the entire crystal instead of aligning individual atoms and shows a higher susceptibility.

Magnetic nanoparticles are commonly defined as magnetic materials whose one dimension is less than 200 nm. Magnetic nanowire, nanospheres or beads and magnetic shells are a few examples of the magnetic nanoparticles successfully synthesized. These magnetic particles possess more than magnetism if these particles are attached with other functional structures such as flurophores, surfactants, biomolecules or other reactive moieties as shown in Figure 9.4. The biomolecules to be detected are initially labeled with magnetic nanoparticles. When they pass over an array of complementary or non-complementary molecules, the sensors could detect the presence of the magnetic labels by the change in sensor resistance. The compound structures after proper functionalization show magnetism which is just one of the traits that are usually needed for a particular application, such as biosensors. Ferrofluids are suspensions of superparamagnetic particles in carrier liquids (Odenbach, S., 2002; Skumiel, A. 2003). Magnetic data storage technologies (Sellmyer 2006) are a few non-biological applications of magnetic nanoparticles.

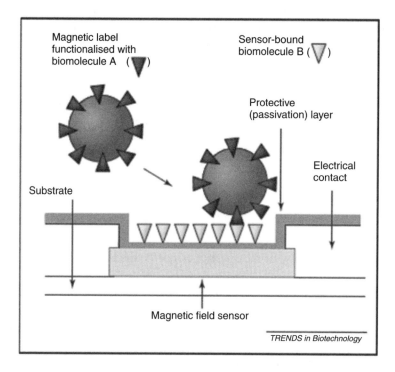

Figure 9.4 Magnetically labeled biomolecule detection in a biosensor. Magnetic label functionalized molecule A is bound to a magnetoresistive sensor. The magnetic moment of the label induces a fringe field resulting in a change in resistance and voltage of the sensor. (Graham *et al.* 2004). Reprinted from: Graham D.L., Ferreira H. A. and Freitas P.P. (2004). Magnetoreisitive based biosensors and biochips, *Trends in Biotechnology*, **22**, 9, 455–462, with permission from Elsevier

Due to the development of magnetic separations of biomolecules, many biomedical and biotechnological applications have been developed. The basic technique of magnetic separation involves coupling of magnetic particles to a biorecognition agent. The magnetic separator allows retaining the particle for preconcentration, washing and chemical processing of the bound targets. These techniques are explained in Sections 9.2 and 9.3.

9.1.1 Sensor Arrays and Integrated Biochips

One of the main benefits of miniaturization, integration and automation by the LoC devices is high throughput. The system design for miniaturization includes the development of miniature micropumps for flow actuation and various detection schemes. In magnetic detection, the biomolecule to be detected (target or analyte) is magnetically labeled first. The labeling is mainly immobilizing the biomolecule on a magnetic material. The labeled analyte is passed over an array of specifically patterned complementary molecules, which are immobilized on the chip magnetic field sensors as shown in Figure 9.4. The change in resistance is usually a measure of the detection (Graham 2004).

Figure 9.5 presents the schematic diagram of a high density magnetoresistive biosensor array integrated with a CMOS chip for DNA hybridization developed by Han *et al.* (2006). The hybridized DNA can be measured using the associated electronic circuits.

Figure 9.5(a) presents the block diagram of the magnetic biochip with an array of 1008 giant magnetoresistive (GMR) sensors arranged in 16 subarrays in an area of 120 μm × 120 μm. The signal read-out circuits are based on time division multiplizing (TDM) as well as frequency division multiplexing (FDM) with a throughput of 16 outcomes/readings. The details of the GMR sensors are presented in Section 9.2.

9.1.2 Manipulation of Biomolecules

The study of dynamic interaction between the biomolecule and the surrounding medium is important in order to understand its mechanism and functions. Separation of molecules and its manipulation is necessary to study this dynamic behavior. Mechanical methods such as microneedles (Essevaz-Roulet 1997), laser traps (Shivashankar and Libchaber 1998; Wuite 2000) and molecular glues (Rief and Grubmuller 2002) are a few examples of the manipulation of molecules to study their interaction. Nanotechnology offers many promising results for the detection, manipulation and identification of biomolecules. A few emerging nanotechnology approaches are: mass spectrometry analysis (Wright *et al.* 2005; Pan *et al.* 2005); the bio-bar code method for amplification (Nam *et al.* 2002–2004; Thaxton *et al.* 2005; Georganopoulou *et al.* 2005); nanowire gated transistors (Mirkin *et al.*1996; Hahn and Lieber 2004; Melosh *et al.* 2003; Beckman *et al.* 2004; Yousaf *et al.* 1999; Cui *et al.* 2001); and nanocantilever methods (Ziegler 2004; Lee *et al.* 2005; Mukhopadhyay *et al.* 2005; Alvarez *et al.* 2004; Weeks *et al.* 2003; Ji *et al.* 2004). Optical detection of a single molecule is emerging as a new tool for biological applications due to the advancements in frequency modulated absorption as well as laser induced fluorescence (Ambrose *et al.* 1995; Nie and Zare 1997). Biomolecular detection is also achieved by optical label-free methods combined with the advantages of ellipsometry and surface plasmon resonance (Westphal and Bornmann 2002), using quantum dots (Chan *et al.* 2002; Han *et al.* 2001) and fiber optic based detection systems (Tazawa *et al.* 2007).

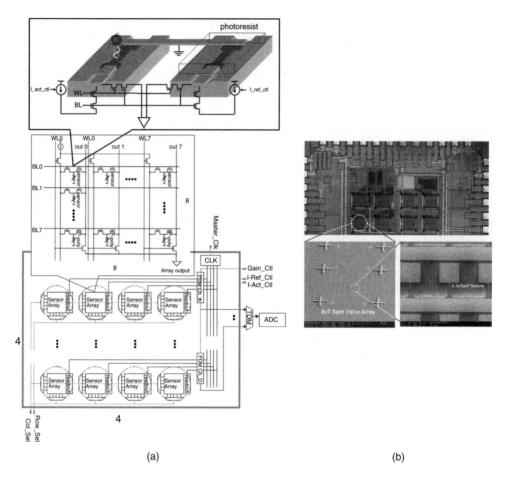

(a)   (b)

Figure 9.5 (a) Schematic diagram of the magnetoresistive biosensor architecture; (b) micrograph of the processed die and the SEM images of the biosensor. (Han *et al.* 2006)

The first magnetic separation was achieved in 1973 using a silica coated and cellulose coated magnetic iron oxide to immobilize a-chymotrypsin and b-galactosidase for the design of biosensors (Robinson *et al.* 1973). The magnetic separation involves the attaching of magnetic particles to a biorecognition agent and a magnetic separator unit that allows washing and preconcentration of bound targets. The binding of the biomaterial can be done by affinity interaction and ion exchange interaction.

Magnetic nanoparticles are of particular interest due to their lesser susceptibility to negative effects of high viscosity, the possibility of using various densities of mixtures and their convenience of use with automated systems such as lab-on-a-chip devices. Magnetic nanoparticles can also be used to concentrate target molecules instead of doing ultra-filtration and precipitation (Alche and Dickinson 1998). Magnetite (Fe_3O_4) nanoparticles are typically used for magnetic separations due to magnetite's magnetizable component. Although these superparamagnetic nanoparticles are attracted by the magnetic field, they will not retain the residual magnetism after the field is removed, which is very important for resuspension and prevention of aggregation after the field is removed. Table 9.1 presents a number of commonly used magnetic particles and their properties (Magnani *et al.* 2006).

Table 9.1 Commonly used magnetic micro- and nanoparticles Source: Magnani, M., Galluzzi, G. and Bruce, I.J (2006). The use of magnetic nanoparticles in the development of new molecular detection systems, *Journal of Nanoscience and Nanotechnology*, **6**, 2302–11

Name	Diameter (µm)	Polymer composition	Activation possibility	Manufacturer/ supplier
BioMag	1	Silica	–COOH	PerSeptive Biosystems
		Charcoal	–NH2	Farmingham, MA, USA
Biosphere	1		–NH2	Biosource International
Dynabeads		Polystyrene	Tosyl- activated	Dynal, Oslo, Norway
M-280	2.8			
M-450	4.5			
M-500	5			
Estapor	1	Polystyrene	–COOH	Prolabo, Fontenay-
			–NH2	sous-Bois, France
Iobeads				Immunotech, Marseille, France
M 100	1–10	Cellulose	–OH	Scigen, Sittingbourne, UK
M 104				
M 108				
MACS MicroBeads	0.05			Miltenyi Biotec, Germany
MagaBeads	3.2	Polystyrene	–COOH	Cortex Biochem, San Leandro
MagaCell	3	Cellulose	–NH2	CA, USA
		Charcoal		
MagAcrolein	3	Acrolein	Epoxy	
MagaCharc	3			
Magarose	20–150	Agarose		Whatman International, UK
Magne-Sphere	1			Promega, Madison, WI, USA
MagneSil	5–8.5	Silica		
Magnetic microparticles	1–2	Polystyrene	–COOH	Polysciences, Warrington,
			–NH2	PA, USA
Magnetic microspheres	1	Polystyrene	–COOH	Bangs Labs., Fishers, IN,
			–NH2	USA

(continued overleaf)

Table 9.1 *(continued)*

Name	Diameter (µm)	Polymer composition	Activation possibility	Manufacturer/ supplier
Magnetic beads	0.8	Latex		ProZyme, San Leandro, CA, USA
Magnetic particles	1	Polystyrene		Boehringer, Mannheim, Germany
Magnetic particles	0.05–24	Polystyrene Dextran	−COOH −NH2	G. Kisker - Products for Biotechnology Steinfurt, Germany
MPG	5	Porous glass	−NH2	CPG, Lincoln Park, NJ, USA
Sera-Mag	1	Polystyrene	−COOH	Seradyn, Indianopolis, IN, USA
SiliMag	0.04–0.14	Silica		Diatheva, Fano (PU), Italy
SPHERO magnetic particles	1–4.5	Polystyrene	−COOH −NH2	Spherotech, Libertyville, IL, USA
XM200 microsphere	1–3.5	Silica Polystyrene	−COOH	Advanced Biotechnologies, Epsom, UK

The MLoC devices in general use the same platform for the detection and manipulation of biological or chemical systems. The handling of very small volumes of liquid samples is very important for analysis using LoC devices. Magnetic markers offer more stability than other methods. Magnetic labs-on-chips are realized by combining magnetic markers with magnetoresistive detectors into a single detecting system (Baselt *et al.* 1998; Tondra *et al.* 1999; Schotter *et al.* 2002). Two-dimensional manipulation of liquids in a microfluidic system is achieved by suspending the drops in silicon oil that contains magnetic microparticles. These particles serve as force-mediators for the magnetic actuation as well as a mobile medium for the molecules (Auroux *et al.* 2002; Vilkner *et al.* 2004; Lehmann *et al.* 2006).

An applied magnetic gradient is used to manipulate the magnetic biomarkers. The analyte molecules can be pulled out from specific binding sites or the binding strength can be tested to distinguish between the specifically bounded and unspecifically bounded molecules. The molecules are detected by measuring the change in resistance due to the presence of an external magnetic field for a fixed sensor current.

(1) Magnetic Separation

Conventional biological and biomedical analyses require concentrated samples. The concentrated samples are prepared from specific biological entities separated from their native environments. The selective separation and purification of biomolecules are generally

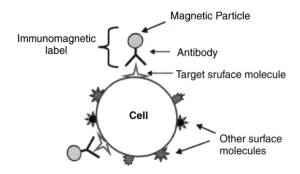

Figure 9.6   Magnetic cell separation. (McCloskey *et al.* 2003a)

achieved by molecular sieves. Magnetic separation is one of the methods to generate the required concentrated samples. The magnetic separation is a two-step process. As shown in Figure 9.6, in the first step, the tagging or labeling of the desired biological entity with magnetic material is done. In the second step, these tagged entities are separated out using a fluid based magnetic separator. The magnetophoretic mobility depends on the intrinsic properties of the magnetic particles, particle size, magnetic susceptibility and the viscosity of the medium.

(2)   Forces Acting on Magnetic Nanoparticles

The separation of nanoparticles for LoC applications can be explained based on vector field theory. The magnetic force acting on a point-like dipole m can be written as (Pankhurst 2003)

$$F_m = (m.\nabla)B \qquad (9.6)$$

which can be explained as differentiation with respect to m. The total moment of a magnetic nanoparticle suspended in a weak diamagnetic medium such as water can be written as $m = V_m M$, where V_m is the particle volume and $M = \Delta\chi H$ is the volumetric magnetization. $\Delta\chi$ can be obtained from $\Delta\chi = \chi_m - \chi_w$ which is the effective susceptibility of the particle relative to the water. For a diluted suspension of nanoparticles in pure water, the magnetic field gradient B can be approximated as $B = \mu_o H$. Equation (9.6) can be written as

$$F_m = \frac{(V_m \Delta\chi)}{\mu_0}(B.\nabla)B \qquad (9.7)$$

Since there is no time varying electric field in the medium, the Maxwell equation can be written as

$$\nabla x B = 0 \text{ and } \nabla(B.B) = 2Bx(\nabla x B)+2(B.\nabla)B = 2(B.\nabla)B. \qquad (9.8)$$

Equation (9.7) can be written as

$$F_m = V_m \Delta\chi \nabla\left(\frac{B^2}{2\mu_0}\right) = V_m \Delta\chi \nabla\left(\frac{1}{2}B.H\right) \qquad (9.9)$$

Tagging in magnetic separation is achieved by chemical modification of the surface of the magnetic nanoparticles with biocompatible molecules. Generally used biocompatible coatings for iron oxide nanoparticles are dextran, polyvinyl alcohol (PVA) and phosopholipids (Molday and MacKenzie 1982; Pardoe 2001). This is a very accurate method of labeling the cells since the antibodies specifically bind to their matching antigen. Magnetic particles coated with immunospecific agents are observed to successfully bind to red blood cells, lung cancer cells, bacteria and golgi vesicles.

The magnetically labeled material is separated from its native solution in a region in which there is a magnetic gradient. This immobilization is achieved due to the magnetic force gradient as explained in Equation (9.9). The hydrodynamic drag force F_d due to the flow of the solution acts opposite to it. The cell separation is achieved only if $F_m > F_d$.

$$F_d = 6\pi \eta R_m \Delta v \qquad (9.10)$$

where η is the viscosity of the medium, R_m is the radius of the nanoparticles and Δv is the difference in velocities of the cell and the medium. The diameter of the particle plays a critical role in hydrodynamic forces. The large size is advantageous to shorten the time frame of cell separation. However, smaller particle sizes reduce the likelihood of interference between the neighboring particles. The maximum flow rate that a particle can withstand under the influence of a magnetic field can be determined from Equations (9.7) and (9.10).

$$\Delta v = \frac{2r^2(B.\nabla)B}{9\mu_0\eta} = \frac{1}{\mu_0}\xi(B.\nabla)B \qquad (9.11)$$

where the magnetophoretic mobility of the particle is given by:

$$\xi = \frac{2r^2\chi}{9\eta} = \frac{V\chi}{6\pi r\eta} \qquad (9.12)$$

(3) Physics of Magnetic Manipulation

For a magnetic nanoparticle particularly a mono-domain nanoparticle, its domain dimension is typically equal to or smaller than the thickness of the magnetic domain wall, which is given by:

$$\delta = \sqrt{\frac{\pi^2 J S^2}{Ka}} \qquad (9.13)$$

where J is the magnetic exchange constant, S the total spin quantum number, a the inter atomic spacing and K the magnetic anisotropy constant of the material (Chikazumi 1964). Substituting the above values for Iron, ($J = 2.16 \times 10^{-21}$ J, $a = 2.86 \times 10^{-10}$ m and $K = 4.2 \times 10^4$ J/m^3 assuming S = 1), the domain wall width will be 42 nm. The important characteristic of a biosensor is the time over which a stable magnetization of the nanoparticles can be established. Mono-domain magnetic nanoparticles become superparamagnetic at room temperature, i.e., their time averaged magnetization without a magnetic field is zero when their magnetic energy is lower than ten times their thermal energy (Gijs 2004). It can be seen that at room temperature, $k_B T = 4.0 \times 10^{-21} J$, the maximum radius required to become superparamagnetic nanoparticles is 6 nm.

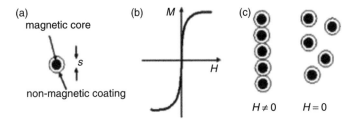

Figure 9.7 (a) Schematic representation of a spherical nanoparticle with internal core diameter s surrounded by a non-magnetic coating for biorecognition; (b) hysteresis loop; and (c) the nanoparticle superstructure in the presence of a magnetic field H. This superstructure decomposes into single particles when the field is removed. (Gijs 2004). Reproduced by permission of Springer Science and Business Media

A typical functionalized nanoparticle consists of a magnetic core of diameter r surrounded by the bioactive coating. As shown in Figure 9.7, this non-magnetic coating is for the selective binding of biomaterials such as protein, DNA or cells. Due to the highly stable nature of the iron oxides such as magnetite and maghemite with diameters in the range of 5–100 nm, they are generally used as core materials. Figure 9.7(b) shows the magnetization curve of such superparamagnetic particles. This hysteresis-free nature is one of the important considerations of the magnetic nanoparticles in biosensing. These suspended superparamagnetic nanoparticles tagged with biomolecules will not form an agglomeration, since they can easily decompose into separated particles when the magnetic field is removed, as shown in Figure 9.7(c). These nanoparticles cause only minimum disturbance to an attached biomolecule while at the same time it has the added advantage of a large surface-to-volume ratio for chemical bonding. However, for large magnetic particles (of the order of 0.5–5 μm), the core consists of a single particle or multiple cores as shown in Figure 9.8(a). The multi-domain microparticles, when exposed to an external magnetic field, acquire a magnetic dipole moment and coalesce under the influence of the dipole interaction and form a supraparticle structure (SPS). It consists of columnar structure along the field direction. The columns of chains are used in microchanels for magnetic separation. The magnetic separation is achieved in nanoparticles in the form of dextran-coated magnetic clusters of 20–100 nm in size.

(4) Separator Design
Different types of magnetic particles have been developed for use in the cell separation process that includes purification and imunoassays. The basic principle of magnetic

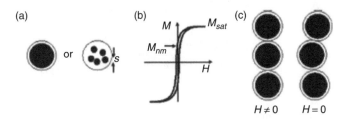

Figure 9.8 (a) Schematic representation of a spherical nanoparticle with single internal magnetic core or with multiple cores; (b) hysteresis loop; and (c) the nanoparticle superstructure in the presence of a magnetic field H. This particle shows remnant moment and the superstructure does not decompose when the field is removed. (Gijs 2004). Reproduced by permission of Springer Science and Business Media

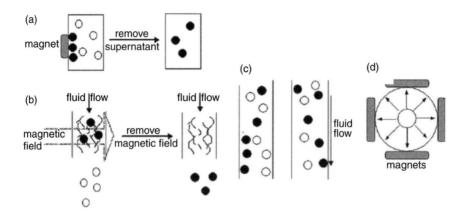

Figure 9.9 Standard method of magnetic separation: (a) the external magnet at the container wall of a solution of magnetically tagged (•) and untagged (o) biomaterials. The tagged particles are collected by the magnet and the unwanted solution is removed; (b) The continuous flow of the solution through a packed column of steel wool can capture tagged particles and can be recovered; (c) a rapid throughput method of magnetic separation with longitudinal and (d) transverse cross-section with four magnets. Pankhurst, Q.A., Connolly, J., Jones, S.K. and Dobson, J. (2003). Applications of magnetic nanoparticles in biomedicine, *Journal of Physics D – Applied Physics*, **36**, R167-R181. Reproduced with permission of the Institute of Physics

separation is similar to the application of a permanent magnet at the sidewalls of a test tube and removal of the aggregated particles followed by the supernatant of the sample. Charge based separation of proteins using lipid coated magnetic particles is found to be very effective in magnetic separation (Buack *et al.* 2003). The magnetic separation is achieved by flowing the sample through a column packed with steel wool in the presence of a magnetic field as shown in Figure 9.9. The magnet collects the tagged particles while removing the unwanted particles.

Many sensing systems prefer to separate the particles using a high gradient magnetic field and catch the magnetic nanoparticles while they float or flow through a capillary

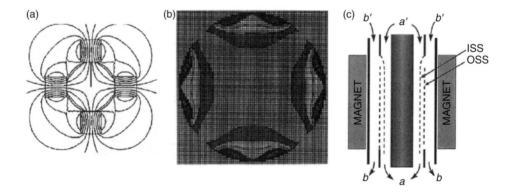

Figure 9.10 (a) Schematic diagram of the quadrupole magnet configuration for cell separation and the magnetic flux lines; (b) the force density distribution; and (c) the cell separator at the center of the field. The sample and the carrier fluid are fed at a' and b' respectively. The inlet splitting surface (ISS) and the outlet splitting surface (OSS) are shown in the figure. The sorted cells are collected at a and b. (Gijs 2004). Reproduced by permission of Springer Science and Business Media

channel. Figure 9.10(a) and (b) presents the schematic arrangement of four magnets to induce maximum magnetic field gradient at the outer side of the sample carrying tube. When the tube containing the sample is inserted in the free space of this quadrupole assembly, the separation takes place due to the laminar flow of the carrier liquid. A magnetic field gradient perpendicular to the direction of the flow is established in the channel. When the sample mixture is entered into the chamber through the flow channel, the particles are carried through the channel due to the flow and the components that interact more strongly with the magnetic field are moved transversely across the channel.

These particles are collected for further analysis. As shown in Figure 9.10(c), when the cylindrical capillary is inserted into the channel with quadrupole magnetic cell sorter, the cell separation sample is inserted at a' and the carrier fluid at b'. Due to the magnetic field gradient, the cell separation takes place at the inner splitting surface (ISS) and outer splitting surface (OSS). The division at the channel outlet using a stream splitter completes the cell separation as shown as b and a. It can be seen from Equation (9.11) that the fluid transport velocity should not exceed a maximum limit (McCloskey 2003a; 2003b). The magnetic cell separation is a function of the antibody binding capacity, which is related to the number of magnetic bead labeling sites of a single cell.

9.1.3 Detection of Biomolecules

A general approach to detect a biomolecule is to attach a label to the target molecule that can produce an observable signal. The detection takes place between the biomolecule and the target molecule and a specific receptor such as an antibody which is tagged with the label. Common labels are fluorescent molecules, enzymes and charged particles. Magnetic labels have recently evolved as labels for biosensing due to their advantages over other labels. The biomolecules which are magnetically labeled or immobilized on a magnetic label are generally detected when they pass over an array of specifically patterned complementary or non-complementary molecules. These probes are immobilized over on-chip magnetic field sensors.

An alternative detection method is based on a secondary detection performed after the interrogation of the probe array with the target molecule. This is done by labeling the target molecules with a small biomolecular label, such as biotin. The biotinylated target molecules which are bound to complementary surface-bound probe molecules are then detected by magnetic labels. Detection of DNA–DNA hybridization can be successfully tested using this approach. Figure 9.11 presents a simplified cross-sectional scheme of detection of magnetically labeled streptavidin on a DNA chip. In step 1, the DNA probes are immobilized over the magnetoresistance sensors that are then hybridized with target DNA. In step 2, magnetically labeled streptavidin is used to detect the hybridized DNA by binding to the biotinylated target DNA.

Figure 9.12 presents the conventional platform of the magnetic nanoTag system for the selective and non-optical detection for DNA microarrays based on magnetic nanoparticles (Wang *et al.* 2005). The magnetic detection is based on spin valve or magnetic tunnel junction detector arrays as shown in Figure 9.12(a). The magnetic particle labeled unknown DNA fragments as shown in Figure 9.12(b) will be tagged to the complementary DNA probes attached to the detection system. Figure 9.12(d) illustrates the detection of the magnetic resistance due to the attachment of the complementary DNA. The system is able to detect one DNA fragment per tag. The magnetic nanoTags are single-domain high-moment nanoparticles with a mean diameter of $100-1000$ Å.

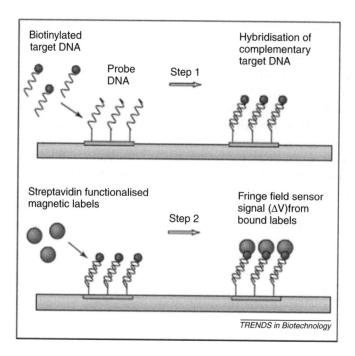

Figure 9.11 Schematic representation of the use of magnetically labeled streptavidin to detect the location of the biotinylated DNA on a chip. The DNA probes are immobilized over the magnetoresistance sensors that are hybridized with target DNA in step 1. In step 2, magnetically labeled streptavidin is used to detect the hybridized DNA by binding to the biotinylated target DNA. (Graham 2003)

9.2   Biochips Based on Giant Magnetoresistance Sensors

Magnetoresistance refers to the change in electrical resistance of a material due to an external magnetic field. This change is very small for most materials. Materials such as permalloy ($Ni_{81}Fe_{19}$) exhibit dramatic changes in magnetoresistance, generally known as anisotropic magnetoresistance (AMR). This is due to the anisotropic scattering of conducting electrons depending on their direction of rotation and magnetization. AMR occurs when the magnetization changes from parallel to transverse, with respect to the direction of current flow. Giant magnetoresistance (GMR) was discovered in 1988 independently in Paris by Baibich *et al.* (Baibich 1988) and Binasch *et al.* (Binasch *et al.* 1988).

GMR has been observed in magnetic multilayered structures where two very close magnetic layers are separated by few nanometer thick spacer layers as shown in Figure 9.13. The first magnetic layer allows electrons only in one spin state to pass through easily if the second magnetic layer is aligned to that spin channel. If the second magnetic layer is misaligned, neither spin channel can get through the structure easily and the electrical resistance becomes high. The electrical resistance of the layers depends on the scattering of the electrons. The orientation of the electron spin with respect to the magnetization direction determines the amount of scattering. Due to the interactions of the layers, the magnetizations of the adjacent layers will normally align in opposite directions and hence this allows the electrons of both spins to scatter equally. However, an externally applied magnetic field forces the alignment of the magnetizations of all layers in one direction.

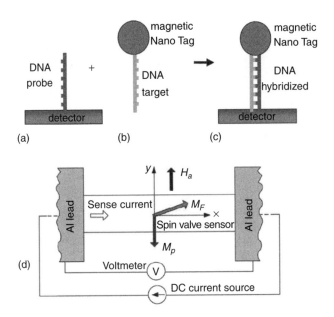

Figure 9.12 Magnetic microarray based on magnetic nanoTag: (a) Magnetoresistance sensors attached with DNA probes; (b) unknown DNA labeled with magnetic nanoTag; (c) highly selective capturing of tagged DNA fragments by the complementary DNA probes and (d) schematic diagram of the magnetoresistance detection system. (Wang 2005)

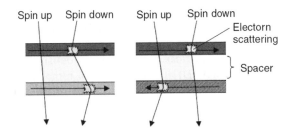

Figure 9.13 Principle of generation of magnetoresistance

This will reduce the scattering of the electrons in one of the spins and hence the resistance of the sensor drops drastically. The resistance of the multilayer architecture decreases drastically when the magnetizations are progressively aligned under the applied field.

An external magnetic field is used to change the relative orientation of the magnetizations of these layers. The electrical resistance is very low as long as they are arranged in aligned fashion and high resistance is observed when they are antiparallel. GMR measures the difference in angle between the two magnetizations in the magnetic layer. It is easy to produce two magnetic layers in parallel by applying a magnetic field. However, an antiferromagnetic layer placed next to the spaced ferromagnetic layer changes the characteristic behavior of the multilayer system, and the sensors with such structures are known as spin valve sensors (Lenssen *et al.* 2000). The details of the spin valve sensors are explained in Section 9.3.

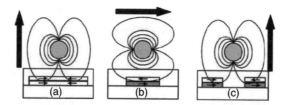

Figure 9.14 Schematic representation of magnetic field arrangement in various GMR sensors: (a) applied field perpendicular to GMR sensor; (b) in plane with modulation across the spin valve and (c) perpendicular to the spin valve. (Megens and Prins 2005)

The resistivity ρ as a function of the angle θ between the magnetization directions can be written as

$$\rho(\theta) = \rho(0^0) + \Delta\rho[1 - \cos\theta]/2 \qquad (9.14)$$

where $\rho(\theta)/\rho(0^0)$ is called the GMR ratio. In general, in AMR sensors the MR ratio is typically about 2%. However, the advancement of new materials have presently shown GMR ratio of several times larger. From Equation (9.14), it can be seen that the angle dependence of the GMR effect has a period of 360°, while that of the AMR effect has a period of only 180°.

The simplest forms of magnetic field arrangement in various GMR sensors are shown in Figure 9.14. In such sensors, the sensitivity depends on the magnitude of the in-plane magnetic field. GMR based biodetection involves labeling biomolecules with magnetic nanoparticles and detecting the magnetic fringe fields of the particle labels by GMR sensors after capturing the target-probe molecules and DNA analysis (Schotter et al. 2002). GMR sensors are not only compatible with the IC fabrication technology and suitable for integration into lab-on-a-chip devices but it is also promising over superconducting quantum interference devices (SQUID) as it is capable of room-temperature operation.

The GMR sensors can be considered as two thin NiFeCo films separated by a very thin Cu spacer layer. All these films should be separated sufficiently so that their magnetizations will not directly couple and align each other. At the same time, these layers should not be separated too far so that the electrons flowing in the sandwich are able to pass from one ferromagnetic layer to the other without losing their spin information. The separation distance must be smaller than the effective scattering lengths of the conducting electrons. Typical layer separations in GMR sensors are 5 nm of NiFeCo, 3.5 nm of Cu and anther 5 nm thick NiFeCo layer.

Figure 9.15 shows a comparison of AMR and GMR read sensors used in memory devices. In AMR read sensors, the magnetically soft adjacent layer (sal) is separated by a spacer layer (spac) from the free ferromagnetic layer (ff). The antiferromagnetic material (af) is at the end of the sensor. As shown in the figure the current is passed parallel to the layers via a pair of current leads. In the GMR sensor, the free ferromagnetic layer is coupled through a spacer layer (spac) to a pinned ferromagnetic layer (pf). Figure 9.15(b) and (c) represents the mode of operation of AMR and GMR sensors respectively.

In the absence of any external field, the maximum rate of change of resistance that can be observed in AMR structure is achieved when the moment of the sense layer ff is at 45° with respect to the direction of the sense current. The introduction of an external field perturbs the structure. The angle of the moment of the sense angle moves away from the balanced 45° and the resistance changes. As a result, the voltage drops across the

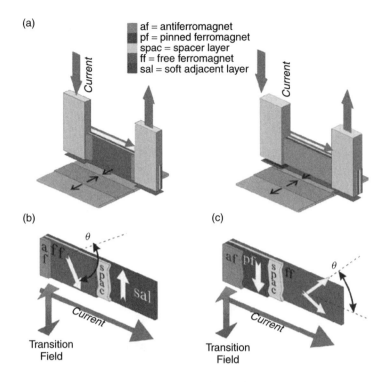

Figure 9.15 (a) Schematic representations of AMR and GMR read sensors. The AMR sensor comprises a magnetically soft adjacent layer (sal) separated via a spacer layer (spac) from the free ferromagnetic layer (ff), which is exchange biased by antiferromagnetic material (af) at the ends of the sensor element only; (b) Mode of operation of AMR sensor; and (c) mode of operation of a GMR sensor. (Parkin *et al.* 2003)

sensor changes in response to the field. The change in resistance of the AMR sensor is proportional to $cos^2(\theta)$.

In the GMR sensor, the magnetic moment of the sense layer ff is the vector sum of the effects of shape anisotropy, coupling of the pinned layer pf and the self-field due to the sense current. The exchange bias from the antiferromagnetic layer af determines the angle of moment of the pinned layer pf. The GMR variation is directionally proportional to $cos(\theta)$ and much greater change is observed than that of AMR sensors. Hence, the GMR sensors are sensitive to much smaller changes in magnetic field and can detect changes in smaller bit patterns than AMR sensors.

Figure 9.16 presents the schematic representation of the principle of MR sensors which makes a table-top genetic screening device. The particles are first functionalized with streptavidin, to enable binding of targets containing biotinyl groups. The sensor surface coated with biomolecular probes will bind with the complementary target species. As shown in Figure 9.16 (5), the magnetoresistive sensors beneath the gene chip can record the capture of the magnetic labels (Ferreira *et al.* 2003).

Magnetoresistive sensors (MRS) are generally magnetic field transducers based on either magnetoresistance of the ferromagnetic materials or on ferromagnetic/non-ferromagnetic heterostructures. The detection in MR sensors is based on the magnetic fringe field of a magnetically labeled biomolecule bound to a magnetic field sensor.

Figure 9.17 presents a schematic diagram of a GMR chip based on a Bead ARray Counter (BARC). The BARC is a sandwich array where the target molecule is bound to

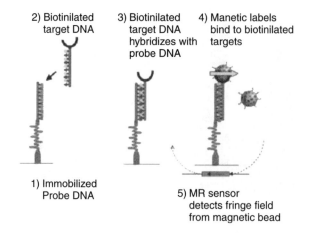

Figure 9.16   Schematic representation of detection of DNA using MR chip. (Ferreira *et al.* 2003)

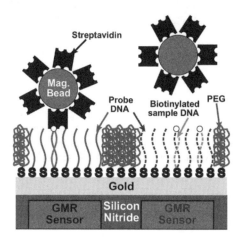

Figure 9.17   The schematic diagram of a GMR BARC chip. (Edelstein *et al.* 2000)

an immobilized probe of the GMR sensor. The specific ligand-receptor interaction binds the magnetic label to the target. The thiolated DNA probes are patterned onto a gold layer directly above the GMR sensor. The non-specific adhesion of the sample DNA as well as the attaching of the magnetic beads on the unmodified areas of the gold are prevented by passivating with thiolated polyethylene gycol (PEG). Biotinylated sample DNA is added to the chip. It hybridizes with the DNA probes on the surface where a complementary sequence is present. The unbounded DNA is removed by washing and the streptavidin coated magnetic beads are injected on to chip. The magnetic beads bound to the biotinylated sample DNA hybridize on the BARC chip. Selective pulling off of the beads that are not bound to the surface is achieved by an external magnetic field gradient using a permanent electromagnet (Lee *et al.* 2000). The magnetic field gradient can also be applied using electromagnets integrated with the chip (Edelstein *et al.* 2000) or a current carrying line placed close to the chip (Lagae *et al.* 2002).

The bound beads are detected by applying a magnetic field perpendicular to the substrate, which imposes a magnetic moment onto the beads. One of the advantages of applying the field normal to the plane of the GMR sensors is that a much larger magnetic field can be applied to the beads without saturating the sensor.

(1) Magnetoresistive Signal Detection

The output signal due to the magnetoresistive changes depends on the magnetic sensitivity, size and ratio of the label and the sensor, magnetic moment of the label, distance between the label and the sensing layer and the sensor current. In general the magnetic moment increases with the increase in applied field until it becomes saturated and the moment no longer increases. Accordingly, the reported sensor signals for single micron-size magnetic labels vary from a few nV to a few $100\,\mu V$ as presented in Table 9.2. It is clear that spin valves of size $1 \times 3\,\mu m^2$ or $2 \times 6\,\mu m^2$ exhibit highest sensitivity with good signal to noise ratios. Magnetoresistive thin films can be easily fabricated using silicon integrated circuit fabrication technology due to the compatibility in patterning and photolithographic techniques. The smaller size and the lower cost are added advantages so that many electronic sensing functions can be integrated into the chips.

Table 9.2 Magnetoresistance detection platforms used by the magnetoresistive sensors and magnetic labels. (Graham 2004)

Sensor Type	Size (μm)	Label size	Sensitivity	Range	Molecular Recognition
Spin valve	2×6	$2\,\mu m$	1	1–6	Yes
		0.5–1.5 mm		1–10	No
		$1\,\mu m$		1–15	No
		250 nm	10 s	10–100 s	Yes
		100 nm	100 s	1000 s	No
		50 nm	1000 s	1000 s	No
Spin valve	1×2.5–3×12	$2.8\,\mu m$	1	<10	No
		11 nm (Co)	>1000 s		No
AMR ring	5 (d)	$4.3\,\mu m$ (NiFe)	1	–	No
Hall sensor	2.4×2.4	$2.8\,\mu m$	1	–	No
GMR spiral	70 (d)	$2.8\,\mu m$	200	<1000	Yes
GMR strip	5×80	$2.8\,\mu m$	1	<100	Yes
GMR Serpentine	200 (d)	$2.8\,\mu m$	10	>1000	Yes

(2) Challenges

GMR sensor development has many remaining challenges. The aggregation of the particle is one of the main problems and it prevents the sensors from working effectively. Freely moving polymer coated iron oxide particles have found a solution to this. However, the magnetic moment is found to be very small. It has been observed that micron-sized particles may impede biomolecular recognition and interaction and hence reduce detection efficiency. Particles of less than 100 nm diameter are found to be difficult to detect individually due their low magnetic moment. Results observed for particle diameters down to 130–250 nm with iron oxide nanoparticles limit the detection.

9.3   Biochips Based on Spin Valve Sensors

It can be seen from Section 9.2 that in GMR sensors, the external field aligns all layers in one direction which reduces the scattering of one of the electron spins. However, a different effect can be observed if an antiferromagnetic layer is placed close to a pair of spaced ferromagnetic layers. The antiferromagnetic layer serves as an exchange bias for the adjacent magnetic layer. Since the other layer is free to rotate, the device shows a linear dependence of magnetic resistance. This device is known as a spin valve, since the applied field effectively acts as a valve for one of the electron spins.

A spin valve is a device in which two conducting magnetic thin films that alternates its electrical resistance depending on the alignment of the layers. It is a sandwich configuration similar to a GMR sensor with one of the two magnetic thin films being connected with an antiferromagnetic layer such as FeMn or CrPtMn. Since these layers are made of different hysteresis materials, the top layer changes its polarity while the bottom layer keeps its polarity and vice versa. The thin magnetic layers align up or down depending on the external magnetic filed. The spin nature of the electrons controls the spin valve functions. In a polarized magnetic thin layer, the unpaired electronics align their spins due to the external magnetic field and they keep that spin while they move through the device. If these electrons encounter a magnetic thin film, which has a field in the opposite direction, the electrons tend to flip their spins and the device shows higher resistance. These technologies are effectively utilized in computer hard-disk drives for high-density magnetic data storage. The device shows higher resistance when the magnetic layers are antiparallel and the resistance is low when they are in parallel. The spin valves are sensitive not only to the magnitude of the applied field, but also to the direction of the field.

Even though the major uses of the spin valve sensors are for the data storage in hard disk drives, due to their ability to detect very weak magnetic fields at room temperature, they are being explored for a number of new applications, particularly in biological sensors (Kurlyandskaya and Levit 2007; Baselt 1998; Ferreira *et al.* 2005; Graham *et al.* 2004, 2005; Freitas *et al.* 2002). Biochips based on magnetoresistive sensors might be able to recognize biomolecules at the single molecule.

In GMR multilayers, the magnetization direction depends on the non-magnetic spacer such as Cu or Ru. However, in spin valves, the layer in between the magnetic layers was engineered to have its magnetization pinned by an antiferromagnetic layer. At the same time, the magnetic dipoles at the other layers are free to rotate. Figure 9.18(a) and (b) presents the schematic representation of the spin valve sensors for DNA detection applications; (a) is the top view and (b) the cross-sectional view. M_f and M_p are the free and pinned magnetization of the spin valve sensors and H_t and H_b are the applied magnetic excitation and bias fields. Figure 9.18(c) presents the schematic representation

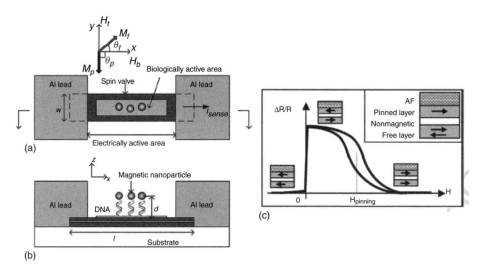

Figure 9.18 Schematic representation of the spin valve sensors (a) top view; (b) cross-sectional view. The conceptual labeling of magnetic nanoparticles bound to the sensor surface using hybridized DNA probes are shown in the active area and (c) the spin valve structure and the magnetoresistance loop at room temperature. (Figures (a) and (b) are adopted from Li *et al.* 2006 and (c) from Schuhl and Lacour 2005). Reprinted from: Li, G. X. Sun, S. H. Wilson, R. J. White, R. L. Pourmand, N. and Wang, S. X. (2006). "Spin valve sensors for ultrasensitive detection of superparamagnetic nanoparticles for biological applications", *Sensors and Actuators* A, **126**, 98–106., with permission from Elsevier

of the spin valve structure and its magnetoresistance loop. The magnetically soft layer is separated by a non-magnetic layer from a second mantic layer. In this figure, $H < 0$ corresponds to a parallel configuration; $0 < H < H_{pinning}$ corresponds antiparallel magnetizations and $H > H_{pinning}$ corresponds parallel alignments.

Figure 9.19 presents the schematic representation of spin valve structure which was introduced in 1991 and is used on most of the read heads in computer hard disks. In the configuration, the magnetic moments in sense and free layers are presented in Figure 9.19(a). This spin valve sensor consists of a magnetically soft layer separated by a non-magnetic layer which has a pinned magnetization by an exchange biasing interaction with an antiferromagnetic layer such as FeMn or IrMn or a ferromagnetic layer. The difference between an antiferromagnetic (AF) and a ferromagnetic (F) layer is that the magnetic dipole inside the AF will tend to align antiparallell to its nearest neighbor. The permanent magnetic dipoles of the F materials interact strongly between each other even without the presence of any applied field. As a result, each ferromagnetic atom will tend to align with its neighbors due to the spin interaction between the nearest atoms. When a ferromagnetic layer is in contact with an AF layer exhibiting exchange anisotrophy, the magnetization field loop is found to be shifted away from the zero applied field (Guedes *et al.* 2006). This is mainly due to the extra magnetic field created by the F layer dipoles due to its coupling. The magnetization of the unpinned or free layer reverses when the magnetic field increases from negative to positive values in the small field range close to $H = 0$. However, due to the magnetization of the pinned layer which remains fixed in the negative direction, the resistance increases sharply, hence the sensitivity of the device is high for a small field. As shown in Figure 9.19, the F1 magnetization is pinned because it

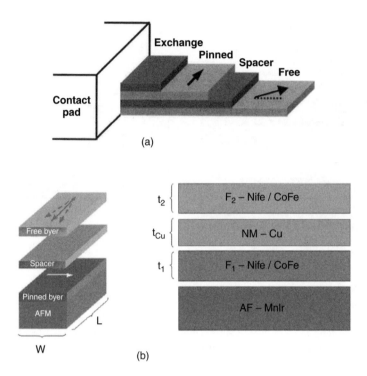

(a)

(b)

Figure 9.19 (a) Schematic diagram of spin valve sensor showing the magnetic moments in the sense (free) and pinned layers (Ferreira *et al.* 2003); (b) structure and arrangement of layers in a simple magnetoresistive spin valve sensor. Due to the coupling to an AF layer, the magnetization of layer F1 is pinned. However, due to the absence of AF-F coupling in layer F2, its magnetization can be rotated freely. (Mendes 2005)

is coupled to the layer AF. Since in layer F2, there is no AF-F coupling, its magnetization is free to rotate. The conducting Cu layer is called a spacer.

Figure 9.20 presents the typical arrangement of layers (in Å units) of a spin valve sensor that consists of two ferromagnetic layers separated by a Cu spacer. In this device, the buffer layer Ta provide a $<111>$ structure that controls the grain size to 10 nm to produce soft free layer properties which leads up to 10 % MR values. Theoretical modeling shows that by decreasing the Cu layer thickness to 15–18 Å on the Ta buffer layer, the MR value is found to be increasing sharply. However, the decrease in Cu thickness affects the interlayer ferromagnetic Neel coupling and hence the usual thickness of the Cu layer is around 20–22 Å (Freitas 2007).

Most widely studied MR sensors use magnetic microspheres of size ranging from 1–3 μm in diameter. Compared to magnetic nanoparticles, the increased volume results in a higher magnetic moment per label in external magnetic field. The uniform size and shape of the microspheres permit quantitative analysis with linearity between the signal and the number of labels to be detected. However, the higher mass of micron-sized particles compared to the biomolecules prevent the label being attached to the sensor surface and the larger diameter of the labels hinders high-density binding for a given area. Magnetic nanoparticles are found to be a solution to these problems. The increased density in binding could be achieved using the nanoparticles due to their smaller size.

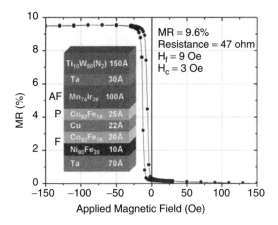

Figure 9.20   Arrangement of various layers along with its thickeness (in Å units) for a spin valve sensor. (Freitas 2007)

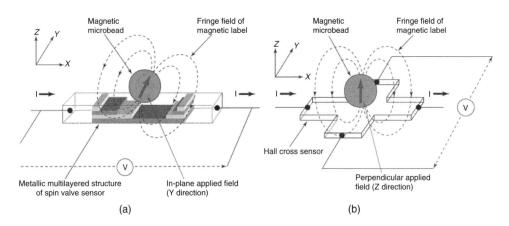

Figure 9.21   Comparison of magnetic labeling using spin valve and Hall effect sensors showing different measurement principles and geometries; (a) in spin valve sensors based on GMR effect, the label fringe field is detected using a sensor biased on in-plane applied magnetic field; (b) in a single layer Hall effect sensor which is based on the AMR effect, detection is possible using an applied magnetic field perpendicular to the sensor surface. (Graham *et al.* 2004). Reprinted from: Graham D.L., Ferreira H. A. and Freitas P.P. (2004). Magnetoreisitive based biosensors and biochips, *Trends in Biotechnology*, **22**, 9, 455–462, with permission from Elsevier

Figure 9.21 presents a comparison of magnetic label measurements for spin valve and Hall effect sensors, which work based on magnetoresistive sensing principles. Figure 9.21(a) is a multilayered magnetoresistive spin valve sensor in which the labeled fringe field is detected with an in-plane sensor. However, as shown in Figure 9.21(b), in the Hall cross sensor, the fringe label field perpendicular to the sensor surface is detected.

Figure 9.22 shows a photograph of a spin valve sensor. The change in electrical resistance of the magnetic materials depends on the relative direction of the current flow and the magnetization. As shown in the figure, the current generally passes through the Cu layer due to its lowest resistivity. However, the flow of current through the Cu layer

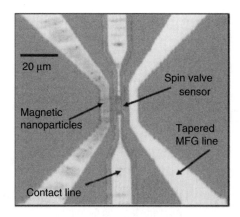

Figure 9.22 Photograph of the spin valve sensor. The figure shows the contact lines for the current flow in the direction of the layers plane and the tapered magnetic field generating lines used to attract the tagged magnetic nanoparticles. (Mendes 2005, p. 26)

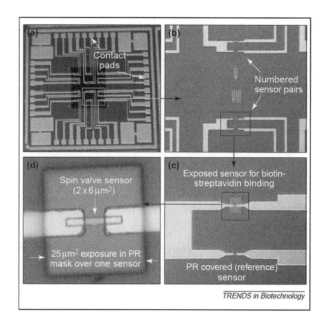

Figure 9.23 Magnetoresistive biochip based on spin valve sensors. The chip has 12 pairs of spin valve sensors for single or differential measurements; (a) fabricated sensors at the center of 8 × 8 mm chip along with connection pads; (b) the input line and output lines; (c) the exposed sensor and a reference sensor; and (d) single 2 × 6 μm² spin valve with opened contacts. (Graham et al. 2004). Reprinted from: Graham D.L., Ferreira H. A. and Freitas P.P. (2004). Magnetoreisitive based biosensors and biochips, *Trends in Biotechnology*, **22**, 9, 455–462, with permission from Elsevier

not only depends on the scattering effects of the electrons in the Cu layer, but also the magnetic interactions of the adjustment layers.

Figure 9.23 shows a photograph of the spin valve sensor fabricated on a silicon wafer (Graham *et al.* 2003). The spin valve stack structure is Ta 65 Å/NiFe 40 Å/CoFe 10 Å/Cu 26 Å/CoFe 25 Å/MnIr 80 Å/Ta 25 Å/TiW(N) 150 Å. These layers are fabricated by magnetron sputtering with a resistance of 15.4 Ω. The structure is defined using laser

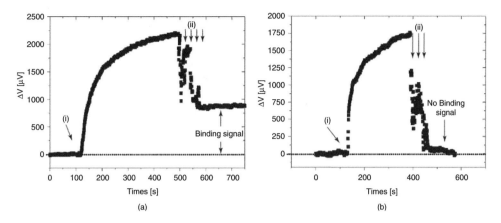

Figure 9.24 Real-time detection of fibrosis related DNA using magnetoresistive sensors with 50mer oligonucleotide probes immobilized on spin valve sensors and hybridized with (a) bi.otinylated complementary target DNA; (b) biotinylated non-complementary DNA. (Graham *et al.* 2004). Reprinted from: Graham D.L., Ferreira H. A. and Freitas P.P. (2004). Magnetoreisitive based biosensors and biochips, *Trends in Biotechnology*, **22**, 9, 455–462, with permission from Elsevier

lithography and ion beam milling techniques with AlCu metal leads as electrical contacts. This chip has 12 pairs of sensors used for the differential signal measurement using the Wheatstone bridge architecture to enable thermal and electrical noise compensation. The device is fabricated using silicon fabrication technologies including lithography and sputtering. Sputtered silicon dioxide of a 2000 Å thick passivation layer is used to protect the chip from corrosion due to the applied fluids during the chemical and biological reactions over the chip.

GMR sensors can be combined with microfluidic flow channels to selectively attach magnetic beads to fabricate miniaturized biosensor arrays. Figure 9.24 presents the measured signals from the magnetoresistive sensors for the detection of a cystic fibrosis related DNA target. A 50mer oligonucleotide probe was immobilized over $2 \times 6 \mu m^2$ spin valve sensor. These sensors were hybridized with biotinylated complementary target DNA in Figure 9.24(a) and biotinylated non-complementary DNA in Figure 9.24(b). It can be seen from the figure that the hybridized DNA was detected by introducing streptavidin functionalized nanoparticles of size 250 nm. The sensor saturation signal level is at (i) and after washing of the signal level is at (ii). After washing, 50% of the sensors are covered with labels in (a) and no binding signal is observed in (b).

9.4 Biochips Based on Magnetic Tunnel Junctions

The ever increasing capacity of magnetic disk drives demands alternatives to GMR spin valve sensors. The maximum useful magnetoresistance provided by spin valve recording sensors is about 16 % – 18 %. It has been proved that much large MR values can be achieved using magnetic tunneling junctions (MTJ). An MTJ is similar to a spin valve device but the metallic Cu spacer is replaced with a very thin layer of insulating tunnel barrier. The sensor current in such a device is passed perpendicularly through the device (Julliere *et al.* 1975; Miyazaki *et al.* 1995; Moodeera *et al.* 1995). Tunneling magnetoresistance (TMR) values as high as 60 % are observed in alumina tunnel barriers at room

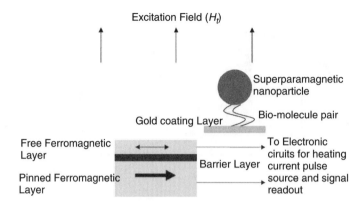

Figure 9.25   Schematic representation of a biosensor based on MTJs. (Wang and Jiang 2007)

temperature (Tsunoda *et al.* 2002). The TMR is defined as the change in resistance divided by the saturated resistance for the parallel magnetization configuration.

Among various magnetic sensors, MTJ devices have greater importance due to their flexibility in designing the required resistances by changing the tunnel barrier thickness. MTJs exploit the asymmetry in the density of states of the majority and minority energy bands in a ferromagnet (Julliere *et al.* 1975; Sloncezewsli 1989; Gallagher *et al.* 1997; Li *et al.* 2006). Tunneling takes place when an electron passes across an insulating barrier. Spin dependent tunneling is related to the relative orientation of two adjustment ferromagnetic films. There is maximum match between the number of occupied states in one electrode and available states in the other when the electrodes are in parallel. The tunneling current is maximum when tunneling resistance is minimum. However, in antiparallel configuration, there is tunneling between the majority states in one electrode and the minority state in the other. This mismatch in the alignment results in minimum current with maximum resistance. In MTJ devices, the directions of magnetizations can be altered by applying an external magnetic field and hence the tunneling resistance is sensitive to the applied field. The perpendicular current flow in such devices is highly attractive to achieve ultra high density since the sensor can be directly attached to the electrical contacts which can also serve as a magnetic shield. In conventional GMR sensors, the current flow is parallel to the layers and the sensor has to be electrically isolated. The isolation layer also occupies space between the shields.

The cross-sectional view of a biosensor-based MTJ is presented in Figure 9.25. The sensor layer is at the bottom ferromagnet, which is pinned in the right direction. It could be possible to freely switch the magnetization orientation of the top layer using an external magnetic field. The sensor area is coated and patterned with a thin layer of gold for the biomolecular attachment. The direction of the in-plane component of the fringe field due to the biolabels can be detected as change in resistance in the MTJ cell.

Figure 9.26 shows the schematic representation of the switching mechanism in MTJs used for the design of MRAM (Butler and Gupta 2004). The top and bottom electrodes are ferromagnetic layers separated by a tunnel barrier. The antiparallel (R_{AP}) and parallel (R_P) alignment layers can be switched using an external magnetic field. The change in magnetoresistance is observed due to relative spin orientation of the electrodes by changing from R_{AP} to R_P. In MRAM, the high and low states correspond to '1' and '0' in non-volatile memory.

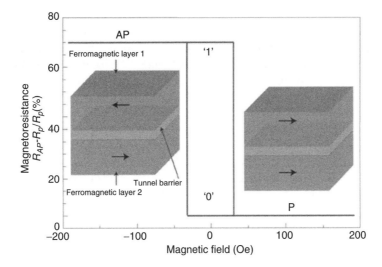

Figure 9.26 Schematic diagram of switching mechanism in MTJ used in MRAM devices. The top and bottom ferromagnetic layers are separated by an insulating tunnel barrier. The change in tunnel resistance is observed due to the switching of the alignments of top and bottom magnetic layers. The high and low resistance states correspond to '1' and '0' in the non-volatile memory device. (Butler and Gupta 2004)

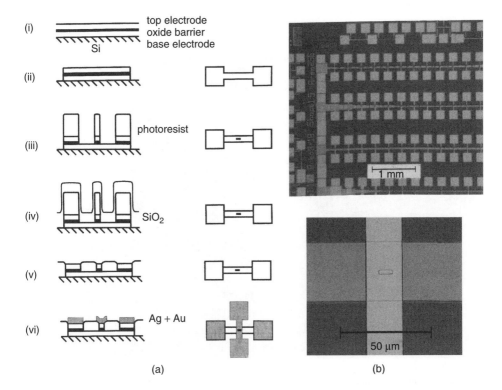

Figure 9.27 (a) Schematic diagram of the fabrication process of magnetic tunnel junctions with cross-sectional (right) and planar (left) views. (b) Layout of the chip and a single tunnel junction. (Gallager *et al.* 1997)

(3) MTJ Fabrication

The performance of an MTJ depends on the size, shape and the geometrical magnetic parameters. The device fabrication starts with the growth of multilayered films using magnetron sputtering on a silicon wafer as shown in Figure 9.27. There are two types of layer structures in an MTJ: structures with Co-free layers and structures with permalloy free layers. The Co-free layer structure (MnFe-Py/I/Co) has the following layer sequence on a Si(100) wafer. Base electrode [200 Å Pt/40 Å Py/100 Å MnFe/80 Å Py]/[oxide (10–30 Å Al_2O_3)]/top electrode (80 Å Co/200 Å Pt), where Py is permalloy (Ni81Fe19). The Py-free layer structure (MnFe–Co/I/Py) has the layer sequences on Si (100) wafer Base electrode [200 Å Cu/40 Å Py/100 Å MnFe/100 Å Co]/oxide (10–20 Å Al_2O_3)/top electrode [200 Å Py/200 Å Pt]. The tunneling barrier is formed by plasma oxidization of 10–30 Å thick Al layer in 100 mTorr of O_2 for 2–7 minutes. The tunneling occurs in both cases between the permalloy and the Co layers. The photolithography process defined as follows is used to pattern the device structures.

The electrodes are defined and patterned using ion milling system as shown in Figure 9.27(a). The junction areas were defined by patterning the top electrode by second ion milling, which stops at the top surface of the base electrode. Using RF magnetron sputtering, the junctions were coated with 1000–1600 Å SiO_2. The junction electrodes were made by a metallization layer of 200 Å Ag/3000 Å Au, deposited and patterned by lift-off process.

The optical micrographs Figure 9.27(b) show the layout of MTJ on a chip and the dimensions of a single junction.

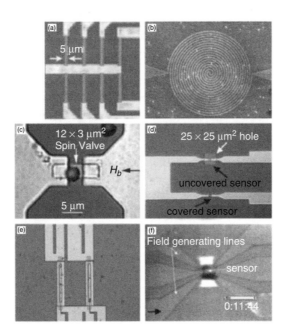

Figure 9.28 Magnetic biosensors developed by various groups; (a) BARC II from Naval Research; (Edelstein *et al.* 2000) (b) Universität Bielefeld Germany; (c) Stanford single bead detector; (d) INESC, Lisbon Portugal for the fluid sensors; (e) Philips Research sensor that can distinguish bulk and surface concentrations; (f) IMEC Leuven, Belgium sensor that combines with an actuator. (Megens and Prins 2005)

9.5 Fully Integrated Biochips

Baselt *et al.* (1998) developed a GMR strip based on 2.8 μm diameter Dynabeads on $80 \times 5 \,\mu m^2$ strips. This concept has been further modified and has resulted in a DNA hybridization sensor array known as BARC as shown in Figure 9.28(a). It consists of 64 sensors with eight sensor strips (Miller *et al.* 2001). Schotter *et al.* developed a similar sensor with a large sensing area as shown in Figure 9.28(b). Figure 9.28(c) presents a single bead sensor developed for the detection of a single DNA fragment by Li *et al.* (Li 2003a; Li 2003b). The sensor developed for binding the magnetic biomarker sample in the fluid is shown in Figure 9.28(d). The details of these sensor developments are presented in Chapter 10.

References

Alche, J.D., and Dickinson H., (1998). Affinity Chromatographic Purification of Antibodies to a Biotinylated Fusion Protein Expressed in Escherichia coli, *Protein Expression and Purification*, **12**, 132–8.

Alvarez, M., Carrascosa L.G., Moreno, M., Calle, A., Zaballos, A., Lechuga, L.M., Martinez, A.C. and Tamayo, J. (2004)., Nanomechanics of the formation of DNA self-assembled monolayers and hybridization on micro-cantilevers, *Langmuir*, **20**, 9663–8.

Ambrose, W.P., Affleck, R.L., Goodwin, P.M., Keller, R.A. and Martin, J.C., (1995). Imaging biological molecules with single molecule sensitivity using near-field scanning optical microscopy, *Experimental Technical Physics*, **41**, 1–12.

Auroux, P.A., Iossifidis, D., Reyes, D.R. and Manz, A. (2002), Micro total analysis systems. 2. Analytical standard operations and applications, *Analytical Chemistry*, **74**, 2637–52.

Baibich, M.N, Brote, J.M. and Fert, A. (1988). Giant mangnetoresistance of (001) iron/(001) chromium magnetic superlattices, *Physical Review Letters.*, **61**, 2472–5.

Baselt, M.N., Lee, G.U., Natesan, M., Metzger, S.W., Sheehan, P.E. and Coltern, R.J. (1998). A biosensor based on magnetoresistance technology, *Biosensors & Bioelectronics*, **13**, 731–9.

Beckman, R.A., Johnston-Halperin, E., Luo, Y., Melosh, N., Green, J. and Heath, J.R. (2004). Fabrication of Conducting Silicon Nanowire Arrays, *Journal of Applied Physics*, **96**, 5921–3.

Binasch, G., Grunberg, P., Saurenbach, F. and Zinn W. (1988). Enhanced magnetoresistance in layered magnetic-structures with antiferromagnetic interlayer exchange, *Physical Review B*, **39**, 4828–30.

Buack, S., Jones, D.A., Laibinis, P.E. and Halotn, T.A. (2003). Protein separations using colloidal magnetic nanoparticles, *Biotechnology Progress*, **19**, 477–84.

Butler, W.H. and Gupta, A. (2004) Magnetic memory: A signal boost is in order, *Nature Materials*, **3**, 845–7.

Chan, W.C.W., Maxwell, D.J., Gao, X., Bailey, R.E., Han, M. and Nie, S. (2002). Luminescent QDs for multiplexed biological detection and imaging, *Current Opinion in Biotechnology*, **13**, 40–6.

Chikazumi, S. (1964). *Physics of magnetism*, R.E. Keiger, Malabar, Florida.

Choi, J.W., Oh, K.W., Thomas, J.K., Heineman, W.R., Halsall, H.B., Nevin, J.H., Helmicki, A.J., Henderson, H.T. and Ahn, C.H. (2002). An integrated microfluidic biomedical detection system for protein analysis with magnetic bead-based sampling capabilities, *Lab on a Chip*, **2**, 27–30.

Cui, Y., Wei, Q., Park, H. and Lieber, C.M. (2001). Nanowire nanosensors for highly sensitive and selective detection of biological and chemical species, *Science*, **293**, 1289–92.

Edelstein, R.L., Tamanaha, C.R., Sheehan, P.E., Miller, M.M., Baselt, D.R., Whitman, L.J. and Colton, R.J. (2000). The BARC biosensor applied to the detection of biological warfare agents, *Biosensors & Bioelectronics* **14**, 805–13.

Essevaz-Roulet, B., Bockelmann, U. and Heslot, F. (1997). Mechanical separation of the complementary strands of DNA, *Proceedings of the National Academy of Sciences of the United States of America*, 94, 11935–40.

Ferreiraa, H.A., Graham, D.L., Freitas, P.P. and Cabral, J.M.S. (2003). Biodetection using magnetically labeled biomolecules and arrays of spin valve sensors, *Journal of Applied Physics*, **93**, 7281–6.

Ferreira, H.A., Feliciano, N., Graham, D.L. and Freitas P.P. (2005). Effect of spin-valve sensor magnetostatic fields on nanobead detection for biochip applications, *Journal of Applied Physics*, **97**, 10Q904.

Freitas, P.P., Silva, F., Oliveira, N.J., Costa, L. and Almeida, N. (2002). Spin valve sensors, *Sensors and Actuators*, **81**, 2–8.

Freitas, P.P., Ferreira, R., Cardoso, S. and Cardoso, F. (2007). Magnetoresistive sensors, *Journal of Physics – Condensed Matter*, **19**, 1–21.

Gallagher, W.J., Parkin, S.S.P., Lu, Y., Altman, R.A., Rishton, S.A., Jahnes, C., Shaw, T. and Xiao, G. (1997). Microstructures magnetic tunnel junctions, *Journal of Applied Physics*, **81**, 3741–6.

Georganopoulou, D.G., Chang, L., Nam, J.-M., Thaxton, C.S., Mufson, E.J., Klein, W.L. and Mirkin, C.A. (2005). Nanoparticle-based detection in cerebral spinal fluid of a soluble pathogenic biomarker for Alzheimer's disease, *Proceedings of the National Academy of Sciences of the United States of America*, 102, 2273–6.

Gijs, M.A.M. (2004). Magnetic bead handling on-chip: new opportunities for analytical applications, *Microfluids and Nanofluids*, **1**, 22–40.

Graham, D.L., Ferreira, H.A., Freitas, P.P. and Cabral, J.M.S. (2003). High sensitive detection of molecular recognition using magnetically labeled biomolecules and magnetoreisiteve sensors, *Biosensors and Bioelectronics*, **18**, 483–8.

Graham, D.L., Ferreira, H.A. and Freitas P.P. (2004). Magnetoresistive based biosensors and biochips, *Trends in Biotechnology*, **22**, 455–62.

Graham, D.L., Ferreira, H.A., Felicino, N., Freitas, P.P., Clarke, L.A. and Amaral, M.D. (2005). Magnetic filed assisted DNA hybridization and simultaneous detection using micron-sized spin valve sensors and magnetic nanoparticles, *Sensors and Actuators B*, **107**, 936–44.

Guedes, A., Mendes, M.J., Freitas, P.P. and Martins, L.J. (2006). Study of synthetic ferrimagnet-synthetic antiferromagnet structures for magnetic sensor application, *Journal of Applied Physics*, **99**, 08B703.

Hahm, J.-I., Lieber, C.M. (2004). Direct ultrasensitive electrical detection of DNA and DNA sequence variations using nanowire nanosensors, *Nano Letters*, **4**, 51–4.

Han, M., Gao, X., Su, J.Z. and Nie, S. (2001). Quantum dot-tagged microbeads for multiplexed optical coding of biomolecules, *Nature Biotechnology*, **19**, 631–5.

Han, S.-J., Xu, L., Yu, H., Wilson, R.J., White, R.L., Pourmand, N. and Wang S.X. (2006). CMOS Integrated DNA Microarray Based on GMR Sensors, *IEEE International Electron Devices Meeting*, 2006, 1–4.

Ji, H.F., Yang, X., Zhang, J. and Thundat, T. (2004). Molecular recognition of biowarfare agents using micromechanical sensors, *Expert Review of Molecular Diagnosis* 47, 859–66.

Julliere, M. (1975). Tunneling between ferromagnetic films, *Physics Letters A*, **54**, 225–6.

Kurlyandskaya, G. and Levit, V. (2007). Advanced materials for drug delivery and biosensors based on magnetic label detection, *Materials Science and Engineering C – Biomimetic and Supramolecular Systems*, **27**, 495–503.

Lagae, L., Wirix-Speetjens, R., Das, J., Graham, D., Ferreira, H., Freitas, P.P.F., Borghs, G. and De Boeck, J. (2002). On-chip manipulation and magnetization assessment of magnetic bead ensembles by integrated spin-valve sensors, *Journal of Applied Physics*, **91**, 7445–7.

Lee, G.U., Metzger, S., Natesan, M., Yanavich, C., Dufrene, Y.F., (2000). Implementation of force diffraction in the immunoassay, *Analytical Biochemistry*, **287**, 261–71.

Lee, J.H., Hwang, K.S., Park, J., Yoon, K.H., Yoon, D.S. and Kim, T.S. (2005). Immunoassay of prostate-specific antigen (PSA) using resonant frequency shift of piezoelectric nanomechanical microcantilever, *Biosensors and Bioelectronics*, **20**, 2157–62.

Lehmann, U., Vandevyver, C., Parashar, V.K. and Martin, A.M.G. (2006). Droplet-based DNA purification a magnetic lab-on-a-chip, *Angewandte Chemie*, **45**, 3062–7.

Lenssen, K.M.H., Adelerhof, D.J., Gassen, H.J., Kuiper, A.E.T., Somers, G.H.J. and Van Zon, J.B.A.D. (2000). Robust giant magnetoresistance sensors, *Sensors and Actuators*, **85**, 1–8.

Li, G. and Wang, S.X. (2003a). Analytical and Micromagnetic Modeling for Detection of a Single Magnetic Microbead or Nanobead by Spin Valve Sensors, *IEEE Transactions on Magnetics*, **39**, 3313–15.

Li, G., Joshi, V., White, R.L., Wang, S.X., Kemp, J.T, Webb, C., Davis, R.W. and Sun, S. (2003b). Detection of single micron-sized magnetic beads and magnetic nanoparticles using spin valve sensors for biological applications, *Journal of Applied Physics*, **93**, 7757–9.

Li, G., Sun, S., Wilson, R.J., White, R.L., Pourmand, N. and Wang S.X. (2006). Spin valve sensors for ultrasensitive detection of superparamagnetic nanoparticles for biological applications, *Sensors and Actuators A*, **126**, 98–106.

Magnani, M., Galluzzi, G. and Bruce, I.J. (2006). The use of magnetic nanoparticles in the development of new molecular detection systems, *Journal of Nanoscience and Nanotechnology*, **6**, 2302–11.

Megens, M. and Prins, M. (2005). Magnetic biochips: a new option for sensitive diagnostics, *Journal of Magnetism and Magnetic Materials*, **293**, 702–8.

Melosh, N.A., Boukai, A., Diana, F., Gerardot, B., Badolato, A., Petroff, P.M. and Heath, J.R. (2003). Ultrahigh-density nanowire lattices and circuits, *Science*, **300**, 112–15.

Mendes, M.J.M.D., Senior Thesis, Universidade Tecnica de Lisboa, 2005.

Mirkin, C.A., Letsinger, R.L., Mucic, R.C. and Storhoff, J.J. (1996). A DNA-based method for rationally assembling nanoparticles into macroscopic materials, *Nature*, **382**, 607–9.

Miyazaki, T. and Tezuka, N. (1995). Giant magnetic tunneling effect in Fe/Al2O3/Fe Junction, *Journal of Magnetism and Magnetic Materials*, **119**, L231–L234.

Moodeera, J.S., Kinder, L.R., Wong T.M. and Meservey R. (1995). Large magnetoresistance at room temperature in ferromagnetic thin film tunnel junctions, *Physics Review Letters*, **74**, 3273–6.

Molday, R.S. and MacKenzie, D. (1982). Immunospecific ferromagnetic iron-dextran reagents for the labeling and magnetic separation of cells, *Journal of Immunological Methods*, **52**, 353–67.

McCloskey, K.E., Chalmers, J.J. and Zborowski, M. (2003a). Magnetic cell separation: characterization of magnetophoretic mobility, *Analytical Chemistry*, **75**, 6868–74.

McCloskey, K.E., Moore, L.R., Hoyos, M., Rodriguez, A., Chalmers, J.J. and Zborowski, M. (2003b). Magnetophoretic cell sorting is a function of antibody binding capacity, *Biotechnology Progress* **19**, 899–907.

Miller, M.M., Sheehan, P.E., Edelstein, R.L., Tamanaha, C.R., Zhong, L., Bounnak, S., Whitman, L.J. and Colton, R.J. (2001). A DNA array sensor utilizing magnetic microbeads and magnetoelectronic detection, *Journal of Magnetism and Magnetic Materials*, **225**, 138–44.

Mukhopadhyay, R., Lorentzen, M., Kjems, J. and Besenbacher, F. (2005). Nanomechanical sensing of DNA sequences using piezoresistive cantilevers, *Langmuir*, **21**, 8400–8.

Nam, J.-M., Park, S.-J. and Mirkin, C.A. (2002). Bio-barcodes based on oligonucleotide-modified nanoparticles, *Journal of the American Chemical Society*, **124**, 3820–1.

Nam, J.-M., Thaxton, C.S. and Mirkin, C.A. (2003). Nanoparticle-based bio-bar codes for the ultrasensitive detection of proteins, *Science*, **301**, 1884–6.

Nam, J.-M., Stoeva, S.I. and Mirkin, C.A. (2004). Bio-bar-code-based DNA detection with PCR-like sensitivity, *Journal of the American Chemical Society*, **126**, 5932–3.

Nie, S. and Zare, R.N. (1997). Optical detection of single molecules, *Annual Review of Biophysics and Biomolecular Structure*, **26**, 567–96.

Odenbach, S. (ed.) (2002). *Ferrofluids: Magnetically controllable fluids and their applications*, Springer, Berlin-New York.

Pan, S., Zhang, H., Rush, J., Eng, J., Zhang, N., Patterson, D., Comb, M.J. and Aebersold, R. (2005). High throughput proteome screening for biomarker detection, *Mol Cell Proteomics*, **4**, 182–90.

Pankhurst, Q.A., Connolly, J., Jones, S.K. and Dobson, J. (2003). Applications of magnetic nanoparticles in biomedicine, *Journal of Physics D – Applied Physics*, **36**, R167–R181.

Pardoe, H., Chua-anusorn, W., St Pierre, T.G. and Dobson, J. (2001). Structural and magnetic properties of nanoscale iron oxide particle synthesized in the presence of dextran or polyvinyl alcohol, *Journal of Magnetism and Magnetic Materials*, **225**, 41–6.

Parkin, S., Jiang, X., Kaiser, C., Panchula, A., Roche, K. and Samant, M. (2003). Magnetically engineered spintronic sensors and memory, *Proceedings of the IEEE,* **91**, 661–80.

Robinson, P.J., Dunnill, P. and Lilly, M.D. (1973). *Biotechnol. Bioeng.* **15**, 603.

Rief, M. and Grubmuller, H. (2002). Force Spectroscopy of Single Biomolecules, *Chemphyschem*, **3**, 255–61.

Schotter, J., Kamp, P.B., Becker, A., Puhler, A., Brinkmann, D., Schepper, W., Brucke, H. and Reiss, G. (2002). A biochip based on magnetoresistive sensors, *IEEE Transactions on Magnetics*, **38**, 3365–7.

Schuhl, A. and Lacour, D. (2005). Spin dependent transport: GMR & TMR, *Comptes Rendus Physique*, **6**(9), 945–55.

Sellmyer, D.J., Zeng, H., Yan, M., Sun, S. and Liu, Y. (2006). *Handbook of Advanced Magnetic Materials*, pp. 211–40.

Shivashankar, G.V. and Libchaber, A. (1998). Biomolecular recognition using submicron laser lithography, *Applied Physics Letters*, **73**, 417–19.

Skumiel, A., Jozefczak, A., Hornwski, T. and Labowski, M. (2003). The influence of the concentration of ferroparticles in ferrofluid on its magnetic and acoustic properties, *Journal of Physics D - Applied Physics*, **36**, 3120–4.

Slonczewski, J.C. (1989) Conductance and exchange coupling of 2 ferromagnets separated by a tunneling barrier, *Physical Review B*, **39**, 6995–7002.

Tazawa, H., Kanie, T. and Katayama, M. (2007). Fiber-optic coupler based refractive index sensor and its application to biosensing, *Applied Physics Letters*, **91**, 113901.

Thaxton, C.S., Hill, H.D., Georganopoulou, D.G., Stoeva, S.I. and Mirkin, C.A. (2005). A biol-bar-code assay based upon dithiothreitol-induced oligonucleotide release, *Analytical Chemistry*, **77**, 8174–8.

Tondra, M., Porter, M. and Liper, R.J. (1999). Model for detection of immobilized superparamagnetic nanosphere assay labels using giant magnetoresistive sensors, *Journal of Vacuum Science & Technology A – Vacuum Surfaces and Films*, **18**, 1125–9.

Tsunoda, M., Nishikawa, K., Ogata, S. and Takahashi, M. (2002). 60 % magnetoresistance at room temperature in Co-Fe/Al-O/Co-Fe tunnel junctions oxidized with Kr-O2 plasma, *Applied Physics Letters*, **80**, 3135–7.

Tudos, A.J., Besselink, G.A.J., and Schasfoort, R.B.M. (2001). Trends in miniaturized total analysis systems for point-of-care testing in clinical chemistry, *Lab on a chip*, **1**(2), 83–95.

Vilkner, T., Janasek, D. and Manz, A. (2004). Micro total analysis systems: recent developments, *Analytical Chemistry*, **76**, 3373–85.

Vo-Dinh, T., Cullum, B.M. and Stokes, D.L. (2001). Nanosensors and biochips: frontiers in molecular diagnostics, *Sensors and Actuators B*, **74**, 2–11.

Wang, S.X., Bae, S.-Y., Li, G., Sun, S., White, R.L., Kemp, J.T., Chris, D. and Webb, C.C. (2005). Towards a magnetic microarray for sensitive diagnostics, *Journal Magnetism and Magnetic Materials*, **293**, 731–6.

Wang, W. and Jiang, Z. (2007). Thermally Assisted Magnetic Tunneling Junction for Biosensing Applications, *IEEE Transactions on Magnetics*, **43**, 2406–8.

Weeks, B.L., Camarero, J., Noy, A., Miller, A.E., Stanker, L. and De Yoreo, J.J. (2003). A microcantilever-based pathogen detector, *Scanning*, **25**, 297–9.

Westphal, P. and Bornmann, A. (2002). Biomolecular detection by surface plasmon enhanced ellipsometry, *Sensors and Actuators B*, **84**, 278–82.

Wright, M.E., Han, D.K. and Aebersold, R. (2005). Mass spectrometry-based expression profiling of clinical prostate cancer, *Molecular & Cellular Proteomics*, **4**, 545–54.

Wuite, G.J.L., Davenport, R.J., Rappaport, A. and Bustamante, C. (2000). An integrated laser trap/flow control video microscope for single biomolecules, *Biophysical Journal*, **79**, 1155–67.

Yousaf, M.N. and Mrksich, M. (1999). Diels-Alder reaction for the selective immobilization of protein to electroactive self-assembled monolayers, *Journal of the American Chemical Society*, **121**, 4286–7.

Ziegler, C. (2004). Cantilever-based biosensors, *Analytical and Bioanalytical Chemistry*, **379**, 946–59.

10

Biomedical Applications of Magnetic Biosensors and Biochips

10.1 Introduction

The advancements in modern technologies especially in microelectronics and biotechnology have revolutionized medical technology. Nanotechnology developments in the manipulation and control of biomaterials and system designs have made far-reaching changes in disease diagnosis, treatment and control. Biological tests to detect the presence of specific molecules become quicker, more sensitive and more flexible when certain nanoparticles are allowed to do the work as tags or labels. Since the dimensions of large biomolecules such as proteins and DNA fall in the 1–1000 nm range, nanotechnology is particularly important in biology. Due to the progress in nanotechnology and the advancements in molecular biology, single molecules can be detected and analyzed. The need for new technologies for medical diagnosis and the requirement for a decentralized diagnosis bring disease diagnosis to a different level. Today, disease diagnosis has been moved from expensive well-established laboratories to cost-effective, disposable, fast and accurate point-of-care testing devices. These portable, accurate and easy-to-use biosensors have become a high priority for many industries due to the recent advancements in detection of single pathogens or molecules. Biochips and other micro-biosensing devices are among the few devices that makes this possible. The integration of biotechnology with microelectronics is becoming the backbone for this evolution. Future diagnostic devices will rely heavily on silicon microelectronic chips and it is clear that the next triumph will be biochips for disease detection, diagnosis, monitoring and analysis.

A nanosensor on a chip is a revolutionary idea to implement total chemical analysis on a single chip. The ability of nanotechnology to make chemical and biological analysis using single or few molecules has fundamentally changed the healthcare, food safety and law enforcement sensor systems. The lab-on-a-chip technologies are micrototal analysis systems in which the sensors are capable of complete processing, including mixing using microfluidic channels, chemical analysis based on few molecules and data outputs and networking. Since nanotechnology has the proven ability to build switchable molecular functions, a completely new approach in designing valves, pumps, chemical separations and detection has been developed. Nanotechnology enabled microfluidic systems can

Nanomedicine: Design and Applications of Magnetic Nanomaterials, Nanosensors and Nanosystems
V. Varadan, L.F. Chen, J. Xie
© 2008 John Wiley & Sons, Ltd

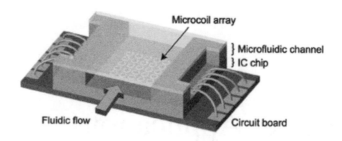

Figure 10.1   Schematic representation of a magnetic biochip. (Lee H *et al*. 2006)

be directed by controlling the surface energy rather than relying on capillary properties of physical channels and the membranes can be replaced with mechanical valves. The requirements for complex signal transductions in conventional laboratories are eliminated by the new approaches in optical, electrical and magnetic detection techniques.

The miniaturization of bioanalytical instruments onto integrated chips has enabled the study of neural signals (Kaul *et al*. 2004; Eversmann *et al*. 2003), tissue dynamics (Bhatia and Chen 1999), electrochemical activities (Hassibi and Lee 2005), the monitoring of ion channels (Fromherz 2005), cell manipulation (Manaresi *et al*. 2003, Lee, H. 2005; Lee, H *et al*. 2004), and the probing of DNA (Cailat *et al*. 1998) using biochips. This miniaturization generally involves the design of sensors, microfluidic flow and mixing channels and signal detection systems. Figure 10.1 shows a typical example of a biochip. This is a hybrid IC microfluidic system for the manipulation of biological cells with a microfluidic channel integrated over the microcoil array circuits to generate spatially patterned microscopic magnetic fields. The magnetic field simultaneously controls the motion of many individual cells which are tagged by the magnetic beads suspended inside the microfluidic system. Efficient and versatile operation of cell manipulation is achieved using the electronically programmable microcoil array circuits using this chip. One of the advantages of this hybrid system is that it can bring together a small number of individual cells with tight spatial control to study cell–cell interactions at the single cell level. Also, this system permits the assembly of a 2-D artificial biological tissue by bringing a large number of cells one by one into a desired geometry at the microscale.

Many different schemes are developed for the detection of biomolecules. Figure 10.2 shows a schematic representation of biomolecule detection schemes using nanoparticles in biochip devices. Gold nanoparticles are widely used as signal reporters for the detection of biomolecules in DNA assay, immunoassay and cell imaging as shown in Figure 10.2(a). Gold nanoparticles are also used in the identification of pathogenic bacteria and DNA microarray technology (Taton *et al*. 2000). In most cases, the metallic properties of the nanoparticles are utilized in particle aggregation, photoemission, electrical and heat conductivity, photo imaging and catalytic activity. These properties are applied in biochips for biomolecule detection, sample preparation, substrate coding and signal transduction and amplification.

Magnetic nanoparticles are found to be a powerful and versatile diagnostic tool in biology and medicine. As shown in Figure 10.2(b), magnetic nanoparticles are used to assist the separation, purification and concentration of different biomolecules. Capturing molecules such as antibodies and oligonucleotides are immobilized on the surface. Magnetic nanoparticles are also used to code biomaterials with different colors during the multiplexing as shown in Figure 10.2(c). The basic approach is to embed a mixture of quantum dots (QD) that emits red, blue and green colors at different ratios into

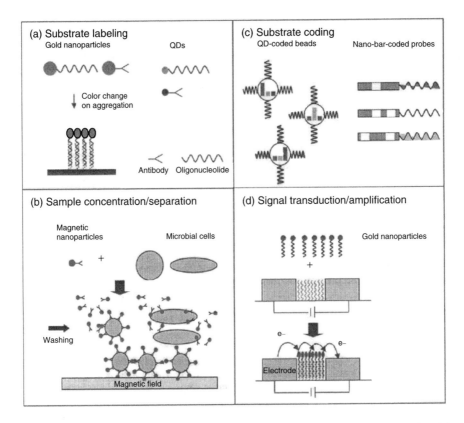

Figure 10.2 Overview of biomolecule detection schemes using nanoparticles in biochips; (a) substrate labeling; (b) sample concentration/separation; (c) substrate coding and (d) signal transduction and amplification. (Liu 2006)

microbeads (Han *et al.* 2001). Theoretically, it is possible to color-code more than a million microbeads using six different QDs and assuming 10 different intensities for each QD with fluorescence labeling (Xu *et al.* 2003). However, the present optical detection systems do not have the required resolution to distinguish all possible intensity combinations. The detection platforms generally suffer from high background fluorescence and, due to the photosensitive nature of the fluorescent labels, it bleaches when exposed to light. Fluorescence-free labeling is possible when the nanoparticles are used as signal transducers in electrical detection systems or signal amplifiers (Park *et al.* 2002) as shown in Figure 10.2(d). Accurate detection of single-nucleotide-polymorphism (SNP) with a target DNA concentration as low as 500 fM is possible using this technique.

Magnetic immunoassay is based on detecting the magnetic fields generated by the magnetically labeled targets. Binding the antibody target molecules or disease causing organisms to a magnetic nanoparticle and measuring that very weak field is the basis of the magnetic immunoassay. This binding gives magnetic signals while exposed to a magnetic field and hence the antibody bound target can be identified. The superconducting quantum interference device (SQUID) based on superparamagnetic nanoparticles is a sensitive, specific and rapid detection technique for biological samples (Chemla *et al.* 2000). Magnetic spheres called 'beads' are frequently used as biomarkers for DNA detection, cell analysis and antibody–antigen interaction studies. These beads are magnetic micro- and nanoparticles of sizes down to 100 nm. The materials applied on the bead shell in the functionalization procedure, for example, streptavidin molecules, will be able to bind specifically to biotin molecules, which in turn can bind selectively to target DNA or proteins. This opens up the possibility of designing biochips to detect the presence of many biomolecules by measuring the magnetic field based on sensors such as magnetoresistance sensors, magnetic tunnel junctions and giant magnetoresistance sensors. This chapter presents the basic principle and designs of various biochips based on magnetic signal transduction for gene and protein analysis, cell analysis, biological warfare and chemical agent detection and environmental applications. Due to its ability to directly translate the changes in magnetization directions to changes in resistances as well as its fabrication compatibility with standard complementary metal oxide semiconductor (CMOS) processing, the complete biochips could be integrated with CMOS electronics devices for signal detection along with flow-controlled microfluidic channels.

10.2 DNA Analysis

In living organisms, the genes are the working subunits of DNA, the chemical information database that carries the complete set of instructions of that organism. The DNA consists of long paired strands spiraled into a double helix. The information in DNA is stored as codes made of four chemical bases: adenine (A), guanine G, cytosine (C) and thymine (T). The order and the sequence of these base codes represent the biological instructions inside the cell. The human genome is composed of 23 distinctive pairs of chromosomes, approximately 3 billion DNA pairs consisting of more than 30 000 genes. Detection of DNA becomes the central theme for the diagnosis and treatment of genetic diseases and forensic analysis. Identification of single-base-mismatched DNA/mRNA strands in complex mixture is very important. It has been proved that most cancers originate from genetic mutations starting with a single-base change in DNA sequences (Cotton 1997). Due to its importance in DNA analysis, development of hybridization assays that permits simultaneous determination of multiple DNA targets becomes of the utmost importance. Multi-target detection using various optical codings that include optical assays (Park et al. 2002; Taton and Mirkin 2001), Raman dye labeled nanoparticles (Cao 2002), quantum dot microbeads (Han 2001), bar-coded nanorods (Nicewarner-Pena et al. 2001) and fiber optic DNA arrays (Ferguson 2000) are found to be very promising in DNA detection. The ultra sensitive electrical detection of DNA using gold nanoparticles (Cheng 2006, Han 2006) and selective molecular hybridization probes (Fuentes 2006; Graham et al. 2003) are materialized due to the characteristic behavior of the nanoparticles. Single stranded DNA (ssDNA) immobilized probes have various biotechnological and medical applications such as DNA-driven nanoassembly (Fritzsche and Taton 2003) and the development of biosensors (Kerman 2004). The choice of the signal transduction medium in a DNA probe is very important and its selection determines the surface chemistry as well as practical use of the conjugates. Most of the established mechanisms are based on solid inorganic supports to capture ssDNA (Riccelli et al. 2001; Peterson et al. 2001).

To understand DNA hybridization and immobilization, it is necessary to evaluate its kinetics. Kinetics of hybridization studies are limited due to the two-dimensional nature of the assemblies where the studies are usually carried out. Magnetic nanoparticles offer special promise due to their ability to direct and manipulate using an external magnetic field as well as their low toxicity and biocompatibility. Gold coated nanoparticles have the added advantage of utilizing the robust chemistry of the gold surfaces along with the unique properties of magnetic particles for immunoassay development (Mikhaylova et al. 2004; Fan et al. 2005) and for the study of the kinetics of the DNA hybridization (Kouassi et al. 2006).

Stability of the DNA probe-support bonds and non-reactivity of oligonucleotide bases to permit hybridization are critical issues for ideal immobilization. Many different methods for chemical immobilization of DNA probes on various supports were developed (Taira and Yokoyama 2005; Fixe *et al.* 2004; Ivanova *et al.* 2004; Walsh *et al.* 2001). The covalent immobilization of aminated probes via an artificially induced amino group at the 3'-or 5' end of the DNA shows a very stable covalent probe-surface bond as well as very low steric hindrances for hybridization (Fuentes *et al.* 2004, 2005). It is established that the DNA probes immobilized on magnetic nanoparticles simplified the target DNA for PCR amplification (Fuentes *et al.* 2006).

Signal detection is one of the key features in all analytical platforms and various detection schemes are developed based on different principles as explained in Section 9.1. Optical, electrical and magnetic detection schemes show very promising results in chemical analysis. However, the portability and higher cost of various optical platforms prevents their use in biochip devices. As explained in Section 9.4, among various types of magnetic sensors, magnetic tunnel junctions (MTJs) have greater importance due to their design flexibility in tuning the resistance while changing the tunnel barrier thickness. Figure 10.3(a) shows the arrangement of the layers in an MTJ. Generally, in a spin valve sensor, two magnetic layers are separated by a non-magnetic conducting spacer, such as Cu. However, in MTJs the conducting spacer is replaced with an insulating tunnel barrier such as aluminum oxide layers. The resistance of the structure depends on the orientation of the magnetization of the two layers. The structure is fabricated on silicon substrate with PtMn (15 nm)/CoFe (2.5 nm)/Ru (0.85 nm)/CoFeB (3 nm)/MgO (0.85 nm)/CoFeB (3 nm) nanolayers as shown in Figure 10.3(a). Using the electron beam and ion beam lithographies, the multilayered structure is patterned to oval-shaped pillars of dimension 200 nm × 100 nm. The antiferromagnetic layer of CoFe and CoFeB at the bottom act as a pinned layer. The magnetization of the top CoFeB layer which acts as a free layer can be changed. Figure 10.3(b) shows the magnetoresistance of the device due to the magnetic field applied at an angle of 30° from the pinned layer.

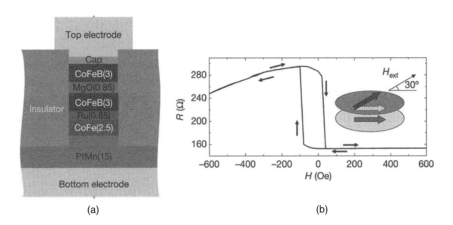

(a)   (b)

Figure 10.3   (a) Schematic diagram of the cross-sectional view of the magnetic tunnel junction diode with the arrangement of the layers. The layer thickness is in brackets. The bottom CoFeB and CoFe layers anti-ferromagnetically couple through the Ru layer which acts as a pinned layer. The top CoFeB layer acts as a free layer, the magnetization of which can be changed. The pinned and free layers are separated by a tunneling MgO barrier; (b) the effect of diode magnetoresistance due to the magnetic field applied at 30° from the pinned-layer magnetization. (Tulapurkar *et al.* 2005)

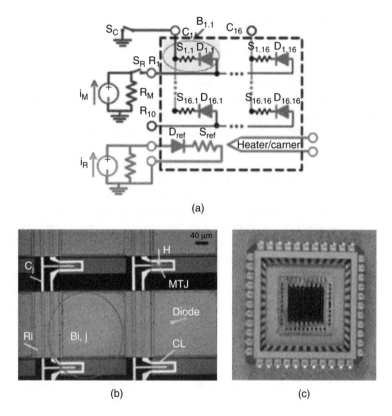

(a)

(b)   (c)

Figure 10.4 The schematic diagram of the biochip for DNA hybridization detection; (b) SEM micrograph of a cell; (c) packaged biochip. (Piedade, M. *et al.* 2006)

Figure 10.4(a) presents the schematic diagram of the biochip matrix array of 16 × 16 MTJ sensors that are connected in series with the thin-film diodes (TFDs). The TFD in this circuit functions as (i) a switching device to enable connection between the column Cjs and row Ris of the matrix; (ii) a temperature sensor for each biosensor. The hand-held platform is integrated with necessary electronics to address the read-out, fluid flow and temperature controller. The MTJs are kept very close to the TFD to sense the planar magnetic field H transverse to its length. This biosensor consists of 16 × 16 cell matrix in which the electric current flows from row conductor Ri through the TFD into the MTJ and finally to the column conductor Cj. The electric current flowing through the row and column addresses the cell and hence establishes the respective connection.

The TFDs are fabricated on hydrogenated amorphous silicon with an aluminum oxide barrier (Lagae *et al.* 2005). The detection site incorporates U shaped carrier lines (CL) which helps to generate a magnetic field to sweep target biomolecules at low frequencies over the immobilized probes to increase the hybridization rate as well as to heat the biochip detection sites.

$200\,\mu m \times 200\,\mu m$ TFDs are formed at the interface of a hydrogenated amorphous silicon (a-Si:H) and an aluminum lead along with an antiferromagnetic layer (MnIr $250\,\mathring{A}$), a fixed ferromagnetic layer (CoFeB $50\,\mathring{A}$), a tunneling insulating barrier layer (aluminum oxide $12\,\mathring{A}$) and a free layer (CoFeB $15\,\mathring{A}$ + NiFe $45\,\mathring{A}$). The transducer shows a tunneling magnetoresistance ratio (TMR) of 27 %. As shown in Figure 10.4(b), a microfluidic

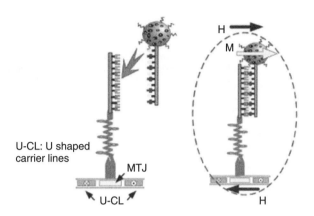

Figure 10.5 DNA hybridization mechanism; (a) magnetically labeled DNA targets at the DNA probe; (b) DNA hybridization detection by sensing the magnetic field created by the labels using MTJ sensors. (Piedade *et al.* 2006)

chamber with a volume of 5 mm × 5 mm × 0.5 mm is mounted over the encapsulated chip. Fluid input and output ports are provided with 2 mm diameter Plexiglas tubing.

The DNA hybridization is detected using MTJ transducers functionalized with a DNA probe as shown in Figure 10.5. The target DNA tagged with paramagnetic nanoparticles is transported to the sensing sites using microfluidic channels. The alternating magnetic field at the sensing site created by the U-shaped CLs is altered by the hybridized DNA targets which are bound to the complementary probe DNA. Finally, the magnetic labels remain bound to the surface of the sensors after washing with buffer solution. The external magnetic field induces a magnetic moment on the nanospheres. Since the probe is connected with the MTJs, the change in magnetic field is detected by the MTJs depending on the number of labels bound to the surface.

DNA sensors can also be materialized using spin valve sensors. The GMR sensors are integrated to a complementary metal oxide semiconductor (CMOS) biochip to form a DNA microarray as shown in Figure 10.6. There are more than 1000 sensing elements in this chip within a 1 mm² area (Han *et al.* 2006; Smith and Schneider 2003). The sensitivity of detection is improved by combining several sensor pixels per biological sample spot. The *in situ* probe synthesis (Mcgall *et al.* 1996) enables large-scale access of genetic information in single pixel per DNA sequences. The light-directed *in situ* DNA synthesis with photo-activatable monomers could achieve densities of the order of 10^6 sequences/cm². This high-density array of oligonucleotide probes is a powerful new tool for large-scale DNA and RNA sequence analysis. 1008 GMR sensors are divided into 16 subarrays with an area of 120 μm × 120 μm. The signal read out from each pixel is achieved by both frequency division multiplexing (FDM) as well as time division multiplexing (TDM) techniques which could easily get 16 outputs/readings.

The performance of the spin valve sensor is highly dependent on the smoothness of the substrate. One of the major obstacles in fabricating a highly sensitive GMR sensor is the planarization of the electrodes' under-layers to obtain the necessary smoothness for successful deposition. Figure 10.7 shows the AFM images of the under-layer before and after chemical mechanical polishing (Smith and Schneider 2003). The surface roughness is found to be changed from 2.9 nm to 0.9 nm after surface polishing.

Figure 10.8(a) shows the schematic diagram of the fabrication procedure of the GMR sensor. The fabrication of the GMR sensor starts with RIE etching of the Sn_3N_4 layer

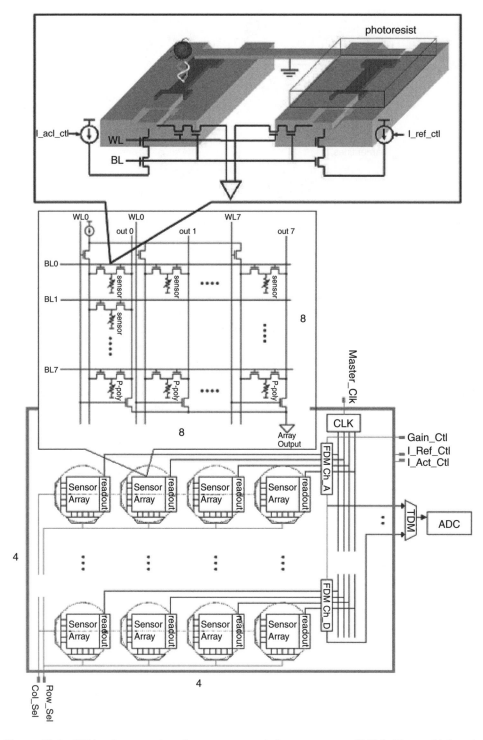

Figure 10.6 DNA microarray based on magnetoresistive sensors on CMOS. The multiple spin valves form the subarray corresponding to 1 DNA spot. 16 subarray share the same control bus. 4 subarrays are frequency division multiplexed to form a channel and 4 channels connect to a multiplexer by time division multiplexing. (Han *et al.* 2006)

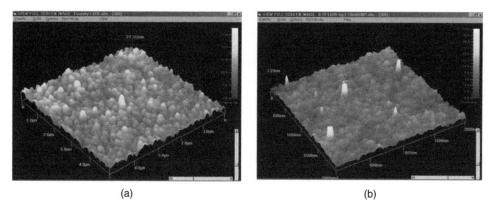

(a)   (b)

Figure 10.7 AFM images of the substrate (a) before planarization, with 2.9 nm roughness; and (b) after chemical mechanical polishing, 0.9 nm surface roughness. (Smith and Schneider 2003)

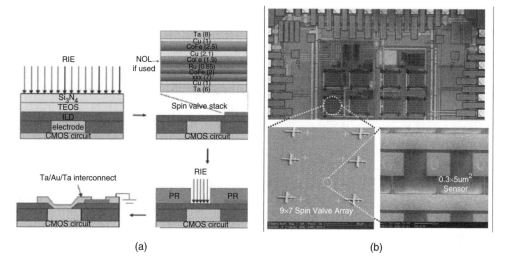

(a)   (b)

Figure 10.8 (a) Schematic diagram of the fabrication procedure of the GMR sensor; and (b) the SEM micrograph of the biosensor array. (Han *et al.* 2006)

to bring the thickness to 400 nm by passivation thinning. The spin valve sensors are patterned using e-beam lithography and lift-off process. Electrical contacts are etched by RIE and Ta/Tu/Ta metal contacts are developed onto it. The antiferromagnetic spin valve sensors are deposited using the ultrahigh vacuum ion beam deposition system. Finally the $SiO_2/Si_3N_4/SiO_2$ layer is deposited to protect the devices. Figure 10.8(b) shows the SEM micrograph of the biosensor array. The chip is fabricated using the 0.25 μm BiCMOS process at the Stanford Nanofabrication Facility.

10.3 Protein Analysis and Protein Biochips

Clinical proteomics is the application of proteomic technologies and informatic tools to clinical evaluation and analysis. The proteome is a dynamic collection of proteins that demonstrate the variation between individuals, between cell types, and between entities of the same type but under different pathological or physiological conditions (Huber 2003). Clinical proteomics demonstrated the promise to identify new targets for treatment and therapeutic interventions. The biomarker detection for diagnosis, prognosis and therapeutic effectiveness is achieved through the comparison of proteome profiles between healthy and disease states (Raj and Chen 2001; Palmer Toy *et al.* 2002). The translational nature of this technology provides unique challenges and many opportunities to transform the way disease is detected, treated and managed. In order to study the responses due to multiple external stimuli over time to a protein as well as its changes within samples from a patient, many technological advancements and new discoveries are required in protein purification, identification and protein interaction with other biomolecules.

Conventional characterization methods for protein-molecule interactions such as surface plasmon resonance (SPR), nuclear magnetic resonance (NMR), affinity chromatography and capillary electrophoresis are time-consuming as well as labor-oriented due to their collection, separation and identification methods. The present methods are based on labor-oriented processing by division of the original sample into multiple divisions. The test will be performed on individual samples for most common substances in each step. This makes the protein analysis more complex.

Protein research has the capability to discover thousands of intact and cleaved proteins in the serum. Disease related protein analysis for disease diagnosis requires separation of protein usually from plasma, urine or saliva and its identification. The tool for this *in vitro* diagnosis (IVD) is a clinical test that has three distinctive development phases as shown in Figure 10.9(a) from identification of the protein to launch of a diagnostic product. The IVD devices should include all necessary reagents, calibrator materials and related instruments for *in vitro* examination in the laboratory after the critical review and approval from the Food and Drug Administration (FDA) before marketing. The array format is very important for the automated platforms for multiple protein detection. The multiplexing capability in protein analysis has the added advantage of flexibility as well as lower cost and time taken for the precision in analysis. As an example, a sample of 12 µl is generally not sufficient for single protein detection by ELISA. However, it is enough for the analysis of 17 cytokines using fluorescent bead-based multiplexed assays. Figure 10.9(b) presents the schematic representation of the strategies behind the design and development of a protein biochip microarray, in which protein immobilization, selection of appropriate functional groups, protein expression and purification and design of novel algorithms for data analysis are involved.

Figure 10.10 shows the protein biochip array technology (PBAT) developed by the Randox Laboratories (Crumlin, UK) to measure 25 analytes simultaneously with a single drop of sample to the biochip. Multiple specific legends are attached at the predefined sites of a 9 mm^2 substrate. The antibody/antigen attachment is done by the silanation methods. Silane reacts with hydroxyl groups on the chip surface followed by the reaction due to the introduction of a hetero-bifunctional linker. This group is capable of reacting with an antibody or peptide or nucleotide. The binding is determined by the chemiluminescence using a charge coupled camera (CCD) imaging system.

The development of functionalized magnetic nanoparticles yields a high throughput assay that has better accuracy, sensitivity and reproducibility. Particularly in the presence of abundant proteins, the nonspecific binding can reduce dramatically the sensitivity of detection (Tang 2004). Hence, there is always a demand for molecular probes with better sensitivity and specificity and nanomagnetic materials are one of the candidate materials for the design of biological assays. The large surface area to volume ratio along with

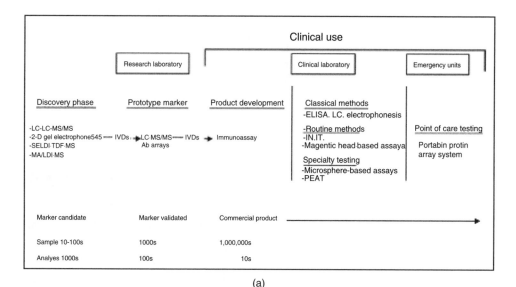

(a)

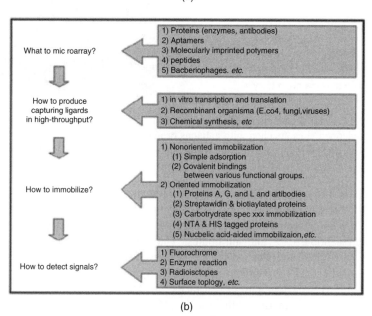

(b)

Figure 10.9 (a) The schematic representation of protein three-step clinical diagnostic system along with exploring technologies at each stage. LC: liquid chiromatography; MS: mass spectrometry; SELDI: surface enhanced laser desroption/ionization; TOF: time of flight; MALDI: matrix assisted laser desorption/ionization; IVD: *in vitro* diagnosis; Ab: antibodies; IN: immunonephelometry; IT: immunoturbidimetry; PBAT: protein biochip array technology (Dupuy, A.M. *et al.* (2005)); (b) schematic representation of microarray fabrication strategy. (Seong and Choi 2003)

the magnetic properties of the magnetic nanoparticles (MNP) demonstrate that antibody conjugated MNP can be used for the separation of target biomarkers with high specificity and sensitivity (Lin 2006; Chou 2005).

Figure 10.11 shows the schematic representation of magnetic nanoparticle based immunoassay developed for affinity mass spectroscopy.

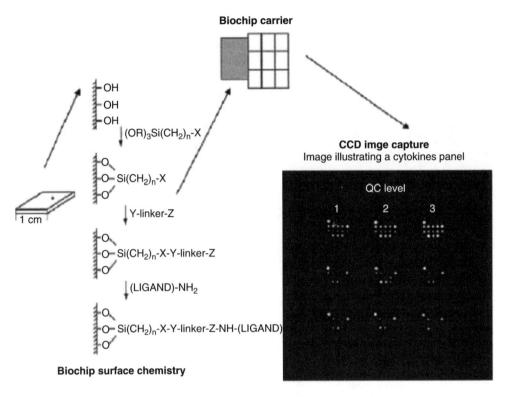

Figure 10.10   Protein biochip array technology (PBAT). This biochip is a 9 mm^2 solid-state substrate with predefined sites attached with antigens or antibodies. The binding is determined by the chemiluminescence and is quantified using a charge coupled camera (CCD). (Dupuy *et al.* 2005)

The antibody conjugated MNPs were first incubated with a biological medium. The antibody of interest is then covalently linked to the nanoparticle surfaces through a cross linker bias (N-hyfroxysccinimide ester). The cross linker can bridge between the aminosilane modified nanoparticles and antibodies. The antibody conjugated nanoparticles are magnetically separated and washed with PBS to remove any excess reactants. MNP1 of 5–15 nm is synthesized by coprecipitation of $FeCl_2$ and $FeCl_3$ under normal conditions (Kang *et al.* 1996). The nanoparticle surfaces are treated to get amino functionality by sol-gel process using tetraethyl orthosilicate (TEOS). This is followed by the addition of 3-aminipropyltrimethoxysilane (APS) to produce MNP2 of size 50 nm. A bifunctional linker suberic acid bis-N-hydroxysuccinimide ester (DSS) is used to crosslink the aminosilane MNPs with anti-serum amyloid P component (anti-SAP) antibody to obtain conjugated MNPs (anti-SAP MNPs).

Mass spectrometers are used to measure the mass-to-charge (m/z) ratio of the ionized atoms or molecules. The sample is introduced into the ionization chamber of the mass spectrometer. After ionization, the samples are extracted into the analyzing region where they are separated according to the mass-charge ratios using an external magnet. The structural information of individual molecules can be obtained from the distinctive fragmentation patterns in a mass spectrum. Today, molecular mass can be measured within an accuracy of 5 ppm or less which is often confirmed by the molecular formula.

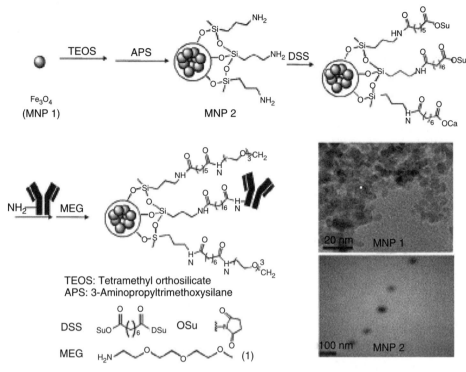

TEOS: Tetramethyl orthosilicate
APS: 3-Aminopropyltrimethoxysilane

Preparation and characterization of MEG-protected antibody-conjugated MNPs.

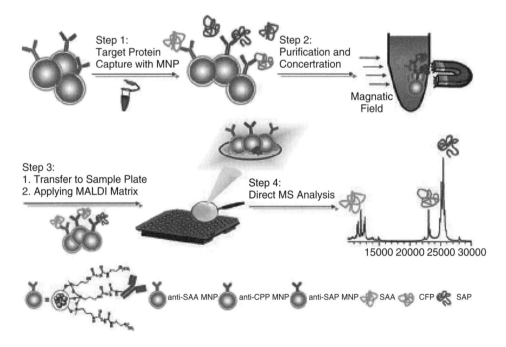

Figure 10.11 Schematic representation of the analytical processing of the multiplexed immunoassay by MEG protected antibody conjugated magnetic nanoparticles and MALDI-TOF MS. MNP: magnetic nanoparticle; CRP: C-reactive protein; SAA: serum amyloid A. (Len *et al.* 2006)

Mass spectrometry is considered as one of the gold standards for the identification of smaller molecules. Among other types of mass spectrometers, matrix-assisted laser desorption/ionization time-of-flight mass spectroscopy (MALDI-TOF MS) is widely used for protein identification due to its high sensitivity (Pasch 2002) as well as its ability to deal with complex mixtures of unknown structures. The molecular mass of large molecules such as biomolecules can be measured within 0.01 % of the total molecular mass of the sample. However, the uses of MALDI-TOF MS in smaller molecules are limited due to interference from the matrix in the low molecular weight region of the spectrum (Sun *et al.* 2006). The detection sensitivity is improved by using methoxyethyl terminated ethylene gycol (MEG) protected antibody conjugated magnetic nanoparticles (Lin *et al.* 2006). Since the MEG protected antibody conjugated magnetic nanoparticles can suppress nonspecific binding during the separation of protein biomarkers in human sera, the development of multiplex immunoassay is possible using the same technology.

The multiplexed assay is prepared using a mixture of different antibody conjugated MNPs and a simultaneous MALDI-TOF MS readout experiment. Figure 10.12 shows the detection of SAP and C-reactive protein (CRP) in human plasma. SAP is a biomarker related to Alzheimer's disease and type-2 diabetes. Healthy human blood contains SAP concentrations of $0–40 \, \text{mgL}^{-1}$. CRP is an inflammatory protein found in human blood, measuring general inflammation in the human body. As shown in the figure, after the immunoaffinity interaction with 1 ml of plasma, the SAP-MNP conjugates were separated by a magnet and the non-antigenic contaminants were removed by subsequent washing.

Figure 10.12 presents the MALDI-TOF MS profiles of healthy individuals and cancer patients analyzed for CRP and SAP in plasma. It can be seen from the results that the CRP peak at m/z = 23042 and SAP peak at m/z = 25462 are significant in the plasma of cancer patients.

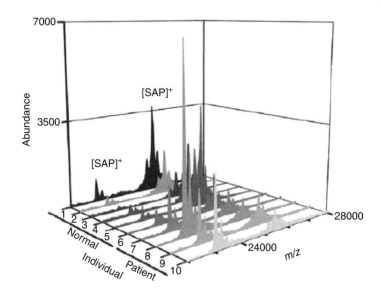

Figure 10.12 MALDI-TOF MS profiles of CRP and SAP from healthy individuals and patients with gastric cancer. (Len *et al.* 2006)

10.4   Virus Detection and Cell Analysis

The detection of viral pathogens is of critical importance in biology and medicine. Viruses have been viewed as infectious agents causing diseases. However, viruses have also been utilized for beneficial purposes including gene therapy (St George *et al.* 2003), vaccine production (Gluck 2002; Polo 2002), and for the use of nucleic acids as protein cages for new material synthesis (Douglas and Young 1998; 1999; Douglas *et al.* 2002). There are many methods dedicated to the detection of viruses; various methods provide different detection and quantification of different viruses. Traditional virus detection and identification has been undertaken in immunological or PCR-based laboratories. Polymerase chain reaction (PCR) utilizes DNA polymerase to amplify the DNA by *in vitro* enzymatic replication. Using the PCR technique, it is possible to amplify a single piece of DNA to a number that is several orders of magnitude higher by several steps and the use of reagents. Enzyme-linked immunoabsorbent assay (ELISAs) is one of the primary immunological techniques for the detection of viruses. ELISA has been widely used as a diagnostic tool in medicine and plant pathology to detect the presence of any antibodies or antigens in the sample. In the ELISA technique, the sample with an unknown amount of antigen is affixed to a surface and is washed with a specific antibody so that it can bind to the surface. The antibody attached biomolecule is then detected by adding a fluorescence marker so that the antigen/antibody complexes can fluoresce in the presence of a light source and the sample can be detected.

Both these techniques have their own inherent limitations. Even though the PCR technique is well advanced in sensitive virus detection, it is prone to failure and is limited in its ability to achieve simultaneous multiple virus identification (Wang 2002). Immunological analysis need specific antisera (blood sample containing specific antibodies) which is both laborious and time-consuming. A rapid and sensitive detection to identify multiple viruses in parallel is sorely needed.

Figure 10.13 presents the schematic representation of virus and protein analysis schemes using microarray technology. The basic design principle is the same for DNA, protein or cells. The specific molecular targets are simultaneously detected by deriving their unique signatures using probes. The probes are chemically attached as array format to a solid substrate for constructing either a DNA or a protein microarray. The microarray technique is also used for the detection of cells, glycans and carbohydrates, due to its significant capability in parallel detection. DNA microarrays for viral analysis can be divided into viral chips and host chips. Both can be applied to detection, identification as well as monitoring the viral populations. The monitoring of viral strains is critical for maintaining the effectiveness of a vaccine. Monitoring is also crucial for the safety of the vaccine as well as to track the attenuation of the viruses due to the vaccine.

Even though many methods have been developed for the detection and quantification of different viruses, different detection schemes may have different sensitivites. For example, the sensitivity of the mouse antibody production test was 10 times higher than that of the viral plaque assay and 10 000 times higher than that of RT-PCR for the detection of MHV-A59 virus. However, the detection of the MMVp virus using PCR is 10^6 times more sensitive than the viral plaque assay and mouse antibody production test. In short, PCR based methods are not always the optimal technique for virus detection. More sensitive and quantitative methods for virological analysis technologies such as electrical biochips (Albers *et al.* 2003; Los *et al.* 2005) and magnetic nanoparticles (Morishita *et al.* 2005; Perez *et al.* 2004; Mornet *et al.* 2004; Hutten *et al.* 2004) have been developed. In general, the electrical technology is based on miniaturized amperometric biosensors measuring the

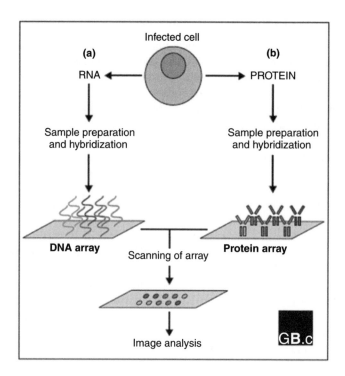

Figure 10.13 Virus and protein analysis using microarray; (a) RNA is first extracted from the infected cell. This RNA is labeled and is hybridized to the chip; (b) protein microarray with either antibodies or antigens as probes. The hybridized chip is scanned and the image is processed to provide the corresponding profiles. (Livingston *et al.* 2005)

redox enzymatic reactions. RNA detections using electrical biochips are reported with a sensitivity of 10^{12} molecules, which is achieved in 25 minutes (Gabig-Ciminska *et al.* 2004).

The interaction of viruses and metallic nanoparticles is used as a unique building block for the design of nanosensors particularly taking advantage of change in the nanoparticle's optical or magnetic properties upon the viral-induced assembly (Huang *et al.* 2007). The nanoparticles act as magnetic relaxation switches due to the spin–spin relaxation time changes by the surrounding water molecules during self-assembly (Perez *et al.* 2002; Allen *et al.* 2005). This property could be used for the detection of biological targets such as nucleic acids and proteins via target-induced self-assembly using MRI and NMR instruments.

When an amine cross-linked iron oxide nanoparticles is placed in an external magnetic field it becomes superparamagnetic. The combined electron spin in such crystals produces a single large magnetic dipole. This local magnetic field gradient inhomogeneity produces an off-resonance when water protons diffuse into the magnetic core due to the dephasing of these spins. This increases the relaxation rate, which is directly proportional to the nanoparticle cross-sectional area. When the individual superparamagnetic nanoparticles are assembled into clusters, their effective cross-sectional area becomes larger and more efficient spin dephasing due to the surrounding water protons could be observed. As shown in Figure 10.14(a) the magnetic nanosensor can act as a magnetic relaxation switch. Figure 10.14(b) shows the Transmission Electron Microscopy (TEM) and Atomic Force

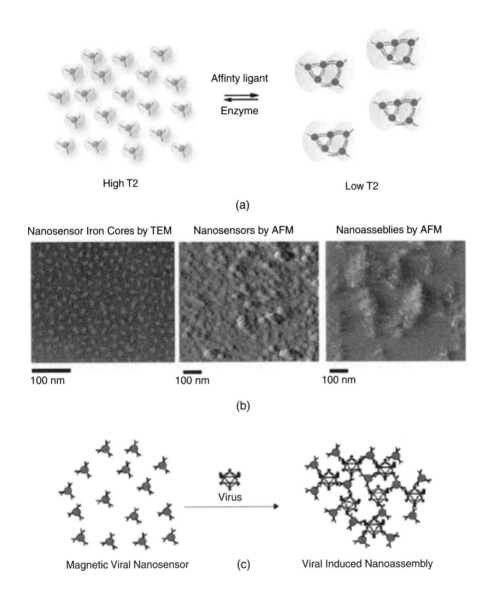

High T2 Low T2

(a)

Nanosensor Iron Cores by TEM   Nanosensors by AFM   Nanoasseblies by AFM

100 nm   100 nm   100 nm

(b)

Magnetic Viral Nanosensor   (c)   Viral Induced Nanoassembly

Figure 10.14 (a) Magnetic nanoparticles act as relaxation switches; (b)TEM and AFM images of nanoassemblies and (c) schematic representation of virus-surface-specific antibodies immobilized on the magnetic nanoparticles to create magnetic viral nanosensors. When exposed to viral particles, clustering of the nanoparticles occurs with a corresponding change in the MR signal. (Perez *et al.* 2003, 2004)

Microscopy (AFM) images of a monolayer of nanosensors. The TEM image shows the iron oxide crystal cores with an average diameter of 8 nm. The AFM image shows a similar monolayer cross-linked dextran shell with an average particle size of 50 nm. The nanoassemblies changes its size to 200–300 nm upon aminated cross-linking.

Figure 10.14(c) shows a magnetic viral sensor consisting of dextran coated superparamagnetic iron oxide nanoparticles, and virus-specific antibodies are immobilized on the surfaces of the dextran coated nanoparticles. The average size of the magnetic nanoparticle is 46 ± 0.6 nm. Anti-herpes simplex virus 1 (HSV-1) virus is attached to the caged dextran via N-succinimidyl-3-(2-pyridyldithio) propionate (SPDP). Addition of 10^4 viral particles to the nanosensor assembly (20 μg Fe/ml), results in two distinct populations of particles corresponding to magnetic nanoparticles of size 46 ± 0.6 nm and viral particles of size 100 ± 18 nm, which is detected by the light scattering. After incubation of 30 minutes, the viral particle population becomes undetectable by the light due to the appearance of a larger population (494 ± 23 nm).

The measurement of magnetic relaxation of these particles for samples of nanosensors with 10 μg Fe/ml shows change in spin–spin relaxation time. This measurement could detect as few as 5 viral particles in the 10 μl of 25 % protein solution without the PCR amplification. This viral-induced nanoassembly of magnetic nanoparticles allows rapid and sensitive detection of the virus in a solution using the measurement of change in relaxation time. Due to its dependency on magnetic relaxation time, this method allows virus detection in complex media such as blood, cell suspension, culture media, lipid emulsions and whole tissues using NMR and MRI instruments. Another advantage of the magnetic nanosensors is that unlike ELISA assays, this method does not require attachment of the virus to a solid surface or a substrate.

Covalent immobilization of oligonucleotides on solid supports is very important in molecular biology. Many different methods have been developed for chemical immobilization of DNA probes (Cohen et al. 1997; Taira and Yokoyama 2005). Immobilization of DNA probes on superparamagnetic nanoparticles is found to improve performance because they permit very high concentration of analysts coupled to their purification (Fredriks and Relman 2000; Corless et al. 2000; Reiss and Rutz 1999; Zhang et al. 2004; Robison et al. 2005). The DNA probes immobilized on magnetic nanoparticles improve the concentration/purification of the target DNA for the PCR amplifications (Fuentes 2005). Nanoparticles containing amino groups are activated by coating with a hetero-functional polymer. It is observed that due to its extreme sensitivity, the detection of hybridized products could be coupled to a PCR-ELISA direct amplification of the DNA bond to the magnetic nanoparticles. Magnetic nanoparticles with immobilized Hepatitis C virus cDNA were able to give a positive result after PCR-ELISA detection with 1 ml of a solution containing 10^{-18} g/ml of HCV cDNA, which is 2 molecules of HCV cDNA.

The immobilization of aminated DNA probes on aminated magnetic nanoparticles is based on aldehyde-aspartic-dextran by the following procedures: (i) initially the DNA probes are immobilized on nanoparticles by a very strong covalent bond to enable them to be used under drastic experimental conditions such as very high temperatures, organic media and alkaline conditions. (ii) The DNA probe is separated from the surface using a long hydrophilic and inert spacer arm which allows a rapid hybridization of the immobilized probe with DNA traces. (iii) The surface of the magnetic nanoparticles is fully inert to avoid nonspecific binding of DNA of the surface. These DNA magnetic nanoparticles are used to detect traces of DNA in the presence of a large amount of non-complementary DNA by PCR-ELISA technique enabling the detection of 10^{-18} to 10^{-19} g of DNA per ml.

As shown in Figure 10.15, the DNA-magnetic nanoparticle conjugates were incubated at a low concentration of complementary DNA in the presence of excess amounts of non-complementary DNA. After hybridization of about one hour, the magnetic nanoparticles were washed and directly added to the PCR reaction. The amplified products were analyzed using the standard ELISA microtiter plates.

The DNA-magnetic nanoparticle assays show specific detection of traces of target DNA in the presence of a large amount of non-complementary DNA by PCR-ELISA. Detection of 10^{-19} g of DNA per ml might be possible using this method. Also, the magnetic nanoparticles containing DNA probes immobilized by the above procedure are very sensitive for the purification/detection DNA/RNA from biological samples.

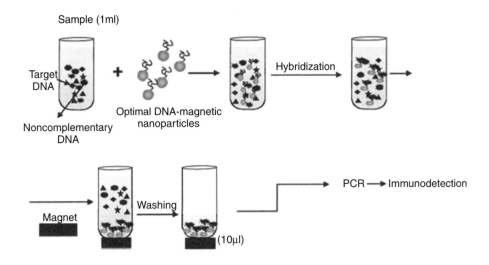

Figure 10.15 Schematic representation of ELISA-PCR assay using magnetic nanoparticles for DNA detection. (Fuentes *et al.* 2006)

10.5 Study of the Interactions Between Biomolecules

One of the fundamental goals of medicine is to study the complex spatio-temporal interaction of biomolecules with drugs at the cellular level. The biomolecular interaction is generally studied using fluorescent labels, both *in vivo* cellular imaging as well as *in vitro* assay detection. However, the intrinsic photochemical properties of these labels and low bleaching thresholds limit their use for long-term imaging. Also, simultaneous detection of multiple signals without the use of complex instrumentation and processing is challenging. Array-based bioassay is a highly promising analytical tool for the study of interaction between biomolecules (Lee *et al.* 2005). The use of magnetic nanoparticles offers many advantages for assays due to the possibility of measuring the analytical signals in terms of magnetization intensity. The use of magnetic nanoparticles also simplifies the preparation of samples because it is easy to separate any trace amount of molecules from the solution. The magnetism-based interaction capture (MAGIC) allows the identification of molecular targets based on induced movement of the superparamagnetic nanoparticles inside the living cell (Won *et al.* 2005). Magnetic nanoparticles are very attractive because of their potential use as contrast agents for magnetic resonance imaging (Saini *et al.* 1987; Stark 1988; Suzuki *et al.* 1996; Mornet *et al.* 2004), heating mediators for cancer thermotherapy (Ito 2005; Yih and Wei 2005), drug targeting (Alexion *et al.* 2000) and drug discovery (Saiyed *et al.* 2003). The interaction of drug molecules with a living cell is very important in drug discovery to understand their therapeutic and adverse effects and to develop second-generation therapeutic remedies.

Self-assembled monolayers (SAMs) have become an important tool for highly accurate analysis of biomolecular interactions due to their ability to position selectively immobilized biomolecules (Banno *et al.* 2004; Houseman *et al.* 2002). The nonspecific interaction of biomolecules can also be prevented using the SAM. Tightly packed organosilane monolayers can be fabricated on a silicon substrate for immobilizing oligonucleotides at both micro- and nanometer scale (Niwa *et al.* 2004). The magnetic detection of biomolecular interaction can be studied using these monolayer modified substrates along with magnetic nanoparticles (Arakaki *et al.* 2004).

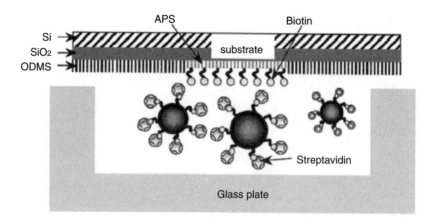

Figure 10.16 Schematic representation of biotin-streptavidin reaction on SAM modified substrate (Arakaki *et al.* 2004). Reproduced by permission of John Wiley & Sons. Inc.

Figure 10.16 presents the schematic illustration of the biotin-streptavidin bioconjugation reaction on SAM modified substrate. The fabrication starts with deposition of a 20 nm SiO_2 layer on a Si(100) substrate. This wafer is placed in octadecyltrimethoxysilane (ODMS) liquid and heated for 8 hours at 110 °C, resulting in deposition of a 20 Å thickness organosilane monolayer. The ODMS monolayer is then spin coated with photoresist and patterned using 350 nm UV light. The patterned substrate is selectively etched using 200 W oxygen plasma and at O_2 flow of 80 sccm for 1 minute. A 3-aminopropultriethoxysilane (APS) monolayer is formed on the exposed clean oxide surface of the patterned region. The SAM modified substrate is immersed in phosphase-buffered saline (PBS, pH 7.2) which has 5 mg/L sulfo-NHS-LC-LC-biotin for one hour. The substrate is sonicated to remove the unbounded suflo-NHS-LC-LC-biotin. It is then incubated for one hour and washed with PBS buffer. Figure 10.17 shows the fluorescent images of biotin modified and unmodified substrates after reaction with Cy-5 labeled streptavidin. The biotin immobilization is visualized using a fluorescence microscope through the reaction with Cy-5 labeled streptavidin. The dots are clearly seen after the reaction as bright spots. As shown in Figure 10.17(b), no pattern is observed in the unmodified substrate in which the biotin is not immobilized. It is clear from this observation that biotin is successfully introduced on the dots of APS. The specific reaction between the biotin and streptavidin could be observed using the monolayer modified substrate. The hydrophobic nature of the ODMS monolayer minimizes any nonspecific interactions of streptavidin in the sample.

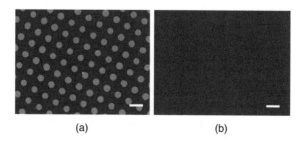

(a)   (b)

Figure 10.17 Fluorescent images of biotin modified and unmodified substrates (Arakaki *et al.* 2004). Reproduced by permission of John Wiley & Sons. Inc.

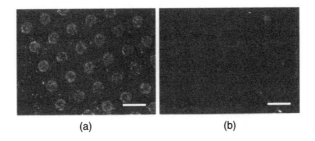

Figure 10.18 Optical microscope images of magnetic nanoparticles immobilized on substrate (Arakaki 2004). Reproduced by permission of John Wiley & Sons. Inc.

The reactivity measurements show that 0.6 molecules of streptavidin are reacted in an area of $10 \times 10\,nm^2$, which has a coverage of 47 %.

Magnetite (Fe_3O_4) nanoparticles of 200 nm diameter are dispersed in APM solution ($NH_4OH:H_2O_2:H_2O = 1:1:5$) and is heated to 100 °C for 20 minutes. The magnetic particles are then collected using a magnet and are washed and dried at 60 °C. These particles are suspended in 5 % (v/v) APS solution and are sonicated for 25 hours at 65 °C. Using the sulfo-NHS-LC-LC-biotin solution, the magnetic nanoparticles are modified and the particles are suspended in 1 ml PBS buffer containing 400 μg/ml streptavidin and incubated for 1 hour. This solution is loaded into a channel on the glass plate as shown in Figure 10.16. The SAM modified substrate is placed on top of this glass plate with the biotin modified surface facing down to avoid any nonspecific sedimentation of particles. The substrate is incubated and rinsed with water. The sample is ready for testing after washing and rinsing with 2-propanol, 1-butanol and dried at 60 °C. Figure 10.18 shows the optical images of magnetic nanoparticles immobilized on the substrate by the interaction between the biotin and streptavidin. It can be seen that the biotin dots are clearly identifiable, as shown in Figure 10.18(a). However there is no reaction when biotin modified particles are used as shown in Figure 10.18(b).

Figure 10.19 shows images of magnetic force microscopy and scanning electron microscopy of the streptavidin modified magnetic nanoparticles immobilized on the dot pattern. Figure 10.19(a) shows the SEM image of a single dot of 4 μm size. Figure 10.19(b) and (c) show topographic and magnetic images of the particles respectively on a 4 μm dot. The biomolecular interaction between biotin and streptavidin is successfully detected using the magnetic nanoparticles and the SAM modified substrates.

10.6 Detection of Biological Warfare Agents

Detection of chemical and biological warfare agents has significant merits in many areas related to public health, food and environmental control. It is very important to develop biological warfare detection systems with low cost, very high sensitivity and specificity, short analysis time and easy operation. The recent proliferation of chemical and biological (CB) agents as instruments for terrorism has triggered the development of an early warning system for CB agents. The early detection of an infectious agent and the identification of individuals exposed to such CB molecules will help to save lives. The early detection of a CB agent provides a better opportunity to respond and to find solutions before the agent becomes uncontrollable. It is possible to minimize the exposure if such an agent is detected immediately after its release and to provide medical treatment against those CB agents. It is clear from the nature of such an agent that, the longer the time after exposure, the

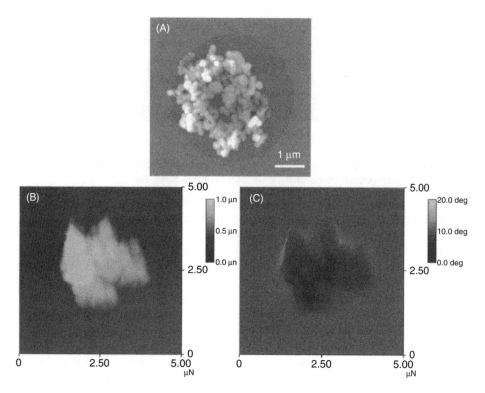

Figure 10.19 Images of magnetic nanoparticles immobilized on patterned substrates by specific interaction (Arakaki 2004). Reproduced by permission of John Wiley & Sons. Inc.

smaller the chances to reverse its damage and chemical effects. The spectroscopic studies at the millimeter wave and submillimeter wave frequencies have indicated that DNA and other cellular materials have their own unique resonance to CB agents due to their localized phonon modes (Woodlard 1999).

Organophosphates (OP) are chemical compounds that are produced by the reaction of alcohols and phosphoric acid. In the 1930s, OPs were used as insecticides and today out of almost 900 different bug killers (pesticides) used in agriculture and household applications, some 37 of them belong to the OP family. Dazinon, disulfoton, azinphos-methyl and fonofos are commonly used in agriculture and household applications. These chemicals kill insects by disrupting their brain and nervous system. These chemicals can stop the key enzyme in the nervous system called cholinesterase from working and kill the insect. The mechanism of action is irreversible inhibition of acetylcholinesterases (AchE), which is found in red blood cells in the nicotinic and muscarinic receptors in the nerve, muscle and the gray matter of the brain. The decrease in plasma cholinesterase results in a decrease of activity in the central, parasympathetic and sympathetic nervous systems due to the accumulation of neurotransmitter acetylcholine in the nervous system. OP esters can also produce delayed neurotoxicity, which takes at least 10 days to develop following a single acute exposure. The presence of OPs in industrial and agricultural drain waters, spills, drifts and its release in the event of chemical warfare, pose great risks to human life.

Among various approaches to decontaminate the OP nerve agents, catalytic destruction (CD) is one of the important methods of decommissioning nerve agent stockpiles, counteracting nerve agent attacks and remediating the OP spills. Even though considerable efforts have been directed towards the development of an OP degrading enzyme, its unavailability in sufficient quantities as well as its relatively low stability caused the majority of practical CD technologies to focus on acid- or base-catalyzed hydrolysis (Yang *et al.* 1992). The use of lanthanide (Ln) cations facilitates the hydrolysis of phosphodiesters and hence it serves as Lewis acid to bind and neutralize the phosphodiesters $P-O^-$ charges. However, the formation and precipitation of lanthanide-hydroxide hindered its use above pH 4 (Bracken K *et al.* 1997). The development of nanosized carriers with a powerful α-nucleophile, an oximate group immobilized on the surface as a CD agent which can disperse in an aqueous medium and could maintain its ability to decompose OP at neutral pH, is important. A high gradient magnetic separation (HGMS) can be used to separate the carriers composed of superparamagnetic iron oxide nanoparticles from aqueous suspensions (Moeser *et al.* 2002). A multi-analyte biosensor counter chip utilizing magnetic microbeads and GMR sensors to detect and identify the biological warfare agent shows promising results (Edelstein *et al.* 2000). The design details of this bead array counter (BARC) system are presented in Section 9.2.

Magnetite nanoparticles complexed with oximate containing moieties result in a colloidally stable aqueous environment at the neutral pH. These nanoparticles of diameter ~100 nm can be effectively removed from water by a portable HGMS device for reuse.

The OP agents may also cause damage to the liver as well as to the nervous system. The most harmful chemicals in this group include sarin, soman, tabun and VX. The chemical structures of the nerve poisons sarin, soman and diisopropyl fluorohosphate (DEF) are shown in Figure 10.20. Due to its wide use in agriculture and household applications, there is considerable concern about reducing the prolusion caused by the OP esters, which can accumulate in the biosphere and in organisms due to their resistance to biodegradation. OP compounds in water are neutralized and destroyed by adding bleach or other chemicals (Yang *et al.* 1992). Even though several environmentally friendly decontamination systems have been developed, significant reduction of the cost and of the environmentally harmful nature of the decontaminant has not been achieved. The use of large volumes of bleaching fluid for decontamination causes logistic challenges and additional environmental problems.

Nanocrystalline metal oxides such as MgO, Al_2O_3, CaO, BaO, SrO, Fe_2O_3, ZnO, TiO_2 and others exhibit destructive chemisorption capability towards acid gases, polar organics and chemical and biological agents (Wagner and Yang 2001; Lucas *et al.* 2001). Due to the possibility of adding surface hydroxides to the surface chemistry of these metal oxides, metal cations, oxide anions and surface bound OH groups can be produced. The nanocrystals absorb several fold more molecules per nm^2 of crystal surface than their microcrystal counterparts at a given temperature and pressure due to the higher fraction of corner and edge sites in the nanocrystalline structure. Nanocrystals also exhibit more active surfaces for the sorbed polar organic molecules. The reaction of chloroethylethyl

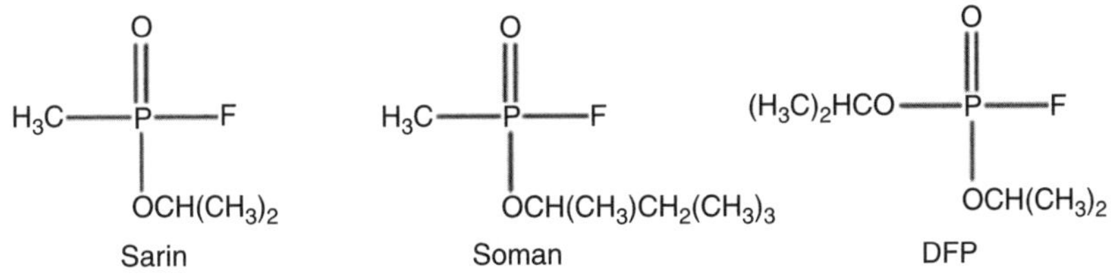

Figure 10.20 Chemical structure of nerve poisons such as Sarin, Soman and DEF. (Bromberg and Hatton 2005)

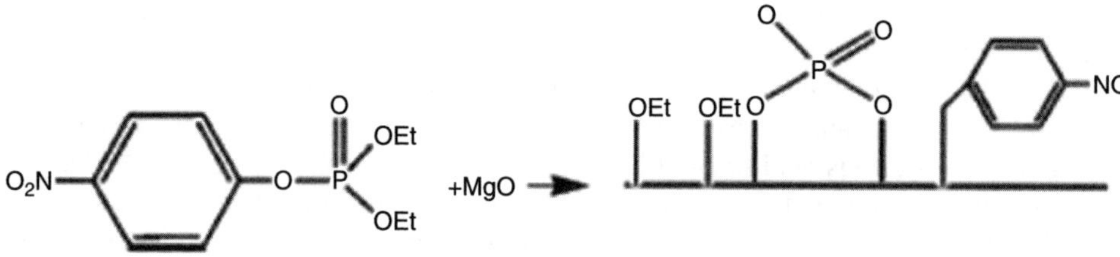

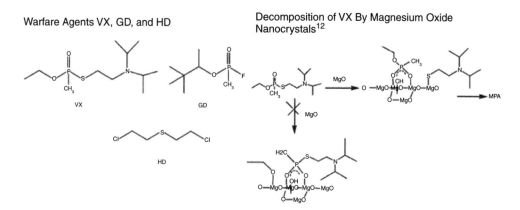

Figure 10.21 Destructive adsorption of paraxon using nanocrystalline MgO. (Bromberg 2007)

sulfide (Ch$_3$Ch$_2$SCH$_2$Ch$_2$Cl, a mimic of mustard gas) with microcrystalline MgO does not occur. However, it is very reactive with nanocrystalline MgO (Lucas *et al.* 2001). Furthermore, as shown in Figure 10.21, the destructive adsorption of paraoxon by nanocrystalline MgO is superior in rate and capacity to that of activated carbon. This is due to the polar nature of the metal oxide surface coupled with a reactive surface which has a high percentage of edge and corner crystal sites, vacancies and defects. The incoming adsorbents are encouraged to adsorb into the nanocrystalline structure due to its large pore openings and the accessibility of these adsorbents to the internal nanocrystalline surfaces.

The reactions of OP nerve agents such as VX, Soman and mustard gas with nanocrystalline MgO and CaO is shown in Figure 10.22, (Wagner *et al.* 1999, 2000). The decomposition of OP agents by MgO and CaO is observed via hydrolysis. The hydrolysis reactions are analogous to those observed in solution, except that the resulting products are from surface-bound complexes. However, the spreading stops when the pores are filled with liquids that form a wet spot. This causes the reactions to slow down and diffusion becomes limited until it reaches a steady state. As shown in Figure 10.22, the reaction with the surface is in stoichiometric, and OP evaporation appears to be the main mechanism. Since the molecules in the entrapped liquid reach the fresh surface to react, the reactions are limited by the evaporation rate or the vapor pressure of the liquid.

One of the solutions to overcome this problem is to use a hybrid molecule. Heterogeneous hydrolysis of toxic compounds catalyzed by hybrid inorganic nanomaterials and polymeric latexes shows promising results (Bromberg and Hatton 2005). The catalytic

Figure 10.22 The reaction and decomposition of VX nerve agent with nanocrystalline MgO. (Wagner *et al.* 1999)

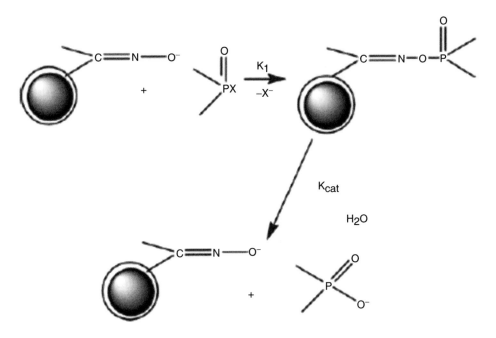

Figure 10.23 Hydropysis reaction of OP by an immobilized nucleophile and the formation of water soluble organophosphoric acid. (Bromberg and Hatton 2007)

decomposition agents are obtained by the surface modification of the magnetic nanoparticles with chemicals and polymers that promote OP decomposition and maintain colloidal stability. The surface modifications of the magnetite nanoparticles are obtained with α-nucleophiles such as oxime and iodosobenzoate groups. Polymer modified magnetic nanoparticles of size 6 to 10 nm aggregates and clusters, resulting in a composite particle of size 100–200 nm, which is optimal for particle removal from the water by HGMS (Ditsch *et al.* 2005). The oximate and α-nucleophiles ion bound to the particle decomposes the O-P bonds such as diisopropylflurophosphate (DFP), resulting in the formation of water soluble organophosphoric acid as primary metabolites as shown in Figure 10.23, (Bromberg and Hatton 2007). Since the phosphoric acids are not efficient inhibitors of the acetylcholine esterase, this hydrolysis allows a drastic reduction of OP toxicity. The inhibition of the enzyme acetylcholine esterase is the cause of OP toxicity in the nervous system.

The presence of the oxime group in the copolymer can be achieved by using acrolein that can convert the acrolein oxime by oximation with hydroxylamine. Given the water insolubility of polyacrolein, oximation was obtained by a seeded copolymerization of acrolein with an amphiphilic copolymer of 1-vinylimidazole (VIm) and acrylic acid (AA) as shown in Figure 10.24. The resulting copolymer is capable of stabilizing magnetic nanoparticles and catalyzing the hydrolysis of OP nerve agents while retaining the product by ion exchange adsorption.

The use of magnetic nanoparticles to detoxify contaminated military personnel or civilians after a CB agent attack is also projected. Functionalized nanoparticles to bind with the foreign toxin are injected into the body and are drawn out using a field gradient magnetic field. Nanoparticles with very high magnetic moment are required to tag the toxin and drag it out.

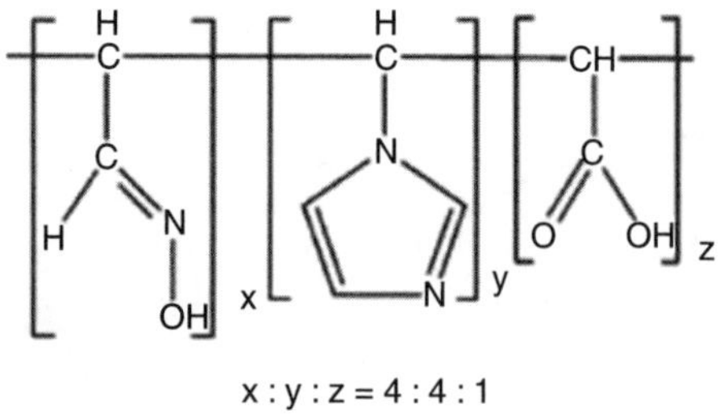

x : y : z = 4 : 4 : 1

Figure 10.24   The structure of p(VIm-AcOx-AA) copolymer. (Bromberg and Hatton 2007)

10.7   Environmental Monitoring and Cleanup

Environmental applications of nanotechnology generally fall into three categories: (i) environmentally benign sustainable products; (ii) remediation of materials contaminated with hazardous substances; and (iii) sensors for detecting environmental agents; microbial agents as well as biomaterials will also fall into these categories. Even though many applications in nanotechnology have been proposed that have direct environmental implications, remediation of contaminated water using nanoparticles is one of the promising technologies with many potential benefits (Masciangioli and Zhang 2003). The large surface area and high surface reactivity of iron nanoparticles provide enormous flexibility for *in situ* applications. Two distinctive redemption technologies in conventional systems, namely, absorptive/reactive and *in situ/ex situ*, are also generally applied in nanotechnology. Absorption technologies remove contaminants by sequestration, while the degradation of the contaminants into harmless products is achieved with reactive technologies. *In Situ* technologies involve the treatment of contaminants in place whereas *ex situ* refers to the treatment of contaminated material by removing the contaminant into more secure locations, similar to pumping the ground water to the surface and treating it with the reactors.

In Situ degradation of contaminants is often preferred due to its cost effective nature. However, *in situ* remediations require delivery of the treatment contaminant to sites. Nanotechnology has special relevance due to its potential ability to inject nanosized particles into contaminated porous media such as soils and sediments as shown in Figure 10.25. In such cases it is possible to use either *in situ* reactive zones where the nanoparticles are relatively immobile, or reactive nanoparticle plumes to contaminated zones if they are sufficiently mobile.

Though many nanoparticles are applicable to *in situ* remediation, zero-valent iron (nZVI) is of particular interest due to its particle morphology, reactivity and mobility. When particle size is smaller than ~ 10 nm, quantum confinement arises because bandgap increases as particle size decreases. The increased reactivity of the nanoparticles is due to the larger overall surface area, greater density of the reactive sites on particle surfaces and their higher intrinsic surfaces site reactivity (Navrotsky 2004). Due to these effects, nZVI exhibits (i) degradation of contaminants that do not react with larger particles; (ii) rapid degradation of contaminants that are already reacted with larger particles; and (iii) environmentally benign products from the contaminants that are rapidly degraded by larger materials.

Nanosized iron particles are very effective for transformation and detoxification of a wide variety of contaminants such as chlorinated organic solvents, organochlorine pesticides and polychlorinated hydrocarbons (PCBs) (Elliot and Zhang 2001). Figure 10.26

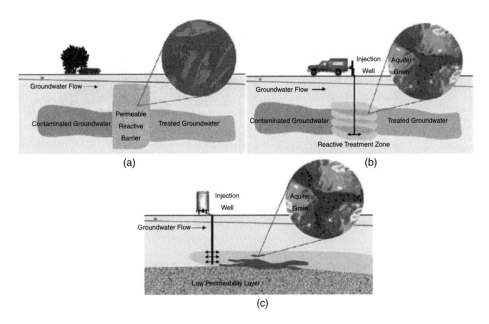

Figure 10.25 Schematic representation of the three approaches of Fe particles for groundwater redemption; (a) conventional permeable reactive barrier made with millimeter sized construction grade granular Fe; (b) reactive treatment zone formed by sequential injection of nanosized Fe to form overlapping zone particles that absorb the native aquifer materials; and (c) treatment of non-aqueous phase liquid (DNAPL) contamination by injection of mobile nanoparticles. The nanoparticles are shown as black dots and the zones that are affected are shown as pink. (Tratnyek and Johnson 2006)

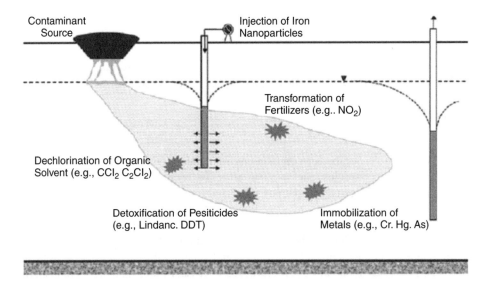

Figure 10.26 Schematic representation of the use of nanoscale iron particles for inexpensive and nontoxic *in situ* remediation. (Zhang 2003)

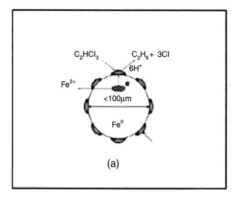
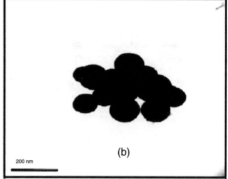

Figure 10.27   (a) Nanoparticle mediated reduction of trichloroethene in water; (b) SEM photograph of nanoparticle cluster. Size bar is 200 nm. (Zhang *et al.* 2003)

shows the schematic representation of iron nanoparticles for site remediation. Direct subsurface injection under gravity-fed or pressurized conditions can be used to transport the nanoparticles to the ground water. The nanoparticles can remain in suspension for an extended period of time in the *in situ* treatment zone. They can also be anchored onto a solid matrix such as activated carbon or zeolite for the treatment of water and waste water.

Figure 10.27 represents the nanoparticles-mediated reduction of trichloroethene in water. The nanoparticle sample is prepared by deposition of two or three droplets of dilute Fe nanoparticle containing ethanol solution onto a carbon film.

Zero-valent iron (nZVI) is a moderate reduction reagent. Usually it can react with oxygen in water as shown in the following equations:

$$2Fe^0_{(s)} + 4H^+_{(aq)} + O_{2(aq)} \rightarrow 2Fe^{2+}_{(aq)} + 2H_2O_{(1)} \tag{10.1}$$

$$Fe^0_{(s)} + 2H_2O_{(aq)} \rightarrow Fe^{2+}_{(aq)} + H_{2(g)} + 2OH^-_{(aq)} \tag{10.2}$$

This corrosion reaction can be accelerated by manipulating the surface chemistry of the particles. Contaminants such as tetrachloroethene (C_2Cl_4) can readily accept the electrons from iron oxidation and be reduced to ethane as follows:

$$C_2Cl_4 + 4Fe^0 + 4H^+ \rightarrow C_2H_4 + 4Fe^{2+} + 4Cl^- \tag{10.3}$$

Table 10.1 shows a number of common environmental contaminants that can be easily treated by iron nanoparticles due to their effective reduction and catalystic activities including chlorinated organic compounds and metal ions (Zhang 1999). The hydrogenated hydrocarbons can reduce benign hydrocarbons by the iron nanoparticles.

The techniques and devices for the detection of chemical and biological agents and environmental pollutions have to be efficient, safe and rapid for first responders to confirm their presence at an incident and to identify and quantify them. The current chemical agent detection is based on spot papers or in a few cases a more sensitive chemical vapor detection system using ion mobility spectroscopy (IMS). IMS usually gives a first warning and subsequently the contaminant is confirmed typically after 6 to 48 hours depending on the analysis using gas chromatography-mass spectroscopy (GC-MS). False positive reduction is generally achieved by combining analysis using two or more systems. One

Table 10.1 Common environmental contaminants that can be removed by nanoscale iron particles. (adapted from Zhang 2003)

Chlorinated methanes	**Trihalomethanes**
Carbon tetrachloride (CCl_4)	Bromoform ($CHBr_3$)
Chloroform ($CHCl_3$)	Dibromochloromethane ($CHBr_2Cl$)
Dichloromethane (CH_2Cl_2)	Dichlorobromomethane ($CHBrCl_2$)
Chloromethane (CH_3Cl)	**Chlorinated ethenes**
Chlorinated benzenes	Tetrachloroethene (C_2Cl_4)
Hexachlorobenzene (C_6Cl6)	Trichloroethene (C_2HCl_3)
Pentachlorobenzene (C_6HCl_5)	cis-Dichloroethene ($C_2H_2Cl_2$)
Tetrachlorobenzenes ($C_6H_2Cl_4$)	trans-Dichloroethene ($C_2H_2Cl_2$)
Trichlorobenzenes ($C_6H_3Cl_3$)	1,1-Dichloroethene ($C_2H_2Cl_2$)
Dichlorobenzenes ($C_6H_4Cl_2$)	Vinyl chloride (C_2H_3Cl)
Chlorobenzene (C_6H_5Cl)	**Other polychlorinated hydrocarbons**
Pesticides	PCBs
DDT ($C_{14}H_9Cl_5$)	Dioxins
Lindane ($C_6H_6Cl_6$)	Pentachlorophenol (C_6HCl_5O)
Organic dyes	**Other organic contaminants**
Orange II ($C_{16}H_{11}N_2NaO_4S$)	N-nitrosodimethylamine (NDMA) ($C_4H_{10}N_2O$)
Chrysoidine ($C_{12}H_{13}ClN_4$)	TNT ($C_7H_5N_3O_6$)
Tropaeolin O ($C_{12}H_9N_2NaO_5S$)	**Inorganic anions**
Acid Orange	Dichromate ($Cr_2O_2^{-7}$)
Acid Red	Arsenic (AsO_3^{-4})
Heavy metal ions	Perchlorate (ClO^{-4})
Mercury (Hg^{2+})	Nitrate (NO^{-3})
Nickel (Ni^{2+})	
Silver (Ag^+)	
Cadmium (Cd^{2+})	

of the advancements in GC-MS, termed matrix-assisted laser desorption ionization time of flight mass spectrometry (MALDI-TOF-MS), has become more widely adopted. However advances in miniaturization of a hand-held system have been hindered due to poor mass resolution.

10.8 Outlook

Developments in nanotechnology for the manipulation and control of biomaterials have revolutionized disease diagnosis and treatment. Biological tests to detect the presence of specific molecules become quicker, more sensitive and more flexible when certain nanoparticles are allowed to do the work as tags or labels. Portable, accurate and easy-to-use biosensors have become a high priority for many industries due to the recent advancements in the detection of single pathogens at the molecular level. This chapter presents the basic theory of biochips in general and magnetic biochips in particular for gene and protein analysis, virus detection and the study of the interaction between biomaterials and magnetic nanoparticles.

The detection of chemical and biological warfare agents has significant importance in many areas related to public health, food and environmental control. The recent proliferation of chemical and biological (CB) agents as instruments for terrorism has led to the development of an early warning system for CB agents. The early detection of a CB agent provides greater opportunities to respond and to find solutions before it becomes uncontrollable. Minimization of exposure to such an agent is possible if it is detected immediately after release and medical treatment against that CB agent can be provided. It is clear from the nature of such agents that, the longer the time that elapses after exposure, the lower the chance to reverse the damage and chemical effects. The last two sections in this chapter present an overview of sensors for CB agent detection using magnetic nanoparticles as well as environmental applications of magnetic nanomaterials.

References

Albers, J., Grunwald, T., Nebling, E., Piechotta, G. and Hintshe, R. (2003). Electrical biochip technology – a tool for microarrays and continuous monitoring, *Analytical and Bioanalytical Chemistry*, **377**, 521–7.

Alexion, C., Arnold, W., Klein, R.J., Parak, F.G., Hulin, P., Bergemann, C., Erhardt, W., Wagenpfeil, W. and Lubbe, A.S. (2000). Locoregional cancer treatment with magnetic drug targeting, *Cancer Research*, **60**, 6641–8.

Allen, M., Bulte, J.W.M., Basu, G., Zywicke, H.A., Frank, J.A., Young, M. and Douglas, T. (2005). Paramagnetic viral nanoparticles as potential high-relaxivity magnetic resonance contrast agents, *Magnetic Resonance in Medicine*, **54**, 807–12.

Arakaki, A., Hideshima, S., Nakagawa, T., Niwa, D., Tanaka, T., Matsunaga, T., Osaka, T. (2004). Detection of biomolecular interaction between biotin and streptavidin on a self-assembled monolayer using magnetic nanoparticles, *Biotechnology and Bioengineering*, **88**, 543–6.

Banno, N., Nakanishi, T., Matsunaga, M., Asahi, T. and Osaka, T. (2004). Enantioselective crystal growth of leucine on a self-assembled monolayer with covalently attached leucine molecules, *Journal of the American Chemical Society*, **126**, 428–9.

Bhatia, S.N. and Chen, C.S. (1999). Tissue engineering at the microscale, *Biomedical Microdevices*, **2**, 131–44.

Bracken, K., Moss, R.A., Ragunathan, K.G. (1997). Remarkably rapid cleavage of a model phosphodiester by complexed ceric ions in aqueous micellar solutions, *Journal of the American Chemical Society*, **119**, 9323–4.

Bromberg, L. and Hatton, T.A. (2005). Nerve agent destruction by recyclable catalytic magnetic nanoparticles, *Industrial & Engineering Chemistry Research*, **44**, 7991–8.

Bromberg, L. and Hatton, T.A. (2007). Decomposition of toxic environmental contaminants by recyclable catalytic superparamagnetic nanoparticles, *Industrial & Engineering Chemistry Research*, **46**, 3296–303.

Cailat, P., Belleville, M., Clerc, F. and Massit, C. (1998). Active CMOS biochips: an electro-addressed DNA probe, *Digest of Technical Papers for IEEE International Solid-State Circuits Conference 1998*, 272–3.

Cao, Y.W.C., Jin, R.C. and Mirkin, C.A. (2002). Nanoparticles with Raman spectroscopic fingerprints for DNA and RNA detection, *Science*, **297**, 1536–40.

Cheng, T.L., Tsai, C.Y., Sun, C.C., Uppala, R., Chen, C.C., Lin, C.H. and Chen, P.H. (2006). Electrical detection of DNA using gold and magnetic nanoparticles and bio-bar code DNA between nanogap electrodes, *Microelectronic Engineering*, **83**, 1630–3.

Chemla, Y.R., Crossman, H.L., Poon, Y., McDermott, R., Stevens, R., Alper, M.D. and Clarke, J. (2000). Ultrasensitive magnetic biosensor for homogeneous immunoassay, *Proceedings of the National Academy of Sciences of the United States of America*, **97**, 14268–72.

Cohen, G., Deutsch, J., Fineberg, J. and Levine, A. (1997). Covalent attachment of DNA oligonucleotides to glass. *Nucleic Acids Research*, **25**, 911–12.

Cotton, R.G.H. (1997). *Mutation*, Oxford University Press, UK.

Corless, C.E., Guiver, M. and Borrow, R. (2000). Contamination and sensitivity issues with a real-time universal 16S rRNA PCR, *Journal of Clinical Microbiology*, **38**, 1747–52.

Chou, P.-H., Chen, S.-H., Liao, H.K., Lin, P.C., Her, G.R., Lai, A.C., Chen, J.C., Lin, C.C. and Chen, Y.-J. (2005). Nanoprobe-Based Affinity Mass Spectrometry for Selected Protein Profiling in Human Plasma, *Analytical Chemistry*, **77**, 5990–7.

Ditsch, A., Laibinis, P.E., Wang, D.I.C. and Hatton, T.A. (2005). Controlled clustering and enhanced stability of polymer-coated magnetic nanoparticles, *Langmuir*, **21**, 6006–18.

Douglas, T. Young, M. (1998). Host-guest encapsulation of materials by assembled virus protection cages, *Nature*, **393**, 152–5.

Douglas, T. and Young, M. (1999). Virus particles as templates for materials synthesis, *Advanced Materials*, **11**, 679–81.

Douglas, T., Strable, E., Wilits, D., Aitouchen, A., Libera, M. and Young, M. (2002). Protein engineering of a viral cage for constrained nanomaterials synthesis, *Advanced Materials*, **14**, 415–18.

Dupuy, A.M., Lehmann, S. and Cristol, J.P. (2005). Protein biochip systems for the clinical laboratory, *Clinical Chemistry and Laboratory Medicine*, **43**, 1291–302.

Edelstein, R.L., Tamanaha, C.R., Sheehan, P.E., Miller, M.M., Baselt, D.R., Whitman, L.J. and Colton, R.J. (2000). The BARC biosensor applied to the detection of biological warfare agents, *Biosensors and Bioelectronics*, **14**, 805–13.

Elliott D. and Zhang, W.X. (2001). Field assessment of nanoparticles for groundwater treatment, *Environmental Science & Technology*, **35**, 4922–6.

Eversmann, B., Paulus, C., Hofmann, F., Brederlow, R., Holzapfl, B., Fromherz, P., Brenner, M., Schreiter, M., Gabl, R., Plehnert, K., Steinhauser, M., Eckstein, G., Schmitt-Landsiedel, D. and Thewes, R. (2003). A 128x128 bio-sensor array for extracellular recording of neural activity, *Digest of Technical Papers for IEEE International Solid-State Circuits Conference* 2003, 222–3.

Fan, A.P., Lau, C.W. and Lu, J.Z. (2005). Magnetic bead-based chemiluminescent metal immunoassay with a colloidal gold label. *Analytical Chemistry*, **77**, 3238–42.

Ferguson J.A., Steemers F.J. and Walt, D.R. (2000). High-density fiber-optic DNA random microsphere array, *Analytical Chemistry*, **72**, 5618–24.

Fixe, F., Dufva, M., Telleman, P. and Christensen, C.B. (2004). Functionalization of poly(methyl methacrylate) (PMMA) as a substrate for DNA microarrays, *Nucleic Acids Research*, **32**, e9.

Fromherz, P. (2005). Joining ionics and electronics: semiconductor chips with ion channels, nerve cells, and brain tissue, *Digest of Technical Papers for IEEE International Solid-State Circuits Conference* 2005, 76–7

Fuentes, M., Mateo, C., Garcia, L., Tercero, J.C., Cuisan J.M. and Fernandes-Lafuente, R. (2004). Directed covalent immobilization of aminated DNA probes on aminated pates, *Biomacromolecules*, **4**, 883–8.

Fuentes, M., Mateo, C., Guisan, J.M. and Fernandez-Lafuente, R. (2005). Preparation of inert magnetic nanoparticles for the directed immobilization of antibodies, *Biosensors and Bioelectronics* **20**, 1380–7.

Fuentes, M., Mateo, C., Rodriguez, A., Casqueiro, M., Tercero, J.C., Riese, H.H., Lafuente, R. and Guisan J.M. (2006). Detecting minimal traces of DNA using DNA covalently attached to superparamagnetic nanoparticles and direct PCR-ELISA, *Biosensors and Bioelectronics*, **21**, 1574–80.

Fredricks, D.N. and Relman, D.A. (2000). Application of polymerase chain reaction to the diagnosis of infectious diseases, *Clinical Infectious Diseases*, **29**, 475–86.

Fritzsche, W. and Tanton, T.A. (2003), *Nanotechnology*, **14**, 3194–8.

Gabig-Ciminska, M., Holmgren, A., Andresen, H., Bundvig, B.K., Wümpelmann, M., Albers, J., Hintsche, R., Breitenstein, A., Neubauer, P., Lose, M., Czyzf, M., Wegrzyn, G., Silfversparre, G., Jurgen, B., Schweder, T. and Enfors, S.-O. (2004). Electric chips for rapid detection and quantification of nucleic acids, *Biosensors and Bioelectronics*, **19**, 537–46.

Gluck, R. and Metcalfe, I.C. (2002). New technology platforms in the development of vaccines for the future, *Vaccine*, **20**, B10–B16.

Graham, D.L., Ferreira, H.A., Freitas, P.P. and Cabral, J.M.S. (2003). High sensitivity detection of molecular recognition using magnetically labeled biomolecules and magnetoresistive sensors, *Biosensors and Bioelectronics*, **18**, 483–8.

Han, M., Gao, X., Su, J. and Nie, S. (2001). Quantum dot tagged microbeads for multiplexed optical coding of biomolecules, *Nature Biotechnology*, **19**, 631–5.

Han, S.J., Xu, L., Yu, H., Wilson, R.J., White, R.L., Pourmand, N. and Wang, S.X. (2006). CMOS integrated DNA microarray based on GMR sensors, *Proceedings of IEEE International Electron Devices Meeting, 2006*, 1–4.

Hassibi, A. and Lee, T.H. (2005). A programmable electrochemical biosensor array in 0.18 μm standard CMOS, *Digest of Technical Papers for IEEE International Solid-State Circuits Conference* 2005, 564–5.

Houseman, B.T., Huh, J.H., Kron, S.J. and Mrksich, M., (2002), Peptide chips for the quantitative evaluation of protein kinase activity, *Nature Biotechnology*, **20**, 270–4.

Huang, X., Bronstein, L.M., Retrum, J., Dufort, C., Tsvetkova, T., Aniagyei, S., Stein, B., Stucky, G., McKenna, B., Remmes, N., Baxter, D., Kao, C.C. and Dragnea, B. (2007). Self-Assembled Virus-like Particles with Magnetic Cores, *Nano Letters*, **7**, 2407–16.

Huber, L.A. (2003). Is proteomics heading in the wrong direction? *Nature Reviews – Molecular Cell Biology*, **4**, 74–80.

Hutten, A., Sudfeld, D., Ennena, I., Reiss, G., Hachmannb, W., Heinzmann, U., Wojczykowski, K., Jutzi, P., Saikaly, W. and Thomas, G. (2004). New magnetic nanoparticles for biotechnology, *Journal of Biotechnology*, **112**, 47–63.

Ivanova, E.P., Pham, D.K., Brack, N., Pigram, P. and Nicolau, D.V. (2004). Poly(l-lysine)-mediated immobilisation of oligonucleotides on carboxy-rich polymer surfaces, *Biosensors and Bioelectronics*, **19**, 1363–70.

Ito, A., Honda, H. and Kobayashi, T. (2005). Cancer immunotherapy based on interacellular hyperthermia using magnetic nanoparticles: a novel concept of heat controlled necrosis with heat shock protein expression, *Cancer Immunology Immunotherapy*, **55**, 320–8.

Kang, Y.S., Risbud, S., Rabolt, J.F. and Strove, P. (1996). Synthesis and characterization of nanometer-size Fe_3O_4 and γ-Fe_2O_3 particles, *Chemistry of Materials*, **8**, 2206–11.

Kaul, R.A., Syed, N.I. and Fromherz, P. (2004). Neuron-semiconductor chip with chemical synapse between identified neurons, *Physical Review Letters*, **92**, 38102.

Kerman, K., Kobayashi, M. and Tamiya, E. (2004). Recent trends in electrochemical DNA biosensor technology, *Measurement Science and Technology*, **15**, R1–R11.

Kouassi, G.K. and Irudayaraj, J. (2006). Magnetic and gold-cated magnetic nanoparticles are DNA sensor, *Analytical Chemistry*, **78**, 3234–41.

Lagae, L., Wrix-Speetjens, R., Liu, C.X., Laureyn, W., Boeck, J.D., Borghs, G., Harvey, S., Galvin, P., Graham, D.L., Ferreira, H.A., Freitas, P.P, Clarke, L.A. and Amarl M.D. (2005). Magnetic biosensors for genetic screening of cystic fibrosis, *IEE Proceedings: Circuits, Devices & Systems*, **152**, 393–400.

Lee, H. (2005). Microelectronic/microfluidic hybrid system for the manipulation of biological cells, Ph.D. dissertation, Harvard Univ., Cambridge, MA, 2005.

Lee, H., Purdon, A.M. and Westervelt, R.M. (2004). Manipulation of biological cells using a microelectromagnet matrix, *Applied Physics Letters*, **85**, 1063–5.

Lee, H., Liu, Y., Westervelt, R.M. and Ham, D. (2006). IC/microfluidic hybrid system for magnetic manipulation of biological cells, *IEEE Journal of Solid-State Circuits*, **41**, 1471–80.

Lee, P.J., Hung, P.J., Shaw, R., Jan, L. and Lee, L.P. (2005). Microfluidic application integrated device for monitoring direct cell–cell communication via gap junctions between individual cell pairs, *Applied Physics Letters*, **86**, 223902.

Len, P.C., Chou, P.H., Chen, S.H., Liao, H.K., Wang, K.Y., Chen, Y.J. and Lin, C.C. (2006). Ethylene glycol-protected magnetic nanoparticles for a multiplexed immunoassay in human plasma, *Small*, **2**, 485–9.

Lien, H. and Zhang, W. (1999). Reactions of chlorinated methanes with nanoscale metal particles in aqueous solutions, *Journal of Environmental Engineering*, **125**, 1042–7.

Liu, W.-T. (2006). Nanoparticles and their biological and environmental applications, *Journal of Bioscience and Bioengineering*, **102**, 1–7.

Los, M., Los, J.M., Blohm, L., Spillner, E., Grunwald, T., Alberts, J., Hintsche, J. and Wegrzyn, G. (2005). Rapid detection of viruses using electrical biochips and ani-virion sera, *Letters in Applied Microbiology*, **40**, 479–85.

Livingston, A.D., Campbell, C.J., Wagner, E.K. and Ghazal, P. (2005). Biochip sensors for the rapid and sensitive detection of viral disease, *Genome Biology*, **6**, 112.1–5.

Lucas, E., Decker, S., Khaleel, A., Seitz, A., Fultz, S., Ponce, A., Li, W., Carnes, C. and Klabunde, K.J. (2001). Nanocrystalline metal oxides as unique chemical reagents/sorbents, *Chemistry – A European Journal*, **7**, 2505–10.

Manaresi, N., Romani, A., Medoro, G., Altomare, L., Leonardi, A., Tartagni, M. and Guerrieri, R. (2003) A CMOS chip for individual cell manipulation and detection, *IEEE Journal of Solid-State Circuits*, **38**, 2297–305.

Masciangioli, T. and Zhang, W.-X. (2003). Environmental technologies at the nanoscale. *Environmental Science and Technology*, **37**, 102A–108A.

McGall, G., Labadie, J., Brock, P., Wallraff, G., Nguyen, T. and Hinsberg, W. (1996). Light-directed synthesis of high-density oligonucleotide arrays using semiconductor photoresists, *Proceedings of the National Academy of Sciences of the United States of America*, **93**, 13555–60.

Mikhaylova, M., Kim, K.D., Berry, C.C., Zogorodni, A, Topark, M., Curits, J. and Mohammed, M. (2004). *Chemistry of Materials*, **16**, 2344–54.

Moeser, G.D., Roach, K.A., Green, W.H., Laibinis, P.E., and Hatton, T.A. (2002). Water-based magnetic fluids as extractants for synthetic organic compounds, *Industrial and Engineering Chemistry Research*, **41**, 4739–49.

Mornet, S., Vasseur, S., Grasset, F. and Duguet, E. (2004). Magnetic nanoparticles design for medical diagnosis and therapy, *Journal of Materials Chemistry*, **14**, 2161–75.

Morishita, N. Nakagami, H., Morishita, R., Takeda, S., Mishima, F., Terazono, B., Nishijima, S., Kaneda, Y. and Tanaka, N. (2005). Magnetic nanoparticles with surface modification enhanced gene delivery of HVJ-E vector, *Biochemical and Biophysical Research Communications*, **334**, 1121–6.

Navrotsky A. (2004). *Encyclopedia of Nanoscience and Nanotechnology*, Marcel Dekker, New York, NY.

Nicewarner-Pena, S.R., Freeman, R.G., Reiss, B.D., He, L., Pena, D.J., Walton, I.D., Cromer, R., Keating, C.D. and Natan, M.J. (2001). Submicrometer metallic barcodes, *Science*, **294**, 137–41.

Niwa D., Omichi, K., Motohashi, N., Homma, T., Osaka, T. (2004) Formation of micro and nanoscale patterns of monolayer templates for position selective immobilization of oligonucleotides using ultraviolet and electron beam lithography, *Chemistry Letters*, **33**, 176–7.

Park, S.J., Taton, T.A. and Mirkin, C.A. (2002). Array based electrical detection of DNA with nanoparticles probes, *Science*, **295**, 1503–6.

Pasch, H. and Schrepp, W. (2002). *MADLI-TOF Mass Spectrometry of Synthetic Polymers*, Springer, New York.

Peterson, A.W., Heaton, R.J. and Georgiadis, R.M. (2001). The effect of surface probe density on DNA hybridization, *Nucleic Acids Research*, **29**, 5163–8.

Perez, J.M., Josephson, L., O'Loughlin, T., Hogemann, D. and Weissleder, R. (2002). *Nature Biotechnology*, **20**, 816–20.

Perez, J.M., Simeone, J.F., Saeki, Y., Josephson, L. and Weissleder, R. (2003). Viral-Induced Self-Assembly of Magnetic Nanoparticles Allows the Detection of Viral Particles in Biological Media, *Journal of the American Chemical Society*, **125**, 10192–3.

Perez, J.K., Josephson, L. and Weissleder, R. (2004). Use of Magnetic Nanoparticles as Nanosensors to Probe for Molecular Interactions, *Chembiochem*, **5**, 261–4.

Piedade, M., Sousa L.A., de Almeida, T.M., Germano, J., de Cost, B.A., Lemos, J.M., Freitas, P.P., Ferriera, H. and Cardoso, F.A. (2006). A new hand-held microsystem architecture for biological analysis, *IEEE Transactions on Circuits and Systems.*, **35**, 2384–95.

Palmer Toy, D.E., Kuzdzal, S. and Chen, D.W. (2002). Protemic approaches to tumor marker discovery in *Tumor Makers: Physiology, pathology technology and clinical applications*, (ed. E.P. Diamandies) AACC Press, Washington DC, 391–400.

Polo, J.M. and Dubensky, T.W. (2002). Virus based vectors for human vaccine applications, *Drug Discovery Today*, **7**, 719–27.

Raj, A.J. and Chen, D.W. (2001). Clinical proteomics: new developments in clinical chemistry, *Labratoriums Medizin*, **25**, 399–403.

Reiss, R.A. and Rutz, B. (1999). Quality control PCR: a method for detecting inhibitors of Taq DNA polymerase. *Biotechniques*, **27**, 920–6.

Riccelli, P.V., Merante, F., Leung, K.T., Bortolin, S., Zastawny, R.L., Janeczko, R. and Benight, A.S. (2001). *Nucleic Acids Research*, **29**, 996–1004.

Robison, D.B., Persson, H.J., Zeng, H., Li, G., Pourmand, N., Sun, S. and Wang, S.X. (2005). DNA-functionalized MFe2O4 (M = Fe, Co, or Mn) nanoparticles and their hybridization to DNA-functionalized surfaces, *Langmuir*, **21**, 3096–103.

Saini, S., Stark, D.D., Hahn, P.F., Wittenberg, J., Brady, T.J. and Ferrucci J.T. (1987). Ferrite particles: a superparamagnetic MR contrast agent for the reticuloendothelial system, *Radiology*, **162**, 211–16.

Saiyed, Z.M., Telang, S.D. and Ramchand C.N. (2003). Application of magnetic techniques in the field of drug discovery and biomedicine, *BioMagnetic Research and Technology*. **1**, 23–33.

Seong, S.-Y. and Choi, C.-Y. (2003). Current status of protein chip development in terms of fabrication and application, *Proteomics*, **3**, 2176–89.

Smith, C.H. and Schneider, R.W. (2003). Very dense magnetic sensor array for precision measurement and detection, *Proceedings of Sensors EXPO, 2003*.

St George, J.A. (2003). Gene therapy progress and prospects: adenoviral vectors, *Gene Therapy*, **10**, 1135–41.

Stark, D.D., Weissleder, R., Elizondo, G., Hahn, P.F., Saini, S., Todd, L.E., Wittenberg, J. and Ferrucci, J.T. (1988). Superparamagnetic iron oxide: clinical application as contrast agent for MR imaging of the liver, *Radiology*, **168**, 297–301.

Suzuki, M., Honda, H., Kobayashi, T., Wakabayashi, T., Yoshida, J. and Takahashi, M. (1996). Development of a target directed magnetic resonance contract agent using monoclonal antibody conjugated magnetic particles, *Brain Tumor Pathology*, **13**, 127–32.

Taira, S. and Yokoyama, K. (2005). Immobilization of single-stranded DNA by self-assembled polymer on gold substrate for a DNA chip. *Biotechnology and Bioengineering*, **89**, 835–8.

Tang, N., Toronatore, P. and Weinberger, S.R. (2004). Current developments in SELDI affinity technology, *Mass Spectrometer Reviews*, **23**, 34–44.

Taton, T.A., Mirkin, C.A. and Lestinger, R.L. (2000). Scanometric DNA array detection with nanoparticles probes, *Science*, **289**, 1757–60.

Taton, T.A., Lu, G. and Mirkin, C.A. (2001). Two color labeling of oligonucleotide arrays via size selective scattering of nanoparticle probes, *Journal of the American Chemical Society*, **123**, 5164–5.

Tratnyek, P. and Johnson, R.L. (2006). Nanotechnologies for environmental cleanup, *Nanotoday*, **1**(2), 44–8.

Tulapurkar, A.A., Suzuki, Y., Fukushima, A., Kubota, H., Maehara, H., Tsunekawa, K., Djayaprawira, D.D., Watanabe, N. and Yuasa, S. (2005). Spin-torque diode effect in magnetic tunnel junctions, *Nature*, **438**, 339–42.

Walsh, M.K., Xwen, W. and Weimer, B.C. (2001). Optimizing the immobilization of single-stranded DNA onto glass beads, *Journal of Biochemical and Biophysical Methods*, **47**, 221–31.

Wang, D., Coscoy, L., Zylberberg, M., Avila, P.C., Boushey, H.A., Ganem, D. and DeRisi, J.L. (2002). Microarray based detection and genotyping of viral pathogens, *Proceedings of the National Academy of Sciences of the United States of America*, **99**, 15687–92.

Wagner, G.W., Bartram, P.W., Koper, O. and Klabunde, K.J. (1999). Reactions of VX, GD, and HD with nanosize MgO, *Journal of Physical Chemistry B*, **103**, 3225–8.

Wagner, G.W., Koper, O.B., Lucas, E., Decker, S. and Klabunde, K.J. (2000). Reactions of VX, GD, and HD with nanosize CaO: autocatalytic dehydrohalogenation of HD, *Journal of Physical Chemistry B*, **104**, 5118–23.

Wagner, G.W. and Yang, Y.-C. (2001). Universal decontaminating solution for chemical warfare agents. U.S. Patent 6,245,957, 2001.

Woodlard, D., Kaul, R., Suenram, R., Walker, A.H., Globus, T. and Samuels, A. (1999). Terahertz electronics for chemical and biological warfare agent detection, *IEEE MTT-S International Microwave Symposium Digest, 1999*, 925–8.

Won, J., Kim, M., Yi, Y.W., Kim, Y.H., Jung, N. and Kim, T.K. (2005). A magnetic nanoprobe technology for detecting molecular interactions in live cells, *Science*, **309**, 121–5.

Xu, H., Sha, M.Y., Wong, E.Y., Uphoff, J., Xu, Y., Tredway, J.A., Truong, A., O'Brien, E., Asquith, S., Stubbins, M., Spurr, N.K., Lai, E.H., and Mahoney, W. (2003). Multiplexing SNP genotyping using the Qbead TM system: a quantum dot-encoded microsphere assay, *Nucleic Acid Research*, **31**, e43.

Yang, Y.C., Baker, J.A. and Ward, J.R. (1992). Decontamination of chemical warfare agents, *Chemical Reviews*, **92**, 1729–43.

Yih, T.C. and Wei, C. (2005). Nanomedicine in cancer treatment, *Nanomedicine: Nanotechnology, Biology, and Medicine*, **1**, 191–2.

Zhang, W.-X. (2003). Nanoscale iron particles for environmental remediation: An overview, *Journal of Nanoparticle Research*, **5**, 323–32.

Zhang, D., Chen, Y., Chen, H.Y. and Xia, X.H. (2004). Silica-nanoparticle based interface for the enhance immobilization and sequence-specific detection of DNA, *Analytical and Bioanalytical Chemistry*, **379**, 1025–30.

Appendix

Units for Magnetic Properties

Physical quantity	Symbol	Gaussian & cgs emu[a]	Conversion factor, C[b]	SI & rationalized mks[c]
Magnetic flux density, magnetic induction	B	gauss (G)[d]	10^{-4}	tesla (T), Wb/m^2
Magnetic flux	Φ	maxwell (Mx), G·cm^2	10^{-8}	Weber (Wb), Volt·second (V·s)
Magnetic potential difference, magnetomotive force	U, F	gilbert(Gb)	$10/4\pi$	ampere (A)
Magnetic field strength, magnetizing force	H	oersted (Oe),[e] Gb/cm	$10^3/4\pi$	A/m[f]
(Volume) magnetization[g]	M	emu/cm^{3h}	10^3	A/m
(Volume) magnetization	$4\pi M$	G	$10^3/4\pi$	A/m
Magnetic polarization, intensity of magnetization	J, I	emu/cm^3	$4\pi \times 10^{-4}$	T, Wb/m^{2i}
(Mass) magnetization	σ, M	emu/g	1 $4\pi \times 10^{-7}$	A·m^2/kg Wb·m/kg
Magnetic moment	m	emu, erg/G	10^{-3}	A·m^2, joule per tesla (J/T)
Magnetic dipole moment	j	emu, erg/G	$4\pi \times 10^{-18}$	Wb·m[i]

Nanomedicine: Design and Applications of Magnetic Nanomaterials, Nanosensors and Nanosystems
V. Varadan, L.F. Chen, J. Xie
© 2008 John Wiley & Sons, Ltd

Physical quantity	Symbol	Gaussian & cgs emu[a]	Conversion factor, C[b]	SI & rationalized mks[c]
(Volume) susceptibility	χ, κ	dimensionless, emu/cm^3	$4\pi\ (4\pi)^2 \times 10^{-7}$	Dimensionless henry per meter (H/m), Wb/(A $\cdot$ m)
(Mass) susceptibility	χ_ρ, κ_ρ	cm^3/g, emu/g	$4\pi \times 10^{-3}$	m^3/kg
			$(4\pi)^2 \times 10^{-10}$	H $\cdot$ m^2/kg
(Molar) susceptibility	χ_{mol}, κ_{mol}	cm^3/mol, emu/mol	$4\pi \times 10^{-6}$	m^3/mol
			$(4\pi)^2 \times 10^{-13}$	H $\cdot$ m^2/mol
Permeability	μ	dimensionless	$4\pi \times 10^{-7}$	H/m, Wb/(A $\cdot$ m)
Relative Permeability[j]	μ_r	not defined	-	Dimensionless
(Volume) energy density, energy product[k]	W	erg/cm^3	10^{-1}	J/m^3
Demagnetization factor	D, N	dimensionless	$1/4\pi$	Dimensionless

[a]Gaussian units and cgs emu are the same for magnetic properties. The defining relation is $B = H + 4\pi M$.

[b]Multiply a number of C to convert it to SI (e.g. $1\,\mathrm{G} \times 10^{-4}\mathrm{T/G} = 10^{-4}\mathrm{T}$).

[c]SI (*Système International d'Unités*) has been adopted by the National Bureau of Standards. Where two conversion factors are given, the upper one is recognized under, or consistent with, SI and is based on the definition $B = \mu_0 H + J$, where the symbol I is often used in place of J.

[d]1 gauss $= 10^5$ gamma (γ).

[e]Both oersted and gauss are expressed as $\mathrm{cm}^{-1/2} \cdot \mathrm{g}^{-1/2} \cdot \mathrm{s}^{-1}$ in terms of base units.

[f] A/m was often expressed as 'ampere-turn per meter' when used for magnetic field strength.

[g]Magnetic moment per unit volume.

[h]The designation 'emu' is not a unit.

[i]Recognized under SI, even though based on the definition $B = \mu_0 H + J$. See footnote c.

[j] $\mu_r = \mu/\mu_0 = 1 + \chi$, all in SI. Relative permeability μ_r is equal to Gaussian μ

[k]$B \cdot H$ and $\mu_0 M \cdot H$ have SI units J/m^3; $M \cdot H$ and $B \cdot H/4\pi$ have Gaussian units erg/cm^3.

Source: R.B. Goldfarb and F.R. Ficket, U.S. Department of Commerce, National Bureau of Standards, Boulder, Colorado 80303, March 1985. NBS Special Publication 696 for Sale by the Superintendent of Documents, U.S. Government Printing Office, Washington D.C. 20402.

Index